AF335729

J. C. Rajapakse, L. Wang (Eds.)

Neural Information Processing: Research and Development

Springer

Berlin
Heidelberg
New York
Hong Kong
London
Milano
Paris
Tokyo

Studies in Fuzziness and Soft Computing, Volume 152

Editor-in-chief
Prof. Janusz Kacprzyk
Systems Research Institute
Polish Academy of Sciences
ul. Newelska 6
01-447 Warsaw
Poland
E-mail: kacprzyk@ibspan.waw.pl

Further volumes of this series
can be found on our homepage:
springeronline.com

Jagath Chandana Rajapakse
Lipo Wang (Eds.)

Neural Information Processing: Research and Development

Springer

Prof. Dr. Jagath Chandana Rajapakse
Prof. Dr. Lipo Wang
Nanyang Technological University
School of Computer Engineering
Nanyang Avenue
639798 Singapore
Singapore
E-mail: elpwang@ntu.edu.sg

ISSN 1434-9922
ISBN 3-540-21123-3 Springer-Verlag Berlin Heidelberg New York

Library of Congress Cataloging-in-Publication-Data

Neural information processing : research and development /
Jagath Chandana Rajapakse, Lipo Wang (eds).
p. cm. -- (Studies in fuzziness and soft computing, ISSN 1434-9922 ; v. 152)
Includes bibliographical references and index.
ISBN 3-540-21123-3 (alk. paper)
1. Neural networks (Computer science) I. Rajapakse, Jagath Chandana. II. Wang, Lipo.
III. Series
QA76.87.N4745 2004
006.3'2--dc22

Springer-Verlag is a part of Springer Science+Business Media
springeronline.com

Typesetting: camera-ready by editors
Cover design: E. Kirchner, Springer-Verlag, Heidelberg
Printed on acid free paper   62/3020/M - 5 4 3 2 1 0

Preface

The field of neural information processing has two main objects: investigation into the functioning of biological neural networks and use of artificial neural networks to solve real world problems. Even before the reincarnation of the field of artificial neural networks in mid nineteen eighties, researchers have attempted to explore the engineering of human brain function. After the reincarnation, we have seen an emergence of a large number of neural network models and their successful applications to solve real world problems.

This volume presents a collection of recent research and developments in the field of neural information processing. The book is organized in three Parts, i.e., (1) architectures, (2) learning algorithms, and (3) applications.

Artificial neural networks consist of simple processing elements called neurons, which are connected by weights. The number of neurons and how they are connected to each other defines the architecture of a particular neural network. Part 1 of the book has nine chapters, demonstrating some of recent neural network architectures derived either to mimic aspects of human brain function or applied in some real world problems.

Muresan provides a simple neural network model, based on spiking neurons that make use of shunting inhibition, which is capable of resisting small scale changes of stimulus. Hoshino and Zheng simulate a neural network of the auditory cortex to investigate neural basis for encoding and perception of vowel sounds. Masakazu, Mori, and Mitarai propose a convolutional spiking neural network model with population coding for robust object recognition. Kharlamov and Raevsky formulate a class of neural network, using neurobilogically feasible multilevel information processing premises, that realizes the temporal summation of signals. Kitano and Fukai introduce a computational neural model to investigate the underlying mechanism of synchrony of neurons in the primary motor cortex to improve the predictive power.

Huang, King, Lyu, and Yang present a novel approach to construct a kind of tree belief network, which improves the approximation accuracy and recognition rate. Chiewchanwattana, Lursinsap, and Chu present an architecture capable of time-series forecasting by using a selective ensemble neural network. Miyajima, Shigei, and Kiriki propose a higher-order multi-directional associative memory with an energy function, which has an increased memory capacity and higher ability for error correcting. Maire, Bader, and Wathne describe a new indexing tree system for high dimensional codebook vectors, by using a dynamic binary search tree with a fat decision hyperplanes.

Neural networks are large parametric models where parameters are stored as weights of connections. Part 2 of this book investigates the recent developments in learning algorithms in seven chapters on adapting weights of neural networks.

Roy attempts to define some external characteristics of brain-like learning and investigate some logical flows of connectionism. Geczy and Usui establish a classification framework with superlinear learning algorithm to permit independent specification of functions and optimization techniques. Chaudhari and Tiwari investigate some approaches for adapting binary neural networks for multi-class classification problem. Ozawa and Abe present a memory-based reinforcement learning algorithm to prevent unlearning of weights. Takahama, Sakai, and Isomichi propose a genetic algorithm with degeneration to solve the difficulties by optimizing structures of neural networks. Wanas and Kamel present an algorithm to independently train the members of an ensemble classifier. Verma and Ghosh present a learning algorithm, by using different combination strategies, to find the optimal neural network architecture and weights.

Artificial neural networks and learning algorithms are increasingly being applied today to solve real world problems. Part 3 of this book contains nine chapters, each describing a recent application of artificial neural networks.

Neskovic, Schuster and Cooper use a neural network for the detection of cars from real-time video streams. Yang, King, Chan, and Huang uses non-fixed and asymmetrical margin setting with momentum in support vector regression for financial time-series prediction. Hu and Hirasawa present a neural network for control of non-linear systems. And Ricalde, Sanchez, and Perez provide an application of recurrent neural network for control of a robot manipulator. Ishikawa presents gesture recognition technique based on self-organizing feature maps (SOMs) using multiple sensors. Hussin, Bakus and Kamel present a technique based on SOMs for phase-based document clustering. Kasabov and Dimitrov discover gene regulatory networks from gene expression data with the use of evolving connectionist systems. Harati and Ahmadabadi use neural networks to solve the multi-agent credit assignment problem. Kim, Lee, Shin, and Yang present an implementation of visual tracking system using an artificial retina chip and a shape memory alloy actuator.

We would like to sincerely thank all authors who have spent time and effort to make important contributions to this book. Our gratitude also goes to Professor Janusz Kacprzyk and Dr. Thomas Ditzinger for their most kind support and help for this book.

Jagath C. Rajapakse
Lipo Wang

Contents

Scale Independence in the Visual System

Raul C. Muresan

Nivis Research, Gh. Bilascu Nr. 85, 3400 Cluj-Napoca, Romania

Abstract. We briefly present some aspects of information processing in the mammalian visual system. The chapter focuses on the problem of scale-independent object recognition. We provide a simple model, based on spiking neurons that make use of shunting inhibition in order to optimally select their driving afferent inputs. The model is able to resist to some degree to scale changes of the stimulus. We discuss possible mechanisms that the brain could use to achieve invariant object recognition and correlate our model with biophysical evidence.

Keywords. Object recognition, scale independence, spiking neurons, shunting inhibition, visual system

1. Introduction

For a better understanding of this chapter, we will begin with a brief description of the major parts of the mammalian visual system, followed by a short review of the current theories on visual processing.

1.1. The architecture of the visual system

Understanding the amazing processing abilities of our visual system is a challenging multi-disciplinary effort. However, the starting point of any modeling attempt is the study of the brain's organization.

Our visual system is a huge, both parallel and hierarchical structure, involving many levels of processing. Everything begins with the projection of the light on the retinal photoreceptors. At this stage, the 100 million photoreceptors transform the light into electrical impulses that are transmitted along the axonal fibers of the optic nerve. The cells that carry the information to higher brain levels are the ganglion cells. These cells exhibit very interesting properties when stimulated by the bipolar and horizontal cells (bipolar and horizontal cells are intermediate cells between the photoreceptors and the ganglion cells). Perhaps the most striking feature of the ganglion cells is their so-called "center-surround" response profile. Some ganglion cells respond vigorously when stimulated with a spot of light in the center and dark surround (ON-center cells), others are exactly the opposite, responding better to dark center and light surround (OFF-center cells). Such a type of response profile is called "center-surround".

Once the visual information has been transformed into neural code (the way information is encoded is still a matter of debate), the spikes propagate along the optic nerve and after passing some intermediate stages (such as the optic chiasm and the thalamic nuclei), they project onto the primary visual cortex. The primary visual cortex, also called area V1, is the first true cortical processing stage. Physiological studies showed that the visual cortex microcircuit is organized in 6 layers and that different areas have different functions. Within each area (such as V1), all the six layers are present. In V1, functionally similar neurons are grouped together in the so-called micro-columns according to some criteria like: orientation selectivity, ocular dominance, etc. One of the most important properties of V1 cells is their orientation selectivity, that is, their ability to respond vigorously to contrasts of a given orientation. For example, some cells are selective to 0 degrees orientation others to 25 degrees orientation and so on. However, V1 cells do not respond only to their preferred orientation, they rather respond to a range of orientations, but as the distance from the preferred orientation increases, the cell's response decreases. This is called orientation tuning (cells that respond to a very limited range besides the preferred orientation are called "sharp-tuned" cells, others that respond to a broad range of orientations are called "broad-tuned" cells). The V2 area is very similar to V1, one of the major differences being the increased number of V2 cells that respond to illusory contours.

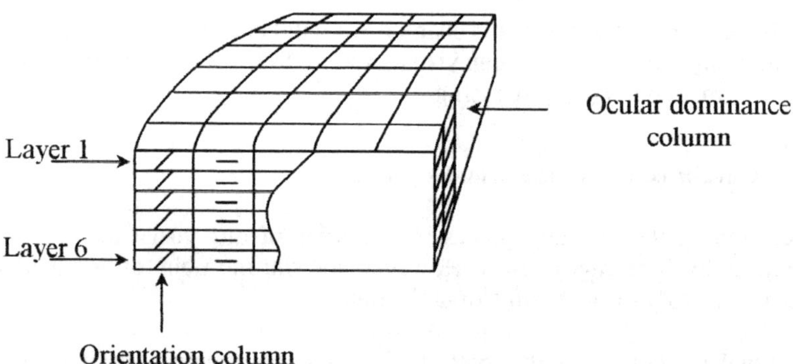

Fig. 1. Organization of the visual cortex (here the primary visual cortex)

The receptive field of a cell in the visual system is considered the area (measured in degrees of the angle formed by the light at the extremities of the area) in the visual field that influences directly the firing rate of the cell. The receptive fields of V1 cells are relatively small (up to a few degrees) and the V1 area has a predominantly retinotopic organization (retinotopy is the property of mapping adjacent visual inputs from the retina onto adjacent neuronal areas in the higher levels).

Although the primary visual cortex has been extensively studied (mainly by Hubel and Wiesel [8]) the exact function of this primary filtering operation is not yet fully understood. Axons from the V1 area project to the deeper levels of the visual cortex. The recognition pathway is thought to be the ventral system, which continues with the V3 and V4 areas. V4 neurons have larger receptive fields than V1 cells and exhibit non trivial responses to contour features (such as angles and curves) [24] and object parts. The hypothesis that V4 cells encode "object-parts" has been speculated by Biederman in his RBC model [1]. However, current experimental evidence does not suffice in order to assign V4 cells a filtering-matching role.

The next stage on the ventral pathway is the infero-temporal cortex (IT). Axons from V4 mainly project to the IT area. Infero-temporal neurons show object specific tuning, that is, they are able to respond selectively to different types of objects (hands, trees, faces). For this reason, IT cells are considered to be among the final cells on the recognition pathway. The receptive field of IT cells covers most of the visual field and their response seems to be transformation invariant (scale-, translation- and to some degree rotation-independent).

1.2. Problems of recognition in the visual system

The visual system performs continuously amazing tasks such as object recognition and it seems to achieve these tasks effortlessly. The last two decades of research revealed some of the mechanisms involved, both at cellular and microcircuit structure level. Numerous studies have been performed on the retina, optic nerve and the first levels of visual processing, such as the primary visual cortex (V1) [8]. However, to this date, we have no complete or correct understanding of the complex processes that take place at higher areas in the brain such as area V4 or the infero-temporal cortex. It is not yet clear how the brain is capable of achieving attentional and recognition tasks.

Many hypotheses have been formulated trying to explain the way our brain can recognize objects. One such theory is based on the RBC (Recognition By Components) model developed in 1987 by Biederman [1]. The RBC model constructs intermediate primitives from primary features. These primitives have a finite number, complex objects being later constructed and described in terms of these primitive decompositions. Attractive as it may be, this theory is still far from solving the problems. It has been thought that the primitive based recognition would eliminate the combinatorial problem of hierarchic recognition. However, it is quite obvious that the RBC model only transfers the problem to higher areas that have to combine primitives. Primitive combination still suffers from the combinatorial explosion in reconstructing real-world complex visual scenes. Also, it is unclear how the visual system might achieve primitive decomposition. One of the hardest problems the visual system has to solve is scale-independent recognition. Should it be primitive extraction or direct recognition, scale-independence must still be achieved.

The temporal correlation hypothesis does not constrain the recognition process. It provides some basic mechanisms in order to solve the "binding problem" formulated by Malsburg and Milner [13,14,15,16,17]. The correlation hypothesis states that synchronization and correlated activity might signal aspects of the visual world that "belong together". Assemblies of such synchronized neurons emerge dynamically leading eventually to perceptual grouping and neural routing to achieve recognition. However, the hypothesis has been widely criticized and the main argument for opposing it has been the difficulty of replicating the experiments that favored it [7].

There are several problems that have no straightforward solution, each approach to solving them proving that the method quickly becomes intractable. Probably the most striking feature of our visual system is the way it combines information. Ultimately we are reduced to a context problem. Simple cells in the lower areas of the visual cortex, such as V1, have small receptive fields. They integrate information from a limited, small area of the visual scene. The problem is how successive layers in the processing hierarchy integrate these separate, small spatial contexts. Combining different feature detectors could rapidly lead to a combinatorial explosion (the combinatorial problem). Riesenhuber and Poggio [25] created a model (HMAX) based on Fukushima's "neocognitron" [6]. These models combined simple feature detector responses into more and more complex representations. The HMAX model combined the features according to a maximum response function, selecting optimal responses for later stages. HMAX is in fact a winner-take-all model that segments the visual scene into approximated representations that are later combined. To some degree, such a hierarchic approach resembles the RBC model. Olshausen et al. provided attention based dynamical routing models in order to achieve spatial context integration [23]. However, all these models are only partial solutions. The HMAX can hardly capture scale variance while Olshausen's model of dynamic routing does not clearly provide an answer for preattentive visual search (a priori knowledge might be required for correct routing).

The visual system is a very powerful processing unit that is able to solve all of the visual tasks with limited resources. One of the most striking aspects of our visual processing is its speed. We are able to segment the visual scene in a few tens of milliseconds, to modulate attention and achieve recognition under 150 milliseconds [28]. All these findings impose even greater constraints on models. The speed of visual processing seriously questions rate-based coding (which assumes that the firing-rate of neurons encodes the information) of information in neural systems. Temporal coding might be, in this context, a much better candidate [29] (in temporal coding, each individual neural spike and the delays between spikes count).

2. Visual scale independence

One very little studied aspect of visual processing is scale independence. Resizing an object leads to big variations in the local organization of the stimulus. Locally speaking, like translation and rotation, scaling effectively changes "where" features will be located. Unlike translation and rotation, scaling also changes "what" local detectors perceive. The same object, at different scales, generates a whole different pattern of stimulation at the lower stages of visual processing (Fig. 2). We might say that only the global "characteristic" of the stimulus is preserved.

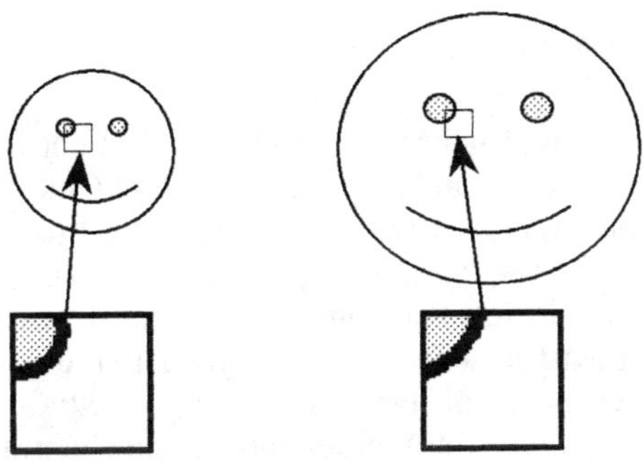

Fig. 2. Local feature variation due to stimulus scaling

Recognizing objects in a scale independent way is a very hard problem. The visual processor has to solve both "what" and "where" problems. There are at least two possible approaches to invariant recognition:
1. the creation of invariant representations;
2. the inference of stimulus characteristics.

Although the second approach is more likely to be used by our visual system, we do not understand yet how it is able to "infer" stimulus characteristics. Accepting that inference processes take place in the visual system would radically change our understanding of visual processing. Most of the present models ignore feedback projections and lateral interactions in visual areas. However, these processing pathways are at least equally important as the feed-forward ones. Quantitatively we have evidence of strong feedback and lateral interactions. It just might be that large, distributed, recurrent and dynamical neural systems can achieve inference. Object representations, in this context, would not be clearly separable. It is much likely that objects and relationships between them generate "mental states". The "liquid state machine" and "far from equilibrium" systems can offer nontrivial processing capabilities, while largely respecting the wiring rules of our cortical microcircuits [12]. Also, the "coherence theory", under development, could offer a new perspective on visual processing and object recognition based on the belief that visual recognition is an inference process [22].

In this chapter, we will focus mainly on the first approach, the creation of invariant representations. To some degree, invariant representations are biologically plausible and are easier to understand. We will provide neural mechanisms that could account for scale invariant recognition. Also, we will describe a model, based on spiking neurons that is fairly biologically plausible and reproduces some known experimental evidence. The neural coding paradigm applied is rank-order coding, while shunting inhibition as an underlying mechanism is extensively used.

3. Neural mechanisms

Studies on the mammalian visual cortex had shown that cortical simple cells might project to complex and hipercomplex cells. Since Hubel and Wiesel [8] it has been thought that such a hierarchical composition might lead to transformation invariant recognition. On the other hand, the speed of processing in the visual system [28] leads us to the conclusion that some feed-forward mechanisms should exist that could achieve, at least to some degree of precision, object categorization and recognition. The speed problem imposes a strong constraint on the neural code that visual neurons might use. Rank-order coding and shunting inhibition could be the answer. Next, we will present the neural mechanism that is able to encode a huge amount of information in a very short time.

Evidence have been accumulated that neurons are able to select their synaptic inputs on the basis of afferent timing. Inhibition could be an important mechanism that contributes to timing sensitivity. Some researchers treat inhibition as having a stabilizing function [26,27]. Although this is probably an important function of inhibition, it can also generate shunting and hyperpolarizing effects that shape the temporal structure of neural activity on a millisecond time scale [2,3,5,11]. Shunting inhibition in general and fast shunting inhibition in particular could offer a strong mechanism for temporal coding rather than firing-rate coding.

The primary question behind all neural systems is how they encode information. At this time, there is widely accepted that neurons encode information by the means of their firing rate. In this framework, we speak about stability and fixed points in the oscillatory processes. However, numerous recent studies seriously question such a neural code because it is unable to account for ultra-rapid brain processes. It is true however that the firing rate of a neuron is directly linked to the stimulus and excitatory as well as inhibitory afferent activity. But we can understand firing rate as a side effect of millisecond time scale processes.

Thorpe and colleagues [29] have proposed a new scheme for information coding in the cortex: rank-order coding. Rank-order coding is based on shunting inhibition to shape the response of a target neuron as a function of afferent spike timing. In principle, considering a target neuron and a pool of inhibitory interneurons, each time an afferent spike arrives, the inhibitory pool shunts down the efficacy of spike integration of the target neuron by the means of shunting inhibition. Simple implementations of such a neural coding proved their power in dealing with large amounts of information [4,18,19,21]. The enormous encoding power of this scheme has been outlined by S.J. Thorpe, showing that rank order codes can be used, in principle, to transmit up to $\log_2(N!)$ bits of information (where N is the number of neurons).

The rank order code scheme presented above has two main advantages. First, due to progressive shunting effects, a normalization process is automatically generated. As afferent activity increases, the shunting effect is also increased and the target neuron desensitised so that its response will be rectified. Secondly, because each time an afferent neuron spikes, the target neuron is desensitised, the response of the target neuron can be made dependent on the timing of afferent spikes (using different combinations of synaptic strengths) [29].

4. A scale invariant recognition model

Next, we will present a simple model, based on a biologically plausible framework that can construct scale invariant (to some degree) representations of objects. The model uses spiking neurons as processing units and has a retinotopic hierarchic organization. A pathway is constructed from the retina, to V1 simple cells and "end-stopped" bar detectors, up to V4 and infero-temporal cortex. Depending on the available computing resources (number of units) used, the model is able to construct scale invariant representations of complex objects with different degrees of precision.

4.1. Building blocks of the model

Spiking integrate-and-fire neurons were used in the recognition system. We have applied a similar model as Thorpe and colleagues for the individual neurons used in simulations. Each neuron is characterized by a small set of parameters:
 – membrane potential (U);
 – instantaneous sensitivity (S);
 – synaptic modulation factor (M).

The update rule adds the current post-synaptic potential (PSP) to the current membrane potential. The PSP is computed by multiplying the synaptic weight (W) by the sensitivity factor (S). After each afferent spike the sensitivity factor is decreased by multiplying it with the synaptic modulation ($M \in (0..1)$). For simplicity, no refractory period or leakage are included in the model.

$$U^{(t+1)} = U^{(t)} + W \cdot S^{(t)} \tag{1}$$

$$S^{(t+1)} = S^{(t)} \cdot M \tag{2}$$

Neurons are organized in maps. Each map has a given, well-determined functional role. Maps are interconnected with each other using synaptic kernels. Because we used retinotopic mapping, each pair of neural maps that are interconnected has exactly one connection kernel associated. This connection kernel is used as a "rule" of interconnection between maps. Each time spikes are being processed, the appropriate connection kernel is selected and used in computing the PSP's. Simulation is iterative and event-driven at the same time. In each iteration a set of spikes that are emitted is computed for every map. In the next iteration, these spikes are integrated according to the position of the spike events. Only those neurons that are affected by the previous spikes are updated. The update is spike-event driven. The simulator we constructed is called RetinotopicNET and is able to simulate large populations of spiking simple neurons [20].

4.2. The architecture of the model

The architecture of the model contains 7 levels of processing, following the retinal, V1, V4 pathway up to the infero-temporal cortex. The seven layers of processing correspond to an ascending feed forward processing with lateral interactions at some levels (Fig. 4). The key feature of the model is the use of extensive competition between different elements of the objects to be recognized. The only information used at this time is contour information but blob-type cells could also be included to account for color or intensity patches as well.

4.2.1. Retinal processing

At the first level of processing the retinal ganglion cells process the incoming image intensities (only intensity 8 bit grayscale images were used). The ON-OFF effect has been achieved by using a classical difference-of-gaussians (DOG), center-ON-surround-OFF and vice versa filter with a ratio of standard deviations: 1 to 3 (eq. 3). Then, the image intensity for the two maps has been converted into spike latency and spikes were fed into the "RetinotopicNET" simulator.

$$DOG(x, y) = \frac{e^{-\frac{x^2+y^2}{2 \cdot \sigma_1^2}}}{\sqrt{2 \cdot \pi} \cdot \sigma_1} - norm \cdot \frac{e^{-\frac{x^2+y^2}{2 \cdot \sigma_2^2}}}{\sqrt{2 \cdot \pi} \cdot \sigma_2} \tag{3}$$

where: - x, y are the coordinates in the kernel space

- σ_1 and σ_2 are the standard deviations ($\sigma_2 = 3 \cdot \sigma_1$)
- *norm* is a normalization constant

Spike latencies are subsequently computed using a linear mapping. The pixel with the highest contrast (highest intensity after applying DOG filtering) generates the first spike (zero latency). The lowest contrast (0) generates the last spike. Intermediate latencies are computed using a linear approximation from the contrast level.

4.2.2. V1 Area

The second layer of processing corresponds to the V1 primary cortex area where different orientation channels are selected by oriented Gabor-like receptive fields (eq. 4). These are the corresponding simple cells, which detect different orientation contrasts.

$$G(x,y) = e^{\frac{-(x^2+y^2)}{2\sigma^2}} \cdot \sin(2\pi f \cdot (x \cdot \cos(\phi) + y \cdot \sin(\phi))) \tag{4}$$

where: - x and y are the coordinates in the kernel space
- $G(x,y)$ represents the Gabor filter kernel value
- f is the spatial frequency of the filter
- ϕ the angle that influences the orientation selectivity of the filter

One key feature is the lateral connection within each orientation map. We have used a butterfly-like lateral connection, which has the property of improving contours. This is a form of primitive contour-integration, but due to the lack of iterative loops only a feed forward contour completion is used. Important work on this topic had been conducted by Zhaoping Li [9]. Further improvement on the system may be achieved by implementing a stronger contour integration mechanism. The Gabor patches were all at the same scale and had a spatial frequency of 0.5 pixels. They covered the range of 0 to 180 degrees with a total of 8 orientations.

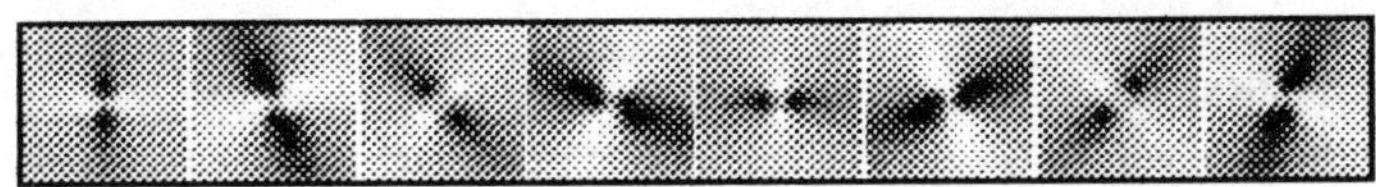

Fig. 3. Lateral "butterfly-like" receptive field profiles. White denotes excitatory areas and black inhibitory ones.

Approximately 2 to 5% of V1 neurons responded to illusory contours. While the contour integration effect is well known to occur in the V2 area, V1 neurons are also known to have some degree of contribution to the integration of contours [10]. Since we haven't modeled the V2 area, our simple model of V1 cells includes also a restricted contour-integration function. Our observations showed that the performance of the model had improved by including lateral interactions between the V1 neurons.

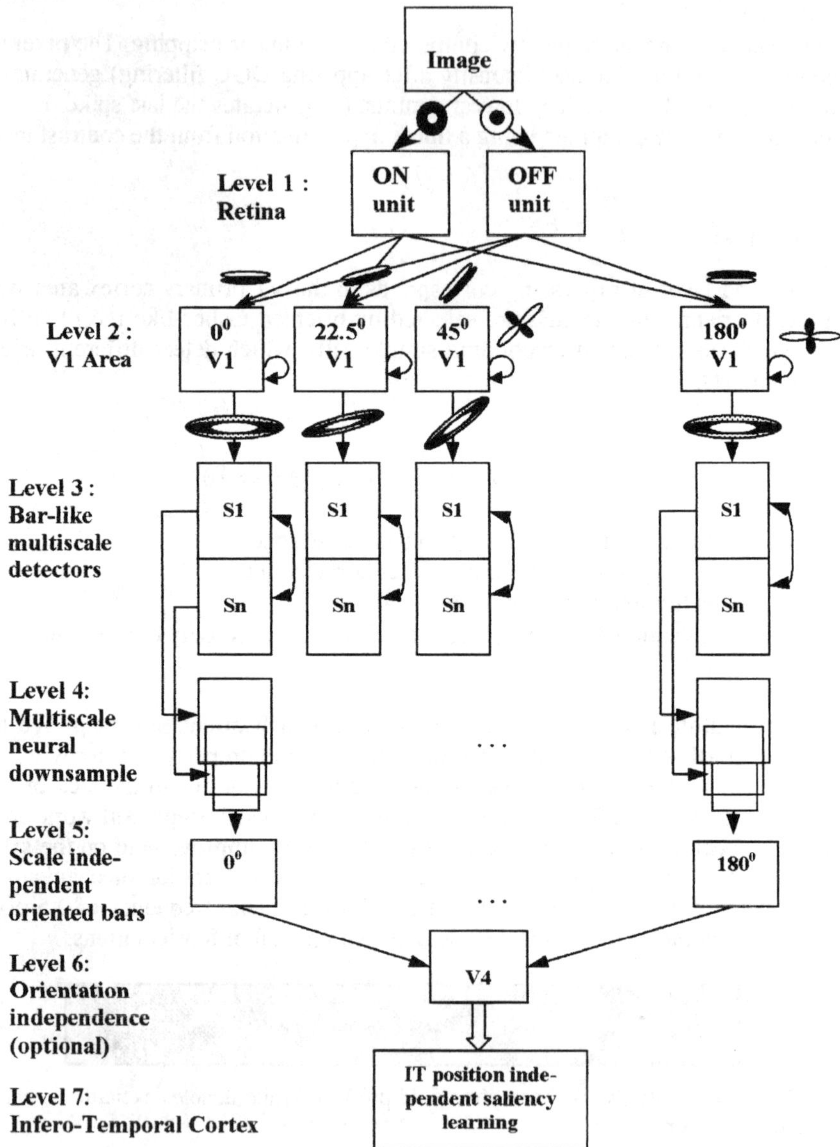

Fig. 4. Model architecture.

4.2.3. Bar-like detectors

For each orientation, a corresponding set of different scaled-receptive-field maps was used to extract the bar-like feature at the corresponding position. Bar like detection, across different length channels, provides the solution to the "what" problem. It is in fact a classification and measurement at the same time.

Each receptive field had an oriented bar-like, end-stopping type. The central, elongated bar, corresponds to excitation. The surrounding area corresponds to weak inhibition proportional to the level of blackness (Fig. 5). This type of receptive field tunes the neuron to the bar that best matches its excitatory size. For the same orientation, multiple scaled bar-detectors were used.

Considering one set of oriented scaled maps (multiscale maps for the same orientation), lateral inhibition has been introduced from large to small sizes, generating a size competition at that orientation. The priority in terms of timing varied from large to small bars. In other words, the maps with a large receptive field had the chance of firing first and, by the means of inter-map lateral inhibition, the smaller bar detectors were inhibited. Such a mechanism ensures that the largest size possible is always detected (instead of composing it from multiple smaller size bars).

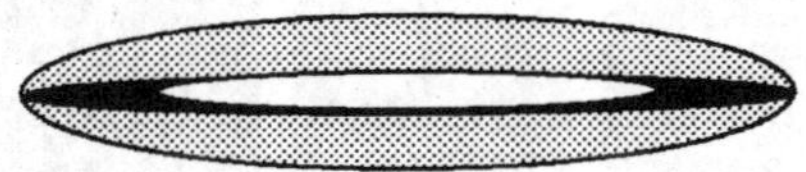

Fig. 5. Bar-like detector for 0^0 orientation. The white bar in the center corresponds to excitatory synapses; the gray and black areas correspond to inhibitory synapses. The black denotes the strongest inhibition, the gray denotes intermediate strength of inhibition.

4.2.4. Multiscale downsample

The fourth layer of processing is responsible for bringing every detail to the same level of spatial importance. We deal here with the "where" problem. Once the constructing elements were detected (bars), we have to create invariance to distance changes between these features that occur due to scaling. In other words, if two long lines are detected, the distance between them has to be brought down to the same distance as the one between the scaled down versions of the same lines (in an object scaling operation). The key mechanism of scale independence is exactly the equivalence of feature distance with feature size. This level is the most important one for scale independence and we will describe the mechanism in detail.

Let us consider, as an example, a simple object, formed of just two lines, oriented at 0^0, as shown in figure 6.

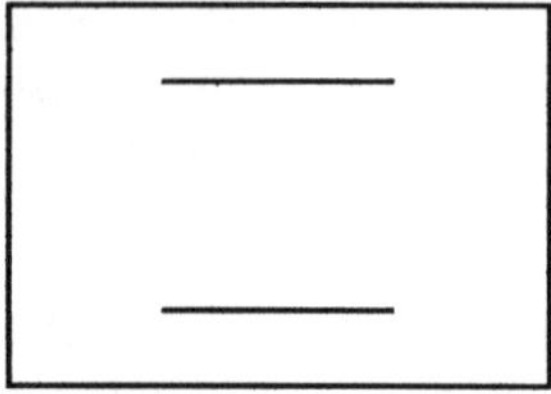

Fig. 6. An object formed of 2 lines, of size 20 pixels each.

For explanation purpose, let us consider that our scaled bar detectors range over 10 to 30 neurons (pixels) and that the lines in the original image (Fig. 6) have a size of 20 pixels. We have 3 maps of 0^0 orientation with bar detectors at 10, 20 and 30 neurons (pixels). The response of the 3 maps to the original image is presented in Fig. 7.

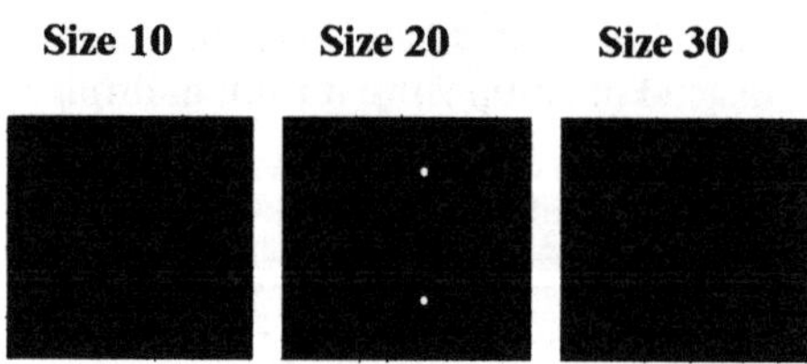

Fig. 7. The response of the 3 maps to the original object with lines of size 20.

Because of the strong lateral inhibition and competition, neurons in the map of size 10 can't fire. Neurons in the map of size 30 have not enough stimulation in the excitatory area to be driven by the 2 lines. Thus, only in the map of size 20 the activity will exist. Now let us scale down by a factor of two the original image (Fig. 8).

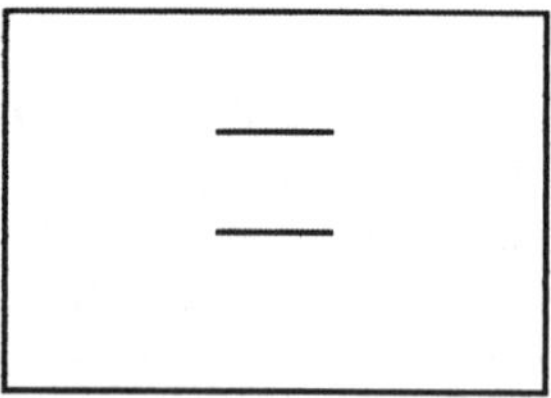

Fig. 8. The original object scaled down by a factor of 2.

The result of the down scaling is that activity will move down to lower sized detectors by a distance proportional to the ratio between the size of different detectors of the maps (Fig. 8).

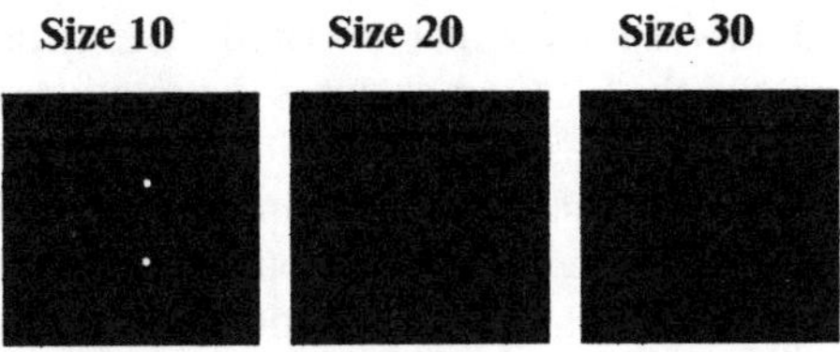

Fig. 9. The response of the 3 maps to the scaled object with lines of size 10.

We take a look at the distance between the two lines: by scaling the object down, the distance between its parts is also scaled down yielding a cortical response with scaled distances between bar detector neurons. All the system has to do, in this simple case, is to scale down the second map by a factor of 2 and the third by a factor of 3 and feed all of these resulting maps into a scale invariant map (Fig. 10).

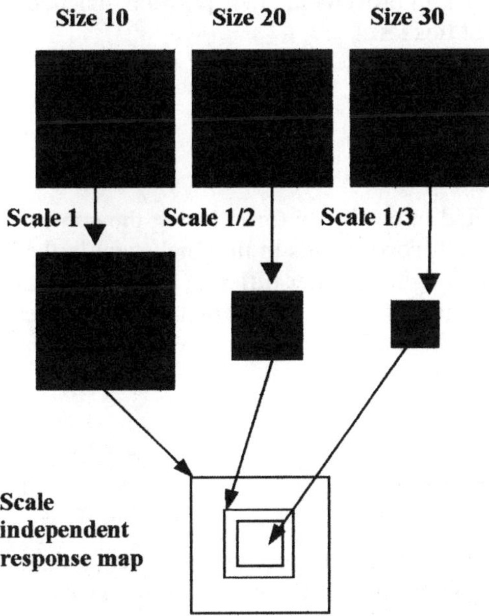

Fig. 10. Neural downsampling and combination to achieve scale independence.

Some may argue that for an object that has lines which will be detected by different sized detectors at the equivalent locations (as dictated by the scale of detection), the neural responses will overlap each other in the final scale invariant map. But we have to take into account that for that specific location, the final neuron will have a multiplied feeding and thus an increased firing rate (the multiplication depends on the number of responses that overlap). The firing rate captures the equivalence information and the location captures the relative position of features.

The neural down sampling is achieved by using window-like receptive fields which could be associated with center-ON-surround-OFF receptive fields in area V4, taking into account the fact that the surround-OFF is a very small silent inhibition which could be used for stability and normalization purposes.

4.2.5. Scale independence maps

At the next layer, at each orientation, the downsampled oriented maps are combined into a scale independent map corresponding to that specific orientation (Fig. 10). The mechanism of combining them should take into account the scaling center of the object. Further improvement, like position independence of features, could be implemented at this level.

4.2.6. Orientation independence

This layer is optional, and in used only for reducing the number of synapses with the infero-temporal map. It corresponds to the final stage in the V4 area. There is no reason for which one might consider different orientations as being equivalent but this type of combination can be used to simulate the hypercomplex cells. Orientation equivalence can be used to improve the generalization capability of the recognition in the IT cortex.

4.2.7. Infero-Temporal Cortex

The infero-temporal cortex is responsible for object recognition. In the architecture presented, learning is performed by increasing the synapse strength with the current sensitivity value as resulted from successive modulator effects generated by the firings in the level 6 map. Every neuron in the final IT map has a retinotopic type of receptive field, covering most of the level 6 map. The synaptic strengths are shared among all neurons, yielding good position independence.

4.2.8. Simulation results

Using the "RetinotopicNET" simulator we calibrated the system for face detection (and recognition). The test database had been generated using a QuickCam web camera and consisted of the faces of three different persons with different face expressions. Image sizes were fixed at 92 x 112 grayscale bitmaps and the faces were scaled in a range 1 to 0.58 the original scale (Fig. 11).

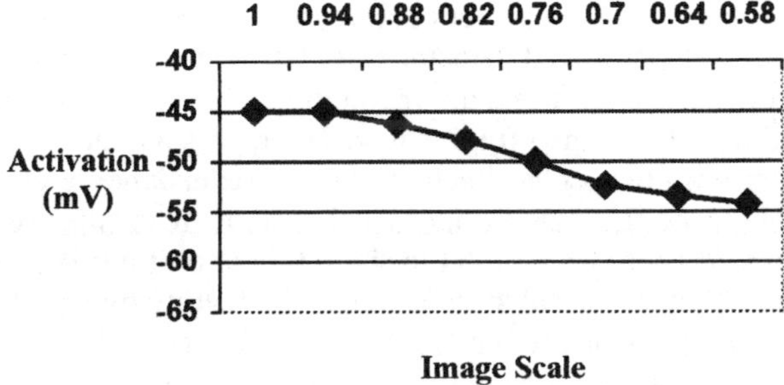

Fig. 21. Activation of infero-temporal cortex for different image scales. The IT layer has been trained to reach the exact threshold activation of -45 mV for the face at the original size.

The number of scales used at level 4 was 7, bar-like detectors ranging from a length of 7 to 13 neurons (7,8,...13). At the infero-temporal level, a strong shunting inhibition had been used to provide enhanced selectivity on learning (for the face recognition case).

The selectivity map of the trained infero-temporal neuron is shown in figure 12 and different sized details can be observed at different positions.

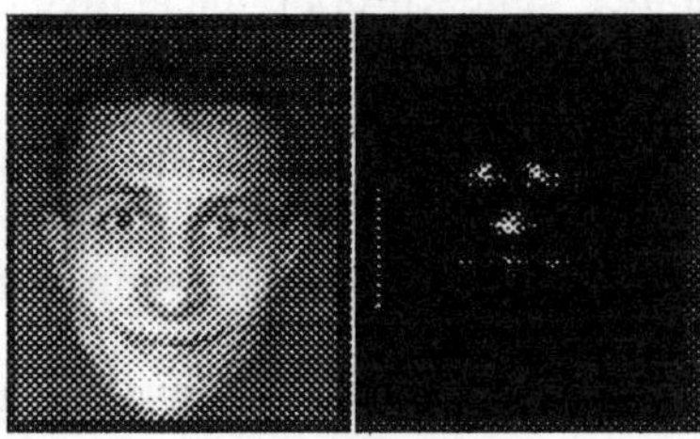

Fig. 32. Training image and the selectivity of the infero-temporal cortex

The given results are surprisingly good, taking into account the limited scale levels used and the size of the details in the image (the details are by far larger than the detectors used). Increasing the size of the bar-detectors and their number (to cover a wider field of scales) can increase the accuracy of recognition. At the same time, we expect that the usage of more orientations can increase accuracy because of the better localized bar detection at the third level of the model.

5. Discussion

We have seen that neural mechanisms like shunting inhibition can contribute to spatial competition processes between different feature detector neurons. However, the model presented in this chapter is clearly not very plausible biologically. The first problem this model encounters is the great redundancy that is generated by the requirement for specialized detector cells. There are many types of features that the visual cortex extracts from the incoming visual signals. Thus, having a good approximation of this signal (in this model) requires an enormous amount of specialized feature-detection cells. Such enormous redundancy and waste of resources is unlikely to be the case in our visual system. It is supposed that many cells change their selectivity from broad to sharp tuning on millisecond time scale. It might be the case that the visual system adapts the feature tuning of neurons dynamically, under coherent attentional feedback supervision. Such adaptation is much likely to occur in the V4 area where complex feature detectors and encoders might exist [24].

The second problem of our model is that at the fifth layer, the neural downsampling should be performed around the center of the object. Such a mechanism would be possible only with attentional involvement. It is not clear however, how attention could modulate the responses and reroute the information flow in order to achieve object based attention as a step in object recognition processes. Nonetheless we have evidence that attention is very important, maybe essential, also in the object recognition process.

Probably the biggest pitfall of the described mechanisms is that they are too algorithmic. Our brain continuously processes information, dynamically reallocating resources and having parallel, multisensorial, integrative capabilities. Algorithmic approaches are good for modeling but unlikely to be used by our visual system.

Although many aspects stand against such simplified models of vision, optimality selection, competition and limited-context scale independence might actually be based on similar mechanisms. Ultimately, such simple analyzers might dynamically arise in a huge hierarchical and recurrent neural system. Pairs of content-context pools might be generated during adaptation to stimulus. Attentional modulation could play an important role, sometimes mixed with recurrent feedback projections that might contribute not only to stability but also to inference processes [22]. The presence of a particular object and the relationship to the overall scene would be captured by the transitions of the dynamical system (as mental states), yielding a characteristics-based recognition capability. Mechanisms similar to our model could be, at least in part, involved in such a complex dynamical picture of visual processing.

As a final statement, we have to mention our belief that vision is not about filtering or matching, *vision is about thinking.*

6. References

1. Biederman I (1987) Recognition by components: A theory of human image understanding. Psychol. Rev. 94:115–147.
2. Bush P, Sejnowski TJ (1996) Inhibition synchronizes sparsely connected cortical neurons within and between columns in realistic network models. J Comput Neurosci 3: 91–110.
3. Chrobak JJ, Buzsaki G (1998) Gamma oscillations in the entorhinal cortex of the freely behaving rat. *J Neurosci* 18:388–398.
4. Delorme A, Thorpe SJ (2001) Face identification using one spike per neuron: resistance to image degradations. Neural Networks 14:795-803.
5. Freeman WJ (1975) Mass Action in the Nervous System. Academic Press, NewYork.
6. Fukushima K (1980) Neocognitron: A self-organizing neural network model for a mechanism of pattern recognition unaffected by shift in position. Biol Cybern 36:193–202.
7. Gray CM (1999) The Temporal Correlation Hypothesis of Visual Feature Integration: Still Alive and Well. Neuron 24:31-47.
8. Hubel D, Wiesel T (1965) Receptive fields and functional architecture in two nonstriate visual areas (18 and 19) of the cat. J Neurophysiol 28:229–289.
9. Li Z (1998) A neural model of contour integration in the primary visual cortex. Neural Comput 10(4):903-40.
10. Li Z (1999) Visual segmentation by contextual influences via intra-cortical interactions in the primary visual cortex. Network: Comput Neural Syst 10:187–212.
11. Lytton WW, Sejnowski TJ (1991) Simulations of cortical pyramidal neurons synchronized by inhibitory interneurons. J Neurophysiol 66:1059–1079.
12. Maass W, Legenstein RA, Markram H (2002) A new approach towards vision suggested by biologically realistic neural microcircuit models. In Buelthoff HH, Lee SW, Poggio TA, Wallraven C (eds) Biologically Motivated Computer Vision. Proc of the Second International Workshop BMCV 2002 Tübingen, Germany, vol. 2525 of Lecture Notes in Computer Science, Springer, Berlin, pp 282-293.
13. von der Malsburg C (1981) The correlation theory of brain function. MPI Biophysical Chemistry Internal Report 81–2.
14. von der Malsburg C (1985) Nervous structures with dynamical links. Ber Bunsenges Phys Chem 89:703–710.
15. von der Malsburg C (1986) Am I thinking assemblies? In: Palm G, Aertsen A (eds) Proceedings of the Trieste Meeting on Brain Theory. Springer, Berlin.
16. von der Malsburg C (1999) The What and Why of Binding: The Modeler's Perspective. Neuron 24:95-104.
17. Milner P (1974) A model for visual shape recognition. Psychol Rev 81:521–535.
18. Mureşan RC (2002) Complex Object Recognition Using a Biologically Plausible Neural Model. In: Mastorakis NE (eds) Advances in Simulation, Systems Theory and Systems Engineering. WSEAS Press, Athens, pp 163-168.
19. Mureşan RC (2002) Visual Scale Independence in a Network of Spiking Neurons. ICONIP '02 Proceedings, Singapore, 4:1739-1743.
20. Mureşan RC (2003) RetinotopicNET: An Efficient Simulator for Retinotopic Visual Architectures. ESANN'03 Proceedings, Bruges, pp 247-254.

21. Mureşan RC (2003) Pattern Recognition Using Pulse-Coupled Neural Networks and Discrete Fourier Transforms. Neurocomputing 51C:487-493.
22. Mureşan RC (2003) The Coherence Theory: Simple Attentional Modulation Effects. CNS'03 Proceeding, in press.
23. Olshausen BA, Anderson CH, van Essen DC (1993) A Neurobiological Model of Visual Attention and Invariant Pattern Recognition Based on Dynamic Routing of Information. J Neurosci 13(11):4700-4719.
24. Pasupathy Anitha, Connor CE (1999) Responses to Contour Features in Macaque Area V4. J Neurophysiol 82: 2490-2502.
25. Riesenhuber M, Poggio T (1999) Hierarchical models of object recognition in cortex. Nat Neurosci 2(11):1019-1025.
26. Shadlen MN, Newsome WT (1994) Noise, neural codes and cortical organization. Curr Opin Neurobiol 4:569-579.
27. Shadlen MN, Newsome WT (1998) The variable discharge of cortical neurons: implications for connectivity, computation, and information coding. J Neurosci 18:3870-3896.
28. Thorpe SJ, Fize D, Marlot C (1996) Speed of processing in the human visual system. Nature 381(6582):520-522.
29. Thorpe SJ, Gautrais J (1998) Rank order coding. In: Bower J (eds) Computational neuroscience: Trends in research. Plenum Press, New York, pp 113-118.

Dynamic Neuronal Information Processing of Vowel Sounds in Auditory Cortex

Osamu Hoshino[1] and Meihong Zheng[2]

[1] Department of Human Welfare Engineering, Oita University, 700 Dannoharu, Oita 870-1192 Japan `hoshino@cc.oita-u.ac.jp`
[2] Department of Applied Physics and Chemistry, The University of Electro-Communications, Chofu, Tokyo 182-8585 Japan `zmh@glia.pc.uec.ac.jp`

Abstract. By simulating a hierarchical neural network model of the auditory cortex, we investigated neuronal bases for encoding and perception of vowel sounds. We demonstrate that information about vowels is encoded by specific dynamic cell assembles, and the perception of an applied vowel sound is mediated by the collective activation of the cell assembly corresponding to the input. We tried to explain the roles of time-varying formant frequencies, which are one of the notable characteristics of natural (vocalized) vowel sounds in humans. We demonstrate that the time-varying frequency stimulation of neurons is advantageous for propagating neuronal excitation throughout the cell assembly corresponding to the applied stimulus as compared with constant frequency stimulation. We suggest that the time-varying change in formant frequencies of vowel sounds may be essential for the enhancement of cognitive performance on vowel sound perception.

Keywords. Vowel sound representation, dynamic cell assembly, time-varying formant frequency

1 Introduction

Regardless of acoustic differences in vocalization among individuals, we can invariantly perceive speeches independent of speakers. We hear to someone speaking and understand what he or she is saying, even though they are vocalizing the same word, phrase, or sentence in different acoustic waveforms. We can also understand speeches that are spoken by people whom we have met for the first time. It is interesting to know how the brain processes such invariant perception of speech sounds.

Vowel is one of the fundamental elements for human speech sounds and is characterized by the so-called formant frequencies, which are peak frequencies in the spectra of vocalized vowel sounds. Based on a psychological experiment,

Peterson and Berney [1] made two-dimensional coordinates (F1-F2) for the first (F1) and second (F2) formant frequencies, and drew ten elliptical enclosures relevant to the English vowels. If, in the F1-F2 coordinates, different formant pairs (F1, F2) of vowels point at loci that are within the same enclosure, the spoken vowels are classified as the same vowel. Different formant pairs belonging to the same vowel imply that we vocalize the same vowel in different acoustic waveforms. Suga [2] made similar two-dimensional coordinates for the Japanese vowels, on which five elliptical enclosures relevant to vowels, /a/, /i/, /u/, /e/ and /o/, were drawn. One question, then raised there, is how auditory systems place a variety of formant pairs (F1, F2) that belong to the same vowel into one relevant vowel category and in what manner the auditory systems use it for the perception of vowel sounds.

We propose here a neural network model for encoding and invariant perception of the Japanese vowels. The model has hierarchical structure. The first network (N_T), which is tonotopically organized, detects formant frequencies of vowel sounds. The second network (N_V) receives inputs from the N_T network in a convergent manner. The neurons of the N_V network are combination-sensitive and detect combinatory information about the formant pairs of vowel sounds.

Dynamic cell assemblies that encode categorical information about the five Japanese vowels are formed in the N_V network dynamics through Hebbian learning, whereby neurons within (across) the cell assemblies are connected through positive (negative) synapses. By this synaptic formation, the same cell assembly is selectively activated when stimulated with formant pairs that belong to the same vowel, during which the activities of the other cell assemblies are completely suppressed. To investigate neuronal bases of invariant perception of vowel sounds, we stimulate the model with various formant pairs, record and analyze the activities of neurons.

A recent experiment [3] has suggested that time-varying spectral changes in formant frequencies are important for the perception of vowel sounds. In that experiment, human subjects were presented with natural and synthesized vowel sounds, between which cognitive performance was compared. The natural vowels had time-varying spectral changes in formant frequencies. For example, vocalization of vowel /e/ started at (F1, F2) = (500Hz, 2000Hz) and continuously changed toward (380Hz, 2200Hz), while a synthetic sound of /e/ consisted of constant frequencies (497Hz, 1982Hz). The researchers showed that the natural (vocalized) vowels are more detectable than the synthesized vowels. Although the experiment has clearly demonstrated the importance of time-varying spectral changes in formant frequencies for vowel sound perception, its neuronal mechanisms have still remained to be seen. To investigate the neuronal significance of the time-varying formant frequencies to the perception of vowel sounds, we stimulate the model with various time-varying formant frequencies, analyze the dynamic behaviors of neurons, and compare these results to those with constant-frequency stimulation.

20

2 Auditory Processing and Neural Network Model

2.1 Auditory Processing

Auditory information is received by the cochlea and sent toward the auditory cortex relayed through multiple central auditory stages, such as the cochlea nucleus, superior olivary complexes, lateral lemniscus, inferior colliculus and medial geniculate body [4]. At the first stage of audition, i.e., at the cochlea, information about sounds is decomposed into frequency components. When an auditory sound is presented, the sound pressure distorts the basilar membrane of the cochlea and makes a traveling wave along the basilar membrane. Since the receptor neurons are distributed over the basilar membrane, the neurons located at a point where the amplitude of the traveling wave (or the amount of displacement of the basilar membrane) is maximal tend to fire more action potentials than others.

The basilar membrane is wider at the apical end than at the basal end, where the traveling wave proceeds from the basal to apical end. Because of this specific shape of the basilar membrane, the location of the maximal displacement of the basilar membrane shifts from the apex to the base as the frequency of a pure tone changes from low (e.g., 60Hz) to high (e.g., 2KHz). This mechanical system enables each receptor neuron to respond to a specific frequency, and the neurons sensitive to lower frequencies are located at the apical end and those to higher frequencies are at the basal end. This orderly arrangement of the frequency-specific neurons is called "tonotopic organization".

It is well known that this tonotopic expression of auditory information is kept throughout the central auditory system toward the primary auditory cortex, and that the cortex is organized in columns of neurons. Each column is perpendicular to the surface of the cortex and contains neurons that tend to respond to the same frequency. As will be explained in section 2.2 (Fig. 1a), we model the primary auditory cortex (as the "N_T" network of Fig. 1a) that has such tonotopic organization and columnar structure.

In a nonprimary auditory cortical area, Langner and colleagues [5] found neurons that are responsive specifically to the German vowels. Using synthetic vowels composed of formant frequencies, they demonstrated that these neurons respond to the specific combinations of formant frequencies (F1, F2). Note that vowel sound spectra contain multiple peaks in frequencies. These peaks are called format frequencies (F1, F2, F3, and so on), where the amplitudes of F1, F2 and F3 are the highest, the second and the third, respectively. An example of the first (F1) and second (F2) formant frequencies of a vowel sound will be spectrally shown in section 2.2 (the bottom of Fig. 1a). Langner and colleagues [5] have suggested that combination-sensitive neurons (i.e., sensitive to specific combinations of formant frequencies) may contribute to processing auditory information about vowels. As will be explained in 2.2, we make a nonprimary auditory cortical area (as the "N_V" network of Fig. 1a) that consists of such combination-sensitive neurons.

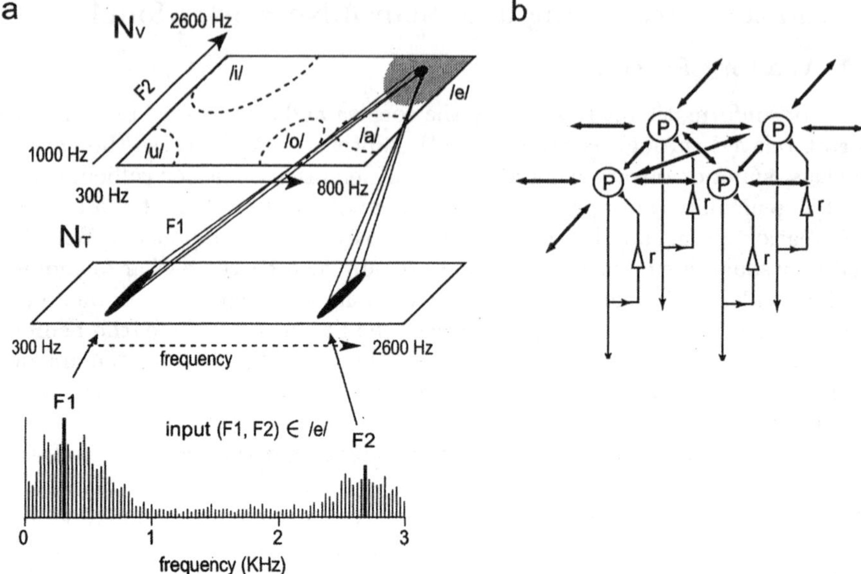

Fig. 1. (a) Structure of the neural network model. The N_T network (middle) and the N_V network (top) are composed of frequency-sensitive columnar neurons and combination-sensitive neurons, respectively. A formant pair of a vowel, e.g., /e/ (bottom), activates the specific N_T neuronal columns ("black ellipses"), which activate one specific N_V neuron ("black circle") and then the cell assembly ("gray region"). The dotted regions indicate the cell assemblies for the other vowels, /a/, /i/, /u/ and /o/. (b) Neuronal architecture of the N_V network. Projection neurons (P) are mutually connected with each other via an excitatory or an inhibitory synapse. Each projection neuron sends axonal signals to its accompanying interneuron (r) via an excitatory synaptic connection, and receives feedback signals from the interneuron via an inhibitory synaptic connection.

2.2 Structure the Neural Network Model

The structure of the model is shown in Fig. 1a. The model is hierarchical. The N_T network is tonotopically organized and detects spectral peaks of vowel sounds, i.e., the first (F1) and second (F2) formant frequencies. The N_V network is a two-dimensional tonotopic map, and receives inputs from the N_T network in a convergent manner ("solid" lines). The N_V neurons are combination-sensitive and detect combinatory information about the formant frequencies.

The N_T network consists of neuronal columns. Positive (negative) synaptic connections are made within (across) the columns. The N_V network consists of neuron units (Fig. 1b). Each unit has a pair of a projection ("P") neuron and an interneuron ("r"). The interneuron (IN) receives an excitatory input from the projection neuron (PN) and sends an inhibitory output to the PN.

The PNs are connected to each other via either an excitatory or an inhibitory synapse. The PNs of the same cell assembly (each of the five "enclosed" regions of the top of Fig. 1a) are recurrently connected through positive synapses. The PNs of different cell assemblies are mutually inhibited through negative synaptic connections. A Hebbian learning process that precedes the present simulations constructs this synaptic structure, which will be shown in section 3.1.

Stimulation of the N_T network with a formant-pair (F1, F2) of a vowel (e.g., /e/) activates specific neuronal columns (the "black" columns of Fig. 1a). These columnar neurons then send axonal outputs to the N_V network and activate one neuron ("black" circle) of the N_V cell assembly ("gray" region) corresponding to /e/. Due to the recurrent positive synaptic connections between the PNs of the same cell assembly, the activation of the PN propagates throughout the cell assembly, and the whole neurons of the cell assembly tend to be activated.

To make the combination-sensitive N_V neurons, we made convergent projections from the N_T to N_V network. This is our assumption and such orderly projections have not been fully confirmed yet in the auditory cortex but in other brain areas. For example, it has been shown [6, 7, 8] that anterior part of the inferior temporal cortex has such convergent projections to the perirhinal cortex. Damasio [9, 10] has proposed a neuronal architecture called "convergence zone" that integrates different modality features distributed over multiple cortical (or subcortical) areas into single perceptual units (e.g., perceptual objects or events). The convergence zone receives convergent inputs from lower cortical areas. We consider that the present N_V network functions as a convergence zone for vowel sound information processing.

2.3 Overall Performance of the Model

The same cell assembly is always activated when stimulated with formant-pairs that belong to the same vowel category. That is, the dynamic cell assembly encodes categorical information about that vowel. The selective activation of one specific dynamic cell assembly could be interpreted as the perception of the applied vowel sound. These dynamic cell assemblies form stable firing patterns (or point attractors) in the N_V network dynamics, as schematically shown in Fig. 2.

For simplicity, we assigned one neuron to process information about one specific formant pair. That is, when the N_T network is stimulated with a given formant pair, only one N_V neuron that is specific to that pair can respond to. However, it should be mentioned that in real cortical neural networks, a population of neurons is likely to respond to the same stimulus. In this respect, the single N_V neuron could be regarded as a population of neurons in the cortex.

As will be shown in section 3.2, one of the notable properties of the present model is that the activities of N_V neurons are not stationary under the ongoing

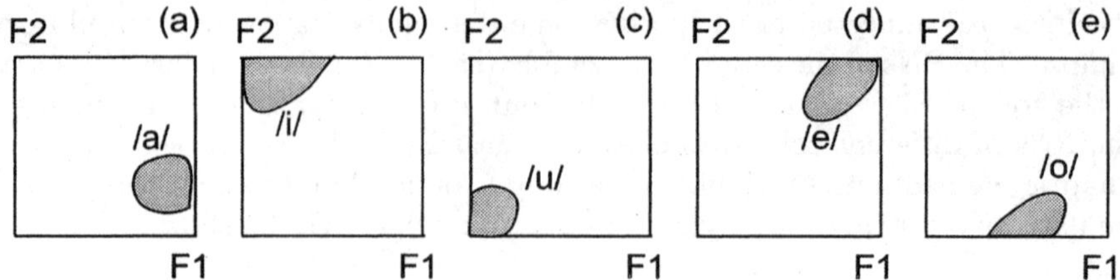

Fig. 2. Representation of vowels by the stable firing patters of the N_V network for vowels /a/ (a), /i/ (b), /u/ (c), /e/ (d) and /o/ (e). The shaded regions denote the collective firings of PNs, or cell assemblies. Note that the PNs within (across) these regions are connected through positive (negative) synapses.

background state but they are changing and show self-generated spontaneous activity like real cortical neurons [11, 12]. The ongoing state of the N_V network is an itinerant state among the five dynamic cell assemblies (or point attractors) [13, 14, 15, 16, 17]. In the present model, the perception of an applied vowel sound is processed as a phase transition from the itinerant state to the dynamic cell assembly (or point attractor) corresponding to the input stimulus [18], as schematically shown in Fig. 3.

The synaptic dynamics plays an important role in establishing the ongoing background state in the N_V network dynamics. The self-inhibition of PNs within the cell assemblies through their accompanying interneurons destabilizes the currently appearing dynamic cell assembly (or attractor), by which the N_V network state is able to make transitions among the dynamic cell assemblies even without external stimulation. The lateral-inhibition between the cell assemblies allows only one dynamic cell assembly to appear. That is, when one cell assembly is appearing, the other cell assemblies tend to be suppressed. This enables each dynamic cell assembly to respond to specific sensory stimulation.

2.4 Model Description

Membrane potentials of N_T neurons are described by

$$\tau_T \frac{du_{p,i}^T(t)}{dt} = -u_{p,i}^T(t) + \alpha_i I_i(t), \tag{1}$$

where τ_T is the time constant of membrane potential $u_{p,i}^T(t)$ of neuron i at time t. The N_T network contains 120 tonotopic neuronal columns, each of which contains 6 neurons that are sensitive to the same frequency. The columns are orderly arranged along the tonotopic axis, in which frequency ranges from 300Hz to 2600Hz with an interval of 20Hz (see Fig. 1a). $I_i(t)$ represents the amount of auditory stimulation and takes on a value of 1 or 0. The intensity of stimulation is determined by a positive coefficient α_i.

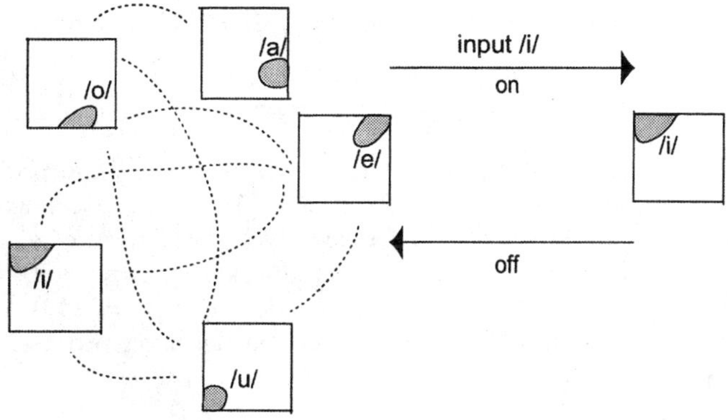

Fig. 3. A schematic drawing of the perception of an applied vowel sound. When the N_V network is stimulated with a formant pair that belongs to a given vowel, e.g., /i/, the state of the N_V network changes from the randomly itinerant motion among the five dynamic cell assemblies (left) to one dynamic cell assembly (right) corresponding to the vowel /i/, where the PNs of the cell assembly (/i/) continuously fire action potentials during the input period. After switching off the input, the state of the network returns to the itinerant state.

Membrane potentials of N_V projection neurons (PNs) and interneurons (INs) are described by

$$\tau_p^V \frac{du_{p,i}^V(t)}{dt} = -u_{p,i}^V + \sum_{j=1}^{M_T} J_{ij}^{V,T} \cdot V_j^T(t)$$

$$+ \sum_{j=1}^{M_V} w_{pp,ij} \cdot V_{p,j}^V(t) + w_{pr} \cdot V_{r,i}^V(t), \tag{2}$$

$$\tau_r^V \frac{du_{r,i}^V(t)}{dt} = -u_{r,i}^V(t) + w_{rp}^V \cdot V_{p,i}^V(t), \tag{3}$$

where τ_p^V and τ_r^V are the time constants of the membrane potentials of PNs and INs, respectively. $u_{p,i}^V(t)$ and $u_{r,i}^V(t)$ are the membrane potentials of PNs and INs, respectively. The N_V network contains 25 (F1-axis) × 80 (F2-axis) neuron units. The neuron units are orderly arranged along the F1- and F2-axis, in which frequency ranges, respectively, from 300Hz to 800Hz and from 1000Hz to 2600Hz with an interval of 20Hz. M_T and M_V are the numbers of N_T neurons and N_V neuron units, respectively. The convergent projections from the N_T to N_V network are made by properly setting $J_{ij}^{V,T}$ to 1 or 0. $w_{pp,ij}$ is the strength of synaptic connection from jth to ith PN. $V_j(t)$ is the axonal output of neuron j. w_{pr} (w_{rp}) is the strength of synaptic connection from IN to PN (from PN to IN).

Outputs of neurons are determined by

$$Prob[V_{z,i}^{X} = 1] = f_z[u_{z,i}^{X}(t)], \qquad (X = T, V; \ z = p, r) \tag{4}$$

$$f_z[y] = \frac{1}{1 + e^{-\eta_z(y - \theta_z)}} \tag{5}$$

Eq. (4) defines the probability of neuronal firing, that is, the probability of $V_{z,i}^{X}(t) = 1$ is given by f_z. η_z and θ_z in Eq. (5) are the steepness and the threshold of the sigmoid function f_z, respectively, for z kind of neuron.

Values of the network parameters used here are $\tau_p = \tau_r = 50$ msec, $\theta_p = 0.1$, $\theta_r = 0.9$, $\eta_p = 14.0$, $\eta_r = 21.0$, $\tau_w = 100$ sec. $w_{pr} = -1.0$, $w_{rp} = 1.0$. The value of $w_{pp,ij}$ is determined by a Hebbian learning process, as will be described in section 3.1.

3 Simulation Results

3.1 Auditory Map Formation

As pre-wired synaptic structure for $w_{pp,ij}$, adjacent PNs were reciprocally connected via excitatory synapses. To make the cell assemblies sensitive to specific vowels, we let the N_V network learn the Japanese vowels according to the Hebbian learning rule as defined by the following equation.

$$\tau_w \frac{dw_{pp,ij}(t)}{dt} = -w_{pp,ij}(t) + \epsilon \cdot (2V_i(t) - 1)V_j(t), \tag{6}$$

where τ_w is a time constant, and ϵ a rate of synaptic modification.

During the learning process, the spectral components of each vowel sound is applied to the corresponding N_T neurons, as shown in Fig. 4. Due to the lateral inhibitory synaptic connections between the neuronal columns, the two columnar activities (the two sets of the six "filled" circles of the N_T network) corresponding to the formant pair (F1-F2) are sharpened and the other columns are suppressed. The simultaneous activation of the two neuronal columns then activates a specific N_V neuron ("black" circle) that then activates its adjacent neurons ("shaded" circles) via the pre-wired positive synaptic connections.

When the N_T network is stimulated with vowel /i/ spoken by subject A, as denoted by $/i/_A$ in Fig. 5a, a population of N_V neurons is selectively activated ("filled" region). The vowels spoken by subjects B and C ($/i/_B$ and $/i/_C$ of Fig. 5b-c) activate other populations of N_V neurons. When the region of an activated population overlaps with the previously activated regions ("open" regions), these neurons are activated together (the "filled" region of Fig. 5d). After learning of /i/ that is spoken by ten different subjects A~J, one specific cell assembly corresponding to /i/ is formed ("filled" region of Fig. 5e). The cell assembly tends to be selectively activated when stimulated with formant pairs that fall within the "filled" region of the N_V map.

Note that if the model learned less numbers of spoken vowels, multiple dynamic cell assemblies would be created for each vowel. This means that different vowel sounds belonging to the same vowel category would activate different dynamic cell assemblies, and therefore the unified perception of the vowel sounds would be impossible. This problem can be overcome when the network is presented with additional vowel sounds until these multiple dynamic cell assemblies overlap and merge into one large dynamic cell assembly.

Through Hebbian learning for the five vowels (/a/, /i/, /u/, /e/, /o/), the state of the N_V network becomes itinerant among the five dynamic cell assemblies (or point attractors) as shown in Fig. 6. The itinerancy of the network state among relevant dynamical attractors such as point attractors, limit-cycle attractors or chaotic attractors, have been proposed as one of the plausible neuronal representations for cognitive sensory mapping and called "Dynamical map" (for detail see ref. [13]). The detail of the dynamic structure of the present itinerant network state will be demonstrated in section 3.2.

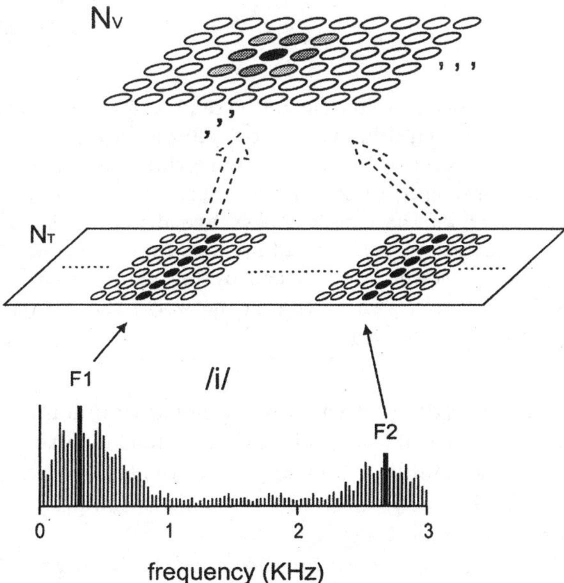

Fig. 4. Responses of the N_T (middle) and N_V (top) networks to vowel (/i/) sound stimulation (bottom). The level of darkness of the shaded circles of the N_V network denotes the level of activation of neurons.

3.2 Dynamic Properties of the Model

The N_V network has ongoing spontaneous neuronal activity as revealed by the raster plots of action potentials of PNs (see Fig. 6), in which the dynamic

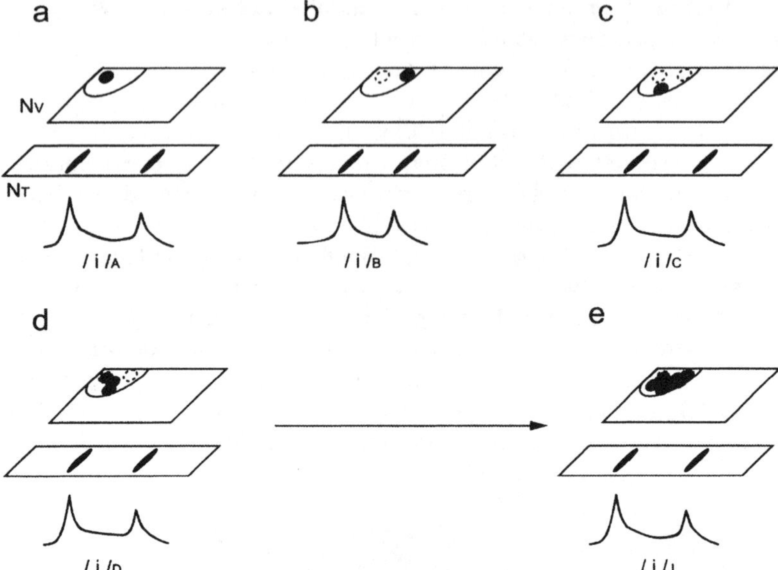

Fig. 5. Schematic drawing of dynamic cell assembly formation through Hebbian learning. (a) The N_T neuronal columns ("filled" ellipses) are activated by $/i/_A$ stimulation that is vowel $/i/$ spoken by subject A. During the stimulation period, the synaptic connections between PNs are modified according to the Hebbian rule. Hebbian learning takes place for the vowels spoken by subjects B (b), C (c) and D (d). The "open" regions indicate populations of neurons that were activated in the past. (e) After learning of the vowel ($/i/$) spoken by ten subjects (A-J), a specific dynamic cell assembly ("filled" region of the N_V network) that represents categorical information about the vowel is formed.

cell assemblies corresponding to the vowels emerge randomly. The temporal formation of the dynamic cell assemblies arises from the mutual (PN-PN) excitation within the cell assemblies. The emergence of each dynamic cell assembly is brief, because the PNs are inhibited by their accompanying interneurons (INs) as soon as they are activated.

Note that the dynamic cell assemblies tend to be segregated from each other under the ongoing spontaneous state. That is, when one dynamic cell assembly is appearing, the other cell assemblies tend to be suppressed. This is due to the mutual inhibition between the cell assemblies via PN-PN inhibitory connections that have been made through the Hebbian learning process.

The present ongoing spontaneous state has an important dynamic aspect. Namely, the ongoing motion among the dynamic cell assembly is random but the activities of the PNs constituting the same cell assembly are coherent. The clear peak centered at around time lag $= 0$ of a cross-correlation function of action potentials between two PNs of the same cell assembly (Fig. 7a) indicates

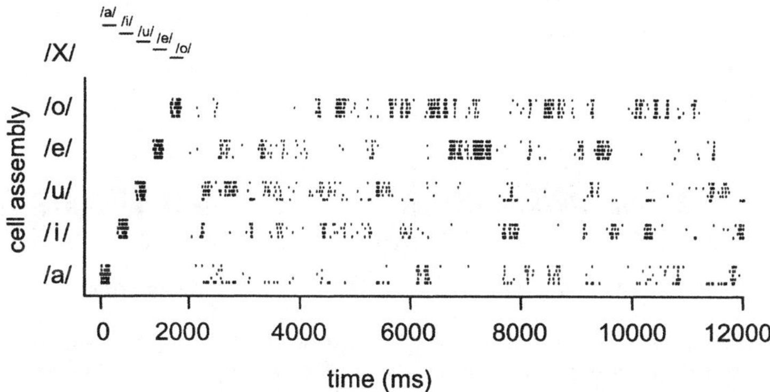

Fig. 6. Raster plots of the PNs of the N_V network during and after the Hebbian learning process. The vowels are separately presented ("horizontal" bars) and the synaptic strengths are modified according to the Hebbian learning rule (see Eq. (6)). After learning, the state of the N_V network becomes randomly itinerant among the five dynamic cell assemblies (or the five point attractors).

that coherent neuronal activity prevails under the ongoing spontaneous state. A cross-correlation function between a PN and its accompanying IN, as shown in Fig. 7b, indicates that the suppression of the PN by the IN is occurring.

The cross-correlation functions of action potentials between neurons are defined by

$$CF(\tau) = \frac{1}{(T_2 - T_1)} \int_{T_1}^{T_2} V(x;t)V(y;t+\tau)dt, \qquad (7)$$

where T_1 and T_2 are the initial and final time of a period that is used to calculate $CF(\tau)$, respectively. $V(x;t)$ and $V(y;t+\tau)$ denote action potentials of neurons x and y, respectively.

The basic cognitive property of the present model is shown in Fig. 8a. When the N_T network stimulated with a vowel sound, whose duration is indicated by a "horizontal bar", the PNs of the cell assembly corresponding to the applied vowel is activated and fire action potentials. After switching off the input, the state of the N_V network returns to the ongoing motion. Figure 8b indicates that coherency in neuronal activity between the PNs improves (as compared with that of Fig. 7a) when the network is stimulated with a vowel (/i/) sound (time = 2000-3000).

The positive synaptic connections between the PNs are necessary to improve cognitive performance. If the positive synaptic connections between PNs within the cell assemblies are weakened, the cell assemblies show similar responsiveness (Fig. 9a) but the activities of these PNs loose coherency (Fig. 9b). One question that arises is what the significance of such neuronal coherency to cognitive performance is.

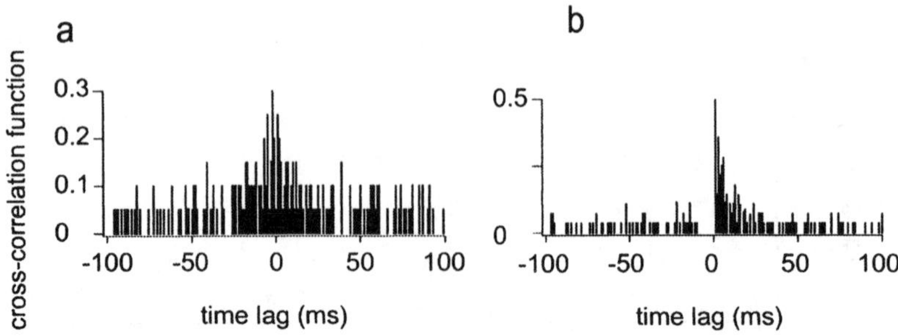

Fig. 7. Cross-correlation functions of neuronal activity during the ongoing spontaneous state. (a) A cross-correlation function of action potentials between two PNs of the same cell assembly (/i/). (b) A cross-correlation function between a PN and its accompanying IN.

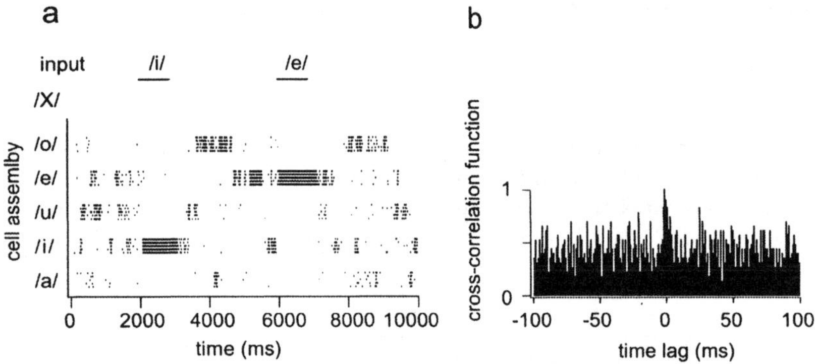

Fig. 8. Cognitive property of the N_V network. (a) Responses of the cell assemblies to vowel sound stimulation. (b) A cross-correlation function of action potentials between the two PNs (the same neurons as used in Fig. 7a) during the stimulation period (time = 2000-3000).

As revealed by Fig. 10, the coherency within the dynamic cell assembly is essential for the enhancement of cognitive performance, or the recognition of the input stimulus. The positive synaptic connections between the PNs enhance reaction time to the input stimulus as shown in Fig. 10a. In contrast, the performance deteriorates for that with weakened connections, increasing reaction time as shown in Fig. 10b. It has been suggested [19] that synchrony in neuronal activity is likely to not only influence the network performance itself but also has a great impact on the neurons of subsequent stages, effectively activating these neurons and thus enhance their neuronal information processing. We consider that the coherent activation of neurons within the dynamic cell assemblies may be essential for neural information processing in the auditory cortex as well.

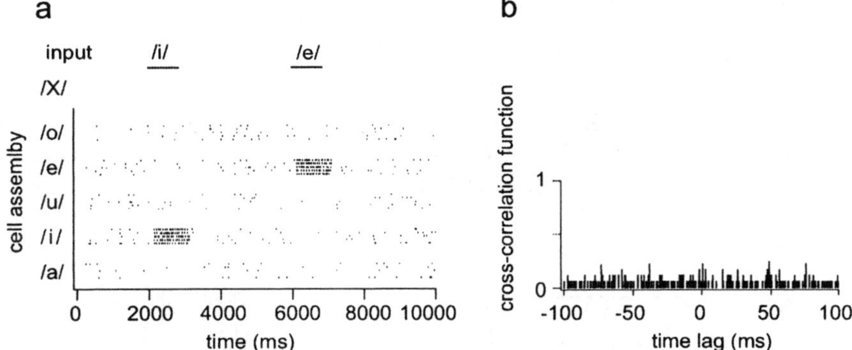

Fig. 9. Cognitive property of the N_V network, in which the synaptic connections between the PNs within the cell assemblies are weakened. (a) Responses of the cell assemblies to vowel sound stimulation. (b) A cross-correlation function of action potentials between the two PNs of the cell assembly (the same neurons as used in Figs. 7a and 8b) during the stimulation period (time = 2000-3000).

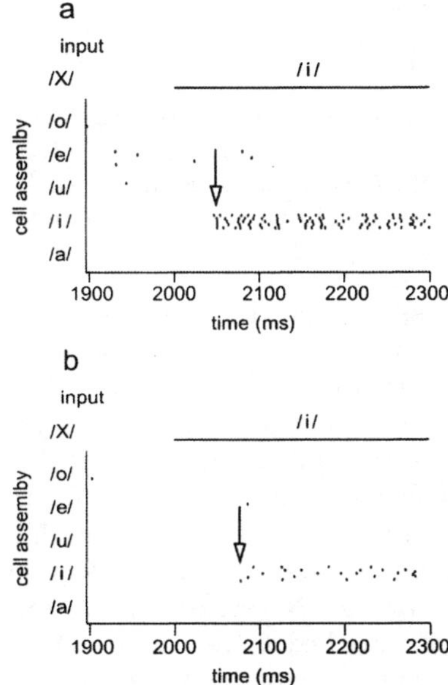

Fig. 10. Reaction time to vowel sound stimulation. (a) Responses of the PNs, where the positive synaptic connections within the cell assemblies are intact. (b) Responses of the PNs, where the positive synaptic connections within the cell assemblies are weakened. The "arrows" indicate the time at which the cell assembly starts to respond.

3.3 Perception of Simple Vowel Sounds

When the N_T network is stimulated with formant-pairs $((F1, F2)_A, (F1, F2)_B, (F1, F2)_C)$ that are spoken by different subjects (A, B, C) but belong to the same vowel /e/ (Fig. 11a), the dynamic cell assembly corresponding to /e/ is always induced in the N_V network dynamics (Fig. 11b). That is, the three different formant-pairs are invariantly perceived as vowel /e/.

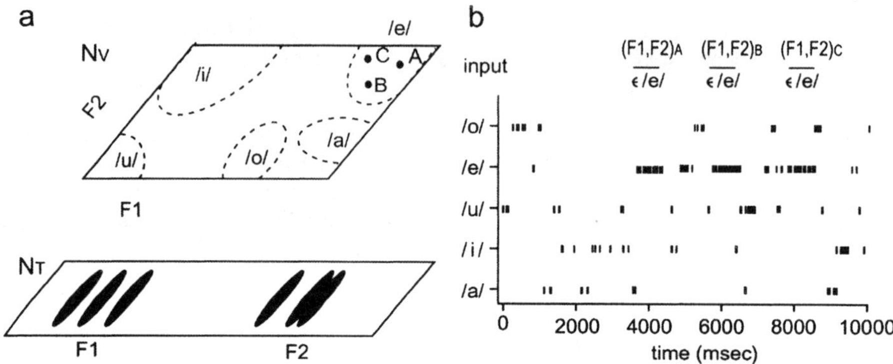

Fig. 11. (a) Stimulation of the neurons ("filled" circles) of the cell assembly (/e/) with different formant-pairs $((F1, F2)_A, (F1, F2)_B, (F1, F2)_C)$. (b) Response of the cell assembly to the stimuli. A vertical bar is drawn at /X/-row (X = a, i, u, e, o) when the cell assembly corresponding to vowel /X/ emerges.

It is interesting to ask whether the network can recognize unknown formant pairs, that is, the formant pairs that the model has not learned previously. When an unknown format pair of vowel /e/ was applied, the same cell assembly (/e/) was activated if the formant pair falls within the region corresponding to /e/ (see the "filled' region of the N_V network of Fig. 5e) or was not otherwise (not shown). The key process for the activation of the cell assembly is to stimulate a member of the assembly. Therefore, the perception of an applied vowel sound does not depend on whether its formant pair has been learned previously or not. Nevertheless, it seems to be required for reliable perception that the applied vowel sound contains formant frequencies similar to those that have been experienced, or learned previously.

To investigate the significance of the time-varying spectral property of natural vowel sounds, we stimulated the N_T network with various time-varying formant frequencies. Figure 12 shows how a certain cell assembly (e.g., /e/) responds to a pair of time-varying formant frequencies, where (F1, F2) changes linearly from (500Hz, 2000Hz) to (380Hz, 2200Hz), as indicated by the "arrows" in Fig. 12a. Figure 12b indicates that the dynamic cell assembly (/e/) is induced by the stimulation.

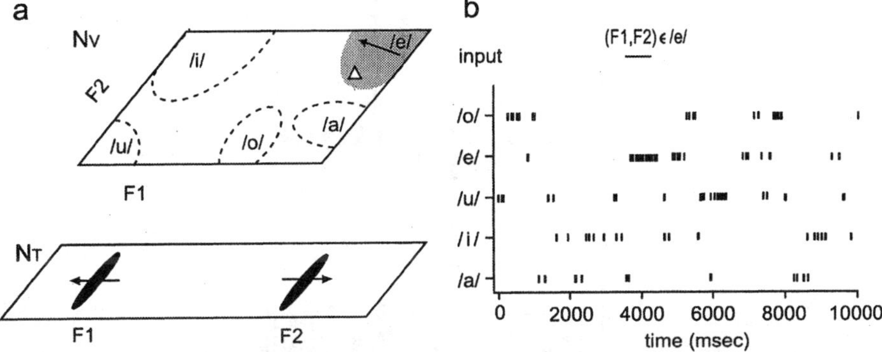

Fig. 12. (a) Stimulation of the cell assembly (/e/) with time-varying formant frequencies, F1 and F2. The "arrows" indicate the changes of individual frequencies, F1 and F2 (bottom), and the change of the locus F1-F2 (top). The "triangle" denotes the neuron whose activity was recorded (see text). (b) Response of the cell assemblies to the time-varying formant frequencies.

We assessed the cognitive performance, reaction time, by measuring the response latency of a N_V projection neuron from the onset of stimulation. Figure 13 is the activity of a PN ("triangle" of Fig. 12a) after the onset of stimulation with $(F1, F2)_A$. Note that the PN does not receive any direct input from the stimulus but is activated indirectly by recurrent inputs from other members of the cell assembly. Stimulation with the constant formant frequencies (Fig. 13a) activates the PN more slowly than that with the time-varying formant frequencies (Fig. 13b).

The difference in reaction time may be due to difference in activity propagation of PNs. That is, when the cell assembly is stimulated with the constant formant-pair (Fig. 11b), the active site could gradually propagate into other PNs of the assembly via their positive synaptic connections. In contrast, the stimulation with the time-varying formant pair could effectively stimulate the PNs, and therefore enhances the propagation of neuronal activation throughout the cell assembly.

3.4 Perception of Complex Vowel Sounds

To understand the basic properties of the present model, we used a simple neuronal representation of vowel information as shown in Fig. 11a (top), where the cell assemblies do not overlap. However, some neurons of the auditory cortex show responsiveness to more than one vowel [5]. For example, neurons of the field L of mynah birds, to which tonotopically organized primary auditory areas project, have responsiveness to both /a/ and /o/ of the German vowels.

To further the present study, we have made another neural network model that contains bimodal neurons that have specific sensitivity to two vowels,

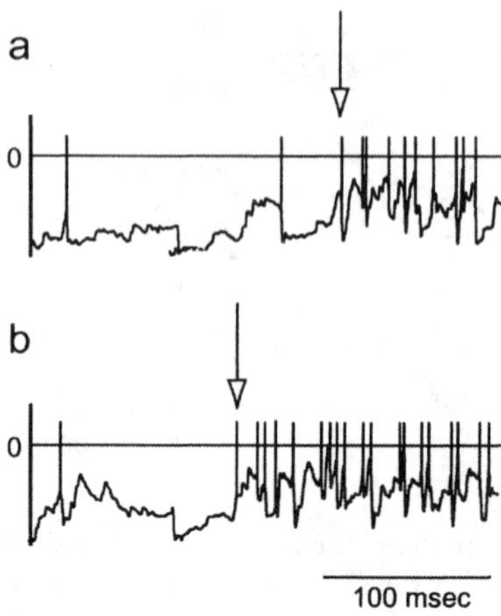

100 msec

Fig. 13. The activity of the PN neuron indicated by the "triangle" in Fig. 12a after the onset of stimulation. The action potentials and the membrane potential of the neuron are superimposed in the same graph. (a) Response to the constant formant-frequency stimulation $((F1, F2)_A$ of Fig. 11). (b) Response to the time-varying formant-frequency stimulation $((F1, F2)$ of Fig. 12). The "arrows" point to the time when neuronal bursting starts.

as expressed by "/a/ ∩ /o/" of Fig. 14a. This is based on a "psychological map" for vowel sound perception in humans [2]. These bimodal neurons are sensitive to both /a/ and /o/. We stimulated one of the bimodal neurons with a constant format-pair (F1, F2). The active site of the N_V network is indicated by the "filled" circle. As shown in Fig. 14b, stimulation of the bimodal neuron induces the two dynamic cell assemblies, /a/ and /o/. These dynamic cell assemblies compete against each other during the stimulation period. This result may imply that the applied formant-pair is not distinguishable.

In a next simulation, we changed the frequencies of the formant-pair to move away from the bimodal region /a/ ∩ /o/ toward the single-modal region /a/, as indicated by the "arrow" of Fig. 15a (top). Just after the stimulation, the two cell assemblies (/a/ and /o/) compete, but the cell assembly /a/ is singled out as the frequencies move toward the /a/ region (Fig. 15b).

Figure 16 shows that the cell assembly /o/ emerges when the frequencies of the formant-pair moves toward the opposite direction, i.e., from /a/ ∩ /o/ to /o/. In this case, the cell assembly /o/ competes against /a/ at the beginning, but is finally singled out (Fig. 16b). That is, the formant-pair is perceived as vowel /o/. The neuronal significance of the time-varying formant frequencies, which is one of the notable characteristics of natural (vocalized) vowel sounds in humans, will be discussed in section 4.

34

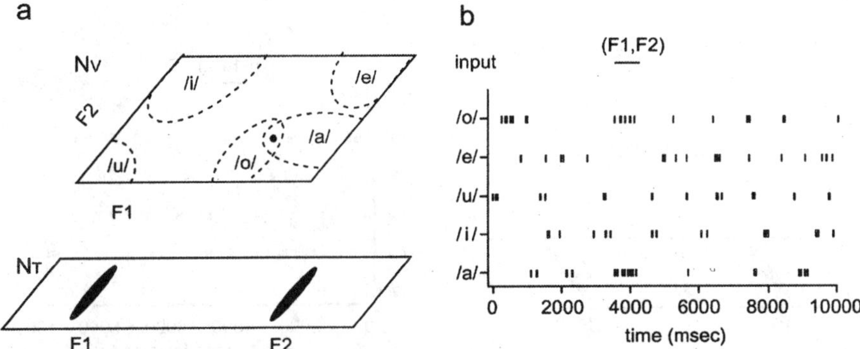

Fig. 14. (a) Stimulation with constant formant frequencies F1 and F2 that falls within the overlapping region, "/a/ ∩ /o/". (b) Responses of the cell assemblies.

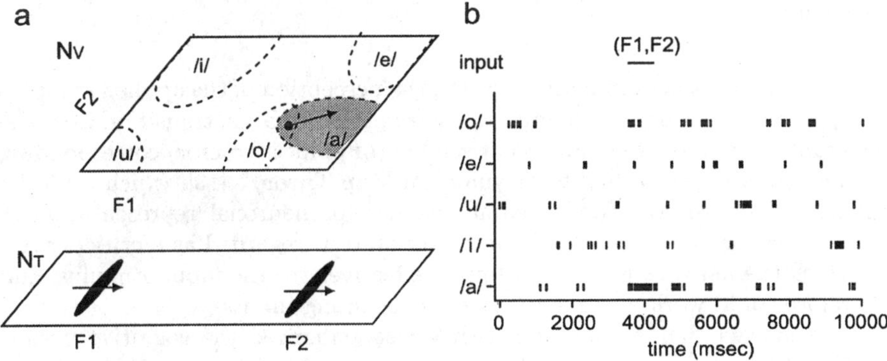

Fig. 15. (a) Stimulation with time-varying formant frequencies that sweep from /a/∩/o/ toward /a/. The arrows indicate the changes of individual frequencies, F1 and F2 (bottom), and the change of the locus F1-F2 (top). (b) Responses of the cell assemblies.

4 Discussion

To explore fundamental neuronal mechanisms for encoding and perception of vowel sounds, we made a hierarchical neural network model. In the model, the lower and higher networks processed spectrally decomposed information about vowels and combinatory information about the formant frequencies of the vowels, respectively. Specific dynamic cell assemblies encoded categorical information about vowels. If a member of a certain cell assembly was stimulated with the formant pair of a vowel sound, the whole members of the cell assembly corresponding to the vowel was activated, and thus the applied vowel sound was perceived.

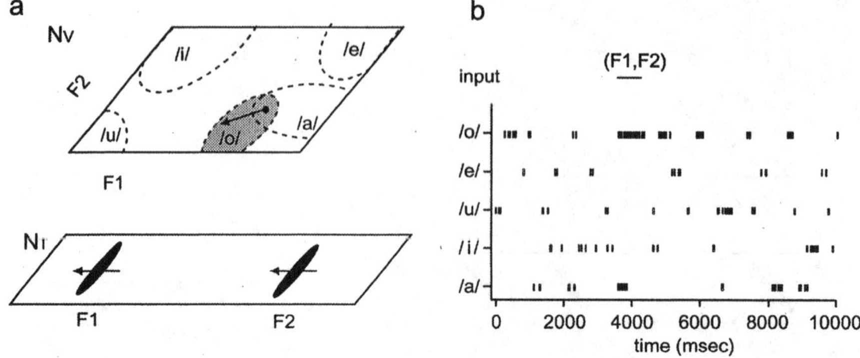

Fig. 16. (a) Stimulation with time-varying formant frequencies that sweep from /a/∩/o/ toward /o/. The arrows indicate the changes of individual frequencies, F1 and F2 (bottom), and the change of the locus F1-F2 (top). (b) Responses of the cell assemblies.

In the present neuronal architecture, the perception of the applied stimulus was processed through a dynamic phase transition, that is, transition from the itinerant state to the dynamic cell assembly (or point attractor) corresponding to the stimulus. According to "Dynamical Map Theory" [13], which could be considered one of the fundamental schemes for neuronal representation of cortical sensory information, the itinerant state is regarded as a critical state at which the network is able to respond effectively to the input stimulus, and therefore could work as a ready state for incoming inputs.

We showed that stimulation with vowels enhanced the cognitive performance of the neural network when the frequencies of formant-pairs varied in time. Since the neurons within the cell assemblies were recurrently connected with each other through positive synapses, the time-varying stimulation of these neurons might be advantageous for propagating neuronal excitation into the other members of the cell assembly, whereby the whole members of the cell assembly could be activated more rapidly and efficiently than stimulation with constant formant-frequencies. The present results may provide some insight into underlying neuronal mechanisms for the enhanced perception of natural vowel sounds in humans [3].

We showed that when bimodal neurons (i.e., sensitive to two different vowel categories) were stimulated with constant formant-frequencies, the two distinct cell assemblies corresponding to the two vowel categories emerged (see Fig. 14), and therefore the stimulus was indistinguishable. This problem could be overcome by changing the frequencies of the formant-pair (see Figs. 15 and 16) in such a way that the frequencies swept in opposite directions, that is, they swept away from each other on the two dimensional (F1, F2) N_V cognitive map. There have been evidence [3] that human subjects change formant frequencies in opposing directions for vowels that are close together in F1 and F2 coordinates (e.g, /e/ and /ε/ of the English vowels). We suggest that such time-varying spectral changes of formant frequencies in opposite directions may be essential for the brain to distinguish vowels that are similar in spectral property.

We assumed here two-dimensional structure for the N_V network. Although a similar two-dimensional neuronal model has been proposed by Langner and colleagues [5], no such auditory cortical area has been found. What we would like to suggest here is that combination-sensitive neurons (i.e., sensitive to pairs of formant frequencies), which have been found in the auditory cortex as addressed in section 1, may play an essential role in encoding and perception of spoken vowel sounds. We used the two-dimensional structure for modeling a simple auditory cortical map, whereby we could demonstrate essential neuronal mechanisms for vowel sound encoding and perception.

Several neural network models have been proposed for vowel sound encoding and perception. Sussman [20] proposed a multi-layered neural network model. The bottom layer, which has similar structure to the N_V network, detects formant frequencies of vowel sounds that are then transferred to and integrated by upper networks. The top layer detects the combination of the formant frequencies. Langner and colleagues [5] made a model in which cell assemblies represent categorical information about vowels. These cell assemblies consisted of recurrently connected neurons. Liu and colleagues [21] proposed an oscillatory neural network model in which synchronization of oscillatory neurons was essential for vowel sound perception.

The similarity between our and their models is that information about relationships between formant frequencies is processed by combination-sensitive neurons, and information about vowels are expressed by the collective activities of neurons, or cell assemblies. The major difference might be background neuronal activity. The background activity of their models is silent or the neurons continuously fire action potentials at maximal firing rates, unless these neurons are externally driven by random or constant stimulation. This is a typical property of conventional neural networks. Our model has ongoing coherent activity that is self-generative (i.e., no external stimulation is required), whereby we could demonstrate the important dynamic aspects of neuronal information processing of vowel sounds in the auditory cortex.

In the present study, we used the same duration of stimulation for every vowel and obtained similar cognitive performance. However, an experimental study [22] has demonstrated that durations of some vowels greatly affect the cognitive performance of human subjects. For example, the probability that /æ/ is heard as /ε/ increases as the duration of /æ/ is shortened. Similarly, when the duration of /ε/ is lengthened the probability that /ε/ is heard as /æ/ increases. This result may imply that information about different vowels are not equally expressed in the cortex and that some neurons (or cell assemblies) that are responsible for encoding /æ/ and /ε/ interact. Our model cannot explain this psychological result, because in the present model the dynamic cell assemblies have equally expressed the five vowels, and there is less interaction

between these cell assemblies. To investigate neuronal mechanisms of how
the vowel duration affects cognitive performance, the present network model
should be improved.

References

1. Peterson GE, Barney HL (1952) J Acoust Soc Am 24: 175-184
2. Suga N (1988) Auditory function: neurobiological bases of hearing. Wiley, New York
3. Assmann PF, Katz WF (2000) J Acoust Soc Am 108: 1856-1866
4. Yost WA (1994) Fundamentals of hearing: an introduction. Academic Press, San Diego California
5. Langner G, Bonke D, Scheich H (1981) Exp Brain Res 43: 11-24
6. Tanaka K, (1997) Curr Opin Neurobiol 7: 523-529
7. Saleem KS, Tanaka K (1996) J Neurosci 16: 4757-4775
8. Suzuki WA, Amaral DG (1994) J Comp Neurol 350: 497-533
9. Damasio AR (1989) Neural Computation 1: 123-132
10. Damasio AR, Damasio H (1994) Cortical systems for retrieval of concrete knowledge: The convergence zone framework. In: Koch C (ed) Large-scale neuronal theories of the brain. MIT Press, Cambridge MA
11. Tsodyks M, Kenet T, Grinvald A, Arieli A (1999) Science 286: 1943-6
12. Engel AK, Fries P, Singer W (2001) Nature Rev Neurosci 2: 704-17
13. Hoshino O, Usuba N, Kashimori Y, Kambara T (1997) Neural Networks 10: 1375-1390
14. Hoshino O, Kashimori Y, Kambara T (1998) Biol Cybern 79: 109-120
15. Hoshino O, Inoue S, Kashimori Y, Kambara T (2001) Neural Computation 13: 1781-1810
16. Hoshino O (2002) Connection Science 14: 115-135
17. Hoshino O, Zheng MH, Kuroiwa K (2003) Biol Cybern 88: 163-176
18. Hoshino O, Kashimori Y, Kambara T (1996) Proc Natl Acad Sci USA 93: 3303-3307
19. McBain CJ, Fisahm A (2001) Nature Rev Neurosci 2: 11-23
20. Sussman HM (1986) Brain and language 28: 12-23
21. Liu F, Yamaguchi Y, Shimizu H (1994) Biol Cybern 71: 105-114
22. Houde RA (2000) J Acoust Soc Am 108: 3013-3022

Convolutional Spiking Neural Network for Robust Object Detection with Population Code Using Structured Pulse Packets

Masakazu Matsugu, Katsuhiko Mori, Yusuke Mitarai

Canon Inc. Leading Edge Technologies Development Headquarters 5-1 Morinosato-Wakamiya Atsugi 243-0193 Japan

Abstract.
We propose a convolutional spiking neural network (CSNN) model with population coding for robust object (e.g., face) detection. Basic structure of the network involves hierarchically alternating layers for feature detection and feature pooling. The proposed model implements hierarchical template matching by temporal integration of structured pulse packet. The packet signal represents some intermediate or complex visual feature (e.g., a pair of line segments, corners, eye, nose, etc.) that constitutes a face model. The output pulse of a feature pooling neuron represents some local feature (e.g., end-stop, blob, eye, etc.). Introducing a population coding scheme in the CSNN architecture, we show how the biologically inspired model attains invariance to changes in size and position of face and ensures the efficiency of face detection.

Keywords. convolutional neural networks, object detection, face detection, population coding, spiking neural networks, pulse packet

1 Introduction

Object detection in cluttered scenes is generally a challenging task especially when we require size, rotation, and translation invariance as well as robustness against illumination changes. For example, locating faces without the explicit use of color and motion information but with discrimination between face and non-face objects in general is still difficult and open problem. Despite these facts, human can detect faces from a monochromatic picture in approximately 100 msec. Once a face is detected by human vision system, one can also recognize it later in a scene from somewhat different viewpoint and identify it even if its size is different from the one formerly detected. A great number of approaches [7, 37, 27, 15, 11, 30, 31, 35] have been taken for robustness in object recognition as well as detection. However, most existing models do not have excellent robustness altogether stated in the above and many of them use color or motion information to help facilitate the detection process.

Convolutional neural network (CNN) models [18], [16], [22] have been used in pattern recognition specifically for face as well as hand-written numeral recognition. A typical CNN model is shown in Fig.1. The CNN is a hierarchical network with feature detecting (FD) layers alternated with sub-sampling (or feature *pooling* [29]) layers. Each layer contains a set of planes in which each neuron has a common *local receptive field* (i.e., *weight sharing*). Each plane in FD layers can be considered as a feature map with regard to a specific feature class detected using common receptive field structure, for each neuron, as a convolution kernel. Sub-sampling or pooling mechanisms such as local averaging [18] or Max-like operation [29] inside a plane in the previous layer provide the entire system with robustness such as translation and deformation invariance. Thus CNNs involve three structural properties, namely, local receptive fields, shared weights, and spatial sub-sampling (or pooling), which are pioneered in a very similar yet older model, *Neocognitron* [9], known as a biologically inspired model.

Spiking neural network models with temporal coding [19] have also been extensively explored to enhance the capacity of information processing in the domain of Hopfield networks[20], RBF networks [28], and so on. In such models, spiking neurons encode information in terms of spike timing. For example, Natschläger and Ruf (1997) [28], using delay coding and competition among different RBF units, implemented RBF-unit in temporal domain in a manner that spiking neurons output an analog number encoded in spike time.

In this study, based on the convolutional spiking neural network (CSNN) model [22], we combine convolutional architecture with pulse neural networks for robustness and economy in object detection. In the hierarchical network, local patterns defined by a set of primitive features are mainly represented in the time domain structure of pulse signals from feature pooling (FP) neurons in a sub-sampling layer to a feature detecting (FD) neuron in the succeeding layer.

Only one pulse from each FP neuron is necessary to encode the existence of a local feature at particular position, similar in spirit to the rank order coding [34] for rapid processing. The proposed model, however, is different from the rank order coding in that each pulse packet [6] is structured so that the set of pulse modulation in the packet uniquely encodes not only the specification, but the saliency of visual features. In Section2, we describe a modular convolutional network architecture and a new module-based learning scheme using a variant of BP that allows efficient training of feature categories.

In Section 3, we propose a population spike coding scheme for size and rotation invariant representation in the convolutional spiking neural network [22]. In Section 4, we discuss properties of proposed population coding scheme and temporal coding aspects (i.e., structured pulse packet) that add to the convolutional networks for the robust face detection. We also give brief description of our ongoing work for VLSI implementation (analog-digital merged CMOS chip) of the proposed model.

2 Object Representation Using a Hierarchy of Local Features

In the proposed model, internal representation of object (e.g., face) is provided by a hierarchically ordered set of convolutional kernels defined by the local receptive field of FD neurons. For example, face model is represented as a spatially ordered set of local features of intermediate complexity, such as eyes, mouth, nose, eyebrow, cheek, or else, and all of these features are represented in terms of lower and intermediate features. The idea is based on our previous work [23], in which spatial arrangement of specific, elementary local features (figural alphabets) of intermediate complexity (not too primitive and not too complex), represented by distributed activation of lattice nodes, are used and integrated for object recognition.

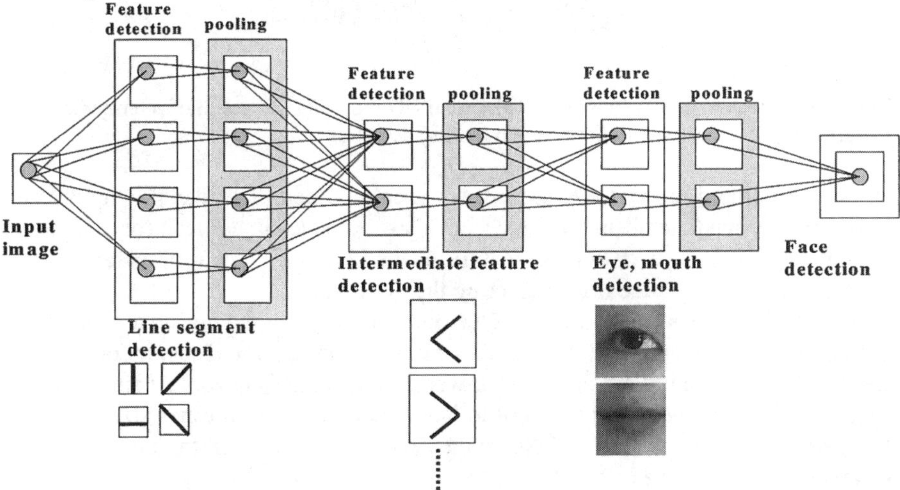

Fig.1. Convolutional architecture (simplified for illustration) for face detection

The lower and intermediate features constitute some form of a fixed set of figural alphabets (similar concept was defined in [8], but not used in this study as the same primitive feature) in our CSNN. Corresponding receptive fields for the detection of these *alphabetical* features [23] are learned in advance to form a local template in the hierarchical network, and once learned, they would never be changed during possible learning phase for object recognition.

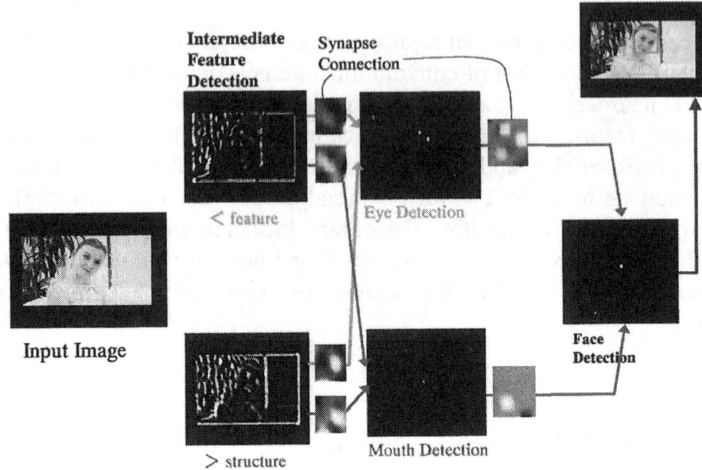

Fig. 2. Face detection by a simple convolutional NN including the result of intermediate features (e.g., '<', '>' end-stops, eye, and mouth) detection

As exemplified in Fig.1, we use a set of local oriented blob-like structures as a basis of intermediate visual features such as '<' shaped corners (or end-stop structures) for an eye or mouth feature as in [23] and a pair of parallel line segments (horizontal blobs) for depicting the eye feature.

Fig. 2 illustrates a process of the face detection in each module in a simplified convolutional network model. Three classes of images, multi-resolution image data, with down-sampling factor of 2 were input to the network. This network with multi-channel coding turned out to have tolerance up to 8 times size changes (i.e., from 35 x 35 to 280 x 280), with each channel having tolerance of size changes up to twice the original.

This set of primary visual features is compatible with Gestalt principles [17] stating that collinear or parallel features tend to be grouped to form a perceptual entity. Other intermediate features such as *blobs* are also incorporated. The proposed CSNN uses two classes of features that constitute face model and those do not. The latter features are used to ensure that non-face objects are discarded by the network. Once a *negative* local feature is detected in a local area where face pattern never holds, whereas in other areas consistent facial features are detected, the corresponding negative signal from the FP neuron reduces the activation level of the object (e.g., face) detecting neuron in the last FD layer, thus false facial pattern is rejected. In the convolutional NN model, FP neurons can perform either maximum value detection or local averaging in their receptive fields of appropriate size.

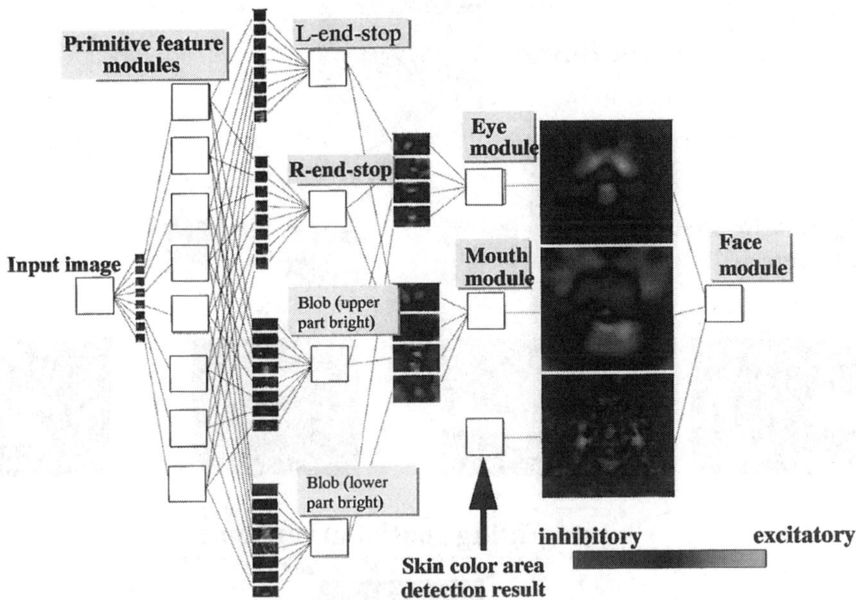

Fig. 3. Receptive field structures (connection weights) of feature detecting neurons in the proposed CNN

Training of the proposed model proceeds module by module only for the FD neurons from intermediate to top layers with connections in the preceding layers fixed. In the first step of training, the two FD layers from the bottom were trained using standard back-propagation by giving small image patches that include specific local features for the neuron in FD layers as a part of training sets. Negative examples that do not constitute the corresponding local (or global in the case of entire face) feature category are also used as false (negative) training data. As the result, the receptive field of bottom FD layer (FD1) neurons formed a sort of line segment detectors with different orientations selectivity as a result of giving teacher signal for each module in FD2 layer. The training data set included three scale levels (channels).

Similarly, in the second step, the next FD layer was trained for the detection of intermediate features of moderate complexity, including '<' or '>' shaped structures, line-pairs. Other local region-based feature (e.g., circular blobs) detectors can also be included. These features (fragments of images) for training are extracted from face images and thereby underwent some Affine transform to ensure robustness to size or orientation change for the detection of respective local features. We do not claim that these are the only local features that constitute the figural alphabets.

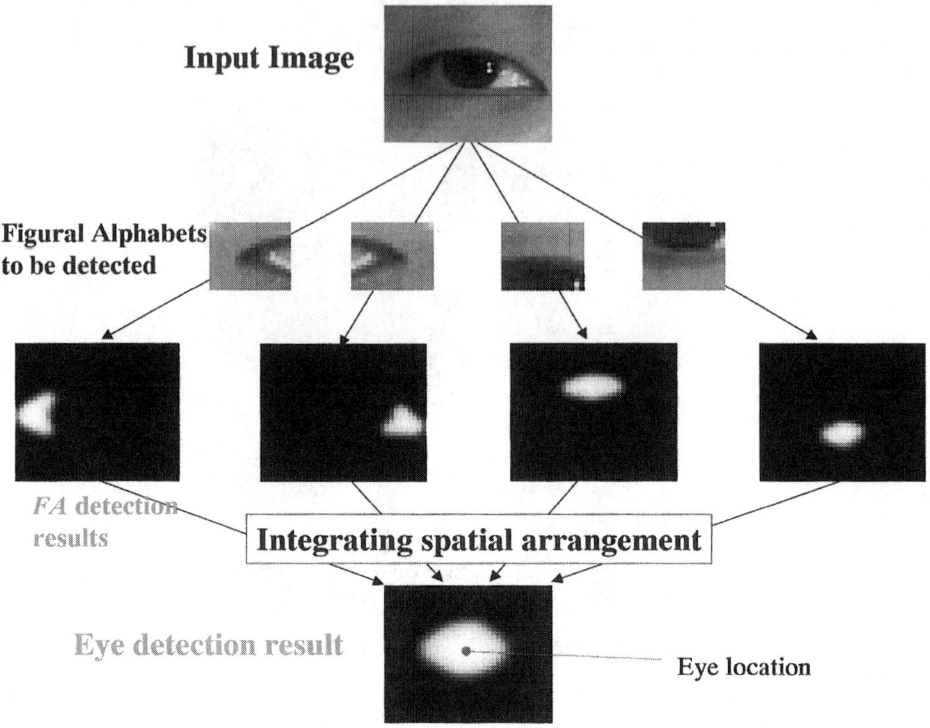

Fig. 4. Eye feature is decomposed of four alphabetical features. The network detects proper configuration of these features by a set of convolutional kernels

Other intermediate alphabetical features can be included in the CSNN model for object recognition, in general. Neurons in the second FD layer and the subsequent FD layers make connections with FP neurons of preceding layer. More complex local feature detectors (e.g., eye, mouth, and nose detectors, but not restricted to these) are trained in the third or fourth FD layer using the patterns extracted from the face image of different persons. The resulting receptive field structures of feature detecting neurons in respective modules are shown in Fig.3. Fig. 4 illustrates a detection process of an eye by integrating the spatial arrangement of a set of local features (figural alphabets) through convolution kernels (receptive field structure) of Fig.3.

3 Population Spike Coding with Structured Pulse Packet

In the following, we propose a conceptual framework to enhance size invariance by means of population codes in pulse packet signal. The original CSNN model

[22], [24] is a feed-forward, pulse-coupled, convolutional network, and synapses of an FD neuron (e.g., integrate-and-fire neuron) exert pulse modulation to pre-synaptic signals.

The amount of modulation, either in phase (PPM: pulse phase modulation) or in width (PWM: pulse width modulation), in each pulse uniquely determines detection strength of a specific local feature that constitutes some more complex pattern. In particular, the presence or absence of a pulse in a specific time-slot represents the detection (existence) or non-existence of the corresponding local feature. Conceptually, a set of phase-modulated pulses from FP neurons of different feature class is combined into a local bus to form a structured pulse packet (SPP) impinging onto an FD neuron (Fig. 5). However, the model in Fig.5 is rather complicated and we do not claim that this building block structure is ideal in terms of parsimony. In fact, we have alternative substrate for implementation as illustrated in Fig.6.

The SPP signals represent visual features as a combination of local features. The packet signal is structured in the sense that each pulse represents the presence and saliency of specific feature class in terms of spike arrival time to the FD neuron.

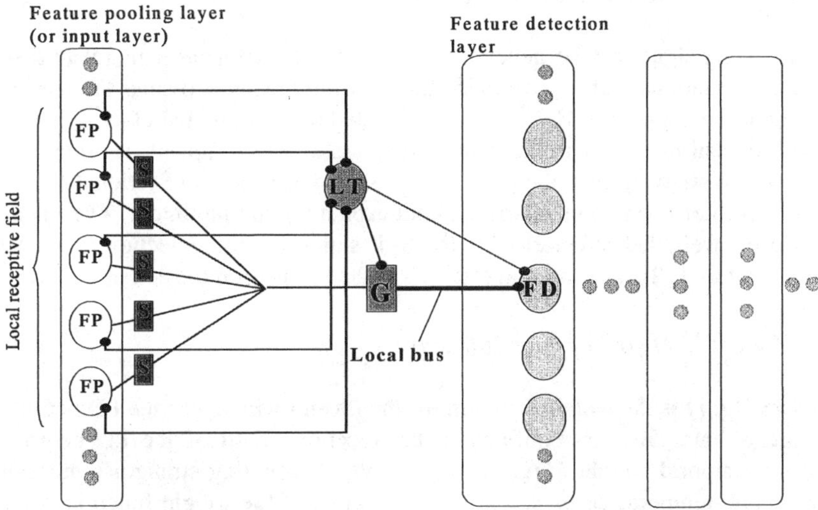

Fig. 5. Conceptual building block of CSNN, S: pulse modulating synapse, LT: local timing inter-neuron (pacemaker neuron)

The SPP signal, in this study, is not restricted to a set of PPM signals, and a set of PWM signals in parallel can also be used. Timing coordination among relevant neurons is important. In Fig.5, a local timing (LT) neuron as a pacemaker makes connections to both FP and FD neurons. Here, the local and distributed timing signal is used to facilitate phase-locking among FP and FD neurons for enhanced precision [22, 33] in the timing structure of SPP.

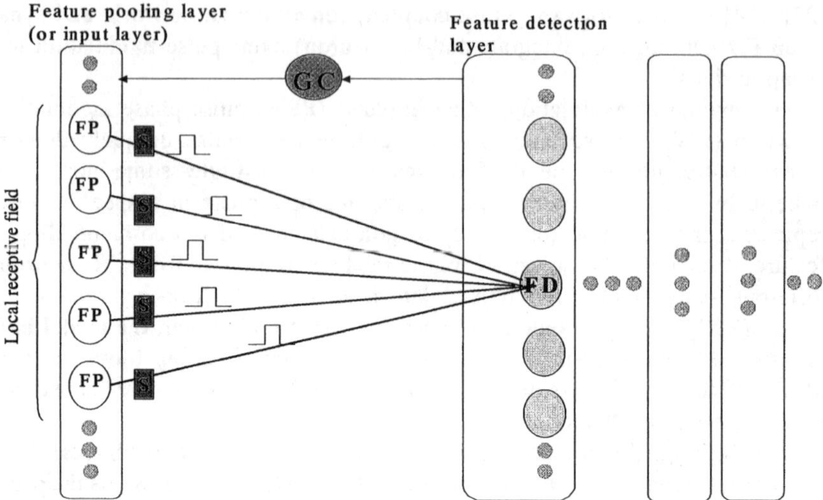

Fig. 6. Reduced building block: one FD neuron and FP neurons in the FD neurons's receptive field are phase-locked to a timing signal (not shown), each S acts as pulse modulator, and GC is a global clock generator

This timing signal can be generated in an event-driven manner in a way that only those LT neurons that receive pulse signals from locally activated FP neurons can generate timing pulses. However, as illustrated in Fig.6, global clock signal can be used to enforce synchronization among relevant group of neurons. In our simulation study, in fact, the latter model is used as shown in Section 4.

Other inter-layer connections are both excitatory and inhibitory. The FD neuron performs weighted integration of the SPP signal for the detection of a complex local feature [22]. The internal state, V, of the FD neuron located at r' in the input image is

$$V = a \iint \sum_j M_j(t - t_j, \mathbf{r} - \mathbf{r'}) u(t) dt d\mathbf{r} \tag{1}$$

where $M_j(t,\mathbf{r})$ is the weight function for the jth sub-window of the FD neuron. The range of integration is confined to the receptive field of the FD neuron. The spatio-temporal weight $M_j(t,\mathbf{r})$ gives a synaptic coupling strength represented in the spatio-temporal domain, and the peak value of the weight function is nothing but the synaptic weight of convolutional network. The weight function is generally monotonically decreasing from the scheduled spike arrival time in the case of PPM coding for the SPP signal. The input spike train, $u(t)$, impinging onto an FD neuron is given by $u(t) = \sum_k \delta(t - s_k)$, and $\delta(\ )$ denotes the Dirac delta

function, s_k is the kth spike arrival time, t_j is the pre-scheduled spike arrival time given by the PPM synapse. Parameters t, s_k, and t_j are all measured from the onset of spike emission by the LT neuron. The weight function $M_j(t,\mathbf{r})$ is internally generated by the FD neuron located at r in the jth feature class, and the time-window integration gives a correlation measure between the basis weight function and the SPP. Such a phase-locked processing stream on top of the time-windowed correlation endows integrate-and-fire systems with a property of coincidence detection.

We employ a population coding strategy for the size and rotation invariance in object detection. Specifically, multi-channel (or multi-resolution) population codes are generated in the CSNN obtaining size invariant signals. Local features are detected by FD neurons with varying size-selectivity.

The population code $q_j(\mathbf{r})$ for FP neuron at $\mathbf{r}$ in the jth feature category (e.g., rotated version of some local feature or the jth class of geometrical structure) is then given by a linear combination of pooled internal states of FP neurons given by $p_{k,j}(\mathbf{r}\text{-}\mathbf{r'})$, each associated with similar feature (e.g., similar figural pattern) with scale index k at position $\mathbf{r}$ of input image.

$$q_j(\mathbf{r}) = \sum_{\mathbf{r}\in C_j}\sum_{k=1} w_{k,j} p_{k,j}(\mathbf{r}-\mathbf{r'}) \tag{2}$$

$$p_{k,j}(\mathbf{r}) = g_{k,j}^m(\mathbf{r}) \Big/ \left\{ D + \lambda \sum_i g_{k,i}^m(\mathbf{r}) \right\} \tag{3}$$

where m is a positive integer (typically, $m=2$), $w_{k,j}$ indicates the weighting coefficient for the pooled internal states of an FP neuron for the class of feature with the kth scale level and jth feature category.

In eqn. (2), C_j denotes receptive field of an FP neuron in the jth feature class. The pooled internal state, $p_{k,j}(\mathbf{r}\text{-}\mathbf{r'})$, of an FP neuron located at $\mathbf{r'}$ with input from an FD neuron located at $\mathbf{r}$ in the jth feature class gives a measure of relative saliency of the jth feature among others in the kth scale level. In eqn (3), $g_{k,j}(\mathbf{r})$ is the output of an FD neuron for the class of feature with the kth scale level and the jth feature category, D and λ are some positive constants (typically, D =1, λ =1). The output $g_{k,j}(\mathbf{r})$ of an FD neuron is typically sigmoid function of its internal state V. In the multi-resolution framework, local features like oriented line segments are detected with n (i.e. the number of resolution or scale in the CSNN) classes of FD neurons, each tuned to selectively respond to the oriented line segments of a specific size (e.g., line segments with the same orientation but with significantly different length).

The integrated representation (2) of specific local features (i.e., similar geometrical structure of different sizes) given by a linear combination of the population code, as in [3], is a scale invariant code of the jth feature. The final output of the FP layer is the pulse signal given by a nonlinear transformation (e.g., sigmoidal activation function) of (2) in the form of either PPM or PWM signals, leading to a set of modulated pulses (SPP). Other class of features with different sizes, but with similar geometrical structures (a subset of feature class of similar figures) can be detected in the same manner.

4 Results

In the training of our CNN, the number of facial fragment images used is 2900 for the FD2 layer, 5290 for the FD3, and 14700 (face) for the FD4 layer, respectively. The number of non-face images, also used for the FD4 layer, is 137. Apparently greater number of facial component images as compared with non-face images is used to ensure robustness to varying rotation, size, contrast, and color balance of face images. So, as shown in Fig.7, we used a fixed set of transformed images for a given training sample image. In particular, we used three different sizes of the fragment images ranging from 0.7 to 1.5 times the original image size (i.e., fragments of facial components for modules in FD2 and FD3, entire face without background for FD4). The performance for face images other than the training data set was tested for over 200 face images with cluttered background and varying image-capturing conditions.

As shown in Fig. 9, the tested images of face are of different size (from 30x30 to 240x240 in QVGA image), pose (up to 30 deg. in head axis rotation and also in-plane rotation), and varying contrasts with cluttered background.

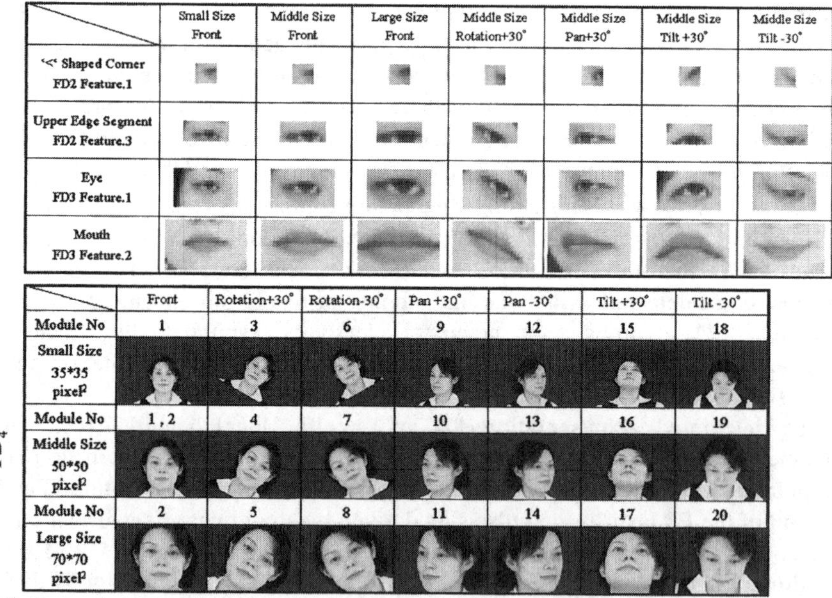

Fig. 7. Example of training datasets (extracted from the database of HOIP Lab, Softpia Japan)

The convolutional network demonstrated robust face detection with 1% false rejection rate and 7% false acceptance rate with quite good generalization ability for 2200 images. As the result of module-based training sequences, using a set of training data (face images) of more than 35 persons with different size, pose, and contrast, the top FD layer can detect and locate faces in complex scenes.

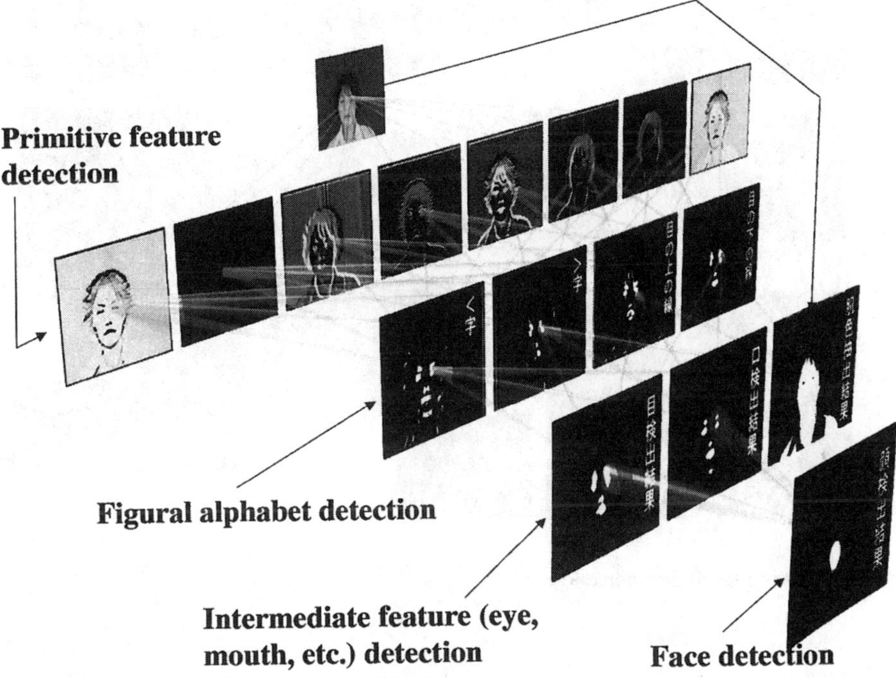

Fig. 8. Face detection process in the proposed convolutional neural network

The network parameters were as follows. The bottom FD layer has 8 classes of feature detecting modules, each of which having 320 x 240 neurons with receptive field size of 13 x 13, The second FD layer has 4 classes of modules for detecting end-stop structures (e.g., '<' or '>') like corners or two classes of horizontally elongated blobs or line segments with opposite gradient direction of gray level. The third FD layer was responsible for detecting eye and mouth, composed of two classes of modules in which each neuron has the receptive field of 31 x 31, and the top layer for detecting faces with neurons having a receptive field size of 77x77. The face detection process for the network as described is shown in Fig.8.

The receptive fields of FP neurons were 13 x 13 from the bottom to the third layer, and in the top layer without FP neurons (i.e., only face detecting FD layer). The activation function of each FD neuron has sigmoidal nonlinearity (e.g., logistic or hyperbolic tangent function) with suitable choice of gradient parameters.

In the simulation, we assumed the existence of some global timing signal instead of local timing neuron, hence enforced phase-locked behavior among locally excited neurons from bottom to top layers.

Fig. 9. Examples of face detection results

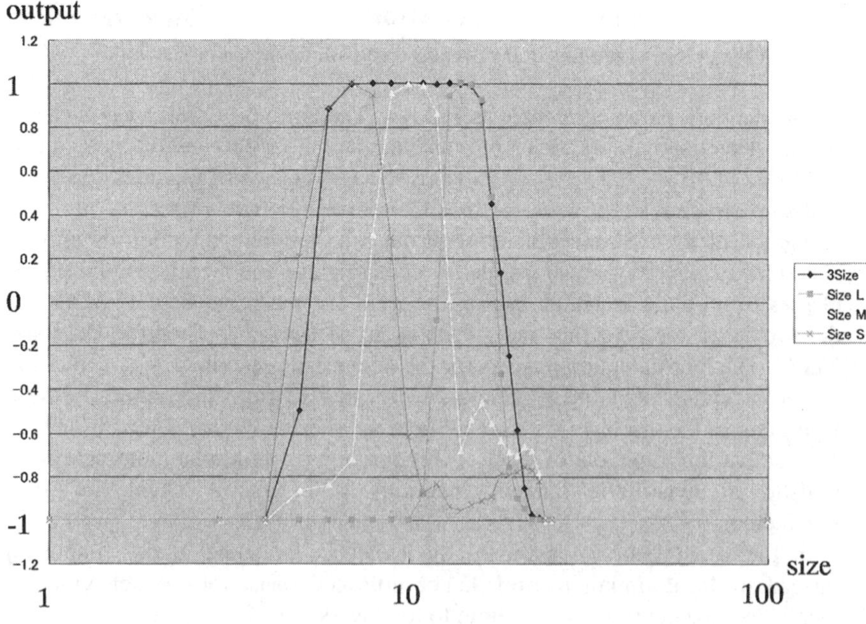

Fig. 10. Population tuning curves for size specific modules (L: large, M: middle, S: small)

Based on this assumption, no spiking dynamics in neurons were considered, since that introduce unnecessary complexity and prohibiting amount of computational cost in our system. The local excitation of neurons in relevant modules for a specific facial feature is ensured by the module-by-module, supervised training.

Fig. 10 shows an example of tuning curves for the size change of face image in three size specific modules. The curve in dark blue denoted as '3 size' give integrated response through population code. The population coding contributed to size as well as orientation invariant response from a set of feature detecting modules, each tuned to a specific class of feature.

The performance for other face images than the training data set was tested using more than 5000 face images with varying image capturing conditions, including size, pose, and contrast. The tested face images are of varying size, pose, and contrast (Fig.9). As long as the size of face is almost the same as that of the training data, the convolutional network demonstrated almost perfect performance (i.e., the FAR and FRR are 1% and 0.7 %, respectively) with good generalization for the database (facial data from HOIP Laboratory, Softpia Japan), part of which we used in the training.

5 Discussion

5.1 Temporal Coding Aspects

Our model used a fixed set of elementary local features (i.e., figural alphabet) as a substrate for general object representation. There are many other existing models for object detection using parts (local feature) based approach [31, 35, 36, 26, 4, 10]. The distinct aspect of the approach presented here, is that we use only a fixed set of local features regardless of object class. In fact, in our preliminary simulation shows that local alphabetical features, not limited to the ones used in this study but by including other class of them, can be used to detect other objects like cars. Moreover, our approach enables to detect proper spatial arrangement of local features by a set of simple convolution operations (i.e., those receptive field structures correspond to convolution kernels).

The proposed framework of one spike (or one volley of spikes) per feature category in the multi-resolution, convolutional, spiking neural network in Section 3, enables efficient and rapid processing in that only one pulse (or one shot of group of pulses) is needed to detect a local feature. What the proposed population coding scheme adds to the convolutional network model is the size/rotation invariant internal representation of object as the result of spatio-temporal population averaging on top of the time domain integration of SPPs. Once some module in the network learns to detect a local feature with a specific size, it is easy to extend the proposed model so that it acquires a set of population code in multi-channel by automatically reproducing neurons, in other channels, with similar receptive field of differing size.

The structured pulse packet as a carrier of visual information accommodates with coincidence detection in temporal coding framework as well as population rate-coding with spatial averaging. The difference between ordinary coincidence detection and our SPP integration in the time domain is that our model neurons in an FD layer detect fine-grained structure of pulse packet signals from a preceding FP layer, whereas in a coarse time scale on the order of the time span of SPP, those neurons detect the coincidence of inputs in multiple SPPs. So, we do not require strict coincidence, but a structured coincidence in that both temporal code and rate code coexist in a meaningful way.

The proposed model is thus compatible with the duality of temporal coding and rate coding in terms of spatial population averaging process as was recently addressed by (Masuda and Aihara 2003) [21]. In fact, coincidence detection can be performed for the SPP using timing signals either from LT neurons or from some center that generate global timing signals like the region generating theta rhythm. The latter aspect conforms with recently proposed hypothesis that the theta rhythm may mediate timing coordination among relevant neurons between hippocampus and neocortex [12], though it was proposed in a different context. The present study addressed global timing scheme (Fig.6) in the context of visual information processing from a rather engineering point of view. On the other hand, a rate code can be derived complementally from spatio-temporal averaging in the number of SPPs for a given timing period.

5.2 Face (Object) Detection and Learning Issues

Face detection or object detection in general, has been extensively studied in the literature (see Yang et al. 2002 [38] for recent survey of face detection approaches; see also Mohan et al. 2001 [26] for object detection approaches). In the proposed model for object (e.g., face) detection, we obtained good robustness to size and changes up to eight times the original object size being used for training. The translation invariance is inherently built into our convolutional architecture so that topological information as well as spatial arrangement of some complex feature (e.g., face) is preserved and the location of the object can be obtained from the firing patterns of top layer.

The limitation of present approach is a tendency toward higher false acceptance ratio (FAR) for greater robustness to variability in size and pose of object. To resolve this and to ensure low FAR while preserving high robustness to variability in object appearance in the image dataset other than those used in training session, we adopt selective activation of face detection modules [25]. In this ongoing approach, an additional FD layer is set on top of the face detection module in the CNN model presented here. Each newly added FD module is sharply tuned to a

specific orientation and a specific size of object (e.g., face). A brief description on
the selective activation of modules can be found in [25].

The supervised training procedure adopted in this study can be considered as a
variant of internal representation learning performed in a multitask learning [3], in
which the weights of the first layer are shared after simultaneous training using all
the tasks (e.g., local feature detection). In the present learning scheme, multitask
learning is equivalent to training multiple feature detectors (planes) in an FD
layer. Once training session for some FD layer is over, the weights of that layer
are fixed and shared by the preceding layer.

The proposed population coding provides explicit rotation as well as size
invariance in the form of linear combination of pooled outputs from detectors of
different local features of specific class of size/orientation.

6 Conclusions

Our module-by-module training of the CSNN in a supervised manner lead to a
set of local feature, including figural alphabets, detectors being hierarchically
formed. The structured pulse packet, represented by a set of pulse modulated
signals, is compatible with a sort of synfire chain [1, 2, 14] coding of visual
information in the hierarchical neural network. The integrated representation of
the population-spike-codes can provide a substrate for economical representation
of size/rotation-invariance in local features as well as complex patterns, which
enables unified perception of objects regardless of their size and pose.

Finally, part of the proposed model can be applied to hardware implementation
of the convolutional spiking neural network, which is currently our ongoing work
[16]. In view of the efficiency of neuronal spike code using a set of pulse packets,
it is natural to apply the original framework [22] for VLSI implementation using
pulse width/phase modulation. In the preliminary study by (Korekado et al. 2003)
[16], a set of PWM signals from neurons in a preceding layer are integrated using
time-domain weight function followed by non-linear transform in a CMOS circuit,
which is essentially discrete time domain, parallel integration of our structured
pulse packet. In that work, for simplified implementation, the synchronization
among relevant neurons are ensured by introducing global clock signal instead of
distributed local timing signal generator [32] like the one proposed in this study.

Acknowledgement

We are indebted to Ms. M. Ishii for providing us with Fig.1 and some photos in
Fig.9.

References

1. Abeles M (1991) Corticonics: Neural circuits of the cerebral cortex. Cambridge: Cambridge University Press.
2. Aviel Y, Mehring C, Abeles M, Horn, D (2003) On Embedding Synfire Chains in a Balanced Network. Neural Comput 15:1321-1340
3. Baxter J (1997) A Bayesian/information theoretic model of learning to learn via multiple task sampling. Machine Learning, 28:7-40 [2b]
4. Chen X, Gu L, Li S Z, Zhang H-J (2001) Learning Representative Local Features for Face Detection. Proc of Computer Vision and Pattern Recognition [2a]
5. Denev S, Lathan PE, Pouget A (1999) Reading population codes: a neural implementation of ideal observers. Nature Neuroscience 2:740-745
6. Diesmann M, Gewaltig MO, Aertsen A (1999) Stable propagation of synchronous spiking in cortical neural networks. Nature 402:529-533
7. Földiák P (1991) Learning Invariance from Transformation Sequences. Neural Comput 3:194-200
8. Fujita I, Tanaka K, Ito M, Cheng K (1992) Columns for visual features of objects in monkey inferotemporal cortex. Nature 360:343-346
9. Fukushima K (1980) Neocognitron: A self-organizing Neural Network Model for a Mechanism of Pattern Recognition Unaffected by Shift in Position. Biol Cybern, 36:193-202
10. Garg A, Agarwal S, Huang T S (2002) Fusion of Global and Local Information for Object Detection. Proc of the 16th Int Conf on Pattern Recog [6a]
11. Gu L, Li SZ, Zhang H (2001) Learning Probabilistic Distribution Model for Multi-View Face Detection. Proc of Computer Vision and Pattern Recognition.
12. Hasselmo M (2003) Theta theory: Requirements for encoding events and task rules explain theta phase relationships in hippocampus and neocortex. Proc of International Joint Conf On Neural Networks
13. Hopfield J J (1995) Pattern recognition computation using action potential timing for stimulus representation. Nature, 376:33-36
14. Ikeda K (2003) A synfire chain in layered coincidence detectors in random synaptic delays. Neural Networks 16:39-46
15. Konen W K, von der Malsburg C (1993) Learning to generalize from single examples in the dynamic link architecture. Neural Comput 5:1019-1030
16. Korekado S, Morie T, Nomura O, Matsugu M, Iwata A (2003) A Convolutional Neural Network VLSI for Image Recognition Using Merged/Mixed Analog-Digital Architecture. Proc. of Seventh International Conference on Knowledge-Based Intelligent Information & Engineering System.
17. Krüger N (1998) Collinearity and Parallelism are Statistically Significant Second-Order Relations of Complex Cell Responses. Neural Processing Lett 8:117-129
18. Le Cun Y, Bengio T (1995) Convolutional networks for images, speech, and time series. In: Arbib MA (ed.) The Handbook of Brain Theory and Neural Networks. MIT Press, Boston, pp255-258
19. Maass W (1999) Computing with Spiking Neurons. In: Maass W, Bishop C M (ed.) Pulsed Neural Networks. Cambridge: MIT Press, pp55-85
20. Maass W, Natschläger T(1997) Networks of spiking neurons can emulate arbitrary Hopfield nets in temporal coding. Network: Computation in Neural Systems, 8:355-372

21. Masuda N, Aihara K (2003) Duality of Rate Coding and Temporal Coding in Multilayered Feedforward Networks. Neural Comput 15:103-125
22. Matsugu M (2001) Hierarchical Pulse-coupled Neural Network Model with Temporal Coding and Emergent Feature Binding Mechanism. Proc International Joint Conf On Neural Networks. pp802-807
23. Matsugu M, Iijima K (2000; filed in 1994) Object Recognition Method. (in Japanese) Japanese Patent No P3078166
24. Matsugu M, Mori K, Ishii M, Mitarai Y (2002) Convolutional Spiking Neural Network Model for Robust Face Detection, Proc 9th International Conf On Neural Info Processing. pp 660-664
25. Mitarai Y, Mori K, Matsugu M (2003) Robust Face Detection Systems Based on Conolutional Neural Networks Using Selective Activation of Modules. Proc 2nd Forum for Information Technology (in Japanese)
26. Mohan A, Papageorgiou C, Poggio T(2001) Example-Based Object Detection in Images by Components. IEEE Trans on Pattern Analysis and Machine Intelligence, 23:349-361
27. Murase Y, Nayar S (1997) Detection of 3D objects in cluttered scenes using hierarchical eigenspace. Pattern Recog Lett 36:375-384
28. Natschläger T, Ruf B (1997) Learning radial basis functions with spiking neurons using action potential timing.
29. Riesenhuber M, Poggio T (1999) Hierarchical models of object recognition in cortex. Nature Neuroscience, 2:1019-1025
30. Rowley H, Baluja S, Kanade T (1998) Rotation Invariant Neural Network-Based Face Detection. Proc of Computer Vision and Pattern Recognition pp38-44
31. Schneiderman H, Kanade T (2000) A Statistical Method for 3D Object Detection Applied to Faces and Cars. Proc of Computer Vision and Pattern Recognition
32. Tanaka H, Hasegawa A, Mizuno H, Endo T (2002) Synchronizability of Distributed Clock Oscillators. IEEE Trans on Circuits and Sys I, 49:1271-1278
33. Tiesinga PHE, Sejnowski TJ (2001) Precision of pulse-coupled networks of integrate-and-fire neurons. Network: Comput In Neural Sys 12:215-233
34. Van Rullen R, Gautrais J, Delorme A, Thorpe S (1998) Face Processing Using One Spike per Neurone. BioSystems 48:229-239
35. Viola P, Jones M (2001) Rapid Object Detection using a Boosted Cascade of Simple Features. Proc Computer Vision and Pattern Recognition
36. Weber M, Welling M, Perona P (2000) Unsupervised Learning of Models for Recognition. Proc European Conf Computer Vision, vol 1, pp18-32
37. Wallis G, Rolls ET (1997) Invariant Face and Object Recognition in the Visual System. Prog in Neurobiol 51:167-194
38. Yang M-H, Kriegman D J, Ahuja N (2002) Detecting Faces in Images: A Survey. IEEE Trans on Pattern Analysis and Machine Intelligence, 24:34-58

Networks constructed of neuroid elements capable of temporal summation of signals

Alexander A. Kharlamov, Vladimir V. Raevsky

Institute of highest nervous activity and neurophysiology, Russian Academy of Science, 5a Butlerova Str., 117865 Moscow, Russia

Abstracts

In this chapter some neurobiological premises are presented that permit to simulate correctly neurons, which realize the temporal summation of signals and networks constructed from them. The formalism of these networks is presented that describes their characteristics with respect to the structural multilevel processing of information. It is also shown that such a statistical analysis is used for textual information processing in the program TextAnalyst® (by Microsystems, Moscow), designed for the automatic semantic analysis of texts.

Key words: neuroid element with temporal summation of signals, neuronal network, associative accessing, structural processing of information, learning, reproduction, recognition, text processing.

Introduction

Among other characteristics of nervous centers Ch. Sherrington postulated their capability of summation. It manifests itself in the fact that a reflex response may not occur to a single stimulus, whereas a series of such stimuli elicits a response. Two types of summation are known: temporal or successive and spatial or simultaneous. Temporal summation is considered to be a summation of stimuli arriving one by one and separated by short time intervals. Spatial summation expresses itself in the reflex response to simultaneous excitation of several spatially-distributed receptors that belong, however, to the same receptive field. The summation in a nervous center is based on the processes in single neurons. In response to a subthreshold stimulation nerve cells generate changes of membrane potential (local response). Since the duration of local responses is in the range of milliseconds, in the framework of the classical views the summation may occur only when stimuli are interspaced by short intervals.

An adaptive behavior of an organism is based on the formation of a memory trace of the dynamics of past events, which implies the fixation not only of causal relations but of temporal relations of external events as well. Spatial summation that reflects a convergence of different excitations may be a candidate for a mechanism of integrative activity of nervous centers that is responsible for a formation of memory. In this case, several stimuli should not necessarily coincide in time. First, the effect of the first stimulus may be prolonged due to the reverberation of excitation in Lorente de No's trap. Second, the first stimulus may not have an ionotropic effect (i.e. change the permeability of a membrane thus eliciting a local response). In case of metabotropic action, causing the change of neuron's functional state indirectly by chemical reactions, the after-effects of a stimulus may last long enough. This variant seems most likely, since temporal relations among various stimuli may then be successfully encoded by the speed of chemical processes.

Retrieval of a memory trace implies the onset of the response, reflecting the succession of past events, to the very first stimulus. Hence the realization of this function on the basis of spatial {temporal?} summation seems rather unlikely. We assume that the pattern of a signal (volley of action potentials) appearing in response to the first stimulus acts as a code for the retrieval from memory of the whole dynamics of the following events. This mechanism is probably realized on the basis of temporal summation of action potentials in a nervous center or on the target neuron.

In 1969 A.N.Radchenko published the monograph "Modeling of the principal brain mechanisms" [1]. The book suggested an interpretation of information processing within networks composed of neurons with temporal summation of signals. It has been shown that addressing the information stored in such networks is characterized by associativity, while structural unit of a network may act as an element of associative memory. Hierarchy of such networks makes it possible to realize the structural processing of information that results, at each hierarchical level, in automatic forming of dictionaries of elements at the given level while the connections of words of these dictionaries in the flow of input information are passed to the output of a level. Networks of the described type may serve as a basis for the realization of a uniform computing environment for the integral representation of information of various types.

Since that time both theory describing such networks [2], [3] and their practical application in information processing systems [4], [5], [6], [7] have been significantly improved.

Below some neurobiological premises are presented that permit to simulate correctly the neurons with temporal summation of signals and networks constructed from them. The formalism of these networks is presented that describes their characteristics with respect to the structural multilevel processing of information. It is also shown that such a statistical analysis is used for textual information processing in the program TextAnalyst® (by Microsystems, Moscow), designed for the automatic semantic analysis of texts.

The major difference between the proposed model and the one described by A.N. Radchenko is that neuron's memory is supposed to be realized on the basis of synapses' plasticity [8] but not on the basis refractory characteristic. In the pragmatic applications based on this model the memory element is completely removed beyond the borders of a neuron.

The suggested model is limited to the case of the binary coding. k-based coding, not increasing the volume of code space, makes the description of the model more complicated, though the model realizing the k-based representation looks more like a natural neuron.

1. Neuronal network for structural information processing

Some neuroid elements are known to have a very complex structure [9]. However, the most popular modifications of neuronal models are based on the spatial summation of signals [10].

Networks constructed from such neuroid elements can realize numerous diverse functions [11] but the most apparent is one of them – network can partition a multidimensional signal space into areas thus giving opportunity for post-processing – to identify the static objects of perceptive space by attributing them to one of these areas [12]. Attempts were made to use these networks to solve the task of memorizing and discrimination of dynamic images, but these attempts resulted either in creation of highly specialized devices requiring manual tuning to the specific narrow class of information [13] or in unjustifiable expenditure of computing resources [14].

Unlike the model of neuron with the spatial summation of signals, the model taking into account the temporal characteristics of signal sequence allows constructing networks that have intrinsic capability to store and retrieve dynamic information and to restore automatically the internal structure of input information.

1.1. Neuroid element capable of temporal summation of signals

In addition to the components characteristic for the model of a neuron with spatial summation of signals, the model of a neuron with temporal summation of signals includes the so-called generalized dendrite [15] that is represented by a chain of summing cells and delays (see fig. 1). It is functionally equivalent to a neuroid element with spatial summation the inputs of which receive information through the binary register of shift (see fig. 2).

Such neuron calculates the convolution of a fragment of input sequence *(a(t-n+1),a(t-n+2),...,a(t))* with neuronal address – *(b₁,b₂,...,bₙ)* – a set of '1' and '0' corresponding to the distribution of excitatory and inhibitory synapses in a register of shift (in generalized dendrite):

$$y(t+1) = f(\sum_{i=1}^{n} a(t+i-n-1)b_i - b_c(t)),\qquad(1)$$

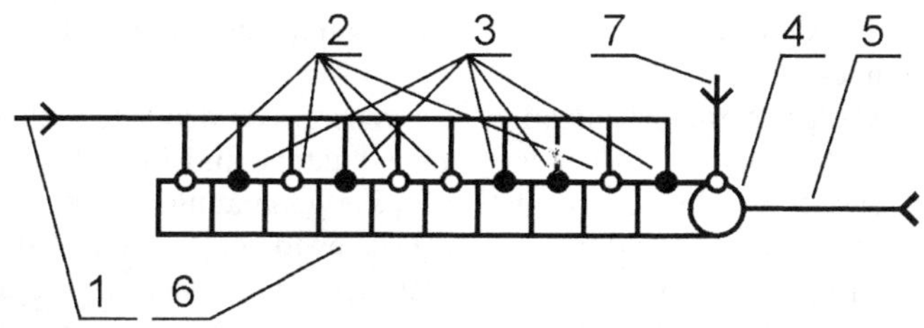

Fig. 1. Neuroid element with temporal summation of signals. 1 – input, 2 – excitatory and 3 – inhibitory synapses, 4 – cell's body realizing threshold transformation, 5 – output connection; 6 – accumulating shifting register, 7 – controlling connection.

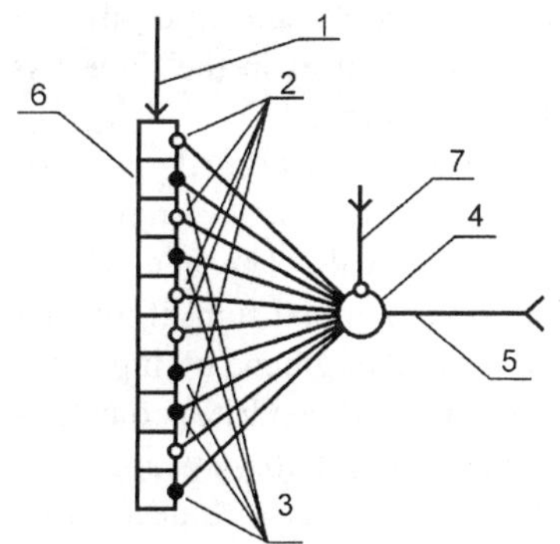

Fig. 2. Equivalent scheme of a neuron with temporal summation of signals. 1 – input, 2 – excitatory and 3 – inhibitory synapses, 4 – cell's body realizing threshold transformation, 5 – output connection; 6 – binary register of shift, 7 – controlling connection.

where f is threshold transformation. If neuron's threshold equals the number of '1' in the address, it will respond strictly to its address.

Neuroid element can be controlled. Synchrosynapse (controlling synapse on neuron's soma) adds its weight with its sign $b_c(t)=\pm k$. This is equivalent to a threshold change.

Neuroid element with temporal summation of signals is A.N. Radchenko's logical development of W. Rall's model [15] which, in turn, was derived from the views of D.A. Sholl [16].

When trying to describe statistically the system of connections within the cerebral cortex, D.A. Sholl suggested a method for the numeric estimation of the number of dendrite branching – their quantitative model. He encircled neuron's soma with concentric spheres and calculated the number of branching foci enclosed between two neighboring concentric surfaces.

Using this model for a starting point, W. Rall formulated the concept of the "equivalent cylinder" (generalized dendrite). He calculated the number of synaptic contacts between two concentric surfaces and substituted the dendrite tree of a neuron with a generalized dendrite the weight of connection in the cell of generalized dendrite equaled the total weight of connections on a part of dendrite tree that is enclosed between the surfaces and that is simulated by the given generalized dendrite contact.

The knowledge of the organization of a generalized dendrite makes it possible to predict the functional consequences of excitatory and inhibitory postsynaptic potentials for the membrane potential of spike-generating region. If the relative distances from the corresponding Shall's surfaces to cell soma as well as time and site of excitatory and (or) inhibitory postsynaptic potentials on the dendrite tree are known, one may calculate the conditions required for neuron's excitation. The summarized postsynaptic potential reaching the cell's soma will be equal to the sum of postsynaptic potentials arriving to the cell's body from the respective cells of a generalized dendrites with respect to their delays. For the cell nearest to soma, this delay equals *dt*, for more distal ones – *2dt, 3dt*, etc.

Joining of all synapses of the generalized dendrite by one presynaptic fiber running from the distal end of dendrite to the proximal one makes it possible to take into account the temporal pattern of a sequence of impulses in the time discretization. The impact of a sequence is enhanced by summation.

Introduction of inhibitory synapses into the generalized dendrite made it possible to form combinations of synapses that help get selective neuronal responses to various temporal patterns of impulses and interspike intervals. The generalized dendrite's cell that is the nearest to soma accumulates the maximal sum, if the distribution of impulses and silence (ones and zeros) in the fragment of input sequence n symbols long (where n is the length of a generalized dendrite) coincides with the distribution of excitatory and inhibitory synapses. Excitatory synapse corresponds to 'one' in the sequence and the inhibitory one – to 'zero'. Indeed, if an impulse and the sum being stored appear in the region of the excitatory synapse synchronously, then the excitatory postsynaptic potential is added to the sum with its weight. If the gap between impulses corresponds to the moment when the sum is in the region of an inhibitory synapse, then inhibitory postsynaptic potential with its weight is not subtracted from the sum. In this case soma receives the maximal sum that is equal to the number of excitatory synapses multiplied by their weight.

1.2. Network consisting of neuroid elements with temporal summation of signals

Joining of a set of 2^n neurons with different address combinations into a single structure (see fig. 3) generates a model of n-dimensional signal space R^n (or, rather a singular hypercube G_e in R^n, if weight of each synapse is either "+1" or "-1"). In this case each neuron simulates one of the nodes of a hypercube. This structure can map any binary sequence as a sequence of nodes G_e – track.

Informational sequence represented in a signal space simulated by neuronal cluster may be stored due to the change in the state of neurons simulating the corresponding nodes of hypercube G_e and later – retrieved.

1.3. Biological premises

The concept of a structure for information processing suggested by A.N. Radchenko gains certain support from neurobiological views, based on anatomy, cytoarchitecture, and electrophysiology of brain.

I.S. Beritov [17] suggested that the main functional unit of the cerebral cortex is a module including layer III pyramids and some stellate cells linking the former together; such module functions as a whole entity influenced by one or

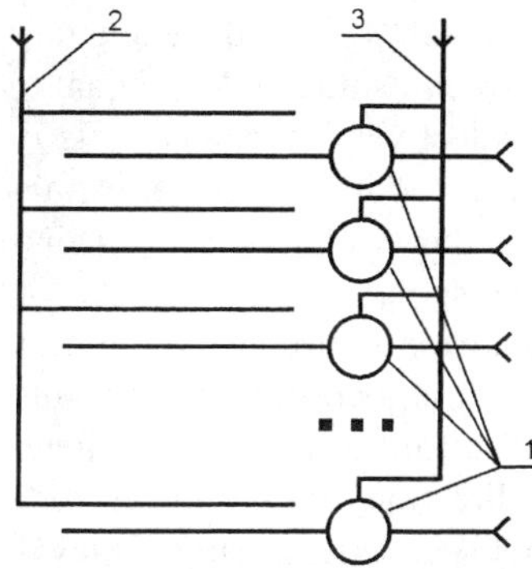

Fig. 3. Neuronal cluster. 1 – neurons of the cluster that have generalized dendrites with different addresses ranging from (000...0) to (111...1), 2 – common afferent fiber, 3 – controlling connection.

few afferents. The cluster of stellate cells that receive innervation from an afferent fiber and have terminals of their own axons on their dendrites, due to such feedback, can consistently respond to the impulsation of an afferent transferring it to the column of pyramids for the further processing.

Pyramids constituting a column have the controlling apical dendrite that receives background impulsation from horizontal cells of layer I, thus facilitating neuronal discharge. This is equivalent to a decrease of neuron's threshold. Layer III pyramids possess (according to G.I. Polyakov [18]) up to 6 basal (informational) dendrites. V.P. Babmindra and T.A. Bragina [19] have demonstrated throughout the human central nervous system (from spinal cord to cerebral cortex) the specific type of innervation of a neuron by a single afferent going from the distal to proximal part of dendrite. The so-called triades have also been demonstrated, i.e. the intercalate inhibitory neuron between terminals of an afferent and dendritic synapse.

Afferents innervating different pyramidal neurons in a column [20] can be linked electrotonically, which creates something like a joint input for all neurons in that structure.

1.4. Associative memory module based on neuroid elements with temporal summation of signals

Let us explicitly add a memory unit to the address part of a neuroid element (generalized dendrite) with constant weights of connections and with threshold transformation also having a constant threshold. Generalized dendrite with a threshold transformation that is the address part of a neuroid element accumulates, at each step, the weighted sum – a convolution of the current n-segment-long fragment of input sequence with address sequence. Neuroid elements differ from each other in distribution of excitatory and inhibitory synapses on their generalized dendrites – i.e. in an address. When the accumulated sum exceeds the threshold, neuron responds to n-segment-long fragment of input sequence. Neuroid element that responded to its address is permitted to process the current symbol of input information, storing included.

2^n neuroid elements of a cluster have addresses ranging from (000...0) to (111...1), where "0" corresponds to inhibitory synapse and "1" – to the excitatory one. Neurons are wired in parallel and each of them (when its threshold is equal to the number of excitatory synapses on its dendrites) selectively responds to its n-segment-long fragment of sequence corresponding to its address. Thus any fragment will be reflected in one of neuroid elements of a cluster and the sequence as a whole – in a sequence of neuroid elements that have responded, i.e. in a track.

Incorporating memory element into a neuroid element we introduce a memory function into nodes of hypercube G_e. In the simplest case, neuroid element with the given address can store a symbol of input sequence following the n-segment-long address fragment. This allows to store the direction of a transfer to a neighboring node, i.e. track, corresponding to an input sequence. During retrieval, addressing the current node enables to define the direction of the transfer to a neighboring node, from the latter – further on and, eventually, to the end of a track.

2. Formalism of neuronal network

The associative memory based on neuroid elements with temporal summation of signals which is principally characterized by associative reproduction with dynamic formation of an address is called below a dynamic associative memory and the system based on this memory will be called dynamic associative storage device. DASD consists of numerous parallel neuroid elements with common input and output and differing from each other only in the sequence of polarities of synaptic connections on generalized dendrites, i.e. in their address. The weight of connections is equal to ± 1. From the informational point of view, DASD simulates an association of the cerebral cortex neurons – a column. DASD transforms the sequence of symbols into a track in multidimensional signal space. It can memorize a sequence of symbols storing the topology of sensory space in a signal space. DASD realizes the stochastic processing of a sequence of symbols and the associative access to the stored information.

Neuroid element (see fig. 1) is a model of a generalized dendrite – an accumulating shifting register. Each cell of a generalized dendrite has an excitatory or inhibitory input, the weight of which in the simplest case is equal to "$+1$" or "-1" (below it is assumed throughout the text). All its inputs are joined by a common afferent fiber. Neuroid element also possesses a threshold transformation and a memory element.

A part of a neuron (without memory element) functions as an address one. Input of a neuroid element receives a succession of impulses that may be represented as an A sequence of "1" and "0": $A=(...,a(-1), a(0), a(1),...,a(t))$, where $a(t) \in \{0,1\}$. Zero signifies the absence of an impulse during a certain tact of time t, whereas "1" stands for the presence of an impulse.

Symbol reaching a neuroid element by presynaptic fiber through synapses at the moment t is received simultaneously by all cells of register (multiplied by its weight coefficients: "+1" or "-1") and is added to the contents of each cell with regard to its weight. Then the contents of each cell is shifted in the direction of neuron's body by one position. Zero is pushed into the cell most distal with respect to neuron's body. After n tacts the contents of the most distal cell is shifted into the cell nearest to the body.

To be able to function as a memory element, neuroid element must change its state during learning, e.g. due to the change of weights of connections [21]. To simplify the interpretation, it is convenient to move a memory element beyond the borders of the address part. Neuroid element memorizes the symbol of a sequence $a(t+1)$ directly following the n-symbol-long fragment $\hat{a}(t) = (a(t-n+1), a(t-n+2),...,a(t))$ (here $\hat{a}(t)$ signifies the fragment of a sequence A in the window with the length equal to n at the current moment of time t).

Let us have 2^n neuroid elements with parallel-connected inputs, each assigned one out of the complete set of addresses:

$$b_0=00...0;$$
$$b_1=00...1;$$
$$...$$
$$b_{N-1}=11...1,$$

where index is the number of a neuroid element in DASD and $N=2^n$. Then any fragment of a sequence A with the length equal to n will induce a response in one of neuroid elements. A change in the state of a neuroid element results in memorizing the next symbol of a sequence. The changed state of a neuroid element makes it possible to reproduce this sequence later.

Let us examine processing and representation of binary internally structured information in the signal space of DASD. This processing making possible to store information effectively, analyze it stochastically (with automatic reconstruction of its internal structure), retrieve it associatively and recognize it.

2.1. Transformation realizing the associative property of accessing the information

Joining 2^n neurons with different address combinations into a unitary structure produces a model of n-dimensional signal space R^n (of a singular hypercube

$\dot{G}_e \in R^n$ if synaptic weights are "+1" and "-1"). In that case each neuron simulates one of hypercube's nodes. Such structure can map any sequence A into a sequence of nodes G_e – track $\hat{A}$:

$$\hat{A} = F(A). \tag{2}$$

Here F is a transformation onto a signal space. Transformation F is the basis for the structural processing of information.

Let us examine transformation F of the binary sequence $A = (..., a(-2), a(-1),$ $a(0), a(1), a(2), ..., a(t), ...)$, where $a(t) \in \{0,1\}$, in n-dimensional space R^n such that each n-segment-long fragment of the sequence $(a(t-n+1), a(t-n+2), ..., a(t))$ is corresponded by a point in R^n - $\hat{a}(t)$ with coordinates, corresponding to the fragment; the whole sequence A is corresponded by a sequence of points: $A = \{..., (a(-n-1), a(-n), ..., a(-2)), (a(-n), a(-n+1), ..., a(-1)), (a(-n+1), a(-n+2), ..., a(0)), (a(-n+2), a(-n+3), ..., a(1)), (a(-n+3), a(-n+4), ..., a(2)), ..., (a(t-n+1), a(t-n+2), ..., a(t)), ...\} = (..., \hat{a}(-2), \hat{a}(-1), \hat{a}(0), \hat{a}(1), \hat{a}(2), ..., \hat{a}(t), ...)$ – track.

Transformation F has a property of associative addressing the points of a track $\hat{A}$ by association with n-segment-long fragment of the sequence A (i.e. with its contents): any set of n symbols directs us to the corresponding point of a track.

It is possible to choose the dimensionality of the space R^n in such a way that a certain finite sequence will be reflected in that space without crossing, i.e. each n-segment-long fragment of the sequence within R^n will be corresponded by exactly one point belonging to the track. In that case there exists a transformation inverse to F – transformation F^{-1} of the track $\hat{A}$ into the original sequence A:

$$A = F^{-1}(\hat{A}). \tag{3}$$

In the general case, among n-segment-long fragments of informational sequence the previously encountered fragment can occur – then the track will pass through the node already belonging to that track, i.e. track will cross itself. From that point on, the track can be continued in more than one way. For the binary sequence no more than two ways are possible.

Associative nature of F transformation makes it possible to preserve the topologic structure of transformed information. Indeed, the similar fragments will be transformed into the same fragment of the track, while different fragments – into different ones.

2.2. Memorizing of information. Retrieval. Auto- and heteroassociativity

Let us have two synchronous sequences A and J. Points belonging to the track $\hat{A}$ corresponding to the sequence A (carrier sequence) within a signal space may be used to store symbols of informational sequence J that is synchronized with A. Let us introduce a memory function M in nodes of a hypercube G_e that establishes the correspondence between each node $\hat{a}(t) \in \ddot{A}$ being passed at the moment t and a binary variable $j(t+1)$ that is a $(t+1)$ symbol of some binary sequence J.

$$M\{\hat{a}(t), j(t+1)\} = [\hat{a}(t)]_{j(t+1)}. \tag{4}$$

Thus we have a track $\hat{A}$ conditioned by the sequence J. [*] means conditionality.

$$[\hat{A}]_J = M\{F(A), J\}. \tag{5}$$

In other words, sequence J is written in the points of track $\hat{A}$ (associated with sequence A).

It is possible to reconstruct the information sequence J on the basis of track $[\hat{A}]_J$ conditioned by it and on the basis of the carrier sequence A:

$$J = M^{-1}\{[\hat{A}]_J, F(A)\}, \tag{6}$$

where in each point $\hat{a}(t) \in \hat{A}$: $M^{-1}([\hat{a}(t)]_{j(t+1)}, \hat{a}(t)) = j(t+1)$. Mapping of the carrier sequence into the track makes it possible to address information stored in the points of that track, i.e. address information sequence. We shall call such way of storage a heteroassociative storage and retrieval, accordingly, – a heteroassociative retrieval.

If the carrier sequence is used as a conditioning one, i.e. points of the track within a signal space store symbols of the same sequence – that there is the case of self-conditioning: if $J \equiv A, M\{\hat{a}(t), a(t+1)\} = [\hat{a}(t)_{a(t+1)}]$,

$$[\hat{A}]_A \equiv [\hat{A}] = M\{F(A), A\}. \tag{7}$$

Analogously to the previous:

$$A = M^{-1}\{[\hat{A}], F(A)\}. \tag{8}$$

In that case the original sequence can be restored starting from any point of the track:

$$A = M^{-1}\{[\hat{A}], \hat{a}(t) \in F(A)\}. \tag{9}$$

Indeed, if we have an n-segment-long fragment of a sequence $\hat{a}(t) = (a(t-n+1), a(t-n+2), ..., a(t))$, we address one of the points $\hat{a}(t)$ of the track $\hat{A}$. This point stores the information $M^{-1}([\hat{a}(t)], \hat{a}(t)) = a(t+1)$, corresponding to the next symbol of the sequence A that has generated the track $\hat{A}$. Adding a new symbol $a(t+1)$ to the $(n-1)$ symbol of n-segment-long fragment, we receive a new n-segment-long fragment $(a(t-n+2), a(t-n+3), ..., a(t+1))$, according to which we can address the next point of the track: $\hat{a}(t+2)$. In that point the next symbol of the sequence is read out: $M^{-1}\{\hat{a}(t+1), a(t+1)\} = a(t+2)$. And so on, till the end of the sequence is reached or till the nearest track branching occurs – (whichever comes first). Such storage is called autoassociative storage and retrieval – autoassociative retrieval.

Memory function M exists everywhere with the exception of branching points where it has to be additionally defined.

Thus using M function together with F transformation that is capable of associative addressing an information, it is possible to realize an associative memory with an ability to store/retrieve information in auto- and heteroassociative manner.

2.3. Development of the stochastic model

Making memory function M a bit more complicated we can construct, along with associative storage/retrieval, a mechanism for stochastic information processing. This may be accomplished by substituting a trigger for registration of the next symbol $a(t)$ of a sequence A by two counters to record the number of passes of a track through the given point in the given direction: C_0 – for transitions to "0" and C_1 – for transitions to "1". We shall also introduce a threshold transformation H that enables, on the basis of the value that function H has in the point of a multi-dimensional signal space defined by its coordinates $\hat{a}(t)$, reconstructing the value of the most probable transition to the next point – to "0" or to "1": $a(t+1)$. Such mechanism is sensitive to the number of passes of the given point in the given direction. It can characterize each point of a track with respect to the number of times the combination $(\hat{a}(t), a(t+1))$ appears in an input information. Additionally, the use of a threshold transformation H enables retrieval of an information with the given level of validity.

This mechanism functions as follows. Transition counters in each point of a fragment of a track store the number of passes. Let us assume that the given fragment of the track is passed twice. If we now apply a threshold transformation H with the value set to $h=2$ to the values stored in the counters of DASD, we will select only those fragments of the track within a signal space that were passed twice. If we set $h=1$, all the information received by DASD will be memorized.

During storage process, counters change their state with respect to the direction of transition:

$$M\{\hat{a}(t), a(t+1)\} = [\hat{a}(t)]_{a(t+1)} = C_{\hat{a}(t)}(t) =$$

$$= \begin{cases} C_0(t) = C_0(t-1)+1, \; C_1(t) = C_1(t-1); & \text{if } a(t+1) = 0; \\ \\ C_0(t) = C_0(t-1), \; C_1(t) = C_1(t-1)+1; & \text{if } a(t+1) = 1. \end{cases} \tag{10}$$

During the retrieval, the state of counters is analyzed and the current symbol is formed with respect to the threshold condition:

$$a(t+1) = HM^{-1}([\hat{a}(t)], \hat{a}(t)) = HM^{-1}\{C_{\hat{a}(t)}(t)\} =$$
$$= \begin{cases} 0, & \text{if } C_1 - C_0 < 0; \\ \\ 1, & \text{if } C_1 - C_0 \geq 0. \end{cases} \tag{11}$$

Besides memorizing – nondecreasing of the values in counters C_0 and C_1 (Eq. 10), forgetting is also possible – uniform decrease of counters' value in time, the speed of decrease being significantly lower than that during memorization:

$$M\{\hat{a}(t), a(t+1)\} = [\hat{a}(t)]_{a(t+1)} = C_{\hat{a}(t)}(t) =$$
$$= \begin{cases} C_0(t) = C_0(t-1)+d_1, \; C_1(t) = C_1(t-1)-d_2; & \text{if } a(t+1) = 0; \\ \\ C_0(t) = C_0(t-1)-d_2, \; C_1(t) = C_1(t-1)+d_1; & \text{if } a(t+1) = 1, \end{cases} \tag{12}$$

where $d_1 >> d_2$.

Adding forgetting capability makes it possible to eliminate any incidental points from a track that receive no confirmation during the latest learning.

2.4. Formation of a dictionary

Memory mechanism sensitive to the number of passes through the given point in the specific direction (mechanism of stochastic processing) is an instrument to analyze an input sequence with respect to its repetitive parts. As it has been demonstrated above, similar fragments of a sequence are converted by F transformation into the same part of track.

Let us have a sequence $A = B_1 * B_2 * B_3 * B_2$, where B_j – some sub-sequences: $B = (a(t_1), a(t+1), ..., a(t+i))$ and $[*]$ stands for concatenation. In a general case, track fragments $\hat{B}_1, \hat{B}_2, \hat{B}_3, \hat{B}_2$ are interspaced by transition fragments $\hat{f}_1, \hat{f}_2, \hat{f}_3$; the starting fragment $\hat{f}_b$ and terminal fragment $\hat{f}_e$ also appear in the track. If we set the value of threshold transformation H to $h=2$, we can select the repeated fragment $\hat{B}_2$ in the track $\hat{A}$, but, setting threshold at $h=1$, we will preserve all information of the original sequence.

If we have a class of sequences $\{A\}$ that contain, in different combinations, sequences $\{B_j\} = \{B_1, B_2, B_3, B_4, B_5\}$, we can form a set of trajectories $\{\hat{B}_j\}$ corresponding to the set of sequences $\{B_j\}$ – a dictionary – by mapping the sequences belonging to $\{A\}$ class into n-dimensional space and applying a threshold transformation.

One might say that the transformation $HM^{-1}MF$, interacting with the input class $\{A\}$, generates a dictionary that describes trajectories corresponding to subsequences of the input class in the space of the given dimensionality.

$$\{\hat{B}\} = HM^{-1}MF(\{A\}).\tag{13}$$

Depending on the value of threshold h of a transformation H, words of a dictionary may be chains or graphs.

2.5. Formation of a syntactic sequence. Multilevel structure

A reconstructed dictionary of frequently occurring events may be used to detect the old information in a flow of the new one. To accomplish that, one has to realize an absorption of the fragments of the input sequence A that correspond to the words already contained in the dictionary, and a free pass for the new (with respect to the contents of the dictionary) information. The result will be an ability to realize the structural approach to information processing.

To solve the task of detection, the transformation F^{-1} is modified to grant it detecting properties. Transformation F^{-1} interacts with an input sequence $\tilde{A}$, containing some new information along with the old one. As a result, the sequence C is formed, in which zeroes substitute those parts of the sequence $\tilde{A}$ that correspond to the parts of the track $\hat{\tilde{A}} = F(\tilde{A})$ coinciding with respective parts of the track $\hat{A}$. In other words, the input sequence $\tilde{A}$ is modified: zeroes substitute those symbols that are corresponded by the points of the track $\hat{\tilde{A}}$ coinciding with the points of the previously formed track $\hat{A} = F(A)$: $C=(..., c(-1), c(0), c(1), ..., c(t), ...)$, where:

$$c(t) = \begin{cases} \tilde{a}(t); & \hat{\tilde{a}}(t) \neq \hat{a}(t), \\ 0; & \hat{\tilde{a}}(t) = \hat{a}(t). \end{cases}\tag{14}$$

where $\hat{a}(t) \in \hat{A}$, $\hat{\tilde{a}}(t) \in \hat{\tilde{A}}$, or, in another form:

$$C = F_c^{-1}(F(\tilde{A}), HM^{-1}(\{\hat{A}\})).\tag{15}$$

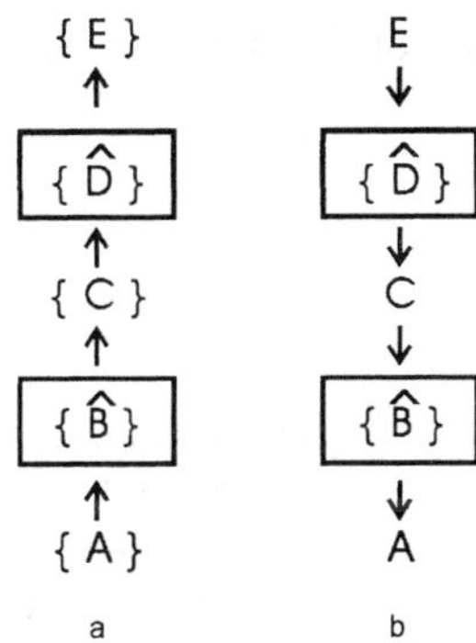

Fig. 4. Standard element of multilevel hierarchic structure. 4a) Analysis mode. Element's input receives a set of sequences $\{A\}$ that form a dictionary of lower level $\{\hat{B}\}$ in DASD, at the output – the set of syntactic sequences $\{C\}$ that serve as an input for DASD of the higher level, where dictionary $\{\hat{D}\}$ is formed on the basis of set $\{C\}$; at the output of the higher level – the set $\{E\}$. 4b) Synthesis mode. At the output of the lower level a synthesized sequence A is formed by substituting sub-sequences B, corresponding to the words of dictionary $\{\hat{B}\}$, to the corresponding place (by association) in the input sequence of abbreviations C that acts as an input of the lower level (and which is the output one from the higher level). Syntactic sequence C is the result of the analogous substitution of sub-sequences, corresponding to the words D of the dictionary of the higher level $\{\hat{D}\}$ to the corresponding places into the syntactic sequence E, fed to the input of the higher level.

If a set of input sequences A has been previously used to compose a dictionary $\{\hat{B}\} = HM^{-1}MF(\{A\})$, the use of transformation F_c^{-1} makes it possible to compose so called syntactic sequence or a sequence of abbreviations C that characterizes the connections between words L of a dictionary $\{B\}$ in a sequence A. Here $\{B\}$ is a set of sub-sequences that correspond to all chains of words $\hat{B}$ of a dictionary $\{\hat{B}\}$: $\{B\} = F^{-1}(\{\hat{B}\})$.

Thus transformation F_c^{-1} makes it possible to remove certain information contained in the dictionary $\{\hat{B}\}$ from the input sequence $\tilde{A}$. This creates the premise necessary to construct a multilevel structure for linguistic (structural) processing of input information. Syntactic sequence C, containing only new information with respect to the given level, becomes the input one for the next level where a new leveled dictionary $\{\hat{D}\}$ and a set of syntactic sequences of the next level $\{E\}$ are formed, as it has been described above, from a set of syntactic sequences $\{C\}$ (see fig. 4a). Here we have a standard element of a multilevel hierarchical structure from DASD: such processing with an extraction of leveled dictionaries can take place at all levels. Dictionary of the next level acts, in this case, as a set of grammar rules for the previous level, since the elements of the former are the elements of inter-word connection of the previous level.

Processes occurring in the described multilevel structure are reversible. Information may be compressed while being processed within the hierarchy in the upward direction, leveled dictionaries being extracted from of this information at each level (information moving upward is refined by removing frequent events with the corresponding frequency of occurrence). Information may be unpacked when processed within hierarchy in the downward direction. If we feed a syntactic sequence C that is an element of a dictionary $\{\hat{D}\}$ to the input of the first level (see fig. 4b), its respective parts will trigger, each at its time, the reproduction of one of the words from a dictionary $\{\hat{B}\}$. Thus at the output of the first level the processes will develop that took part when the syntactic sequence of the first level was formed.

2.6. Defocused transformation

A further development of the concept of F transformation is defocused transformation $F_{(r)}$, that transforms each n-segment-long fragment of symbols of a sequence not into a point – hypercube's node, but into a set of nodes that are located within an area with a radius r (r-area) and the center coinciding with the given node. Thus the original sequence A is transformed into a pipe with the radius r and the axis – track $\hat{A}$:

$$\hat{A}_{(r)} = F_{(r)}(A).\tag{16}$$

The use of defocused transformation $F_{(r)}$ allows to increase the fidelity of reproduction if non-defocused track was formed during learning. In this case, due to noise pollution of input information (e.g. inversion of some symbols) track points storing the necessary information may be found among $C = C_n^r + 1$ points of r-area around the current point of the track. During the retrieval, the mentioned necessary information (information on transition stored in the points of r-area) and the information (with respect to its weight) on the next symbol of retrieval-triggering input sequence are used to make a decision about the next symbol of the sequence being retrieved. Here four variants are possible.

If input sequence coincides with the stored one in the given symbol – then the stored sequence is retrieved. If input sequence does not coincide with the stored sequence in the given symbol, but the weight of the information on transition in the points that belong to the vicinity of the point being addressed is greater than the weight of input information – retrieval follows the stored track. Or the opposite situation: the weight of input information is greater. In that case retrieval leaves the stored track for the empty areas of signal space and stops. Retrieval is resumed when n-segment-long fragment addressing the stored sequence appears in a n-digit register again. And finally, if a decision can not be made on the basis of the stored information, it is determined by the information from the upper level.

2.7. Recognition

Recognition is understood as a process of making decision on the degree of similarity of input and previously stored information. Recognition implies the preceding learning process. Recognition mechanism is based on the comparison of the input sequence $\tilde{A}$ and the most closely matching sequence A from those stored in the DASD, the reproduction of the latter by means of the transformation

$HM^{-1}MF$ starts in response to the input sequence $\widetilde{A}$. The degree of similarity D_x:

$$D_X = \| \widetilde{A} - A \| \tag{17}$$

is calculated by summing up distances (by Hemming) between the corresponding n-segment-long fragments of input and reproduced sequences measured at each step.

$$D_X = \sum_T \| \widehat{\widetilde{a}}(t) - \hat{a}(t) \|, \tag{18}$$

where T is the length of the track. The decision about matching, with the given precision, is based on the comparison with the recognition threshold.

A more simple mechanism of recognition based on the heteroassociative way of storing/retrieval needs special consideration. During learning, sequence A, corresponding to the event being stored, is used as a carrier sequence. The sequence of code symbols corresponding to the event J serves as information sequence. In this case, recognition is understood as the retrieval of the information sequence J – event's label – that is initiated by the input sequence A.

2.8. Topology of a signal space. Resistance to damage

Describing track as a sequence of points in a signal space we imply that two points of signal space following each other in a track are produced by two consecutive fragments of input sequence that differ from each other due to the shift of one symbol. But if we consider a topology of a track in a multidimensional signal space, two neighboring points of a track are not at all geometrical neighbors. Moreover, usually they are located in distant areas of hypercube.

Such topology of a track in a signal space entails two important consequences. First, due to defocusing, each point turns into an exact sphere that does not intersect with the spheres on neighboring points of a track. The second and more important consequence is the following. If signal space is realized in any physical manner, e.g. as a neurochip, a physical damage of a part of a signal space (neurochip) affecting neighboring elements results only in an even decimation of points in a signal space. In other words, DASD is a device providing a high degree of resistance of a stored information to damage.

3. Text processing based on neuro-networking technology

The described approach to information processing proved effective for the structural analysis of information of various kinds: speech, images, texts. And the differences in the analysis are mostly confined to the stage of the preprocessing. It is the preprocessing that is usually the most intricate part of an analysis.

We shall demonstrate the capabilities of the neuro-networking approach using the processing of electronic texts, where the preprocessing is relatively simple, since the algorithms conversion of analog signal into the symbol form are well developed.

The described neuro-networking approach was utilized in the family of program products TextAnalyst® developed by Moscow company Microsystems, Ltd [5], [22] to create the frequency profile of a text.

3.1. Structural processing of information.

Human cerebral cortex may perform structural processing of information that can be explained in terms of formalism described above. Any information reaches sensory organs is processed in this way after the primary processing. Multilevel paradigmatic representation of information is formed in the left hemisphere. Separate elements of the description of the given level (words of a dictionary) are equiprobable candidates that can be used as words in a higher-level description.

Such information processing in the left hemisphere results in the automatic reconstruction of dictionaries of several levels. E.g. in case of processing of verbal information the following dictionaries are recovered: phoneme dictionary, stems and inflexional morphemes dictionary, dictionaries of words, phrases and syntaxems.

Information processing in the right hemisphere results in formation of syntagmatic two-level representations following the principle 'part vs whole' [23]. Elements of syntagms are formed at lower level while inter-elemental connections are fixed at the higher level. The specific characteristic of information processing in the right hemisphere is the redistribution of information according to the level of its significance in the framework of holistic representation with the help of hippocampus [24]. Weight characteristics of elements are recalculated in the organized two-level representations, this recalculation taking into account the density of connections of representation's elements with other elements.

Let us consider one of representation pairs, where the separate words represented at the lower level of the pair are connected to form sentences of a text at the next level. Then elements of the lower level – words – are carriers of lexical component of the meaning of text units, while the inflexional structure of the sentence (with gaps in place of the stems) composed of these words (not filled with meaning of specific words) is the carrier of the grammar component of meanings.

Words (stems), extracted during the previous stage of analysis, are connected with each other by certain links with the help of inflexional structure. To simplify the procedure of analysis of connections, we will analyze the frequency of paired occurrence of words (stems) in a sentence instead of forming the inflexional structure of a sentence. We will assume that words of a text that occurred with other words within one sentence are connected with them.

Now let us combine both text components (words and their pair connections) into the single representation. Then words (stems) appear to be connected into a network. It is worth noting here that initially both dictionaries of stems and inflexions were formed on the basis of the frequency analysis procedure. In other words, both words within the network and their connections are numerically characterized: by weight of words and weight of connections.

Then stored elements are iteratively redistributed with respect to their weight. This reorganization results in the change of initial numeric characteristics of the words. Words within the network that are connected with many other words, including connections via intermediate words, increase their weight; the weight of other words is accordingly uniformly decreased. The resulting numeric characteristic of words – their semantic weight – is an indication of their significance in the text.

3.2. System for the analysis of text TextAnalyst®

The presented realization of the system for textual information processing is based on the usage of the structural characteristics of language and text that may be revealed by statistical analysis realized on the basis of hierarchical structures composed of DASD.

Statistical analysis reveals the most frequently occurring textual elements: words or set expressions as well as interactions among the discovered textual elements. Statistical indexes are converted into the semantic ones. Later semantic weights of network elements are used to discover the most informative parts of the text. Here the following functions are realized: organization of textual base into a hypertext, automatic summarization, clusterization and classification of texts and semantic search.

Software realization of technology

The core of the system includes three major blocks and the base of linguistic information. The block of preprocessing performs sentence segmentation on the basis of punctuation marks and special grammatical words; it eliminates working words and stop-words. It also normalizes grammatical forms of words and of word combinations on the basis of stemming, reducing all wordforms to theirs stems.

Indexation block automatically selects the fundamental concepts of a text (words and word combinations) and their interconnections, calculating their relative significance. It also forms the representation of semantics of a text (set of texts) in the form of a semantic network.

The base of general linguistic knowledge includes the following dictionaries: words-separators, auxiliary words, inflexions, common words.

Indexation block (selection of words and word combinations) was created on the basis of the programmed model of hierarchical structure of DASD and realizes algorithms of automatic extraction of the dictionary of subject-specific words and their expressions with common words.

The number of DASD levels in hierarchic structure determines the a priori set maximal possible length of a concept of a specific field and is equal to twenty.

Hierarchic structure forms the dictionary of words of specific field, that passed all filters of the preprocessing and were not attributed as common words, and the dictionary of word combinations containing both special and common words. The first level of the hierarchic structure represents two-character combinations of special and common words from the dictionaries. The second hierarchical level represents DASD that store dictionaries of three-character words and combinations of characters from special and common words encountered in the text in the form of indexes of elements of corresponding first-level dictionaries with one more character added. The representation of information at further levels is completely uniform – DASD store indexes of elements stored in lower-level DASD, each with one more character added.

During the dictionaries formation the frequency of occurrence of each combination of characters in respective DASD elements is calculated. The frequency of words and word combinations (combinations of characters that are not continued at the next level) is used for the subsequent analysis.

To evaluate the semantic weight of a concept, the weights of all concepts linked with the former, i.e. the weights of the whole "semantic constellation", are used. As a result of a redistribution, those concepts gain the maximal weight that have the most powerful connections and are located, figuratively, in the center of "semantic constellations". Frequency characteristics are converted to semantic weights with the help of iterative procedure:

$$w_i(t+1) = f(\sum_{ji} w_{ji} w_j(t), w_{ij}) = f(\sum_{ipj} w_{ip} w_p(t) w_{pi}), \qquad (19)$$

here $w_i(0) = \ln z_i$; $w_{ij} = z_{ij} / z_j$ and $f(s) = \sigma(s) = 1/(1 + e^{-ks})$, were z_i – frequency of occurrence of i word in the text, z_{ij} – frequency of joint occurrence of i and j words in text fragments, and σ-function is a normalizing one.

The main functions of the system TextAnalyst[®]

Indexation block serves as a basis for the realization of the following functionalitiy of textual information processing:

- formation of hypertext structure;
- navigation through hypertext structure with the help of the semantic network;
- formation of thematic tree;
- summarizing texts;
- automatic partitioning of a set of documents into thematic blocks (clusterization);
- comparison of texts (automatic classification of texts);
- formation of a response to a semantic request of a user – formation of a thematic summary.

Conclusions

Traditional networks based on neuroid elements with spatial summation of signals do not permit effective processing of dynamic information. The major difference between neuroid elements with spatial summation of signals and neuroid elements with temporal summation of signals is the generalized dendrite. The neurons with temporal summation of signals are selectively excited by different fragments of input signal sequences and networks composed of them realize memorizing, storage and retrieving of information with associative accessing to it, as well as the structural processing of information – reconstruction of the inner structure of input information in the form of leveled dictionaries.

Such networks are basic for the realization of application programs for the automatic structural processing of a textual information – TextAnalyst®.

References

[1] A.N. Radchenko, *Modeling of the principal brain mechanisms,* (in Russian), Leningrad: Nauka, 1969.

[2] A.N. Radchenko, *Informational mechanisms of neuronal memory and models of amnesia,* (in Russian), St-Petersburg: Anatolia, 2002.

[3] A.A. Kharlamov, "Associative processor based on neuroid elements for the structural processing of information," (in Russian), Informational technologies, no. 8, 1997, pp. 40 – 44.

[4] A.A. Kharlamov, "Dynamic recurrent neuronal networks for the representation of speech information," (in Russian), Informational technologies, no. 10, 1997, pp. 16 – 22.

[5] A.A. Kharlamov, A.E. Ermakov, D.M. Kuznetsov, "Technology of processing of textual information based on semantic representations on the basis of hierarchical structures composed of dynamic neuronal networks controlled by attention mechanism," (in Russian), Informational technologies, no. 2, 1998, pp. 26 – 32.

[6] A.A. Kharlamov, R.M. Zharkoy, V.I. Volkov, G.N. Matsakov, "A system for recognition of isolated handwritten symbols based on hierarchical structure of dynamic associative storage devices," (in Russian), Informational technologies, no. 5, 1998, pp. 27 – 31.

[7] A.A. Kharlamov, C.A. Allahverdov, E.S. Samaev, "Neurochip – a neuroid element with temporal summation of input signals – an element of neuronal network for the structural analysis of information," (in Russian), Neurocomputers: development and applications, no 2, 2003, pp. 35 – 38.

[8] A.A. Kharlamov, "Neuroid elements with temporal summation of input signal and associative memory devices on the elements basis," (in Russian), in *Cybernetics questions. Devices and systems*, N.N. Evtikhiev, Ed., Moscow: Moscow institute of radiotechniks, electronics and automatics, 1983, pp. 57 – 68.

[9] Ah Chung Tsoi, "Locally recurrent globally feedforward networks: A critical review of architectures," IEEE Transactions on Neural Networks, vol. 5, no. 2, 1994, pp. 229 – 239.

[10] F. Rosenblatt, *Principles of Neurodinamics,* New York: Spartan Books, 1962.

[11] A.A. Frolov and I.P. Muravyev, *Neuronal model. of associative memory,* (in Russian), Moscow: Nauka, 1987.

[12] J.E. Dayhoff, *Neural Network Architectures. An Introduction,* New York: Van Nostrand Reinhold, 1990.

[13] A. Waibel, T. Hanazava, G. Hinton, K. Shikano, K. Lang, "Phoneme recognition: neural networks vs. hidden Markow model," in *International Conference on Acoustics, Speech and Signal Processing 1988*, pp. 107 - 110.

[14] A.J. Robinson and F. Fallside, "Static and dynamic error propagation networks with application to speech coding," in *Neural Information Processing Systems,* D.Z. Anderson, Ed., New York: American Institute of Physics, 1988, pp. 632 – 641.

[15] W. Rall, "Theoretical significance of dendritic trees for neuronal input-output relations," in *Neural Theory and Modelling,* R.F. Reiss, Ed., Stanford: Stanford University Press, 1964 , pp. 73 – 97.

[16] D.A. Sholl, "Dendritic organization in the neurons of the visual and motor cortices of the cat," Journal of Anatomy, no. 87, 1953, pp. 387 – 406.

[17] I.S. Beritov, *Structure and functions of the cerebral cortex,* (in Russian), Moscow: Nauka, 1969.

[18] G.I. Polyakov, *Basics of the systematics of human cerebral cortex neurons,* (in Russian), Moscow: Medicine, 1973.

[19] V.P. Babmindra, T.A. Bragina, *Structural basis of interneuronal integration,* (in Russian), Leningrad: Nauka, 1982.

[20] J. Midtgaard, "Processing of Information from Different Sources: Spatial Synaptic Integration in the Dendrites of Vertebrate CNS Neurons," Trends in Neurosciences, vol. 17, no. 4, 1994, pp. 166 –173.

[21] D.O. Hebb, *The Organization of Behavior: A Neuropsychological Theory*, New York: Wiley, 1946.

[22] Dan Sulliven, *Document Warehousing and Text Mining,* New York: Wiley, 2001.

[23] V.D. Glezer, *Vision and thought,* (in Russian), Leningrad: Nauka, 1985.

[24] O.S. Vinogradova, *Hippocampus and memory*, (in Russian), Moscow: Nauka, 1994.

Predictive synchrony organized by spike-based Hebbian learning with time-representing synfire activities

Katsunori Kitano[1] and Tomoki Fukai[2]

[1] Department of Computer Science, Ritsumeikan University
`kitano@cs.ritsumei.ac.jp`
[2] Department of Information-Communication Engineering, Tamagawa University
`tfukai@eng.tamagawa.ac.jp`

In this chapter, we introduce a computational model to give a theoretical account for a phenomenon experimentally observed in neural activity of behaving animals. Pairs of neurons in the primary motor cortex exhibit significant increases of coincident spikes at times when a monkey expects behavioral events. The result provides an evidence that such a synchrony has predictive power. To investigate the underlying mechanism of such a predictive synchrony, we construct a computational model based on two known characteristics in the brain: one is the synfire chain, the other is spike-timing-dependent plasticity. The synfire chain is a model to explain a precisely firing spike sequence observed in frontal parts of the cortex. Synaptic plasticity, which is commonly believed a basic phenomenon underlying learning and memory, has been reported to depend on relative timings of neuronal spikes. In the proposed model, occurrence times of events are embedded in synapses from the synfire chains to time-coding neurons through spike-timing-dependent synaptic plasticity. We also discuss the robustness of the proposed mechanism and possible information coding in this cortical region.

Keywords: computational neuroscience, information coding, synaptic learning, prediction, synchronization

1 Introduction

Learning of temporal information in the external world is important for organizing behavior. In order that an animal respond appropriately to external events, it is definitely advantageous to learn their causal relationships, to predict their occurrences, and to prepare for them. However, how the predictive aspect in organizing behavior is encoded by a neuronal activity has been an open problem, therefore, many authors have challenged this issue so far. Using a monkey that was trained on a motor response task, Riehle et al. showed that a neuronal activity in the monkey's motor cortex reflects its prediction for an external stimulus [1]. In the task, two successive visual stimuli, a CUE signal and a GO signal, were presented to the monkey: the CUE signal told the monkey the of a trial, while the GO signal, which was presented after fixed time intervals from the CUE presentation, instructed the monkey on a motor response. After the monkey engaged in a number of trials, pairs of neurons in the motor cortex exhibited spike synchronization, which occurred more frequently than by chance, at the timings when the the GO signal was expected to be presented. Since the significant spike synchronization occurred even though the GO signal was not actually presented, such a neuronal activity is thought to have predictive power.

From a general point of view, neural coding of a time lapse must be essential not only for prediction of occurrence times of the external events, but also for various behavioral organizations [2]. One of time-representing neuronal activities that has been experimentally observed is a sustained activity; neurons in cortical areas such as the prefrontal cortex continue to discharge spikes at firing rates of several tens spikes per second while a subject engaging in a delay response tasks is waiting and preparing for a GO instruction following a CUE signal [3, 4, 5, 6]. The sustained activity has been thought to represent a time interval of the preparatory period as well as the information associated with the CUE signal. Alternatively, a precisely firing sequence is another possible candidate for an internal representation of time. Multi-unit recording studies of the frontal (prefrontal and premotor) cortex found a couple of neurons generate a precise spike sequence with a fixed order and fixed intervals through several tens of synaptic connections [7, 8]. These neural activities are theoretically understandable, if synchronous spike packets propagate through a layered feed-forward structure, which was proposed as 'synfire chain' [9]. The synfire activity, in which synchronous spikes tick the layers, can perform as an internal clock. To test plausibility of this synfire hypothesis, stability of the synfire activity has been intensively investigated with both numerical and theoretical methods [10, 11, 12]. In terms of the organization of such a structure, recent modeling studies have reported that spike-based Hebbian plasticity [13], what is called spike-timing-dependent plasticity, contributes to organization of such a network because of the temporal asymmetry of the plasticity [14, 15]. Moreover, collateral spread of the spike-timing-dependent plasticity experimentally observed in hippocampal culture systems is likely to promote such a feedforward structure [16]. However, functional roles of the synfire chain have been discussed only by a few authors [17, 18], therefore, possible information processing by the network structure remains to be studied.

A neural substrate for learning has been attributed to synaptic plasticity since the Hebb's postulate [19]. Recent experimental studies have revealed that the synaptic plasticity depends on relative timing between pre- and post-synaptic spikes; if a pre-synaptic spike precedes a post-synaptic spike, the corresponding synapse is potentiated. Otherwise, the synapse is depressed [13].

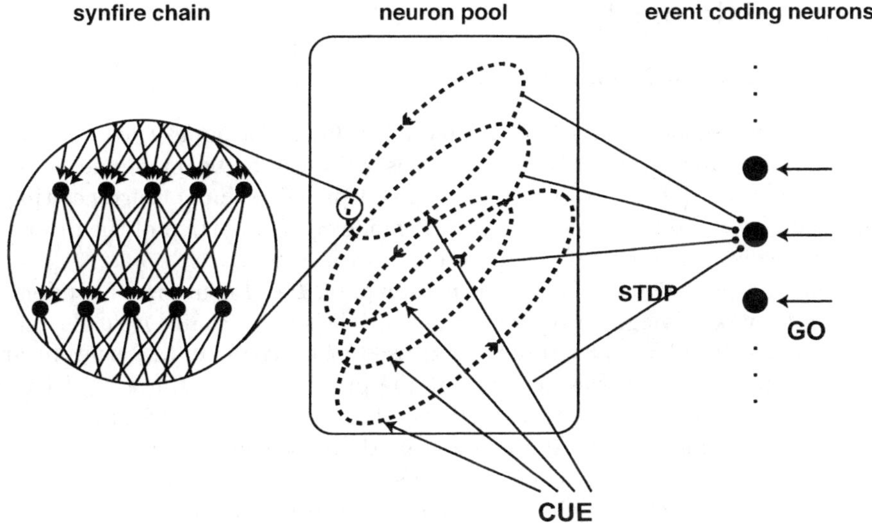

Fig. 1. A schematic picture of the model network. A single closed synfire chain is comprised of randomly chosen neurons out of a neuron pool. The synapses from the neurons in the pool to EC neurons are randomly connected with a ratio c, which means that the EC neurons receive excitatory synaptic inputs both from neurons belonging to some chains and from ones belonging to no chains. Therefore, an arbitrary pair of the EC neurons are projected to by different cell assemblies.

The spike-timing-dependent plasticity (STDP) has a time window in the causal potentiation part and the acausal depression part, respectively [20, 21]. A modeling study implied that the time window of the causal part play a role of coincidence detection [22]. Furthermore, as is theoretically shown, STDP imposes a competition among synapses connecting to an identical post-synaptic neuron so that the synaptic strengths redistribute to regulate a firing activity of the post-synaptic neuron [23].

In this chapter, we demonstrate that multiple synfire chains and STDP cooperatively play important roles of generation of the predictive synchrony. Since every layer in a synfire chain is activated at a precise timing due to the inherent property of the synfire activity, a relative time interval from the presentation of the CUE signal can be represented by a combination of precisely activated layers each of which belongs to different synfire chains. Considering the coincidence function of STDP, we expect that synapses from layers activated at the timing of GO signal are selectively potentiated. Based on this insight, we construct a computational model in which occurrence times of external events are embedded in and detected from the coordinate activities of the multiple synfire chains via STDP [24].

2 Materials and Methods

2.1 Structure of Network

Our network model consists of two modules. One is a pool of neurons that contains multiple synfire chains, which is assumed to be in premotor and prefrontal areas [8]. The other is a group of neurons coding occurrence times of events, event coding neurons, which is modeled as a portion of the primary motor cortex. The schematic structure is shown in Fig.1.

Out of N_{all}=5,000 neurons in the neuron pool, $L \times M$ neurons are randomly chosen to make a single feed-forward network, a synfire chain. A single chain that has a closed loop structure is comprised of L layers, and M neurons are included in each layer. We here repeat the procedure four times in order to make four synfire chains with L_1=90, L_2=110, L_3=130, L_4=150, and M=20 for all the chains. Due to randomness in the procedure, some neurons are involved in more than one chain, whereas some other ones are never chosen. A neuron in a layer of a chain received excitatory synaptic inputs from all the neurons in the preceding layer. Synaptic projections from the neuron pool to event coding neurons (EC neurons) are randomly determined with the probability c. We typically take $c = 0.2$ for demonstrations.

Accordingly, the EC neurons are innervated by the synaptic inputs from the neuron pool. Since we demonstrate only ten EC neurons, mutual connections among the EC neurons are not dominant in comparison with the input from the neuron pool. As a result, whether the mutual connections exist or not, results are not different. Therefore, we neglect such connections in the present model.

2.2 Single Neuron Model

We use a leaky integrate-and-fire model as a dynamics of the membrane potential. For the neuron belonging to synfire chains (SC neuron), the dynamics is described with

$$\tau_m \frac{dV}{dt} = -V + E_{leak} - I_{syn} + I_{bg} \tag{1}$$

$$= -V + E_{leak} - \sum_j g_{c,j} r_j (V - E_{AMPA}) + I_{bias} + \sqrt{2D}\eta, \tag{2}$$

with τ_m=20ms, E_{leak}=-70mV. When the membrane potential reaches a spike threshold of -54mV, the neuron generates a spike and then the membrane potential is reset to -70mV within a refractory period of 1.5ms. The dynamics of the synaptic current $g_{c,j} r_j (V - E_{AMPA})$ is determined by the first order kinetics of the gating variable r_j [25], where we take E_{AMPA}=0mV for excitatory synapses. g_c represents a synaptic conductance of the feedforward projection from the neurons in the preceding layer, measured in a unit of the leak conductance of the neuron. The value is given as a constant of $g_c = 0.06$. The

SC neuron is driven by the background input I_{bg} as well. It is thought that I_{bg} arises from any other cortical inputs than the feedforward projection. In order to reduce heavy computational load to simulate a number of excitatory and inhibitory cortical spikes, we mimic the input with a fluctuating current. We assume that the background input can be decomposed into a constant bias current I_{bias} and Gaussian white noise. We adjusted the intensity of I_{bg} so that SC neurons discharge at about 1H in the spontaneous activity.

On the other hand, the dynamics of the EC neuron is

$$\tau_m \frac{dV}{dt} = -V + E_{leak} - \sum_j g_j r_j (V - E_{AMPA})$$

$$+ I_{GO} + I_{bias} + \sqrt{2D}\eta. \tag{3}$$

The synaptic conductances of the synapses from the neuron pool to EC neurons g_j change according to the STDP rule mentioned later. The values of I_{bias} and $\sqrt{2D}$ are adjusted so as to ensure that EC neurons fire at around 20Hz. Furthermore, the EC neuron is given an excitatory stimulus that arises from the GO stimulus, I_{GO}. The stimulus is transiently supplied as a depolarizing current. The other parameters are the same as ones for the SC neuron.

2.3 Formulation of STDP

The synapses from the neuron pool to the EC neurons (NP-EC synapse) exhibit synaptic plasticity depending on relative spike timings between pre- and postsynaptic neurons. We here use the formulation of STDP by Song et al. [23]. In the formulation, a conductance change Δg is described as follow: if $\Delta t = t_{post} - t_{pre} > 0$, $\Delta g = g_{max} A_p \exp(-\Delta t/\tau_p)$. Otherwise, $\Delta g = -g_{max} A_d \exp(-|\Delta t|/\tau_d)$. Here, t_{post} and t_{pre} represent a spike timing of a post- and a pre-synaptic neuron, respectively. g_{max} is a maximum conductance. A_p and A_d are a ratio of the potentiation and the depression, respectively. τ_p and τ_d are the time constants. We set $A_p = 0.01$, $A_d = 0.0105$ and $\tau_p = \tau_d = 20$ms. g_{max} is set to 0.02. The conductances of the NP-EC synapses change in the range of $0 \leq g \leq g_{max}$. The initial conductances of them are determined by uniform random values in the range.

2.4 Scheme of Simulated Task

The network engages in a simulated task in which two kinds of stimuli are presented as external events. A CUE signal represents the external (usually sensory) stimulus instructing the start of a trial, whereas after the CUE signal, a GO signal is presented at three timings of 1000ms (GO1), 1500ms (GO2), and 1800ms (GO3) as external events. In our model, the CUE activates a specific layer of each chain, consequently initiates spike packet propagation in each chain. Numerical simulations are conducted in the following procedure.

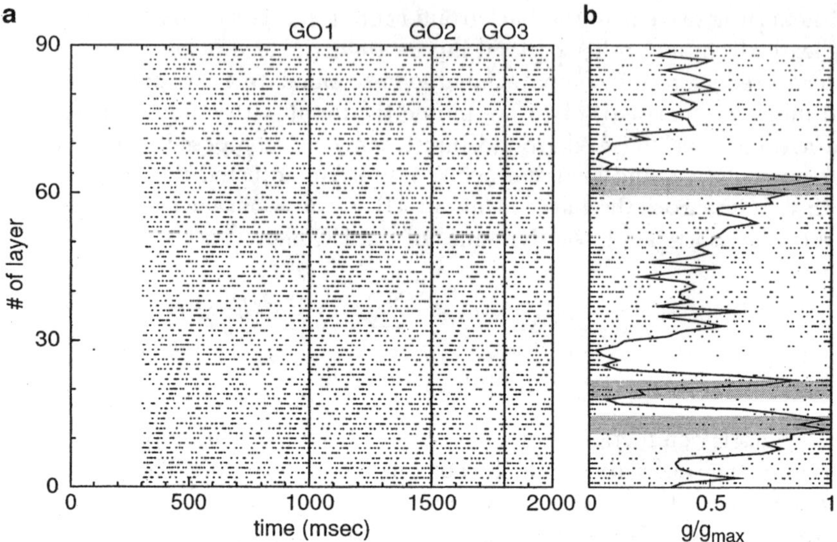

Fig. 2. An example of an activity in a synfire chain and a synaptic projection from the synfire chain achieved through learning trials. (a) A spike raster plot of the neuron involved in the chain with L_1=90 layers. We set the layer activated by the CUE signal to the first layer and arrange the neurons by the index of the layer that each neuron belongs to. The CUE signal is provided at 300ms. The vertical lines indicate the occurrence times of the GO signal, GO1, GO2, and GO3. (b) The individual and the averaged conductances of the synapses from each layer in the chain to the EC neurons are indicated by dots and a line, respectively. The conductances are normalized by the maximum value g_{max}.
The ordinate is common with the one of **a**

Firstly, trials for learning of the times are performed by the network. During the learning trials, the excitatory inputs that are supposed to be derived from the GO stimulus, I_{GO}, are delivered to the EC neurons. After learning, trials without the GO stimulus are conducted in order to see whether salient neural activity relevant to the occurrence times is observed in the EC neurons

3 Results

3.1 Synaptic Configuration after Learning

Fig.2a displays spiking activities of the neurons in the chain with $L_1 = 90$ layers during a trial. The identical layer is activated by the CUE in every trial so that a synchronous spike packet passes through specific layers at the presentation times of the GO signal. As for this chain, the packet passes through around the 20th layer at GO1, the 15th layer at GO2, and the 65th layer at GO3, respectively. At the timings, the EC neurons are activated by the GO stimulus. Therefore, the synapses from the layers activated immediately before the EC neurons' spikes are steadily potentiated over the trials due to one of the functions of STDP, coincidence (causality) detection. Fig.2b shows the synaptic conductances from the chain to the EC neurons averaged over each layer. Gray areas indicate the layers through which the synfire activity passes within 20ms before the GO presentations. As shown in the figure, the layers associated with the presentation of the GO signal by STDP gives strong synapses to the EC neurons. In the similar manner, the synapses from the other chains are organized through STDP. Thus, the times are represented by combinations of the layers that are coincidently activated.

3.2 Unitary Event Analysis

After learning trials, we make the network perform 40 trials of the task without the GO stimulus and use these spike data of 10 EC neurons for unitary event analysis [26, 27]. Firstly, for a pair of the EC neurons, we obtain firing rates of each the neuron in the sliding time window of 100ms in steps of 5ms. The step corresponds to a bin to determine precision of spike coincidences, that is, when spikes of a pair fall within the same bin, the spikes are regarded as coincident. Using the firing rates, we estimated the number of expected spike coincidences, which accidentally occur, in the time window N_{exp}, based on the assumption of independent firing of the pair. Next, the number of empirical spike coincidences, which actually occur, N_{emp} is counted in the time window. The statistical significance of the excessive spike coincidences is tested with the obtained N_{exp} and N_{emp} The P-value in the time window is represented by the area in which the number of the event occurrences n is larger than N_{emp} for a Poisson distribution with a mean N_{exp}. If the P-value is smaller than a significance level α, the epoch, which is represented by the middle of the time window, is turned out significant. We here take $\alpha{=}0.05$.

In Figs.3, a example of the analysis for a pair of the neurons, neuron 4 and 8, is shown. In Fig.3a, dots show conventional raster plots of the pair and black circles represents raw coincident spikes. The average firing rates over trials of both the neurons are about 25Hz. As shown in Fig.3b, unitary events, the coincident spikes with statistical significance $P < \alpha(= 0.05)$, are represented by black circles. The unitary events of the pair occurred around at the occurrence times of the GO stimulus even though the stimulus are not given actually. This is because both of the EC neurons commonly detect the combinational activities of the layers that are associated with the occurrence times of the GO stimulus. Fig.3c shows the joint surprise measure $\log[(1 - P)/P)]$ in each sliding time window. The measure is introduced in order to increase the resolution of large and small value of P. The dashed line corresponds to the significance level, $\log[(1 - \alpha)/\alpha] \approx 2.94$ $(\alpha = 0.05)$. If the measure exceed the dashed line, the epoch is turned out significant. The measure is found to be larger than the significance level around at the timings

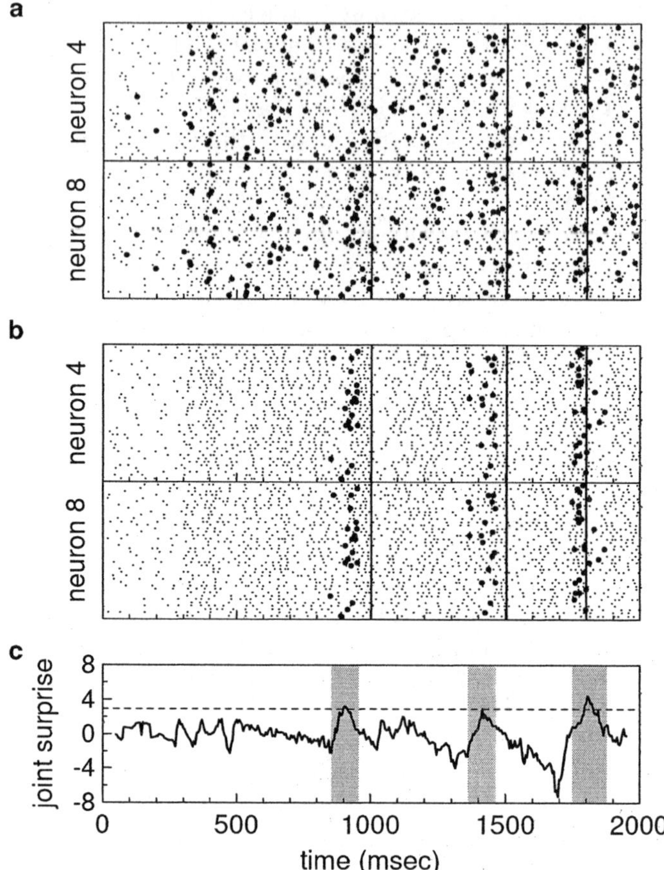

Fig. 3. A typical result of unitary event analysis for the spiking activity of a pair of the EC neurons during test trials without the GO stimulus. Spike data of 40 trials are used for the analysis. (**a**) Dots showed a conventional raster display for spiking activities of neuron 4 and 8. Raw coincident spikes are represented by filled circles. (**b**) Among the coincident spikes, significant synchronous spikes, unitary events, are indicated by the filled circles. (**c**) In the analysis to detect the unitary events, we calculate significance measures of spike coincidences, P-value. In order to increase resolutions for large or small P, we introduce the joint surprise measure, $\log[(1 - P)/P]$, which is indicated by the curve. The significance level $\alpha = 0.05$ is indicated by the dashed line. If the curve of the measure excess the dashed line, it is turned out that the significant spike coincidences occur at the epoch. The corresponding time windows are indicated by gray areas

of the GO presentation. The significant time windows that contain unitary events are indicated by gray areas.

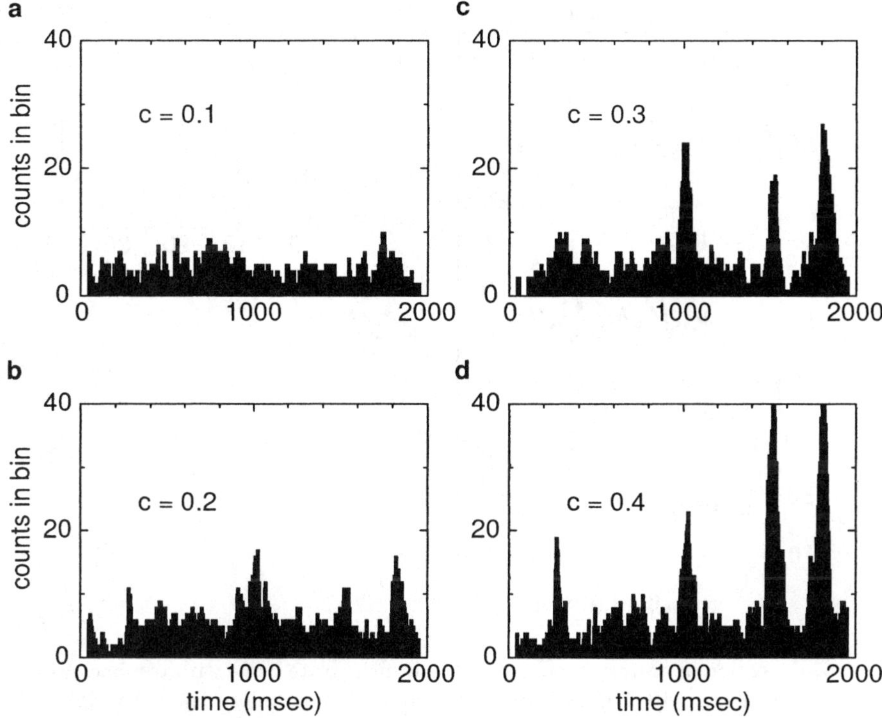

Fig. 4. The generation of the predictive synchrony depends on the connection ratio of the NP-EC projection, c. For all the pair of 10 EC neurons (45 pairs), we carry out the unitary event analysis. When a time window is turned out significant, we count it in the middle bin of a time window. The figures show how many times each bin is determined as statistically significant in the case of different connection ratios **(a)** c=0.1, **(b)** c=0.2, **(c)** c=0.3, and **(d)** c =0.4

3.3 Dependence on Connection Rate c

As a result of the unitary event analysis for all the pairs (45 pairs) of the EC neurons, it is found that only two pairs show unitary events time-locked to all the three timings. Next, we investigate a frequency distribution of the significant epoch, which stands for how many times significant spike synchrony is confirmed in the epoch through the analysis for all the pairs in three simulations with different initial conditions. Figs.4 displays the distributions for different connection rate of c=0.1, 0.2, 0.3 and 0.4. In the case of $c = 0.1$, there is no peak in the distribution. As c becomes larger than 0.1, however, peaks are observed at the timings of the GO presentations, which implies that significant spike synchrony tends to occur at the timings. Since the large c means that NP-EC synapses are rich, the synaptic inputs from the layer

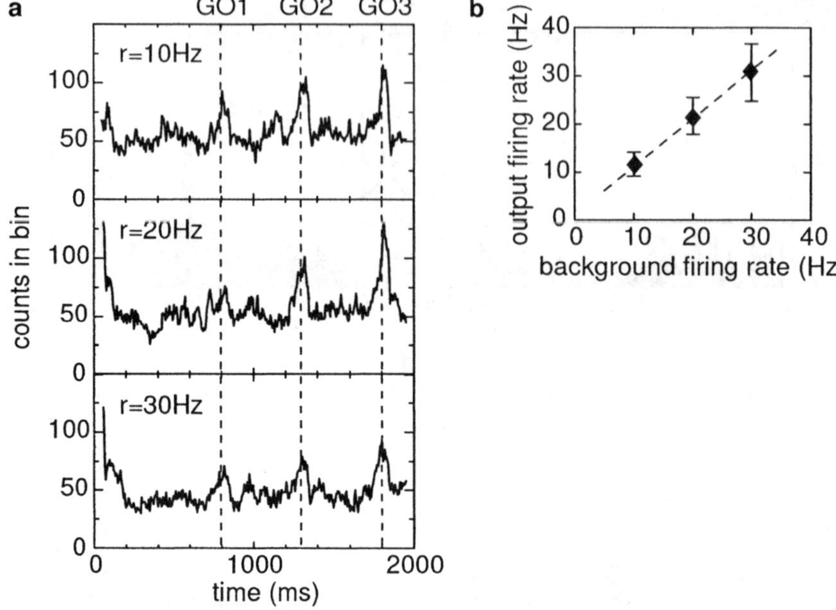

Fig. 5. Detection of unitary events at different firing rate r of excitatory background activity. (a) For r=10, 20, and 30Hz, we counted how many times the bin with significant spike coincidences is determined. Timings of GO stimulus are 800ms, 1300ms, and 1800ms in this numerical simulations. (b) The relationship between the input firing rate of the background activity and the average output firing rate of the EC neurons. Due to the activity regulation function of STDP, the average output firing rate increases only modestly against an increase in the input firing rate of the background

activated at the timings in the synfire chains are intensive enough for the EC neurons to detect. Consequently, "effective" common inputs that the layers provide pairs of the EC neurons become strong so that pairs tend to exhibit spike synchronization significantly at the timings.

3.4 Robustness against Background Activity

As shown by the results mentioned above, the generation of the predictive synchrony relies on "effective" common inputs at the timings of the GO signal. In order that such a predictive synchrony is successfully generated, the effective common inputs must be detectable against the background activities. To investigate robustness of the predictive synchrony against background activities, excitatory and inhibitory Poisson spike trains delivered through synapses are supplied to the EC neurons as the background cortical inputs, instead of applying the fluctuating currents. Concurrently, we replace the spiking activity obtained by the simulation of the multiple synfire chains with time-locked spikes generated by hand for the purpose of reducing the computational load. In this case, both the synapses, synapses mediating the time-locked spikes (NP-EC synapses) and excitatory synapses mediating background activities (BG synapses), exhibit synaptic plasticity in the spike-timing-dependent manner. In this simulation, we assume that the innervations of 1000 excitatory BG synapses, inhibitory 300 synapses, and 300 NP-EC synapses to each EC neuron are independent over the neurons. The inhibitory synapses mediate Poisson spike trains with a fixed firing rate of 10Hz. An excitatory BG synapse delivers Poisson spikes with a mean firing rate of rHz, whereas a NP-EC synapse a superposition of time-locked spikes and random spikes. It is assumed that, in each of NP-EC synapse, time-locked spikes mimicking spiking activities of neurons in the multiple synfire chains and random spikes occur at 2Hz and 3Hz, respectively. Due to a random projection of the NP-EC synapses, the neurons coding a relative time interval from the CUE presentation randomly innervate the EC neurons. Based on the fact, we make the time-locked spikes by hand uniformly through the trial time.

We carry out numerical simulations, changing the firing rate r of the excitatory background activities. The similar plots to Figs.4 are indicated in Fig.5a. In the case of r=10Hz, the background activities are relatively low, so that the network reliably exhibits predictive synchrony at the timings of the GO signal. Although the peaks at the timings become lower with an increase in the firing rate, the three peaks remain detectable. In general, as an input becomes more intensive, an output firing rate steeply increases as long as the rate stays in biologically plausible range. In this case, however, the output firing rate of the EC neurons moderately increases, as shown in Fig.5b. This is because STDP performs the other function, activity regulation. The intensity of the synaptic inputs through the excitatory BG synapses, in the other words, a gain of noise, is regulated through a synaptic competition among the synapses that is introduced by STDP. Consequently, the cooperation of two functions of STDP, coincidence detection and activity regulation, enhances the reliable generation of the predictive synchrony.

3.5 Dependence on Precision of Synfire Activities

Despite any pairs of the EC neurons share no common innervations, the neurons can synchronize due to coincidences in the inputs through NP-EC synapses, which perform as "effective" common inputs. The input coincidences highly rely on the precise propagation of the synfire activity. In the simulation of the synfire chains, spikes evoked by the synfire propagation actually show temporal fluctuations, which are caused by external noise, trial by trial. From our results, denoting the standard deviation of temporal fluctuations of a spike representing a relative time interval from the CUE signal T by $\sigma(T)$, we obtain an approximate relationship of $\sigma(T) \sim k\sqrt{T}$, where a constant k is 0.2. The constant k can be related to the precision of the synfire propagation.

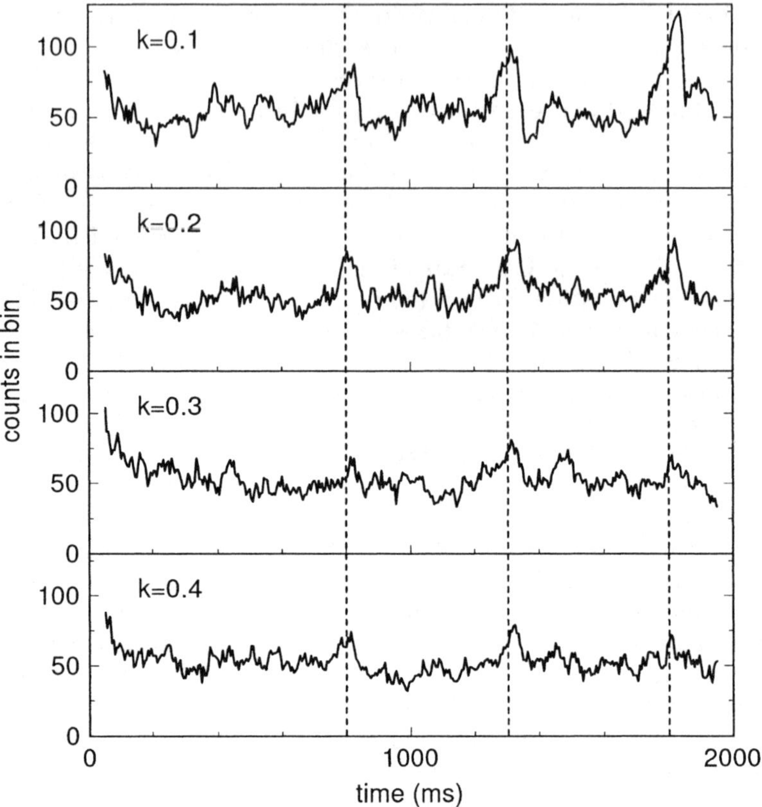

Fig. 6. Fluctuations of timed-locked spikes affect the generation of the predictive synchrony. From numerical simulations of the multiple synfire chains, it is found that spikes representing a relative time interval T from the CUE presentation fluctuate following a relation of $k\sqrt{T}$, where k is a constant. For different values of parameter k=0.1, 0.2, 0.3, and 0.4, the generation of the predictive synchrony is investigated.

To clarify to what degree the generation of the synchrony depends on the precision of the spike timings, we carry out numerical simulations for different values of k. In this simulation, the time-locked spikes introduced in the previous subsection fluctuates in every trial according to such a relationship, i.e., a spike representing a relative time from the CUE, T, shows spike jitters $k\sqrt{T}$ around at the mean time T.

As obviously shown in Fig.6, if k is larger than 0.2, the EC neurons fail to detect the layers' activities around at the timings of the GO signal. In this simulation, we set the CUE and the GO3 to 100ms and 1800ms, respectively. The spikes responsible for the GO3 are supposed to represent a relative time interval of 1700ms. At the timing, σ becomes 8.3ms for k=0.2, whereas 12.4ms

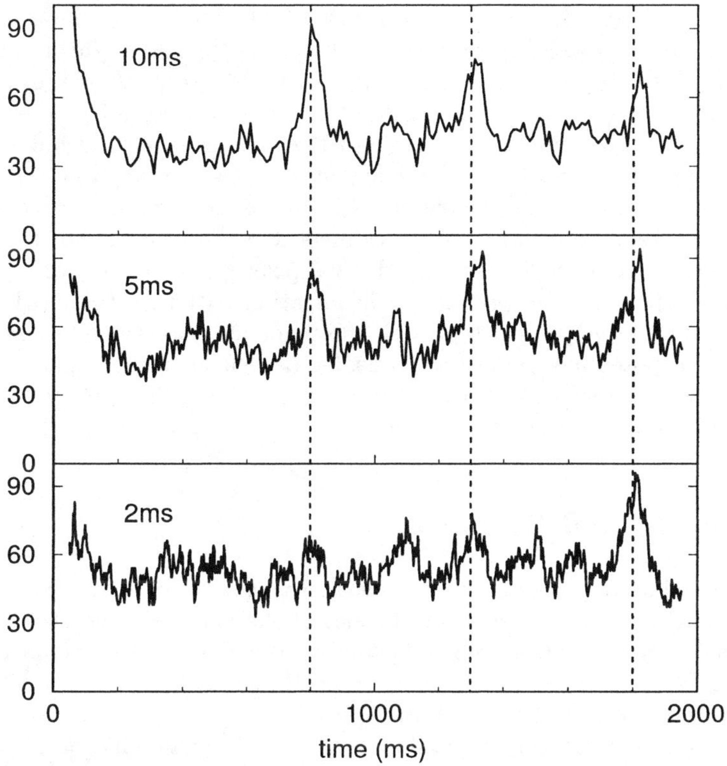

Fig. 7. Temporal characteristics of precision of significant synchrony. The same data is analyzed with different coincidence windows of 2ms, 5ms, and 10ms

for $k=0.3$. For $k=0.2$, the width of the fluctuation 2σ falls within the time window of STDP, 20ms.

As an alternative representation of the time lapse, we model modulations of firing rates instead of time-locked spikes. In the alternative model, a neuron projecting to the EC neurons temporally shows increases in its firing activity at its responsible times. In our numerical simulations, the model fails to generate predictive synchrony (not shown), which also implies that precision of spike timing is necessary for the proposed paradigm.

3.6 Temporal Characteristics of Unitary Events

Further analysis on the same task reported that significant synchrony occurred more precisely as time advances in the preparatory period [28]. Therefore, precision of the spike synchronization is likely to represent an animal's elevated expectation of predictable external events. Although the underlying mechanism of such a phenomenon is still unclear, higher-order cognitive functions is thought to be involved in updating precision of the spike synchrony.

To discuss this issue with the proposed model, we analyze the data obtained in the case $k=0.2$, changing the bin size that determines precision of spike coincidences. Fig.7 is the result of the analysis with the bin of 10ms, 5ms, and 2ms. For the 10ms bin, the first peak corresponding to GO1 is highest, whereas the third peak is lowest. This implies that relatively broad spike coincidences tend to occur at early timings. On the other hand, the analysis with the 2ms bin indicates only the 3rd peak remains high, which means spikes in the pairs exhibit precise synchronization at the 3rd timing. These results are qualitatively consistent with the tendency observed experimentally. Since the proposed model does not consider any effects of the higher-cognitive function, such a temporal characteristics emerges only from a synaptic-level of dynamics under the noisy condition.

4 Summary and Discussion

As presented above, the proposed network was capable of generating the predictive synchrony at the occurrence times of the external events such as the GO signal. The mechanism relied on combinational activities of multiple synfire chains and cooperative functions of STDP, coincidence detection and activity regulation. A combination of layers activated by synfire propagations represented a relative time interval from the CUE signal. The coincidence detection function of STDP associated timings of the GO signal with synapses from layers in the chains activated at the timings to the EC neurons in the learning process (Figs.2). The precise propagation of synfire activities enabled the inputs through the NP-EC synapses to coincidently increase at the timings, therefore, the EC neurons came to exhibit spike synchronization even without the GO presentation (Figs.3). Because of the mechanism, the successful predictive synchrony depended on the inputs through the NP-EC synapses. To depolarize the EC neurons, a sufficient amount of the NP-EC projection was necessary (Figs.4). The other function of STDP, activity regulation, played a crucial role as a gain reduction of the background activity so that the network can detect inputs from synfire chains against the background activity (Figs.5). Since the synchronization of the EC neurons depended on the coincident increase of the inputs through the NP-EC projection, fluctuations of spike timings of neurons in the synfire chains must be small enough to be within a time window of STDP (Fig.6). Finally, the proposed model gives an account for a time course of precision of significant spike synchrony, which simply arose from the synaptic learning under the noisy condition (Fig.7).

In our model, the EC neurons were driven by the inputs from the background cortical activity, which was more dominant than the inputs from the synfire chains. Since the origins of the background inputs can be considered to be many cortical areas such as the recurrent local circuit in the same cortical area and the projection from some other motor-related cortical regions. We assumed that the background activities include inputs from any possible cortical areas except from the synfire chains. Therefore, the modulation of the firing rate in the EC neurons experimentally observed in primary motor neurons must arise from the background inputs. In general, information on motor commands such as torque or directions is though to be encoded by the modulation of the firing rates in the primary motor cortex [29, 30, 31]. As shown in Fig.5b, STDP regulated the background activities so as to maintain almost liner relationship between the input and the output firing rate in the cortical area. In addition to the predictive synchrony treated here, several studies have reported that spike synchronization in the primary motor cortex plays a crucial role in the additional information coding [32, 33]. In conjunction with these evidences, the present study suggest STDP should be one of the potential neural substrates to enable the temporal coding and the rate coding to coexist in the primary motor cortex.

Appendices

A1. Leaky integrate-and-fire model

If it is assumed that a behavior of a neuronal membrane potential depends only on the passive membrane characteristics and external inputs, the neuron is modeled as a leaky integrate-and-fire neuron

$$\tau_m \frac{dV}{dt} = -(V - E_{leak}) + I_{input}. \tag{4}$$

τ_m is a time constant of the neuronal membrane, the value of which depends on cell types. While the neuron receives no external input current ($I_{input} = 0$), the membrane potential V exponentially decays due to a leak current and reaches the reversal potential of the leak current E_{leak}. When the potential depolarized by an excitatory input current ($I_{input} > 0$) excesses a threshold potential E_{th}, the neuron generates an output signal, a spike, and the membrane potential is immediately set to a resting potential E_{rest}. Once a spike is generated, the neuronal membrane shows refractoriness in a certain period. The membrane potential is held at a resting potential in the refractory period, and then the potential is released from the refractoriness after the period.

A2. Kinetic model

A spike generated by a presynaptic neuron is transmitted to postsynaptic neurons through synapses. When the spike reaches a synapse through the axon, a synaptic transmitter is released from the presynaptic terminal. If receptors on the postsynaptic site receive the transmitter, they become open

to allow ionic currents to flow through the membrane. The synaptic current
to a postsynaptic neuron is modeled as

$$I_{syn} = gr(V - E_{syn}), \tag{5}$$

where g is a synaptic conductance, r is a variable representing a fraction of the
"open" receptors, V is a postsynaptic membrane potential, and E_{syn} is the
reversal potential which depends on the kind of synapses. The rate variable r
can be described by the first-order kinetic equation [25]

$$\frac{dr}{dt} = \alpha T(1 - r) - \beta r, \tag{6}$$

where α and β are the raise and decay time constants of this process, respectively. The variable T represents the transmitter release. Since the transmitter
release is a very rapid process, the value of T can be approximated by an impulse. During the transmitter is released or in a fixed period (typically 1ms),
T is set to 1. Otherwise, T takes 0.

A3. Spike-timing-dependent plasticity

Recent experimental studies have reported that synaptic plasticity depends
on the relative timing of the pre- and postsynaptic spikes [13, 20, 21]. If the
postsynaptic spike follows the presynaptic spike, the synapse is potentiated.
Otherwise, the synapse is depressed. The amount of change in the synaptic
strength is well fit to an exponential function of difference of the spike times.
Therefore, such a plasticity rule can be mathematically incorporated [23].
Denoting the difference of the times of the pre- and the postsynaptic spikes
with $\Delta t = t_{post} - t_{pre}$, the amount is represented by

$$\Delta g = \begin{cases} g_{max}A_p\exp(-\Delta t/\tau_p) & (\Delta t > 0) \\ -g_{max}A_d\exp(-|\Delta t|/\tau_p) & (\Delta t < 0) \end{cases}, \tag{7}$$

where A_p and A_d are the ratios of synaptic potentiation and depression, respectively (both are positive). τ_p and τ_d are time constants to determine width
of plasticity windows. According to the rule, synapses are modified between
0 and g_{max}.

References

1. Riehle A, Grün S, Diesmann M, Aertsen A (1997) Science 278: 1950–1953
2. Fuster JM (2001) Neuron 30: 319–333
3. Niki H, Watanabe M (1976) Brain Res 105: 79–88
4. Goldman-Rakic PS (1995) Toward a circuit model of working memory and the
   guidance of voluntary motor action. In: Houk JC, Davis JL, Beiser DG. (eds)
   Models of Information Processing in the Basal Ganglia. MIT Press, Cambridge

5. Fuster JM (1997) The prefrontal cortex: anatomy, physiology, and neuropsychology of the frontal lobe. Raven, New York
6. Funahashi S, Inoue M (2000) Cerebral Cortex 10: 535–551
7. Abeles M, Bergmann H, Margalit E, Vaadia E (1993) J Neurophysiol 70: 1629–1638
8. Prut Y, Vaadia E, Bergman H, Haalman I, Slovin H, Abeles M (1998) J Neurophysiol 79: 2857–2874
9. Abeles M (1991) Corticonics. Cambridge University Press, Cambridge
10. Diesmann D, Gewaltig MO, Aertsen A (1999) Nature 402: 529–533
11. Câteau H, Fukai T (2001) Neural Netw 14: 675–685
12. Aviel Y, Mehring C, Abeles M, Horn D (2003) Neural Comput 15: 1321–1340
13. Markram H, Lubke J, Frotscher M, Sakmann B (1997) Science 275: 213–215
14. Levy N, Horn D, Meilijson I, Ruppin E (2001) Neural Netw 6-7: 815–824
15. Kitano K, Câteau H, Fukai T (2002) NeuroReport 13: 795–798
16. Tao HW, Zhang LI, Bi G-q, Poo M-m (2000) J Neurosci 20: 3233–3243
17. Miller R (1996) Biol Cybern 75: 263–275
18. Arnoldi HM, Englmeier KH, Brauer W (1999) Biol Cybern 80: 433–447
19. Hebb DO (1949) The organization of behavior: a neuropsychological theory. Wiley, New York
20. Bi G-q, Poo M-m (1998) J Neurosci 18: 10464–10472
21. Bi G-q, Poo M-m (2001) Annu Rev Neurosci 24: 139–166
22. Gerstner W, Kempter R, van Hemmen JL, Wagner H (1996) Nature 383: 76–78
23. Song S, Miller KD, Abbott LF (2000) Nature Neurosci 3: 919–926
24. Kitano K, Okamoto H, Fukai T (2003) Biol Cybern 88: 387–94
25. Destexhe A, Mainen ZF, Sejnowski TJ (1998) Kinetic models of synaptic transmission. In: Koch C, Segev I. (eds) Methods in Neural Modeling. MIT Press, Cambridge
26. Grün S, Diesmann M, Aertsen A (2002) Neural Comput 14: 43–80
27. Grün S, Diesmann M, Aertsen A (2002) Neural Comput 14: 81–119
28. Riehle A, Grammont F, Diesmann M, Grün S (2000) J Physiol (Paris) 94: 569–582
29. Georgopoulos AP, Kalaska JF, Caminiti R, Massey JT (1982) J Neurosci 2: 1527–1537
30. Muir RB, Lemon RN (1983) Brain Res 261: 312–316
31. Kalaska JF, Cohen DA, Hyde ML, Prud'homme M (1989) J Neurosci 9: 2080–2102
32. Hatsopoulos NG, Ojakangas CL, Paninski K, Donoghue JP (1998) Proc Natl Acad Sci USA 95: 15706–15711
33. Baker SN, Spinks R, Jackson A, Lemon RN (2001) J Neurophysiol 85: 869–885

Improving Chow-Liu Tree Performance by Mining Association Rules

Kaizhu Huang, Irwin King, Michael R. Lyu, and Haiqin Yang

Department of Computer Science and Engineering
The Chinese University of Hong Kong
Shatin, N.T., Hong Kong
{kzhuang, king, lyu, hqyang}@cse.cuhk.edu.hk

Abstract. We present a novel approach to construct a kind of tree belief network, in which the "nodes" are subsets of variables of dataset. We call this model Large Node Chow-Liu Tree (LNCLT). This technique uses the concept of the association rule as found in the database literature to guide the construction of the LNCLT. Similar to the Chow-Liu Tree (CLT), the LNCLT is also ideal for density estimation and classification applications. More importantly, our novel model partially solves the disadvantages of the CLT, i.e., the inability to represent non-tree structures, and is shown to be superior to the CLT theoretically. Moreover, based on the MNIST hand-printed digit database, we conduct a series of digit recognition experiments to verify our approach. From the result we find that both the approximation accuracy and the recognition rate on the data are improved with the LNCLT structure, when compared with the CLT.

Key words: Classification, Association Rule, Chow-Liu Tree, Large Node, Bayesian Network.

1 Introduction

One of the interesting problems in Machine Learning is density estimation, i.e., given a training dataset, how can we estimate the data distribution? The estimated distribution can be used to perform classification or prediction.

The Naive Bayesian (NB) network demonstrates good performance in using the estimated distribution to construct classifiers, even when compared with the state-of-the-art classifiers, e.g., C4.5 [27]. With a conditional independency assumption among the features or attributes, i.e., $P(A_i, A_j|C) = P(A_i|C)P(A_j|C)$, with A_i, A_j, $1 \leq i \neq j \leq n$ and C representing the attributes and class variable, respectively, NB estimates the joint probability $P(C, A_1, A_2, \ldots, A_n)$ from data and classifies a specific sample into the class with the largest joint probability. Furthermore, this joint probability can be

94

decomposed into a multiplication form based on its independency assumption. Therefore, the decision function can be written as follows:

$$c = \arg\max_{C_i} P(C_i, A_1, A_2, \ldots, A_n)$$

$$= \arg\max_{C_i} P(C_i) \prod_{j=1}^{n} P(A_j|C_i), \tag{1}$$

where, $P(C_i), P(A_j|C_i)$ are usually estimated empirically.

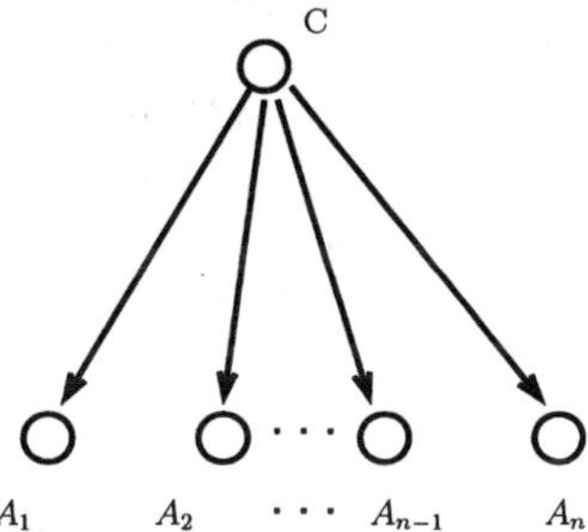

Fig. 1. A Naive Bayesian Classifier. $A_i, 1 \leq i \leq n$, is the attribute. In this figure, the attribute is independent of each other, given the class label C.

The success of the NB is somewhat unexpected since its independency assumption typically does not hold in many cases. A representative example is the so-called "Xor" problem. In this problem, attributes are two binary random variables A and B. When $A = B$, $C = 1$; otherwise $C = 0$. Thus the attribute A is not independent of B, when given the class variable C. NB encounters problems in classifying the "Xor" data. The reason is that $P(C = 0), P(C = 1), P(A = 0|C = 0), P(A = 1|C = 0), P(A = 0|C = 1), P(A = 1|C = 1)$ will all be nearly 0.5 when the data samples are randomly generated. It will be hard to assign any data into the class "0" or "1" since the estimated joint probabilities, according to (1) for both classes, will be about $0.5 \times 0.5 \times 0.5 = 0.125$.

By relaxing the strong assumption, i.e., the independency among the data attributes, of NB, many researchers have developed other types of Bayesian belief networks such as Semi-naive Bayesian networks [18, 12], Selective Naive Bayesian networks [20], and Tree Augmented Naive Bayesian networks [9].

One of the competitive models in this trend is the so-called Chow-Liu Tree (CLT) model [3]. Rather than assuming an independence among the attributes, CLT assumes a tree dependence relationship among the attributes, when given the class variable. The decision function of the CLT constructed from the estimated distribution can be written as a decomposed form:

$$c = \arg\max_{C_i} P(C_i, A_1, A_2, \ldots, A_n)$$

$$= \arg\max_{C_i} P(C_i) \prod_{j=1}^{n} P(A_j | Pa(A_j), C_i), \qquad (2)$$

where $Pa(A_j)$ represents the parent node of A_j in the tree structure. The decomposed item $P(A_j | Pa(A_j), C_i)$ is usually estimated empirically.

When compared with NB, CLT can generate a more accurate distribution [3] and achieve lower error rates in classification tasks [9]. Its advantages are partly due to the relaxed restriction than NB [9], its decomposable ability in approximating distribution, and the resistance to over-fitting problems.

However, there are still problems for the CLT, i.e., the tree dependence assumption on the underlying structure of the training dataset will be still too strong to be satisfied in many cases. For a simple example, see Fig. 2(a). If the underlying structure of a dataset can be represented as a graph as Fig. 2(a), the CLT method will not be able to restore this structure, since Fig. 2(a) is not a tree due to its cyclic characteristic.

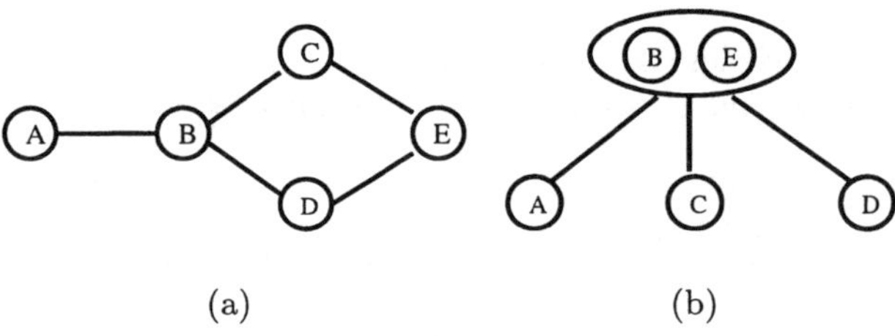

(a)   (b)

Fig. 2. (a): The underlying structure of a dataset (b): A large node tree structure we call "LNCLT"

However, if we combine some nodes as a "large node", then Fig. 2(a) can be represented as a tree. Fig. 2(b) is indeed such a structure, which is compatible with Fig. 2(a), since they both represent "A, C, D are conditionally independent of each other, when given B and E".

Motivated from this finding, we develop a Large Node Chow-Liu Tree (LNCLT), where the large node means a subset of attributes as $\{B, E\}$ in Fig. 2(b). Based on the improved techniques of association rules [1], we propose reasonable combination rules to construct the large node tree structure directly from the draft structure by the CLT. Both theoretical results and experimental results demonstrate the superiority of our novel model over CLT.

One of the important features of our approach is that, implied by its name, the resulting large node tree maintains a tree structure, where the estimated distribution is easily decomposed and therefore naturally enjoys the resistance ability to the overfitting problems.

The main contributions of this paper are described as follows. First, we propose a novel Large Node Chow-Liu Tree, which outperforms Chow-Liu Tree theoretically and experimentally. Second, we develop a theory to determine the threshold used in mining association rules, which is usually set by hand. This will, thus, save the time to adapt the threshold by some intuitive methods such as Cross Validation methods [16].

This paper is organized as follows. In next section, we present the related work. In Sect. 3, we describe the background for this paper including the notations, the CLT algorithm and the concept of the association rule. In Sect. 4, we introduce the main work of this paper, namely, the main theoretical results in guiding the construction of the LNCLT and the practical algorithm. Following that, in Sect. 5, we demonstrate the advantages of the LNCLT based on a serious of experiments. We then conclude this paper with final remarks in Sect. 6. Some of the theoretical and experimental results in Sect. 4 and Sect. 5 have been earlier presented in [11] and are expanded significantly in the current paper, while other sections are new.

2 Related Work

It has been an active research topic to attempt to improve the performance of belief networks by relaxing strong connection assumptions. A number of algorithms are proposed to relax the strong assumption of NB [18, 25, 12, 20, 19, 28, 32]. Similarly, in relaxing the CLT, Malvestuto [22] used acyclic hypergraph and brought out a local heuristic method to search the structure. The similar work to learning hypergraph[1] from data was presented by Srebro et al in [29, 15]. They aimed to solve this problem globally and proposed the approximation method as well. Another school of approaches to extend the CLT is the so-called Bayesian Networks [26]. Instead of assuming a dependence tree structure, this method tried to search the dependence relationship among the attributes from data.

However, the above models suffer from the difficulties in approximating the good distributions from data. As shown by Srebro [29], it is an NP-hard problem to find the optimal hypergraph. Even for the proposed approximation method, it cannot achieve satisfactory result [14]. Furthermore, it is also NP-hard to obtain the optimal Bayesian Networks from data [6]. On the other hand, unrestricted Bayesian Networks do not demonstrate an increase in accuracy even when compared to the simple NB network [9].

Other extensions of the Chow-Liu Tree are also investigated recently. Meila [23] proposed to model distributions as the mixture of the Chow-Liu Trees. Dasgupta and Luby [4] suggested polytree Bayesian networks or trees with oriented edges. Huang et al. invented a discriminative way to training Chow-Liu Trees [13].

---

[1] Srebro et al. named this structure as Markov network or hypertree.

In this paper, we do not aim to find an optimal large node tree structure. Similar to [23], we perform the upgrading directly on the CLT. Instead of using a linear combination of CLTs, we construct a more relaxed graph structure than the CLT, namely, the large node tree, based on the improved techniques from association rules. Moreover, we theoretically prove that the constructed large node tree has a larger log likelihood than that of the CLT and therefore generate a more accurate distribution approximation.

3 Background

In this section, we first describe the notations used in this paper. Next, the concept of CLT and association rules, rather than the details of these two topics will be introduced.

3.1 Notations

The notation here will largely follow that of [23]. Let V denote a set of n random discrete variables and assume A is a subset of V. We denote x_A as one assignment of the variables in A. Moreover we consider a graph $T = (V, E)$ where V is the vertex set and E is a set of undirected edges. If T is a connected acyclic graph, we call T a tree. If the number of edges $|E|$ in a tree T is equal to the number of vertex minus one: $|V| - 1$, we call T a spanning tree. Let V^* denote a set of the subsets of V, where V^* satisfies the following condition:

$$\cup_{U_i \in V^*} U_i = V, \tag{3}$$

$$U_i \cap U_j = \varnothing \quad with \quad U_i, U_j \in V^*, \quad i \neq j. \tag{4}$$

A large node tree $T^*(V^*, E^*)$ is defined as a tree where V^* is the vertex set satisfying the above conditions and E^* is the set of edges among V^*. Here we can see that each vertex of T^* is actually the subset of V and these subsets have no overlapped variables. Figure 3(b) is an example of a large node tree.

According to the tree decomposition, the distribution encoded in the large node tree can be written into:

$$P^*(x_V) = \frac{\prod_{(u,v) \in E^*} P(x_u, x_v)}{\prod_{v \in V^*} P^*_v(x_v)^{deg(v)-1}},$$

where, $deg(v)$ refers to the number of edges which contain v as one vertex. The directed large node tree distribution can be written into:

$$P^*(x_V) = \prod_{v \in V^*} P^*_{v|Pa(v)} P^*(x_v | x_{Pa(v)}).$$

The problem of learning Large Node Chow-Liu Tree can be informally stated as: given the training dataset S with s independent observation $x^1, x^2, \ldots, x^s$, find a large node tree structure that match S well, where x^i is the n-dimensional vector, which can be represented as $\{x_1{}^i, x_2{}^i, \ldots, x_n{}^i\}$.

3.2 Chow-Liu Tree

We here introduce the algorithm to construct the CLT from data. We will not talk much about the Chow-Liu Tree techniques. Readers interested in this method can refer to [3].

(1) a) Calculate all the mutual information denoted as $I(X_i, X_j)$, between any two nodes X_i, X_j, where, the mutual information between two variables X, Y is defined as

$$I(X, Y) = \sum_{x,y} P(x, y) \log \frac{P(x, y)}{P(x)P(y)}. \tag{5}$$

    b) Insert them into a set B.

    c) Initiate tree $T(V, E)$ where $V = \{$all the nodes of a data set$\}$, $E = \{\}$,

(2) Do until E contains $n - 1$ edges (n is the number of nodes)

    a) Find the nodes pair (X_{m_1}, X_{m_2}) with maximum mutual information denoted as Im from B.

    b) If no cycle is formed in T when the vertex X_{m_1} is connected with X_{m_2}, add edge (X_{m_1}, X_{m_2}) in E, and delete $Im(X_{m_1}, X_{m_2})$ from B.

    c) Otherwise, delete $Im(X_{m_1}, X_{m_2})$ from B

    d) Go to (2).

The CLT structure obtained from this algorithm is proved to be the optimal one in the sense of Maximum Likelihood criterion [3].

3.3 Association Rules

Mining association rules is recently under great attentions in data mining [1]. This method can be typically applied in the supermarket database analysis problem. In such a problem, it is interesting to know what other goods customers will buy when they buy a certain type of goods. A representative example is that a large number of customers will buy the butter when they buy the bread. Then *bread → butter* is called one association rule.

The notation of association rule is that: assuming that $I = \{i_1, i_2, ..., i_n\}$ is a set of items and T is a set of transactions, one transaction is a set of items. We use $X \to Y$ $(X \cap Y = \varnothing)$ associated with a confidence $c \in [0, 1]$ to specify an association rule that means customers will buy X item together with Y item with the confidence level c if a fraction c of the transactions consisting of X also consist of Y. The rule has a *support s* in T, if a fraction s of the transactions in T consist of both X and Y. To make the association reliable, this support s has to be greater than a threshold which is called the minimum support. In our problem, T is the dataset and I is the attributes set. Because we are concerned about the classification accuracy, we fix Y and X as the class variable C and a subset of the attributes, respectively. Since we construct each LNCLT or CLT for each class, mining the association rule will be reduced to mining all the frequent itemsets X, whose supports are larger than the minimum support.

With regards to the algorithm to mine association rules, we refer the interested readers to [1, 10], since it is out of the scope of this paper to introduce the algorithm in detail. In this paper, we use the algorithm called Apriori developed in [1].

4 Learning Large Node Chow-Liu Tree

In this section, we first define a concept called combination transformation. We then in Sect. 4.1 present the combination rules and give the theoretical justifications why these rules will improve the performance of the draft structure. Following that, we propose the theory on how to determine the minimum support used in the combination rules. Finally, we detail the practical algorithm in Sect. 4.2.

Definition 1 *A combination transformation is defined to be a transformation in a tree structure T. This transformation combines several nodes into a large node and keep the connection relationship of T.*

Figure 3 is an illustration of combination transformation. In Fig. 3, (a) is a tree structure and (b) is the result after a combination transformation. In (b) when nodes D, B are combined, the edge $\overline{BE}$ in (a) will be kept as the edge $\overline{(BD)E}$ in (b).

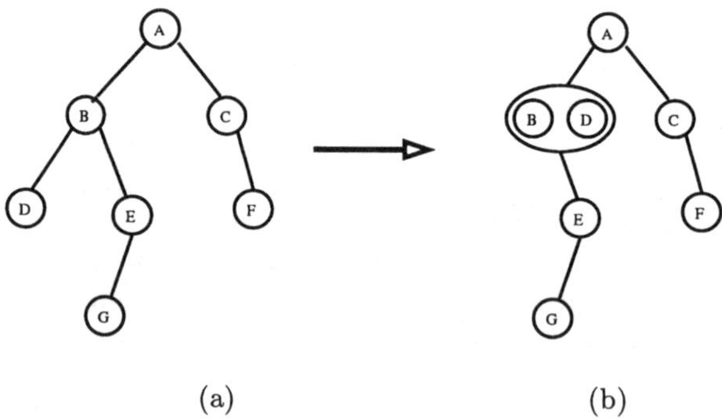

(a)   (b)

Fig. 3. An illustration of combination transformation

4.1 Main Results

In this subsection, we will present theoretical results on combination rules. We first describe the combination rules.

Rule 1 *Sibling rule: The nodes to be combined satisfy that the set of these nodes are sibling relationship, i.e., there exists another node as their common parent.*

Rule 2 *Parent-child rule: The nodes to be combined satisfy that the set of these nodes can be sorted as a sequence based on a certain node as the root, in which each node is the parent node of its sequent node.*

Rule 3 *Association rule: The nodes to be combined satisfy that, under a given confidence level and a minimum support, the set of these nodes denoted by A forms an association rule , i.e., $A \to C$, where C is the class label.*

Rule 4 *Bound rule: The nodes to be combined satisfy that the number of these nodes is fewer than a given integer bound K.*

We theoretically show that the resulting graphical structure after a combination transformation satisfying Rule 1 or Rule 2 will have a larger log likelihood and thus can approximate the dataset more accurately. We give Proposition 1, Proposition 2 and further Corollary 1, Corollary 2 to prove this.

We first present a preliminary lemma on the log likelihood of the CLT.

Lemma 1. *Given a training dataset S and n variables defined as in Sect. 3.1, the log likelihood $l_t(x^1, x^2, \dots, x^s)$ of the observations can be written as the following when the dataset is fit as a maximum weight spanning tree, where the weight is given by the mutual information between two nodes:*

$$l_t(x^1, x^2, \dots, x^s) = \sum_{i=1}^{n} \sum_{k=1}^{s} \log P(x_i^k \mid x_{j(i)}^k), \tag{6}$$

where $j(i)$ represents the parent node of variable i obtained by the ordering based on any certain node as the root in a tree and x^k is an n-dimensional vector $\{x_1^k, x_2^k, \dots, x_n^k$ with $1 \le k \le s\}$. Moreover, this log likelihood is maximized when the spanning tree is obtained with Chow-Liu Tree method [3].

The proof can be seen in [3].

Proposition 1. *Given a spanning tree T, if any two nodes satisfy parent-child relationship based on a certain root, then the graphical structure T^* after a combination transformation of these two nodes is, based on the Maximum Likelihood criterion, superior to the original tree T.*

Proof. Using Fig. 4 as an illustration, we assume that the left part (a) is one sub-part of the spanning tree T and in this subpart we perform the combination of two variables X_1 and X_2. To be simple, we assume X_1 has children: X_2, X_q, and X_2 has only one child : X_m. We have the similar proof if X_1 and X_2 have multiple children. Figure. 4(b) is the structure after these two nodes X_1, X_2 with parent-child relationship are combined. For the spanning

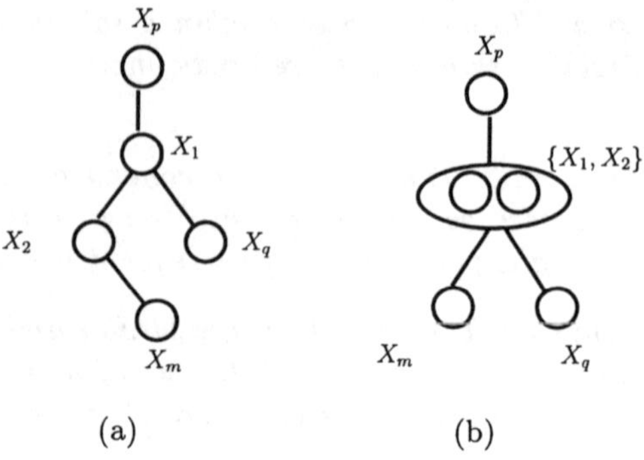

Fig. 4. A parent-child combination (a): The original sub-tree. (b): The result sub-tree after the combination of X_1 and X_2.

tree T, only the subpart (a) is changed into (b) when combining X_1 and X_2 and the other parts of T remain unchanged. We rewrite the log likelihood of the training dataset according to tree T into two parts:

$$l_t(x^1, x^2, \ldots, x^s) = \sum_{i \neq X_1, X_2, X_m, X_q} [\sum_{k=1}^{s} \log P(x_i{}^k \mid x_{j(i)}^k)] +$$

$$+ \sum_{k=1}^{s} [\log P(x_{X_2}^k \mid x_{X_1}^k) + \log P(x_{X_m}^k \mid x_{X_2}^k) +$$

$$+ \log P(x_{X_q}^k \mid x_{X_1}^k) + \log P(x_{X_1}^k \mid x_{X_p}^k)]. \tag{7}$$

The same as (7), we can write the log likelihood encoded in the transformed structure T^* with X_1 and X_2 combined into (8):

$$l_{t^*}(x^1, x^2, \ldots, x^s) = \sum_{i \neq X_1, X_2, X_m, X_q} [\sum_{k=1}^{s} \log P(x_i{}^k \mid x_{j(i)}^k)] +$$

$$+ \sum_{k=1}^{s} [\log P(x_{X_m}^k \mid x_{X_1}^k x_{X_2}^k) + \log P(x_{X_q}^k \mid x_{X_1}^k x_{X_2}^k) +$$

$$+ \log P(x_{X_1}^k x_{X_2}^k \mid x_{X_p}^k)]. \tag{8}$$

Further we can define the second part of (7) as $R(l_t)$ and write it into the entropy form as in (9).

$$R(l_t) = \sum_{k=1}^{s} [\log P(x_{X_2}^k \mid x_{X_1}^k) + \log P(x_{X_m}^k \mid x_{X_2}^k) +$$

$$+ \log P(x_{X_q}^k \mid x_{X_1}^k) + \log P(x_{X_1}^k \mid x_{X_p}^k)]$$

$$= \sum_{k=1}^{s} \log P(x_{X_2}^k \mid x_{X_1}^k) + \sum_{k=1}^{s} \log P(x_{X_m}^k \mid x_{X_2}^k) +$$

$$+ \sum_{k=1}^{s} \log P(x_{X_q}^k \mid x_{X_1}^k) - \sum_{k=1}^{s} \log P x_{X_p}^k +$$

$$+ \sum_{k=1}^{s} \log P(x_{X_1}^k \mid x_{X_p}^k) + \sum_{k=1}^{s} \log P x_{X_p}^k$$

$$= -H(X_2|X_1) - H(X_m|X_2) - H(X_q|X_1) -$$

$$-H(X_1 X_p) + H(X_p). \tag{9}$$

In the same way, we can write the second part of (8) into (10).

$$R(l_{t^*}) = \sum_{k=1}^{s} [\log P(x_{X_m}^k \mid x_{X_1}^k x_{X_2}^k) +$$

$$+ \log P(x_{X_q}^k \mid x_{X_1}^k x_{X_2}^k) + \log P(x_{X_1}^k x_{X_2}^k \mid x_{X_p}^k)]$$

$$= -H(X_2|X_1 X_p) - H(X_m|X_1 X_2) -$$

$$-H(X_q|X_1 X_2) - H(X_1 X_p) + H(X_p). \tag{10}$$

According to the information theory, we have:

$$H(X_2|X_1) \geq -H(X_2|X_1 X_p),$$
$$H(X_m|X_2) \geq H(X_m|X_1 X_2),$$
$$H(X_q|X_2) \geq H(X_q|X_1 X_2).$$

Therefore, we have the following inequality:

$$R(l_t) \leq R(l_{t^*}). \tag{11}$$

From (7), (8), (11) we obtain that:

$$l_t \leq l_{t^*}. \tag{12}$$

Proposition 1 shows that a single parent-child combination transform will increase the log likelihood of a tree T, which means the data fitness will be increased.

Proposition 2. *Given a spanning tree T, if two nodes satisfy sibling relationship based on a certain root, then the graphical structure T^* after a combination transformation of these two nodes is, based on the Maximum Likelihood criterion, superior to the original tree T.*

The proof of Proposition 2 is much similar to Proposition 1, we will not prove it here.

Based on a sequence of combination transformation, We can easily expand Proposition 1 and Proposition 2 into the following Corollary 1 and Corollary 2.

Corollary 1. *Given a spanning tree T, if a subset of nodes can be sorted as a sequence based on a certain node as the root, in which each node is the father of its sequent node, then the graphical structure T_* after a combination transformation of these nodes in this subset is, based on the Maximum Likelihood criterion, superior to the original tree T.*

Corollary 2. *Given a spanning tree T, if all the nodes in a subset are sibling relationship, then the graphical structure T_* after a combination transformation of all the nodes in this subset is, based on the Maximum Likelihood criterion, superior to the original tree T.*

These two corollaries prove that the combination transformation of parent-child relationship and siblings relationship will increase the approximation accuracy on data. Another advantage of combining the nodes with parent-child or sibling relationship lies in that the transformed graphical structure will maintain a tree structure, which is easily decomposed and enjoys the ability to resist the overfitting problem. On the other hand, combining nodes without parent-child or sibling relationship may result in a non-tree structure. Such example can be seen in Fig. 5.

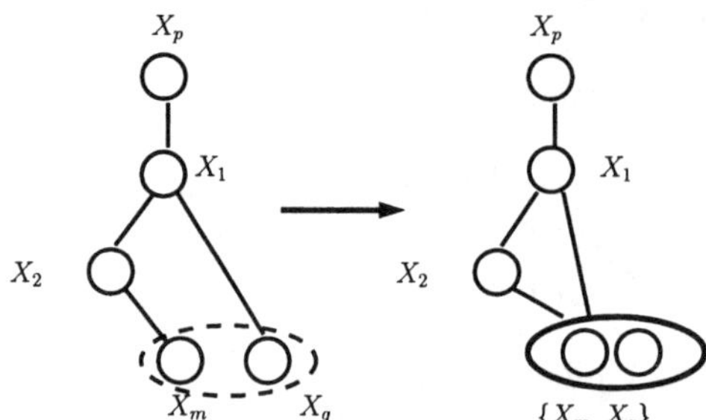

Fig. 5. An example to illustrate that combining nodes without parent-child or sibling relationship may result in a non-tree structure.

Here we argue that Rule 3 is reasonable. Since those attributes with an association rule pointing to class label C will occur with one another more frequently, they should be more dependent on one another than other attributes. Thus they are more like a single node and should be combined with the higher priority.

On the other hand, Rule 4 is also necessary. The bound K cannot be too large, or the estimation of the probability of the large node will be unreliable. An extreme case is that, when K is equal to n, i.e., the number of the attributes, all the nodes will be combined into one large node. In this case the estimated distribution will be the empirical distribution, which is very unreliable to represent the data.

Until now, we do not mention how to set the threshold, i.e., the minimum support in using the associate rules. In the next section, we present how to determine the minimum support theoretically.

4.2 How to Determine the Minimum Support?

Without loss of generality, we begin with the $2-1$ association rule:$X \to Y$, which means X contains just two attributes $X = \{i, j\}$ and Y contains one variable $Y = \{l\}$ (in our problem, l is the class variable). The derivation for the general case will be similar. In the following, we use the Chebyshev Theorem to derive the suitable minimum support.

This theorem gives the lower bound on the probability that the frequency f of an event c after n trials differs from the real probability p within a ε variation:

$$P(|f - p| \leq \varepsilon) \geq 1 - \frac{p(1-p)}{\varepsilon^2 n}. \tag{13}$$

In our problem, the frequency is given by

$$f = \frac{N_{ijl}}{N_{ij}},$$

where, N_{ij} is defined as the number of the occurrence of the item $\{i, j\}$ and N_{ijl} is similarly defined. The value p is defined as the real probability of the event "the itemsets which consist of i, j will also consist of l". If we rewrite the absolute form of (13),we can have the following:

$$f - \varepsilon \leq p \leq f + \varepsilon.$$

In the association rule mining process, it is required that p is greater than the confidence level p_{cf} and p is also has to be less than 1. So we can simply specify that:

$$f - \varepsilon = p_{cf},$$
$$f + \varepsilon = 1.0.$$

From above, we obtain: $\varepsilon = \frac{(1-p_{cf})}{2}$. Combined this with (13), we have the following:

$$P(|f - p| \leq \varepsilon) \geq 1 - \frac{p(1-p)}{\varepsilon^2 n}$$

$$= 1 - \frac{p(1-p)}{\frac{(1-p_{cf})^2}{2}n}$$

$$\geq 1 - \frac{0.5(1-0.5)}{\frac{(1-p_{cf})^2}{2}n}$$

$$= 1 - \frac{1}{(1-p_{cf})^2 n}. \tag{14}$$

In order to obtain reliable association rule, the frequency: $f = \frac{N_{ijl}}{N_{ij}}$ has to be close to the real probability of c event. So the probability that the frequency is close to the real probability must be at least greater than 0.5, which implies:

$$1 - \frac{1}{(1-p_{cf})^2 n} \geq 0.5. \tag{15}$$

Here n is equal to N_{ij}, which at least achieves a number

$$n = N_{ij} \geq s_m N, \tag{16}$$

where N is the number of the cases or samples in dataset, and s_m is the minimum support. To satisfy (15), n should be big enough. Thus its lower bound $s_m N$ should be big enough. At last we obtain the bound of the minimum support:

$$s_m \geq \frac{2}{(1-p_{cf})^2 N}.$$

In a word, the above can be written into a lemma:

Lemma 2. *In order to make the inference in mining association rule reliable, the minimum support of the association rule must satisfy the following inequality:*

$$s_m \geq \frac{2}{(1-p_{cf})^2 N}, \tag{17}$$

where N is the total number of cases in dataset, p_{cf} is the confidence level specified by the user.

4.3 Practical Algorithm

In this section, we describe the detailed algorithm to build up Large Node Chow-Liu Tree from data. Our algorithm consists of three phases. In the first phase we utilize Apriori in [1] to detect all the association rules satisfying

Rule 4. The second phase is basically the Chow-Liu Tree construction algorithm. In the last phase, we combine the attributes, which satisfy combination rules and have higher supports, and upgrade the Chow-Liu Tree structure into the LNCLT structure iteratively.

Phase 1: Detecting all the association rules $X \rightarrow Y$, where Y is specified by the class variable C and X is a subset of attributes set, with the cardinality fewer than a bound K.

(1) Determine a good value of the minimum support, based on (17). Call the Apriori procedure to generate the association rules, whose X's have the cardinality fewer than K.

(2) Record all the association rules together with their supports into list L.

Phase 2(3): Drafting Chow-Liu Tree [3].

Phase 3: Adapting the tree structure based on combination transformation

(4) According to tree T, filter out association rules from L whose X's do not satisfy combination conditions, i.e., Rule 1 or Rule 2 from L. We get the new L'.

(5) Sort L' in descending order based on the supports of the association rules.

(6) Do until L' is $NULL$.

(a) Do the combination transformation based on the first itemset l_1 of L'.

(b) Delete l_1 and any other association rules l_i in L' which satisfy the following condition:

$$l_1.X \cap l_i.X \neq \varnothing,$$

where $l_1.X$ and $l_i.X$ refers to the X part of l_1 and l_i, respectively.

(c) Examine whether the newly generated items satisfy the combination rules. If yes, insert them into L' and sort L'.

(d) Go to (a).

5 Experiments

In this section, we first present the setup information of our experiments. Following that, we describe our pre-processing methods including handling zero-counts problems and feature extraction. In Sect. 5.3, we demonstrate the experimental results.

5.1 Setup

Our experiments are implemented on MNIST datasets [21]. The MNIST datasets consist of a 60000-digit training dataset and a 10000-digit test dataset. Both the training dataset and the test dataset consist of 28×28 gray-level pixels digits. As mentioned before, the bound K in Rule 4 cannot be set to a big value, we set K to 3 in our experiment.

5.2 Pre-Processing Methods

5.2.1 Feature Extraction Methods

We use the same method as [2] to extract 96-dimensional binary features from the digits. Since this method requires the binarization of the images, we first use a global threshold to binarize the training and test dataset. Then we segment the digit images into 2×3 sub-regions uniformly. In each sub-region, we judge whether four configurations given in Fig. 6 and their rotated configurations in other three main directions exist. Each configuration corresponds to a binary feature; therefore, the total number of the features will thus be $2 \times 3 \times 4 \times 4 = 96$.

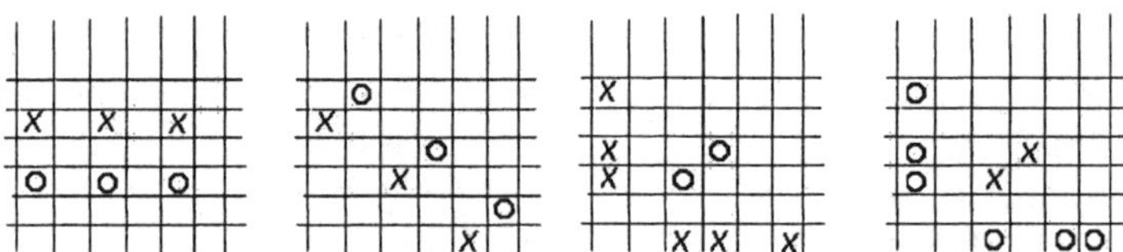

Fig. 6. Four configurations to extract features with $\times$'s and o's representing black pixels and white pixels respectively. These configurations will be rotated clockwise with angles $90^o, 180^o$ and 270°, respectively

5.2.2 Attacking Zero-Counts Problem

Zero-counts problems happen when a given class label and some value of the attribute never occur in the training dataset. This may cause problems in using the estimated probabilities to construct the decision function. For example, in the CLT's decision function, if one value a_j^k for an attribute A_j is never achieved, the empirically estimated $P(A_j = a_j^k | A_l = a_l^p, C = C_i)$ will be zero. Consequently, when $A_j = a_j^k$, the joint probability in the right part of (2) will be 0, whatever its parent is and the other terms are. Similar problems also happen for LNCLT. To tackle this problem, we use the popular Laplace correction method [24]. The modified estimated empirical probability for $P(A_j = a_j^k | A_l = a_l^p, C = C_i)$ is given by

$$(\#(A_j = a_j^k, A_l = a_l^p, C = C_i) + f)/(\#(A_l = a_l^p, C = C_i) + fm_j), \quad (18)$$

instead of the uncorrected one,

$$\#(A_j = a_j^k, A_l = a_l^p, C = C_i)/\#(A_l = a_l^p, C = C_i), \qquad (19)$$

where m_j is the number of values for attribute A_j, $\#(A_j = a_j^k, C = C_i)$ denotes the number of the occurrence that the attribute A_j achieves its k-th value a_j^k and the class label C achieves C_i. Other $\#(\cdot)$'s are similarly defined. We take the same value $1/N$ for parameter f as [5, 17], where N is the number of samples in training database. The technique is similarly used in estimated the probability of large nodes in LNCLT.

5.3 Results

In this subsection, we compare the performance of the LNCLT with the CLT in the tasks of approximating the dataset and performing classification. We built 10 LNCLTs for 10 digits. When examining the performance in approximating the dataset of the LNCLT and the CLT, we use the log likelihood criterion. When performing classification, we calculate the 10 probabilities for the test sample based on 10 LNCLTs and output the digit, whose LNCLT has the maximum probability, as the result.

5.3.1 Log Likelihood

Table 1. Minus Log Likelihood

Digit	Training (bits/digit)		Testing (bits/digit)	
	LNCLT	CLT	LNCLT	CLT
0	30.14	30.87	30.05	31.00
1	13.08	13.75	12.12	12.86
2	33.78	34.68	33.03	34.05
3	34.49	35.51	33.87	34.95
4	27.98	28.70	27.58	28.34
5	32.45	33.17	32.31	33.18
6	26.96	27.63	26.60	27.26
7	25.01	25.83	24.84	25.79
8	34.15	34.94	33.75	34.58
9	26.90	27.52	26.12	26.63

From Table 1, we can see that the log likelihood of the LNCLT is larger than that of the CLT for all the ten digits both in training dataset and test dataset. This result shows that the LNCLT approximates the data more accurately, which is consistent with our theoretical analysis in the previous sections.

5.3.2 Recognition Rate

We first use the 60000-digit training dataset to train the LNCLT and CLT. To test the performance of the LNCLT and CLT, we extract $1,000$ digits from the 10000-digit test dataset randomly as our test dataset. We do the 1000-digit test for 10 times to evaluate the performance difference between the LNCLT and CLT. Table 2 describes the result. From Table 2, it is clearly observed that the LNCLT performs better than CLT in all of 10 test datasets. We note that, when compared with the results of other approaches on MNIST, the recognition rate here is relatively low. The simple binarization method and different feature extraction method may partly explain this phenomenon.

Table 2. Recognition Rate

Dataset	1	2	3	4	5
CLT(%)	83.20	84.70	84.10	83.50	83.70
LNCLT(%)	83.70	85.90	84.70	84.20	84.90

Dataset	6	7	8	9	10
CLT(%)	85.10	84.30	83.30	83.50	83.80
LNCLT(%)	86.00	85.40	83.50	83.90	85.70

6 Conclusion

In this paper, we have described a method for constructing a kind of "tree" belief network: Large Node Chow-Liu Tree. This method can be seen as the extension of Chow-liu Tree algorithm. With the combination of improved association rule techniques, our novel model can partially overcome the disadvantages of Chow-Liu Tree, i.e., the inability to represent non-tree structures and maintain the advantages of Chow-Liu Tree, such as the decomposition ability in estimating the distribution. we demonstrate that the Large Node Chow-Liu Tree is superior to the CLT both theoretically and experimentally.

Two issues need to be checked in the near future. First, although the LNCLT model achieves performance superior to the CLT model, the proposed iterative process of combining nodes into large nodes may be time-consuming. How to reduce the time-complexity thus becomes a part of our future work. Second, the parameter K, namely, the maximum number of nodes which can be combined, is simply set to 3, which is unnecessarily the optimal value. How To investigate parameter selection methods such as [7, 8, 30, 31] and propose efficient algorithms remains one of our research directions.

Acknowledgment

This research is supported fully by grants from the Hong Kong's Research Grants Council (RGC) under CUHK 4407/99E and CUHK 4222/01E.

References

1. R. Agrawal and R. Srikant. Fast algorithms for mining association rules. *Proceedings of International Conference on Very Large Data Bases (VLDB-1994)*, 1994.

2. R. Bakis, M. Herbst, and G. Nagy. An experimental study of machine recognition of hand-printed numerals. *IEEE Transactions on Systems Science and Cybernetics*, SSC-4(2), JULY 1968.

3. C. K. Chow and C. N. Liu. Approximating discrete probability distributions with dependence trees. *IEEE Trans. on Information Theory*, 14:462–467, 1968.

4. S. Dasguota. Learning polytrees. In *Uncertainty in Artificial Intelligence*, 1999.

5. P. Domingos and M. J. Pazzani. On the optimality of the simple baysian classifier under zero-one loss. *Machine Learning*, 29:103–130, 1997.

6. J. Dougherty, R. Kohavi, and M. Sahami. Supervised and unsupervised discretization of continuous features. In *International Conference on Machine Learning*, pages 194–202, 1995.

7. G. Elidan, N. Lotner, N. Friedman, and D. Koller. Discovering hidden variables:a structure-based approach. In *NIPS 13*, 2001.

8. N. Friedman and G. Elidan. Learning the dimensionality of hidden variables. In *Proceedings of Seventeenth Conference on Uncertainty in Artificial Intelligence (UAI)*, 2001.

9. N. Friedman, D. Geiger, and M. Goldszmidt. Bayesian network classifiers. *Machine Learning*, 29:131–161, 1997.

10. J. Hipp, U. Guntzer, and G. Nakhaeizadeh. Algorithms for association rule mining-a general survey and comparison. *ACM SIGKDD Explorations*, 2:58–64, July 2000.

11. K. Huang, I. King, and M. R. Lyu. Constructing a large node chow-liu tree based on frequent itemsets. In Lipo Wang, Jagath C. Rajapakse, Kunihiko Fukushima, Soo-Young Lee, and Xi Yao, editors, *Proceedings of the International Conference on Neural Information Processing (ICONIP-2002), Orchid Country Club, Singapore*, pages 498–502, 2002.

12. K. Huang, I. King, and M. R. Lyu. Learning maximum likelihood semi-naive bayesian network classifier. In *Proceedings of IEEE International Conference on Systems, Man and Cybernetics (SMC-2002), Hammamet, Tunisia*, 2002.

13. K. Huang, I. King, and M. R. Lyu. Discriminative training of bayesian chow-liu tree multinet classifiers. In *Proceedings of International Joint Conference on Neural Network(IJCNN-2003), Oregon, Portland, U.S.A.*, volume 1, pages 484–488, 2003.

14. K. Huang, I. King, and M. R. Lyu. Finite mixture model of bound semi-naive bayesian network classifier. In *Joint 13th International Conference on Artificial Neural Network (ICANN-2003) and 10th International Conference on Neural Information Processing (ICONIP-2003), Long paper, Lecture Notes in Computer Science*, pages 115–122, 2003.

15. D. Karger and N. Srebro. Learning markov networks: maximum bounded tree-width graphs. In *Symposium on Discrete Algorithms*, pages 392–401, 2001.

16. R. Kohavi. A study of cross validation and bootstrap for accuracy estimation and model selection. In *Proceedings of the 14th International Joint Conference on Artificial Intelligence (IJCAI-1995)*, pages 338–345. San Francisco, CA:Morgan Kaufmann, 1995.

17. R. Kohavi, B. Becker, and D Sommerfield. Improving simple bayes. In *Technique report*. Data Mining and Visualization Group, Silicon Graphics Inc., Mountain View, CA., 1997.

18. I. Kononenko. Semi-naive bayesian classifier. In *Proceedings of Sixth European Working Session on Learning*, pages 206–219. Springer-Verlag, 1991.

19. P. Langley. Induction of recursive bayesian classifiers. *In Proceedings of the 1993 European Conference on Machine learning*, pages 153–164, 1993.

20. P. Langley and S. Sage. Induction of selective bayesian classifiers. In *Proceedings of the Tenth Conference on Uncertainty in Artificial Intelligence (UAI-1994)*, pages 399–406. San Francisco, CA: Morgan Kaufmann, 1994.

21. Y. Le Cun. http://www.research.att.com/ yann/exdb/mnist/index.html.

22. F. M. Malvestuto. Approximating discrete probability distributions with decomposable models. *IEEE Transactions on Systems, Man and Cybernetics*, 21(5):1287–1294, 1991.

23. M. Meila and M. Jordan. Learning with mixtures of trees. *Journal of Machine Learning Research*, 1:1–48, 2000.

24. T. Niblett. Constructing decision trees in noisy domains. In *Proceedings of the Second European Working Session on Learning*, pages 67–78, 1987.

25. M. J. Pazzani. Searching dependency in bayesian classifiers. In D. Fisher and H.-J. Lenz, editors, *Learning from data: Artificial intelligence and statistics V*, pages 239–248. New York, NY:Springer-Verlag, 1996.

26. J. Pearl. *Probabilistic Reasoning in Intelligent Systems: networks of plausible inference*. Morgan Kaufmann, CA, 2nd edition, 1997.

27. J. R. Quinlan. *C4.5 : programs for machine learning*. San Mateo, California:Morgan Kaufmann Publishers, 1993.

28. M. Sahami. Learning limited dependence bayesian classifiers. In *Proceedings of the Second International Conference on Knowledge Discovery and Data Mining*, pages 335–338. Portland, OR:AAAI Press, 1996.

29. N. Srebro. Maximum likelihood bounded tree-width markov networks. *MIT Master thesis*, 2001.

30. A. Stolcke and S. Omohundro. Hidden markov model induction by bayesian model merging. In *NIPS 5*, pages 11–18, 1993.

31. A. Stolcke and S. Omohundro. Inducing probabilistic grammars by bayesian model merging. In *International Conference on Grammatical Inference*, 1994.

32. G. Webb and M. J. Pazzani. Adjusted probability naive bayesian induction. In *the eleventh Australian Joint Conference on Artificial Intelligence*, 1998.

A Reconstructed Missing Data-Finite Impulse Response Selective Ensemble (RMD-FSE) Network

Sirapat Chiewchanwattana[1,2,3]
Chidchanok Lursinsap[2]
Chee-Hung Henry Chu[3]

[1] Department of Computer Science, Faculty of Science, Khon Kaen University, Khon Kaen, Thailand `sunkra@kku.ac.th`
[2] AVIC Research Center, Department of Mathematics, Faculty of Science, Chulalongkorn University, Bangkok, Thailand `lchidcha@chula.ac.th`
[3] Center for Advanced Computer Studies, The University of Louisiana at Lafayette, Lafayette, Louisiana, U.S.A. `cice@cacs.louisiana.edu`

Abstract. This chapter considers the problem of time-series forecasting by a selective ensemble neural network when the input data are incomplete. Six fill-in methods, viz. cubic smoothing spline interpolation, k-segment principal curves, Expectation maximization (EM), regularized EM, average EM, and average regularized EM, are simultaneously employed in a first step for reconstructing the missing values of time-series data. A set of complete data from each individual fill-in method is used to train a finite impulse response (FIR) neural network to predict the time series. The outputs from individual networks are combined by a selective ensemble method in the second step. Experimental results show that the prediction made by the proposed method is more accurate than those predicted by neural networks without a fill-in process or by a single fill-in process.

Keywords. Time-series prediction, ensemble of networks, finite impulse response networks.

1 Introduction

An important application of neural networks is in time-series data prediction. Data points from the past are used to predict those in the future; i.e., the value of a system at time n, denoted x_n, is modeled as a time-series function $x_n = f(x_{n-1}, x_{n-2}, \ldots, x_{n-k})$, where k is the number of previous time steps. This model can be used in the forecasting [1] of such natural and social phenomena as hydrological cycles, climate [2], and financial trend [3]. When a neural network is used, the time-series data are partitioned into three phases. The data in the first phase are the training data, to be used by the network learning process. A second phase of the data is used to validate the network's accuracy. The third phase is the data input when the network is used to predict sample values.

The accuracy of the prediction depends upon the related issues of the choice of an appropriate mathematical model and the integrity of the collected data. Data collected in practice can often be incomplete in that some data points are missing due to such reasons as malfunctioned sensors, human errors, and sometimes even natural disasters. If some x_{n-j}, for $1 \leq j \leq k$ are missing from the input, the value of x_n cannot be correctly computed by the time-series function $x_n = f(x_{n-1}, x_{n-2}, \ldots, x_{n-k})$. One approach is that those missing x_{n-j} must be estimated or filled in first and, then, the functional approximation of the complete data is performed. In this case, the estimated value of x_n, denoted by $\hat{x}_n$, must be computed from the existing x_{n-j}, $1 \leq j \leq k$. The key to using this approach is to decide which methods should be used to derive the value of $\hat{x}_n$ such that some distortion measure, such as $(\hat{x}_n - x_n)^2$, is minimized.

Statistical and neural computational methods have been used to handle the problem of missing data values. A well known statistical method for estimating missing data is the expectation maximization (EM) algorithm [4, 5]. In most neural computation approaches, the missing data problem is solved by a supervised neural network. Training any supervised neural network requires both input and target data; previously reported work has considered that the missing data can occur either in the input [6], in the target output [7], or both [8]. A combination of supervised learning and EM algorithm has been used to improve the missing data estimation [9].

While neural networks have been used to perform time-series data prediction without any missing data [10], prediction with a given set of incomplete data has received more recent interest [2]. Most reported techniques [5, 3, 2, 8, 9] use only either EM estimation or neural networks. Even though a single neural network can be efficiently used for the prediction of time-series data, a combination of many neural networks of the same type often shows an improvement in the prediction performance. We present an approach that uses several EM-based algorithms, as well as interpolation methods using the smoothing spline interpolation and the k-segment principal curves, to fill in the missing data values. The ensemble network therefore consists of independently trained neural networks, each drawing an input stream from a fill-in method, which are then combined as a single master network. Each individual network uses a Finite Impulse Response model [12] to perform the prediction. We denote this approach as a *reconstructed missing data-finite impulse response selective ensemble (RMD-FSE)* network. A structure of this approach is illustrated in Figure 1.

The remainder of this chapter is organized as follows. The methods used for reconstructing the missing data points are described in Section 2. The selective ensemble neural network we used and the index for measuring its performance are presented in sections 3 and 4, respectively. The numerical results from two case studies are reported in Section 5 and the conclusion is in Section 6.

2 Techniques for Reconstruction of Missing Data

The six fill-in techniques considered in our experiments are (1) cubic smoothing spline interpolation, (2) The imputation by EM algorithm with random selection, (3)The imputation by EM algorithm with average selection, (4) The regularized EM algorithm with random selection, (5) the regularized EM algorithm with average selection, and (6) k-segments algorithm for finding principal curves. These six methods are used to fill in any missing data points in the set.

In *Cubic Smoothing Spline Interpolation*, a piecewise cubic function $f(t)$ is defined by the observed values $x(t)$. The missing value $x(k)$ can then be obtained by evaluating the interpolated spline $f(t)$ at the appropriate value of $t = k$. The spline is obtained by choosing parameters to minimize the cost function:

$$\kappa \sum_n w_n(x_n - f(n))^2 + (1 - \kappa) \int (D^2 f)^2 \tag{1}$$

where κ is a constant, $0 \leq \kappa \leq 1$, w_n is the weighting factor of datum x_n, $f(t)$ is a cubic spline, and $D = I_{N \times N}$ is an identity matrix, and N is the length of the data set. In our experiments we set $w_i = 1$ for all i and $\kappa = 0.99$.

The next two methods are based on *Imputations by EM Algorithm*. Using the EM algorithm [4], multiple candidates for a missing value are calculated. The two methods differ in how the actual output is chosen. The inputs to the EM algorithm are partitioned and arranged in a time-series form with a window size of k. Each set of inputs is, then, stacked to form an input matrix A:

$$A = \begin{bmatrix} x_1 & x_2 & \cdots x_k \\ x_2 & x_3 & \cdots x_{k+1} \\ \cdots & \cdots & \cdots \cdots \\ x_{N-k+1} & x_{N-k+2} & \cdots x_N \end{bmatrix} \tag{2}$$

Element x_n denotes an input value at time n. The value of each missing datum is set to some special values, such as -9999. The EM method considers matrix A as a table of $N - k + 1$ observations of a k-dimensional variable, which is normally distributed [4]. After the imputation process, all missing data are estimated and a reconstructed matrix $A^{EM} = [a_{i,j}]$ is produced.

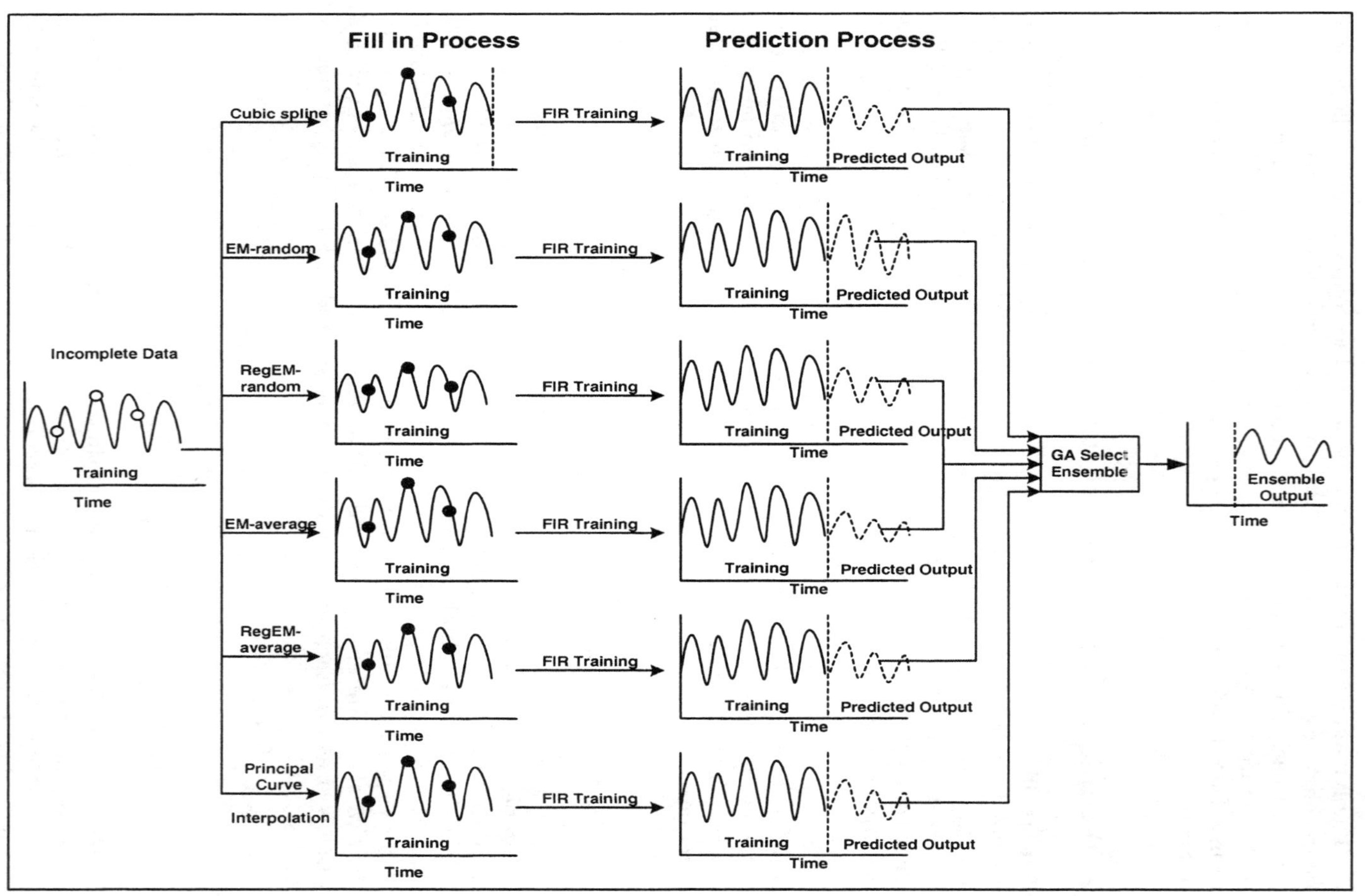

Fig. 1. RMD-FSE network structure.

Suppose x_n is a missing datum; obviously, the number of x_n can occur diagonally from 1 to k times. Let the number of occurrences of x_n in A be k'. After applying the EM algorithm, there are k' possible estimated values of $\hat{x}_n$. In the *Random Selection* method, we randomly select one of the values of $\hat{x}_n$ as the fill-in value. In the *Average Selection* method, the estimated value of $\hat{x}_t$ is computed by averaging all its possible values.

In the *Imputation by the regularized EM algorithm* methods, we used a method to estimate the mean and the covariance matrix of an incomplete data set and fill-in missing values with imputed values by the regularized EM [2]. The regularized EM algorithm is based on the iterative analysis of linear regressions of variables with missing values and regression coefficients estimated by ridge regression. After imputation process, a reconstructed matrix is produced. Similar to our discussion above, in the *Random Selection* method, the estimated value of $\hat{x}_n$ is randomly selected from all of its estimated values. In the *Average Selection* method, the value of $\hat{x}_n$ is performed by averaging all of its estimated values.

The *k-segment principal curve interpolation* method is an incremental one that finds principal curves which are generalizations of principal components [11]. Line segments are inserted and fitted to form polygonal lines on the univariate time-series data. The missing value is obtained by evaluating the interpolation on those combined polygonal segments of time-series data.

3 FIR Selective Ensemble Network

The six data reconstruction methods produce six input data streams, which may differ from each other since the reconstruction methods are not the same. Each stream is fed to an individual network to produce a prediction of the current value of the time series; the selective ensemble of outputs are combined by a single master network. The RMD-FSE network has two parts, the individual FIR networks and the master network used to integrate the outputs.

3.1 Individual Networks

The task for each individual network is to predict the next sample $x(n)$ of a time series based on a number of past samples. Suppose K past samples are to be used. A straightforward approach is to use K input neurons in the network for the delayed input samples $\{x(n-k) : k = 1, \cdots, K\}$. A standard feedforward network can be trained to produce the output sample that minimizes the difference from a target sample in a training sequence. After each step, the input samples are shifted accordingly. For instance, at time n, the ith input neuron $\xi_i^{(0)}$ is fed with $\xi_i^{(0)} = x(n-i)$, where the superscript is used to denote the layer of the network. At time $n+1$, the ith input neuron receives the sample that was fed to the $(i-1)$th neuron at the previous time step so that $\xi_i^{(0)} = x(n - (i-1)) = x((n+1) - i)$. At $i = 1$, the input neuron receives either the previous target sample (during training) or the previous output sample (during operation) as its input. The samples to two neighboring input neurons are therefore delayed by one time unit apart.

A time-delay neural network is a variation on the basic network in that delay units are furnished for the hidden units as well. Consider a time-delay network with a layer of K_i input neurons, a layer of M hidden units, and an output unit. Consider a two-layer network with a single output unit. We can write the mth hidden unit output as $\xi_m^{(1)}$ and the output of the network as $\xi^{(2)}$. Suppose each hidden unit has K_h delay elements, each hidden unit has a set of $K_h + 1$ outputs, so that the mth hidden unit output is written as $\xi_{m,k}^{(1)}$, where $k = 0, \cdots, K_h$. At time step n, the connection of the network is as follows. The output unit $\xi^{(2)}$ is a function of the hidden units $\xi_{m,k}^{(1)}$, $m = 1, \cdots, M$, and $k = 0, \cdots, K_h$. The hidden units $\xi_{m,0}^{(1)}$, $m = 1, \cdots, M$ are functions of the input units $\xi_i^{(0)}$, $i = 1, \cdots, K_i$. The input unit $\xi_1^{(0)}$ is fed with $x(n-1)$, the previous target or output sample. The output $\xi^{(2)}$ is the network's prediction of the sample $x(n)$.

In a time-delay neural network, those neurons that are associated with a delay unit are connected to an upper layer but are not connected to neurons from a lower layer. At the input layer, the input units $\xi_i^{(0)}$, for $i = 2, \cdots, K_i$, receive as input the delayed value of $\xi_{i-1}^{(0)}$. At the hidden layer, the hidden unit $\xi_{m,k}^{(1)}$, for all m and for $k = 1, \cdots, K_h$, receive as input the delayed value of $\xi_{m,k-1}^{(1)}$.

It is straightfoward to derive the number of input samples that are used by the network to predict $x(n)$. At time n, the input layer contains the samples $\{x(n-k) : k = 1, \cdots, K_i\}$. At the hidden layer, the oldest hidden unit outputs are $\xi_{m,K_h}^{(1)}$, $m = 1, \cdots, M$. This set of hidden unit outputs is computed K_h time steps ago, using the samples $\{x(n-k) : k = K_h + 1, \cdots, K_h + K_i\}$. The set of input samples used by the network to predict $x(n)$ is therefore $\{x(n-k) : k = 1, \cdots, K_h + K_i\}$.

In our work, the individual networks used in the ensemble are FIR networks. An FIR network is functionally equivalent to a time-delay neural network. It is a layer-by-layer fully connected feedforward network modeled with tapped-delayed synapses. Neurons in an FIR network are organized as layers, similar to a traditional feedforward network. Consider the connection between a pair of neurons from adjacent layers. In a conventional synapse, the ith unit at the lth layer is connected to the jth unit at the $(l-1)$th layer with an associated weight $w^{(l-1)}(i,j)$. In a tapped-delayed synapse, the jth unit at the $(l-1)$th layer has an output vector of $[\xi_{j,0}^{(l-1)}, \xi_{j,1}^{(l-1)}, \cdots, \xi_{j,K_{l-1}}^{(l-1)}]^T$. A weight vector $\mathbf{w}^{(l-1)}(i,j)$ is associated with the link between the ith unit at the lth layer and the jth unit at the $(l-1)$th layer, so that the ith neuron at the lth layer receives the input

$$s_i^{(l)} = \sum_j \{ w_0^{(l-1)}(i,j)\xi_{j,0}^{(l-1)} + w_1^{(l-1)}(i,j)\xi_{j,1}^{(l-1)}$$

$$+ \ldots + w_{K_{l-1}}^{(l-1)}(i,j)\xi_{j,K_{l-1}}^{(l-1)} \}. \tag{3}$$

The inner product of the weight and the lower layer output vectors $s_i^{(l)}$ is passed through a sigmoidal function γ as the output of the i unit at the lth layer:

$$\xi_i^{(l)} = \gamma(s_i^{(l)}). \tag{4}$$

The input to a unit at a higher level can be viewed as an inner product or as the output of an FIR digital filter, with the weight vector serving the role of the filter coefficients. Similar to a time-delay neural network, the output vector of a unit is shifted so that the most recently computed output is saved while the oldest output value is shifted out.

During the training of the network, the value of each weight is adjusted in order to minimize the specified squared error by using the temporal backpropagation algorithm [12]. Analogous to the traditional backpropagation algorithm, errors at the higher layers are distributed to the lower layers. Because the weights at each synapse are organized as weights of an FIR filter, the errors are filtered through these connections to the lower layers.

3.2 Master Network

Suppose we use an ensemble of N networks to predict a sample value $f_{RMD-FSE}(n)$. The overall ensemble output can be any function of the individual network outputs. The simplest such function is to combine the individual outputs as a weighted sum:

$$x_{RMD-FSE}(n) = \sum_{i=1}^{N} \alpha_i x^{(i)}(n), \tag{5}$$

where $x^{(i)}(n)$ is the output value of the ith network, with an associated weighting parameter α_i. Intuitively, the weights should satisfy $\sum_i \alpha_i = 1$ because each network produces a prediction of the same sample $x(n)$. Additionally, since the predicted values should have the same sign as the sample, it is reasonable to expect $0 \le \alpha_i \le 1$, for all i.

The weights should be set so that if the jth network is more reliable, its associated weight α_i should be larger. Choosing the actual values of α_i is an optimization problem based on the observed mean square error of individual networks. Let $\mathbf{C}$ is the correlation matrix of the errors from the network predictors $x^{(i)}$ and $x^{(j)}$; the (i,j)th element of $\mathbf{C}$ is given by

$$c_{ij} = \frac{1}{|\Gamma|} \sum_{k \in \Gamma} e^{(i)}(k) e^{(j)}(k) \tag{6}$$

where Γ is the training set and $e^{(j)}(k)$ is the error of Network j produced in response to the kth input of the training set. The diagonal terms of $\mathbf{C}$ are the mean square errors of the individual networks while each off-diagonal term is a pairwise correlation of the corresponding networks. In the following we write c_{ii} as c_i^2. Since the ensemble output should have a better performance than the individual networks, a criterion for finding $\boldsymbol{\alpha}$ is to minimize the ensemble output mean square error:

$$E_{\boldsymbol{\alpha}} = \boldsymbol{\alpha}^T \mathbf{C} \boldsymbol{\alpha}. \tag{7}$$

Some insight about the choice of the weights and the output mean square error can be gained by considering the simple case when $N = 2$. The solution for $\boldsymbol{\alpha}$ is to be found by minimizing

$$E_2 = [\alpha_1 \alpha_2] \begin{bmatrix} c_1^2 & c_{12} \\ c_{12} & c_2^2 \end{bmatrix} \begin{bmatrix} \alpha_1 \\ \alpha_2 \end{bmatrix}. \tag{8}$$

Since $\alpha_2 = 1 - \alpha_1$, we can differentiate E_2 with respect to α_1 to solve for α_1 as:

$$\alpha_1 = \frac{c_2^2 - c_{12}}{c_1^2 + c_2^2 - 2c_{12}}. \tag{9}$$

It follows that

$$\alpha_2 = \frac{c_1^2 - c_{12}}{c_1^2 + c_2^2 - 2c_{12}}. \tag{10}$$

With this set of weights, the ensemble output has a mean square error of

$$\begin{aligned} E_2 &= \alpha_1^2 c_1^2 + \alpha_2^2 c_2^2 + 2\alpha_1 \alpha_2 c_{12} \\ &= \frac{c_1^2 c_2^2 - (c_{12})^2}{c_1^2 + c_2^2 - 2c_{12}} \end{aligned} \tag{11}$$

This set of weights illustrates that when Network 1 is more reliable, i.e., $c_1^2 < c_2^2$, its output is weighted more heavily since $\alpha_1 > \alpha_2$. In fact, if Network 1 makes no mistakes during training, i.e., $c_1^2 = 0$, the weight for the other network α_2 is set to zero so that the output of Network 1 is taken as the ensemble output.

When the correlation of the errors by the two networks are low, i.e., when they make mistakes at different input patterns, the ensemble mean square error simplifies to

$$\frac{c_1^2 c_2^2}{c_1^2 + c_2^2}. \tag{12}$$

Suppose Network 1 has a better performance so that $c_2^2 = \rho c_1^2$, where $\rho > 1$. The ensemble mean square error is then

$$\frac{\rho}{\rho + 1} c_1^2, \tag{13}$$

which is less than the smaller of the mean square errors of the two individual networks. In this case, we can see that the improvement, in the mean square sense, of having an ensemble network can be as high as a factor of 2, corresponding to the case of $\rho = 1$, or when the two networks make equal amounts of mistakes in an uncorrelated fashion.

Consider another scenario. Suppose the two networks make about the same amount of mistakes in the training set and that the mistakes are of approximately the same order. Further, suppose that their mistakes overlap, i.e., they make the same mistakes for a subset of the training set. In this case, $c_1^2 = c_2^2$ and $c_{12} = \beta c_1^2$, where β is the amount of overlap of their mistakes, so that $0 \leq \beta \leq 1$. The mean square error of the output is then

$$\frac{c_1^2}{2}(1 + \beta). \tag{14}$$

When the two networks make the same set of mistakes, $\beta = 1$, and the output mean square error is the same as that of each individual network. When the two networks make different mistakes, β approaches 0 so that the ensemble output is improved in the mean square sense by a factor approaching 2.

In the general case when $K > 2$, a genetic algorithm-based selective neural network ensemble method (GASEN) [13] can be used. A genetic algorithm is used to search for a solution of E_{α} by defining $h(\alpha) = 1/E_{\alpha}$ as the fitness function which is to be maximized. The components of a candidate solution α may violate the constraints during its evolution, therefore its elements α_i should be normalized to $\alpha_i / \sum_j \alpha_j$ at each generation.

There are different strategies in using the weights found by the genetic algorithm. In GASEN [13], the weights are used to eliminate networks rather than to be used in a weighted average. We can exclude those networks which are associated with weights less than a threshold value $\lambda > 0$; i.e., exclude Network i if $\alpha_i < \lambda$ After these networks are discarded, the outputs of the remaining networks are averaged to form the ensemble output. This is justified by observing that after eliminating those networks that are particularly unreliable, the remaining individual networks have approximately equal performance.

In our work, we set λ to zero and use the weights found by the genetic algorithm to form a weighted average as the ensemble output. Empirically we found this method to be more desirable than using a simple average with or without discarding networks with particularly unreliable outputs.

4 Performance Index

The prediction performance of a RMD-FSE network is evaluated by measuring the difference between the mean squared error of the test networks and the reference network. We choose a FIR network, which is trained by the complete training set with no missing data as the reference network. Because the occurrence of data sample dropouts is stochastic in nature, we make several runs for each fill-in method. The performance index is defined as

$$P_d = \frac{1}{\rho}\left(\sum_{i=1}^{\rho} E_i^T - \sum_{j=1}^{\rho} E_j^R\right) \tag{15}$$

where E_i^T denotes the mean square error of the tested network, E_j^R denotes the mean square error of the reference network, and ρ denotes the number of runs per fill-in method. The parameter ρ is twenty-five in our experiments. The interpretation of P_d is that if the prediction performance of the reference network is worse than that of the test network, then P_d is less than zero, and the lower value shows the better performance. If the prediction performance of the reference network is better than that of the test network, then P_d is greater than zero. Otherwise P_d is zero.

When the RMD-FSE network is compared with individual networks, the performance index of the RMD-FSE network is defined in terms of the average of the following difference:

$$P_d' = \frac{1}{L \times \rho}\left(\sum_{i=1}^{L \times \rho} E_i^T - \sum_{j=1}^{L \times \rho} E_j^{RF}\right) \tag{16}$$

where L denotes the number of different percentage of missing data, and E_j^{RF} denotes the mean square error of RMD-FSE network. The meaning of P_d' is that if the average prediction performance of RMD-FSE network is worse than that of an individual network, then P_d' is less than zero. If the average of prediction performance of RMD-FSE network is better than that of the individual network, then P_d' is greater than zero. Otherwise, P_d' is zero.

5 Numerical Results

Two different data sets are used in our experiment. The first set is the sunspot data and the second set is the daily gauge height. The prediction results of individual networks and the results of the averaging ensemble network, whose outputs are averaged from those individual networks, are compared with the results of RMD-FSE network. The experimental results of these two data sets are concluded as follows.

5.1 Sunspot data

The structure of each individual network has three layers consisting of one input neuron, two sub-hidden layers of three and two neurons, respectively, and one output neuron. Each layer's neurons from the input to output layer have ten, two and one tapped delayed, respectively. The weights assigned to each individual network of the ensemble network are shown in Figure 2(a). We can see from Figure 2(a) that the weights associated with some individual networks at every percentage of missing input were set to zero by the genetic algorithm. The output of those networks were therefore essentially discarded from the ensembling. The prediction performance is shown in Figure 3(a). When we consider the results, we found some individual networks yield P_d less than zero at 2.5%, 7.5%, 10%, 12.5% and 15% missing data but the averaging ensemble network and RMD-FSE networks yield P_d less than zero for every percentage of missing data. Furthermore, RMD-FSE networks yield the lowest P_d and, consequently, give the better prediction performance than both of the reference network and the averaging ensemble network. We note that when the percentage of missing data is high, viz. at 10%, 12.5% and 15%, the network with the average EM selection gives the better result than the network with the random EM selection. The experimental results in Table 1 show that the minimum and maximum P_d' are 1.025×10^{-4} and 8.686×10^{-4}, respectively. Once the complete data are used as the training set of the FIR network, P_d' is 3.330×10^{-4}. All P_d' values are also greater than zero. Hence, RMD-FSE networks exhibit better performance than the individual networks.

5.2 The daily gauge height at Ban Luang gauging station, Mae Tun stream, Ping river

The structure of the network used in this experiment is the same as that used in the sunspot data experiment. The weights assigned to each individual network of the ensemble network are shown in Figure 2(b). Some individual network are discarded from the ensembling at 7.5 %, 10%, 12.5% and 15 % missing. The prediction performances are shown in Figure 3(b). Some individual networks yield P_d less than zero at 2.5% and 7.5%missing data. The P_d of the averaging ensemble networks are less than zero at 2.5% and 7.5% missing data, while at 10%, 12.5% and 15% missing data, they are greater than zero. In contrast, RMD-FSE networks yield P_d less than zero for every percentage of missing data. Furthermore, RMD-FSE networks yield the lowest P_d and, consequently, give better prediction performance than both of the reference network and the averaging ensemble network. These results signify that RMD-FSE networks also give better prediction performance than the individual networks. The experimental results in Table 1 show that the minimum and maximum P_d' are 1.578×10^{-5} and 22.730×10^{-5}, respectively. When the complete data are used as the training set of the FIR network, P_d' is 1.578×10^{-5}. All P_d' values are greater than zero. We note that the individual networks give worse prediction performance than RMD-FSE networks both in terms of mean square error and in terms of consistency.

5.3 Discussion

From the results of both studies in our experiments, it can be seen that using only one fill-in technique does not achieve desirable performance. Thus, we use several EM-based estimation methods and two interpolation methods for filling in the missing data. The different input data streams are then fed to

Table 1. Average performance index P'_d for sunspots data and the daily gauge height data. We found that all values of P'_d are greater than zero.

Fill-In Methodology	Sunspots $\times 10^{-4}$	Gauge Height $\times 10^{-5}$
Spline	2.928	5.345
EM(random)	8.686	9.009
Reg EM(random)	4.487	22.730
EM(average)	5.129	4.412
Reg EM(average)	3.076	20.362
Principal Component	5.735	2.505
Averaging Ensemble	1.025	2.226
Complete Data	3.330	1.578

an ensemble network for prediction. In almost all cases, RMD-FSE network gave the best prediction performance in our experiments. However, when we compare RMD-FSE network and the averaging ensemble network which was averaged from the individual networks in two problems, the prediction performance of RMD-FSE is better than the averaging ensemble network.

6 Conclusions

Incomplete data sets can be problematic when a neural network is used for time-series prediction. We use a variety of fill-in techniques to generate multiple input streams for an ensemble of FIR neural networks. The new network model, referred to as the RMD-FSE, is a GA-based selective ensembling of the outputs of these FIR networks. We evaluated the accuracy of prediction with a performance index which measures the accuracy of prediction for the desired network with respect to the individual networks. We conducted our experiments using two real benchmark data sets, viz. the annual sunspot and the daily gauge height data collected at the Ban Luang gauging station, Mae Tun stream, Ping river, Thailand. Our results show that RMD-FSE outperforms each individual network and thus RMD-FSE is proposed for highly accurate incomplete time-series prediction.

There are some limitations of the work described here. First, using an ensemble of networks increases the computational resources needed. Secondly, while the quality of the ensemble output is better than that of the individual networks, it can only be improved by having better individual networks. The assessment of how much improvement can be expected by using an ensemble in a general case remains as future work.

References

1. Hu T S, Lam T S, Ng S T (2001) River flow times series prediction with a range-dependent neural network. Hydrological Sciences-Journal-des Hydrologiques 46: 729–745.

2. Schneider T (2001) Analysis of incomplete climate data: Estimation of mean values and covariance matrices and imputation of missing values. Journal of Climate 14:853–887.
3. Abdullah M H L b, Ganapathy V (2000) Neural network ensemble for financial trend prediction. In: Proc. IEEE TENCON 3:157–161.
4. Schafer J L (1997) Analysis of Incomplete Multivariate Data. Chapman & Hall/CRC, Boca Raton.
5. Dempster A P, Laird N M, Rubin D B (1997) Maximum likelihood from incomplete data via the EM algorithm. J. Royal Statistical Society Series B, 39:1–38.
6. Pedreira C E, Parente E (1995) Neural networks with missing values attributes. In: Proc. IEEE International Conference on Neural Network 6:3021–3023.
7. Verikas A, Gelzinis A, Malmqvist K, Bacauskiene M (2001) Using unlabelled data to train a multilayer perceptron. In: Proc. ICAPR 2001, LNCS 2013, 40–49.
8. Tanaka M (1996) Identification of nonlinear systems with missing data using stochastic neural network. In: Proc. 35th Conf. On Decision and Control, 933–934. Japan.
9. Ghahramani Z, Jordan M (1994) Supervised learning from incomplete data via an EM approach. In: Cowan J D, Tesauro G, Alspector J (eds), Advances in Neural Information Processing System 6, 120–127. Morgan Kaufmann, San Francisco.
10. Mozer M C (1993) Neural net architectures for temporal sequence processing. In: Weigend A S, Gershenfeld N A (eds), Time Series Prediction: Forecasting the Future and Understanding the Past, 243–264. Addison-Wesley, Reading, Mass.
11. Verbeek J J, Vlassis N, Krose B (2002). A k-segments algorithm for finding principal curves. Pattern Recognition Letters 23:1009–1017.
12. Wan E A (1994) Time series prediction by using a connectionist network with internal delay lines. In: Weigend A, Gershenfeld N (eds), Time Series Prediction. Forecasting the Future and Understanding the Past, 195–217. Addison-Wesley, Reading, Mass.
13. Zhou Z-H, Wu J-X, Tang W, Chen Z Q (2001) Combining regession estimators: GA-based selective neural network ensemble. International Journal of Computational Intelligence and Applications 1:341–356.

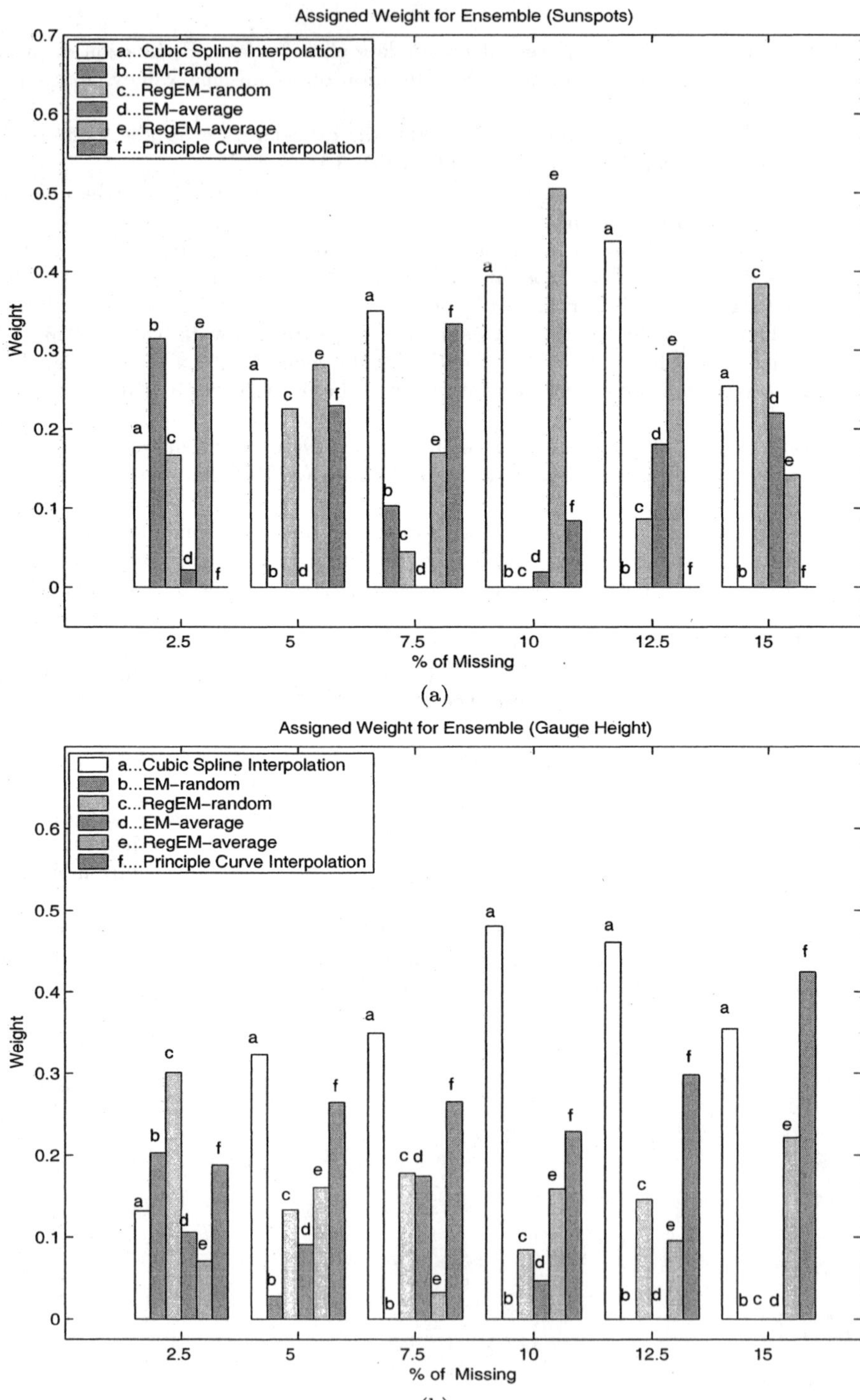

Fig. 2. Weight in the selective ensemble network assigned to different FIR networks for the (a) sunspots and (b) gauge height data sets. They differ depending on the amount of data points missing.

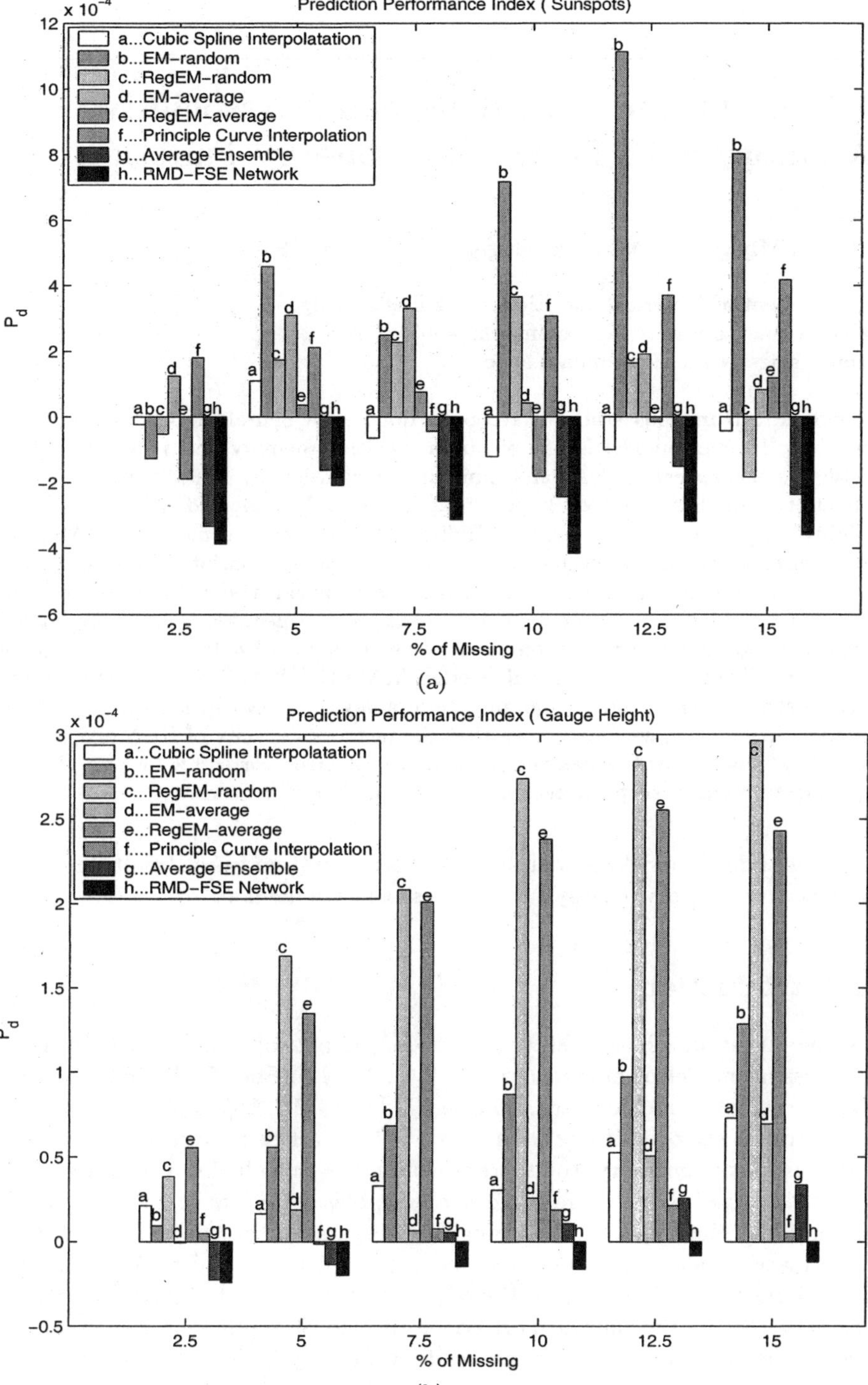

Fig. 3. Prediction performance index for the (a) sunspots and (b) gauge height data sets.

Higher Order Multidirectional Associative Memory with Decreasing Energy Function

Hiromi Miyajima[1], Noritaka Shigei[1] and Nobuaki Kiriki[1]

Department of Electrical and Electronics Engineering,
Kagoshima University, Kagoshima, 890-0065 Japan
{miya,shigei}@eee.kagoshima-u.ac.jp

Abstract. Numerous studies have been done with neural models of associative memory. Hopfield model is one of autoassociative memory and is deeply studied with the theoretical analysis and numerical simulation by introducing the energy functions. Further it is well known that higher order Hopfield model has higher ability than the conventional one. Bidirectional associative memory (BAM) and multidirectional associative memory (MAM) are heteroassociative models and are also deeply studied using the same method as Hopfield model. However, there are few higher order neural models of heteroassociative memory. Specifically, little is known about higher order heteroassociative memory with the decreasing energy function. This paper proposes higher order MAM (HOMAM) with the energy function. From numerical simulation and static analysis, HOMAM is superior to other model. Specifically, we have shown that the memory capacity of HOMAM is about $0.18P/n^2$ and HOMAM has the high ability of error correcting, where P is the number of memorized patterns and n is the number of neurons.

Key words: Hopfield model, higher order neural networks, multidirectional associative memory, energy function, associative memory

1 Introduction

Numerous studies have been done with neural networks as a new information processing model of human brain[1, 2, 3, 11, 12]. Specifically, many models with associative memory are proposed[2]–[6],[13, 14, 18]. Associative memory can store a set of patterns as memory. When a key pattern is presented to the associative memory, the system responds by producing whichever one of the memorized patterns most closely resembles or relates to the key pattern. Hence, the recalling is done through association of the key pattern with the information memorized. The basic concept of using neural networks as associative memory is to interpret the system's evolution as transition of an input pattern toward the one memorized pattern most resembling the input pattern. The models to perform associative memory include (correlation type) autoassociative memory and heteroassociative memory. Hopfield model is one of autoassociative memory, and is deeply studied with the theoretical analysis and numerical simulation by introducing the energy function. Further it is well known that higher order neural networks (HONNs) are a generalized model whose potential is represented as the linear sum of weights and input vectors are more effective in associative memory and combinatorial application problems than the conventional ones[7, 8]. Specifically, Yatsuki[9] and Abbott[10] have shown that HONNs have high memory capacity by using the theoretical analysis. This result shows that HONNs have higher ability than the conventional ones even in associative memory. On the other hand, numerous studies with heteroassociative memory also have been done[3, 15, 16, 17]. Bidirectional associative memory (BAM) as one-to-one model and multi-directional associative memory (MAM) as many-to-one model, which means that plural key patterns are presented to the system and one pattern is output, are well known so far. They also are known as models with decreasing energy functions. Although second order BAM (SOBAM) is a generalized model of BAM to second order neural networks and has high ability in associative memory, the stability of the networks does not hold[17]. So, it is desired to propose a higher order model with the energy function.

In this paper,we propose higher order MAM (HOMAM) as a higher order model with decreasing energy functions and show the effectiveness of this model. Further, we analyze the static ability in associative memory of HOMAM and compare it with the conventional models such as the cascade and k-AM ones.

2 Higher Order Neural Networks and Multidirectional Associative Memory

In this section, higher order neural networks, the second order BAM and multidirectional associative memory are introduced.

2.1 Higher Order Neural Networks

First, let us consider the conventional neural element. The output y for each neuron is given by

$$u = \sum_{j=1}^{n} w_j x_j - \theta \tag{1}$$

$$y = f(u), \tag{2}$$

where x_j is the j-th input, u is the internal potential of the neuron, $f(u)$ is the output function, w_j is the weight of the j-th input, θ is the threshold (See Fig.1.(a)).

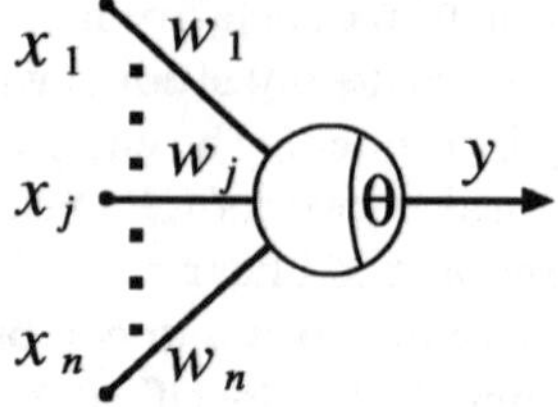

(a) The conventional neural element.

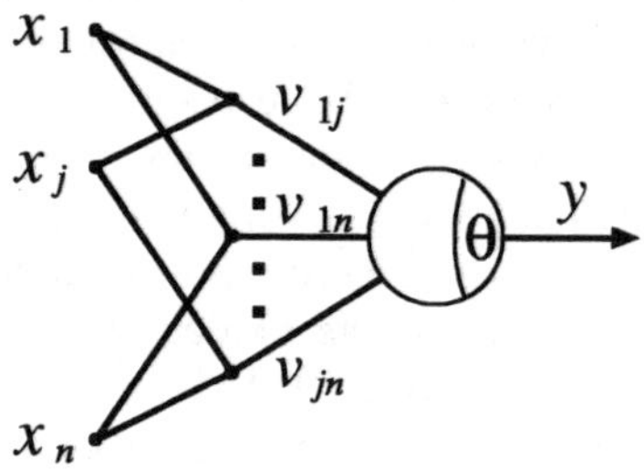

(b) A higher order neural element with $k = 2$.

Fig. 1. Neural elements. The output y of both types of neural elements is calculated by Eq.(2). For the conventional neural element shown in Fig.1.(a), the internal potential u is calculated by Eq.(1). For a higher order neural element shown in Fig.1.(b), u is calculated by Eq.(8) with $k = 2$.

The internal potential for a higher order neural element is represented by

$$u = \sum_{j=1}^{n} w_j x_j + \sum_{j=1}^{n}\sum_{k=1}^{n} v_{jk} x_j x_k + \cdots - \theta, \tag{3}$$

where the second term of the RHS is the potential of the sums of products for combinations of two input, x_j and x_k, and v_{jk} is the weight for them. The k-th term is represented by

$$\sum_{[l_k]} v_{[l_k]} x_{l_1} x_{l_2} \cdots x_{l_k}, \tag{4}$$

$$\sum_{[l_k]} = \sum_{l_1}\sum_{l_2}\cdots\sum_{l_k}, \tag{5}$$

$$1 \le l_a \le n, \tag{6}$$

where $[l_k] = l_1, \cdots, l_k$, $v_{[l_k]}$ is the weight for products of input to the neuron, and $a = 1, \cdots, k$. The Eq.(3) is rewritten as follows:

$$u = \sum_{k'=1}^{k}\sum_{[l_{k'}]} v_{[l'_k]} x_{l_1} x_{l_2} \cdots x_{l_{k'}} - \theta, \tag{7}$$

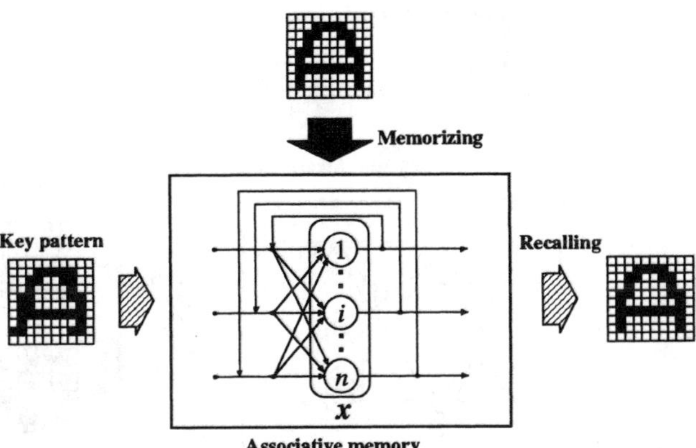

Fig. 2. One-directional associative memory model. The model, which is the well-known Hopfield model, is a single layered neural network with feedback connections. The network state x^t at time t is calculated by using the previous state x^{t-1} as a single transition. The initial network state x^0 is set to a noised pattern A'. After some transitions, the network state x^t becomes the memorized pattern A. This figure shows a network consisting of the conventional neural elements. Higher order neural elements also can be used for this model.

where k is the order of the products of the network. In this paper, the following element with the order k is used (See Fig.1.(b)).

$$u = \sum_{[l_k]} v_{[l_k]} x_{l_1} \cdots x_{l_k} - \theta \tag{8}$$

$$y = f(u). \tag{9}$$

It is called the k-th order neural network whose components are composed of the k-th order neural elements.

2.2 Associative Memory

First, let us explain associative memory for the conventional models. Figs.2, 3 and 4 show the concepts of autoassociative, bidirectional associative and multidirectional (three-directional) associative memory models, respectively. Each model has different structure of the network. It performs associative memory as the following steps:

1. Memorizing of some patterns is performed: In Figs.2, 3 and 4, the patterns A, A and B, and A, B and C are memorized in each model, respectively.

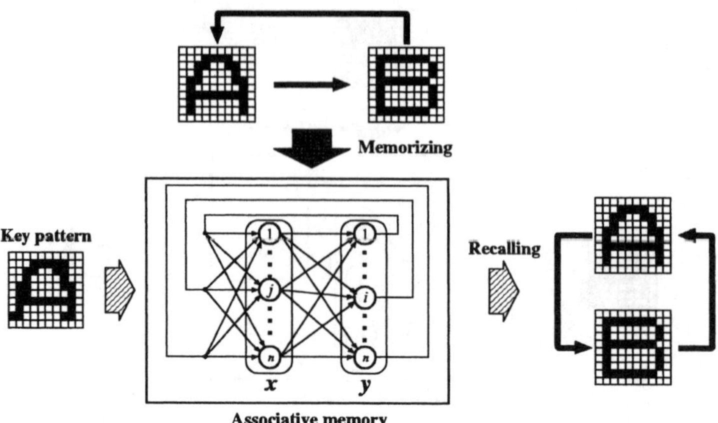

Fig. 3. Bidirectional associative memory model. The network consists of two layers with feedback connections. If the connections between two layers are symmetrical, the feedback connections can be replaced with bidirectional connections. The initial network state x^0 is set to a noised pattern A'. After some transitions, the states of two layers become memorized patterns A and B, respectively.

2. A pattern similar to any memorized pattern or a memorized pattern with noise is input to each model: In Figs.2 and 3, the noised pattern A' and in Fig.4, the noised patterns B' and C' are input to each model.
3. Recalling for the memorized patterns is performed: In Fig.2,

$$A' \to A^1 \to A'' \to \cdots \to A \to A \to \cdots,$$

where the memorized pattern A is obtained finally.
In Fig.3, a series of patterns as follows are produced:

$$A' \to B^1 \to A^1 \to B^2 \to A'' \to \cdots \to B \to A \to B \to A \to \cdots,$$

where the memorized patterns A and B is finally produced by turns.
In Fig.4, sequential patterns produced and A, B and C in this order are produced finally as follow:

$$A' \to C^1 \to A^1 \to B^1 \to C^2 \to A^2 \to B^2 \to \cdots \to A \to B \to C \to A \to \cdots$$
$$\begin{array}{ccccccccccc} \nearrow & & \nearrow & & \nearrow & & \nearrow & & \nearrow & & \nearrow & & \nearrow & & & \nearrow & & \nearrow & & \nearrow & & \nearrow & & \nearrow \\ B' & & B' & & C^1 & & A^1 & & B^1 & & C^2 & & A^2 & & & C & & A & & B & & C \end{array}$$

In this case, two patterns such as A' and B' are input and one pattern such as C^1 is output. Further, C^1 and B' are input and A^1 is output and so on.

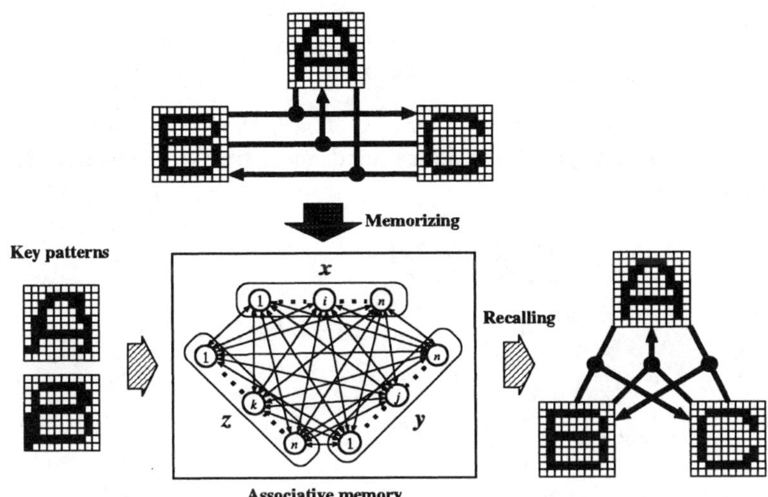

Fig. 4. Three-directional associative memory model. The network consists of three layers with bidirectional connections. The state of a single layer, e.g. z^t, is calculated by using the states of two layers, e.g. x^{t-1} and y^{t-1}, as a single transition. For example, the initial network states x^0 and y^0 are set to two noised patterns A' and B', respectively. After some transitions, the states of three layers x, y and z become memorized patterns A, B and C, respectively.

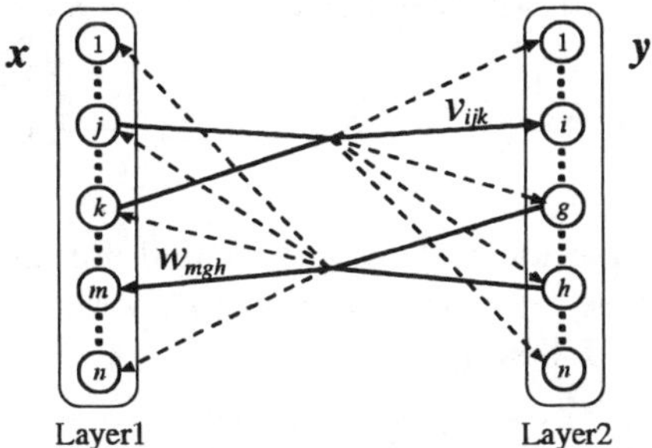

Fig. 5. Structure of SOBAM. SOBAM, which is a two-layered network, is composed of second order neural elements. All the connections are one-directional. Solid lines show two inter-layer connections corresponding to a weight v_{ijk} or w_{mgh}. Dashed lines show inter-layer connections corresponding to a product $x_j \cdot x_k$ or $y_g \cdot y_h$.

2.3 SOBAM

SOBAM is a two-layered network which is composed of the second order neural elements as shown in Fig.5[17]. Let Eq.(10) be the memorized patterns,

$$\begin{cases} \boldsymbol{x}^{(s)} = (x_1^{(s)}, \cdots, x_n^{(s)}) \\ \boldsymbol{y}^{(s)} = (y_1^{(s)}, \cdots, y_n^{(s)}) \end{cases} \quad (s = 1, \cdots, P), \tag{10}$$

where each element of them is 1 or -1. Then, the connection weights between Layer 1 and Layer 2 are determined according to correlative learning as follows:

$$v_{ijk} = c \sum_{s=1}^{P} y_i^{(s)} x_j^{(s)} x_k^{(s)}$$
$$(i=1,\cdots,n,\, j=1,\cdots,n-1,\, k=j+1,\cdots,n), \tag{11}$$

$$w_{mgh} = c \sum_{s=1}^{P} x_m^{(s)} y_g^{(s)} y_h^{(s)}$$
$$(m=1,\cdots,n,\, g=1,\cdots,n-1,\, h=g+1,\cdots,n), \tag{12}$$

where c is a positive number, v_{ijk} is the weight of the i-th element of Layer 2 for the product between the j-th and k-th elements of Layer 1 and w_{mgh} is the weight of the m-th element of Layer 1 for the product between the g-th and h-th elements of Layer 2. Here, let us consider the state transition of network. Given an input $\boldsymbol{x}^t = (x_1^t, \cdots, x_n^t)$ to Layer 1 at time t, then the output of the i-th element of Layer 2 at time $t+1$ is determined as follows:

$$y_i^{t+1} = f\left(\sum_{j=1}^{n-1} \sum_{k=j+1}^{n} v_{ijk} x_j^t x_k^t - \theta_i \right), \tag{13}$$

where f is an output function and θ is a threshold. Further, given the obtained result to Layer 2 at time $t+1$, then the output of the m-th element of Layer 1 is also determined as follows:

$$x_m^{t+2} = f\left(\sum_{g=1}^{n-1} \sum_{h=g+1}^{n} w_{mgh} y_g^{t+1} y_h^{t+1} - \theta_m \right). \tag{14}$$

By continuing the process, associative memory by SOBAM is performed. In order to see the dynamics of the recalling process, numerical simulation is performed for the case of $n = 200$. In this case, each element of memorized patterns is a random variable with the probability $1/2$ for each 1 or -1, and the following output function is used.

$$\mathrm{sgn}(u) = \begin{cases} 1 & (u > 0) \\ -1 & (u \leq 0). \end{cases} \tag{15}$$

Let the pattern ratio r be defined as the number of memorized patterns divided by the number of different weights connected to each neuron. In SOBAM, let the pattern ratio be defined as $r = \frac{P}{\binom{n}{2}}$, $\binom{n}{k}$ means the combination of k from n. Let the direction cosine between memorized pattern s and the pattern x^t be defined as follows:

$$a^t = \frac{1}{n} \sum_{i=1}^{n} x_i^t s_i. \tag{16}$$

a^0 means the distance between the pattern x^0 and the memorized pattern s and, a^1 means the distance between the pattern x^1 after one transition of x^0 and s and so on. Fig.6 shows a result of numerical simulation for $r = 0.15$, $c = 1/\binom{200}{2}$ and $\theta = 0$. As you can see, complicated dynamics for Layer 1 and Layer 2 is formed. Although SOBAM has high performance, symmetric property for weights, such as $w_{ijk} = w_{jki}$, does not hold. This is a weak point for SOBAM.

2.4 MAM

MAM is a three-layered network which is composed of the conventional neural elements shown as Fig.7[3, 15]. Let Eq.(17) be the memorized patterns:

$$\begin{cases} x^{(s)} = (x_1^{(s)}, \cdots, x_n^{(s)}) \\ y^{(s)} = (y_1^{(s)}, \cdots, y_n^{(s)}) \quad (s = 1, \cdots, P), \\ z^{(s)} = (z_1^{(s)}, \cdots, z_n^{(s)}) \end{cases} \tag{17}$$

where each element is 1 or -1. The connection relation from Layer 1 to Layer 2 as shown in Fig.7 is represented by the following correlation matrix U.

$$(U)_{ij} = u_{ij} = c \sum_{s=1}^{P} y_i^{(s)} x_j^{(s)}, \tag{18}$$

where U_{ij} means the (i,j)-th component u_{ij} of the matrix U. And, the connection relation from Layer 2 to Layer 1 is represented by the transpose U^T of the matrix U as follows:

$$(U^T)_{ij} = u_{ij}^* = c \sum_{s=1}^{P} x_i^{(s)} y_j^{(s)}. \tag{19}$$

Likewise, the connection relations from Layer 2 to Layer 3 and Layer3 to Layer 1 are represented by the following correlation matrixes V, W, respectively.

$$(V)_{ij} = v_{ij} = c \sum_{s=1}^{P} z_i^{(s)} y_j^{(s)}, \tag{20}$$

$$(W)_{ij} = w_{ij} = c \sum_{s=1}^{P} x_i^{(s)} z_j^{(s)}. \tag{21}$$

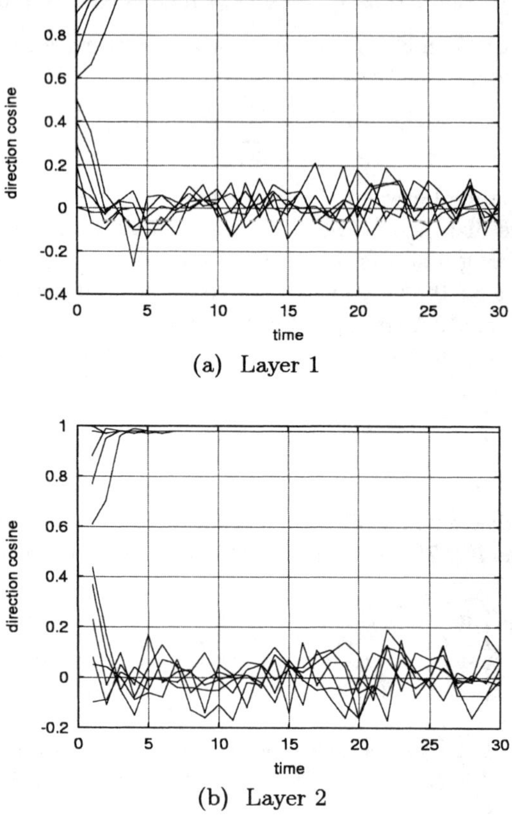

(a) Layer 1

(b) Layer 2

Fig. 6. Dynamic behaviors of recalling processes (SOBAM) ($r = 0.15$, $n = 200$). In Layer 1, given an input pattern whose direction cosine with the memorized pattern s is more than or equal to 0.6, then the input pattern approaches to the pattern s after some transitions, and given an input pattern whose direction cosine with the memorized pattern s is less than or equal to 0.5, then the case seems to be failure in recalling. In Layer 2, the case where the direction cosine is more than 0.6 is successful in recalling and the case where the direction cosine is less than 0.4 seems to be failure in recalling. Remark that the cases of failure in recalling do not have stable equilibrium states.

As well, the transpose of them, V^T and W^T, are defined. Each element of V^T and W^T is represented as v^*_{ij} and w^*_{ij}, respectively. In the recalling process of MAM, each state $x^{t+1}, y^{t+1}, z^{t+1}$ at time $t + 1$ is determined in order as follows:

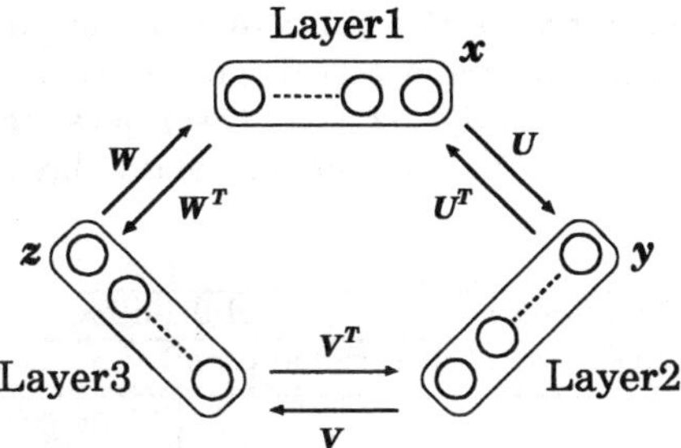

Fig. 7. Multidirectional Associative Memory. Three layers are interconnected. A weight matrix between two layers is the correlation matrix of two memorized patterns. The connection relation from Layer 1 to Layer 2 is represented by the transpose of the matrix from Layer 2 to Layer 1.

$$\begin{cases} x_i^{t+1} = f[\sum_{j=1}^{n}(w_{ij}z_j^t + u_{ij}^*y_j^t) - \theta_i] \\ y_j^{t+1} = f[\sum_{k=1}^{n}(u_{jk}x_k^{t+1} + v_{jk}^*z_k^t) - \theta_j] \\ z_k^{t+1} = f[\sum_{i=1}^{n}(v_{ki}y_i^{t+1} + w_{ki}^*x_i^{t+1}) - \theta_k]. \end{cases} \qquad (22)$$

Each element continues this process until all states of the network reach equilibrium state. The network has the same stable equilibrium states as BAM has[3, 15]. MAM is a three-layered model whose connections between two layers are the same as ones of BAM. Therefore, we can consider SOMAM whose connections are the same as ones of SOBAM. Table 1 shows the relation among the conventional models.

3 Higher Order BAM

Although associative memory with SOBAM or SOMAM has high ability, the stability of the network does not hold. That is, the symmetric structure of weights, such as $w_{ijk} = w_{jki}$, does not hold. Therefore, in order to overcome the problem, we propose a new model as shown in Fig.8. The model is one which generalizes BAM to multidirectional model. We call it HOMAM (with decreasing energy function). In this paper, we will consider the bidirectional model without loss of generality. Let the pattern ratio for HOMAM be defined as $r = \frac{P}{n^2}$.

[**Memorizing process**] Let the Eq.(17) be the memorized patterns. Then each weight is determined as follows:

Table 1. The relation between the proposed model and the conventional models. Associative memories are classified into autoassociative (one-directional case) and heteroassociative (multi-directional case) ones, and are constructed by the conventional and HONNs. The proposed model is 3-directional one by HONNs.

	One-directional case	Multi-directional case k	
		2-directional case	3-directional case
The conventional model (first order)	Autoassociative $X \longleftrightarrow X$	BAM $X \longleftrightarrow Y$	MAM Y / $X \longleftrightarrow Z$
HONNs	Autoassociative $X \longleftrightarrow X$	SOBAM $X \longleftrightarrow Y$	SOMAM Y / $X \longleftrightarrow Z$ — The proposed HOMAM Y / $X \longleftrightarrow Z$

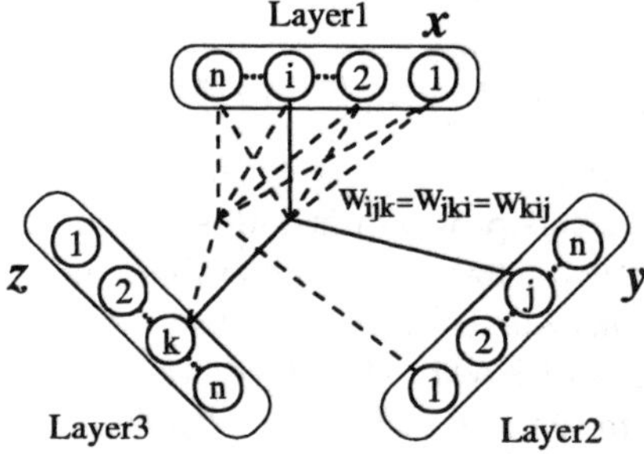

Fig. 8. An example of HOMAM with three layers. Triplet elements from different layers are interconnected by a three-directional connection. When the connection among elements i, j and k directs to element i, the product $y_j \cdot z_k$ is weighted by w_{ijk}. Although this example is for three layers, an HOMAM network may have an arbitrary number layers more than three.

$$w_{ijk} = c \sum_{s=1}^{P} x_i^{(s)} y_j^{(s)} z_k^{(s)} \qquad (i, j, k = 1, \cdots, n). \tag{23}$$

[Recalling process] First, assume that Layer 2 and Layer 3 are input layers and Layer 1 is output layer. Given input patterns y and z to Layer 2, Layer 3, then the output of the i-th neuron at $t + 1$ is determined as follows:

$$x_i^{t+1} = f\Big(\sum_{j=1}^{n} \sum_{k=1}^{n} w_{ijk} y_j^t z_k^t - \theta_i \Big). \tag{24}$$

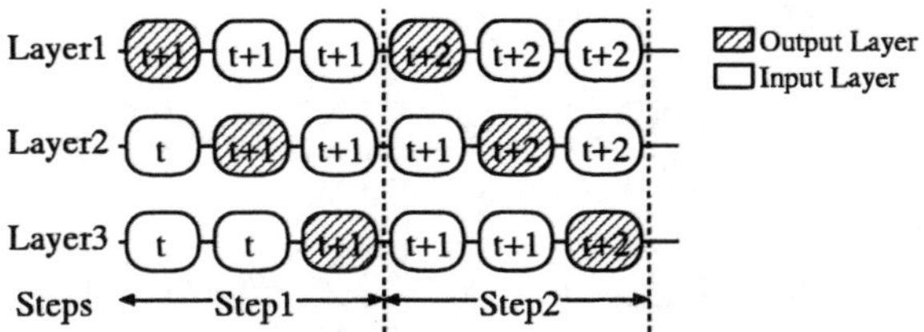

Fig. 9. A state transition of HOMAM. Given input patterns to Layer 2 and Layer 3, then an output pattern of Layer 1 is obtained. Next, given the input pattern of Layer 3 and the obtained pattern of Layer 1, then an output pattern of Layer 2 is obtained. Further, given the obtained patterns of Layer1 and Layer 2, then an output pattern of Layer 3 is determined. Likewise, the same procedure is repeated.

At the next step, input patterns z^t, x^{t+1} are given to Layer 3 and Layer 1 and the output y^{t+1} of Layer 2 is determined. Further, output z^{t+1} of Layer 3 is determined as follows:

$$y_j^{t+1} = f\Big(\sum_{k=1}^{n} \sum_{i=1}^{n} w_{jki} z_k^t x_i^{t+1} - \theta_j \Big), \tag{25}$$

$$z_k^{t+1} = f\Big(\sum_{i=1}^{n} \sum_{j=1}^{n} w_{kij} x_i^{t+1} y_j^{t+1} - \theta_k \Big). \tag{26}$$

In recalling process of this model, input to two layers are given and output of the rest of layers is determined at the first step. At the next step, one of two layers and the obtained output are input, output of the rest of layers is determined and so on. Let one step be defined as the time when all layers transit one time.

Fig.9 shows the case where Layer 2 and Layer 3 are used as the first input layers at step t. In the case of the conventional associative memory, products between elements of one layer are taken. In the proposed model, product between elements between two layers is taken. Therefore, the following relation holds from the definition of Eq.(23).

$$w_{ijk} = w_{jki} = w_{kij} \tag{27}$$

Then, we can prove that the proposed model has decreasing energy function. Assume that f is Eq.(15). Let energy function E for a state S of the network at time t be defined as follows:

$$E(S^t) = - \sum_{i=1}^{n} \sum_{j=1}^{n} \sum_{k=1}^{n} w_{ijk} x_i^t y_j^t z_k^t. \tag{28}$$

When the state changes from t to $t + 1$, the variation of the energy E is represented as follows:

$$\Delta E(S^t) = E(S^{t+1}) - E(S^t)$$

$$= -\sum_{i=1}^{n}(x_i^{t+1} - x_i^t)\sum_{j=1}^{n}\sum_{k=1}^{n}w_{ijk}y_j^t z_k^t$$

$$-\sum_{j=1}^{n}(y_j^{t+1} - y_j^t)\sum_{i=1}^{n}\sum_{k=1}^{n}w_{ijk}x_i^{t+1}z_k^t$$

$$-\sum_{k=1}^{n}(z_k^{t+1} - z_k^t)\sum_{i=1}^{n}\sum_{j=1}^{n}w_{ijk}x_i^{t+1}y_j^{t+1} \tag{29}$$

From the definition of x_i^{t+1}, y_j^{t+1}, z_k^{t+1}, and Eq.(29), the following holds:

$$\Delta E(S^t) = -\sum_{i=1}^{n}(x_i^{t+1} - x_i^t)u_i^t - \sum_{j=1}^{n}(y_j^{t+1} - y_j^{t+1})u_j^t - \sum_{k=1}^{n}(z_k^{t+1} - z_k^{t+1})u_k^t$$

$$\leq 0. \tag{30}$$

Therefore, as energy function of the proposed model decreases with time, the stable state transition is performed. Fig.10 shows numerical simulation for dynamics with $n = 200$, where pattern ratio $r = 0.15$, $c = 1/200^2$ and $\theta = 0$. As you can see, stable equilibrium state is performed. Fig.10 shows the validity of the above argument.

4 The Ability of the Proposed Model

We will analyze the ability of the proposed model by using the analytic method of the static model[2, 8, 18]. Let the Eq.(22) be the memorized patterns, and each element of $\{x^{(s)}, y^{(s)}, z^{(s)}\}$ is the random variable with probability $1/2$ for each 1 or -1, and $\theta = 0$.

4.1 The ability of correct recalling

For input patterns $y^{(r)}$ and $z^{(r)}$, the probability that the output x_i for the i-th element equals to the desired output $x_i^{(r)}$ is computed. The Eq.(31) holds for the inner state u_i of each element x_i.

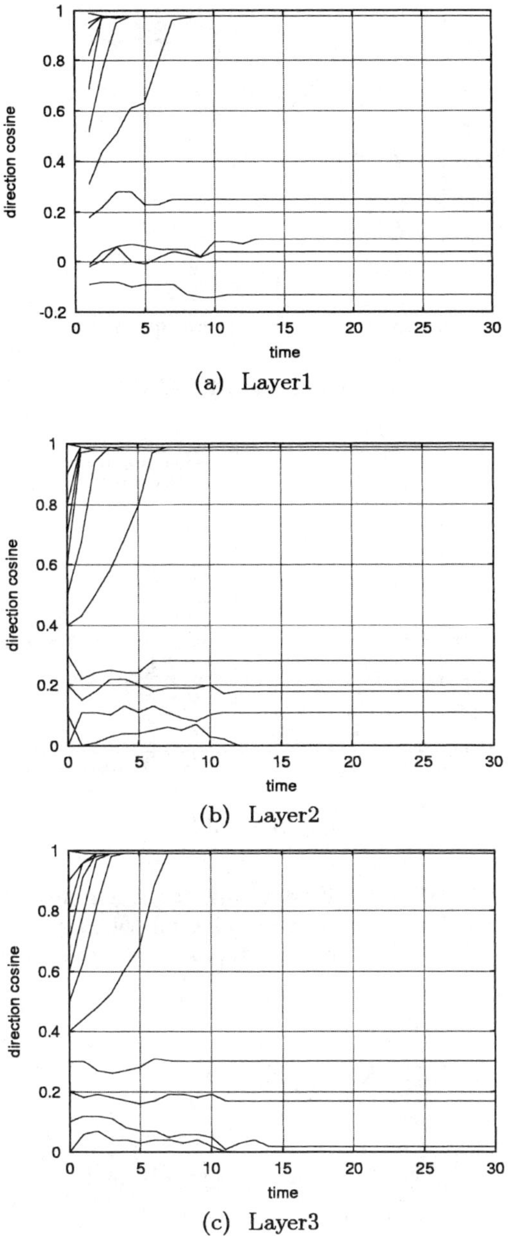

Fig. 10. Dynamic behavior of recalling process (HOMAM) ($r = 0.15$, $n = 200$). The explanation of the Fig.10 is the same as one of the Fig.6. In this case, the critical direction cosines of Layer1, Layer 2, and Layer 3 in successful (failure) in recalling are about 0.3 (0.2), 0.4 (0.3) and 0.4 (0.3), respectively. Remark that the cases of not only successful in recalling but also failure in recalling have stable equilibrium states.

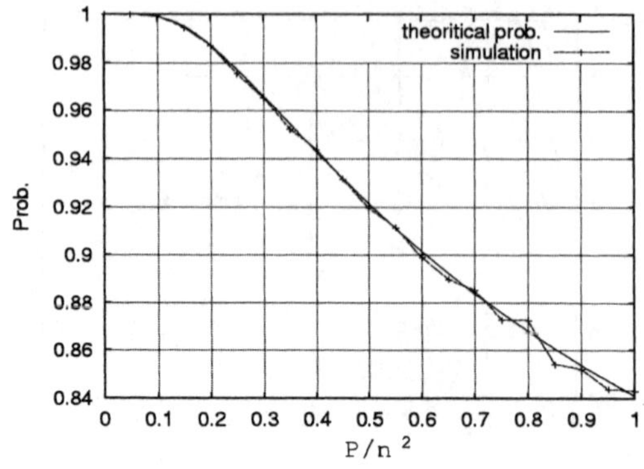

(a) The proposed HOMAM

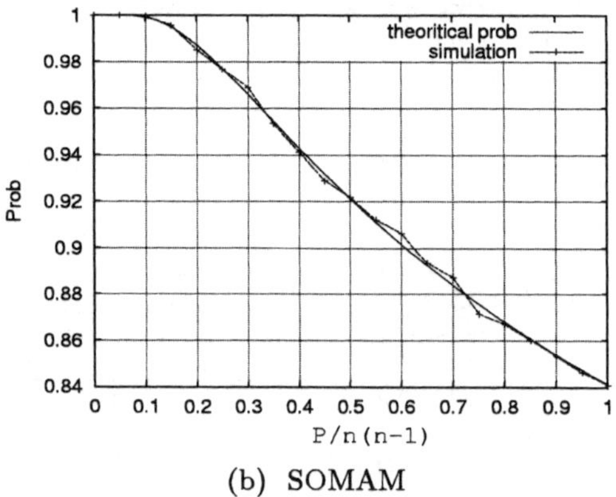

(b) SOMAM

Fig. 11. Probability of correct recalling for each model. Figs.11.(a) and 11.(b) show the probabilities of correct recalling for the Eqs.(40) and (41), respectively. The solid and dotted lines are the graphs for the theoretical and simulation results, respectively.

$$
\begin{aligned}
u_i &= \sum_{j=1}^{n}\sum_{k=1}^{n} w_{ijk} y_j^{(r)} z_k^{(r)} \\
&= \sum_{j=1}^{n}\sum_{k=1}^{n} \left[c \sum_{s=1}^{P} x_i^{(s)} y_j^{(s)} z_k^{(s)} \right] y_j^{(r)} z_k^{(r)} \\
&= c x_i^{(r)} \sum_{j=1}^{n}\sum_{k=1}^{n} (y_j^{(r)})^2 (z_k^{(r)})^2 \\
&\quad + c \sum_{j=1}^{n}\sum_{k=1}^{n}\sum_{s\neq r} x_i^{(s)} y_j^{(s)} z_k^{(s)} y_j^{(r)} z_k^{(r)}.
\end{aligned}
\tag{31}
$$

Now, let **Prob**(correct recalling) be defined as the probability that $\boldsymbol{x} = \boldsymbol{x}^{(r)}$.
As y_j and $z_k = \pm 1$, the following equation holds:

$$\sum_{j=1}^{n}\sum_{k=1}^{n}(y_j^{(r)})^2(z_k^{(r)})^2 = n^2. \tag{32}$$

Let $c = 1/n^2$.

$$u_i = x_i^{(r)} + \frac{1}{n^2}\sum_{j=1}^{n}\sum_{k=1}^{n}\sum_{s\neq r} x_i^{(s)} y_j^{(s)} z_k^{(s)} y_j^{(r)} z_k^{(r)} \tag{33}$$

Each variable of $x_i^{(s)}, \cdots, z_k^{(r)}$ is mutually independent and takes the value 1
or -1 with the probability $\frac{1}{2}$. Let

$$v = \frac{x_i^{(s)} y_j^{(s)} z_k^{(s)} y_j^{(r)} z_k^{(r)}}{n^2}. \tag{34}$$

Then, v is a random variable and takes the value $\frac{1}{n^2}$ or $-\frac{1}{n^2}$ with the probability $\frac{1}{2}$. Let h be defined as the crosstalk term of the Eq.(31) as follows:

$$h = \frac{\sum_{j=1}^{n}\sum_{k=1}^{n}\sum_{s\neq r} x_i^{(s)} y_j^{(s)} z_k^{(s)} y_j^{(r)} z_k^{(r)}}{n^2}. \tag{35}$$

Then, the following equation holds:

$$\mathbf{Prob}(h = \frac{2t-(P-1)n^2}{n^2}) = \left(\frac{1}{2}\right)^{(P-1)n^2}\binom{(P-1)n^2}{t}$$

$$(t = 1, 2, \cdots, (P-1)n^2). \tag{36}$$

From the Central Limit Theorem, the random variable h is distributed according to the normal low $N(0, P/n^2)$ with the average 0 and the covariance
P/n^2. Let us compute the probability that $x_i = x_i^{(r)}$, that is $u_i > 0$ when
$x_i^{(r)} = 1$.

$$\mathbf{Prob}(h > -1) = \frac{1}{\sqrt{2\pi\sigma}}\int_{-1}^{\infty}\exp(-\frac{s^2}{2\sigma})ds$$

$$= \frac{1}{\sqrt{2\pi}}\int_{-\sqrt{\frac{n^2}{P}}}^{\infty}\exp(-\frac{s^2}{2})ds \tag{37}$$

As a result, the following equation holds:

$$\mathbf{Prob}(h > -1) = 1 - \Phi(-\sqrt{\frac{n^2}{P}})$$

$$= \Phi(\sqrt{\frac{n^2}{P}}), \tag{38}$$

where

$$\Phi(u) = \frac{1}{\sqrt{2\pi}} \int_{-\infty}^{u} \exp(-\frac{s^2}{2})ds. \qquad (39)$$

Then, we have the following equation:

$$\mathbf{Prob}(\text{correct recalling for HOMAM}) = \Phi(\sqrt{\frac{n^2}{P}}), \qquad (40)$$

We can also show the Eq.(40) when $x_i^{(r)} = -1$. Fig.11(a) shows the graph of Eq.(40) and the result of numerical simulation. The axis of ordinates is correct recalling probability and the axis of abscissa is pattern ratio $r = \frac{P}{n^2}$. In order to be $\mathbf{Prob}$(correct recalling for HOMAM) > 0.99, $\frac{P}{n^2} < 0.184$ must hold. This case equals to the direct cosine with 0.98, because the direct cosine becomes $2\times\mathbf{Prob}$(correct recalling for HOMAM)-1 from the definition.

By using the same method as the above, the abilities for MAM and SO-MAM are shown as follows:

$$\mathbf{Prob}(\text{correct recalling for MAM}) = \Phi\left(\sqrt{\frac{2n}{P}}\right), \qquad (41)$$

$$\mathbf{Prob}(\text{correct recalling for SOMAM}) = \Phi\left(\sqrt{\frac{n(n-1)}{P}}\right). \qquad (42)$$

Fig.11(b) shows the graph of Eq.(42) and the result of numerical simulation for SOMAM, where the axis of abscissa is $\frac{P}{n(n-1)}$. From the comparison between Eq.(40) and Eq.(42), it seems that both models has the same ability. The difference is that HOMAM is stable network but SOMAM is not.

The another models performing associative memory of sequential patterns like HOMAM are well known. In order to perform them for the conventional model of one-dimensional case, asymmetric property for weights is needed, because symmetric one has one stable equilibrium pattern in dynamics. With associative memory of sequential patterns, plural stable equilibrium patterns are needed, however, in a recalling process. Therefore, it is known that it is difficult to construct the model performing associative memory of sequential patterns[19]. While Meir[20] and Amari[21] have proposed the cascade and k-AM models ($k = 1, 2, \cdots$), respectively. These models consist of plural layers each of which has one stable equilibrium pattern in dynamics as same as HOMAM. However, the memory capacity of the models is about $0.15(P/n)$. Further, as these models do not have symmetric property for weights, analysis of stability by use of energy functions can not do. Compared with the conventional models, as the proposed model has symmetric property for weights, dynamics of the model is easily analyzed by use of energy functions.

4.2 The ability of error correcting

Let us consider how many errors the proposed model can correct. Let $\boldsymbol{y}^{(r)'}$ and $\boldsymbol{z}^{(r)'}$ be defined as the patterns with D_1 and D_2 elements different from $\boldsymbol{y}^{(r)}$ and $\boldsymbol{z}^{(r)}$, respectively. That is, $1 - \frac{2D_1}{n} = d(\boldsymbol{y}^{(r)}, \boldsymbol{y}^{(r)'})$ and $1 - \frac{2D_2}{n} = d(\boldsymbol{z}^{(r)}, \boldsymbol{z}^{(r)'})$. Then, the Eq.(43) holds for u_i.

$$
\begin{aligned}
u_i &= \sum_{j=1}^{n}\sum_{k=1}^{n} w_{ijk} y_j^{(r)'} z_k^{(r)'} \\
&= \sum_{j=1}^{n}\sum_{k=1}^{n} [c\sum_{s=1} x_i^{(s)} y_j^{(s)} z_k^{(s)}] y_j^{(r)'} z_k^{(r)'} \\
&= c x_i^{(r)} \sum_{j=1}^{n}\sum_{k=1}^{n} y_j^{(r)} z_k^{(r)} y_j^{(r)'} z_k^{(r)'} \\
&\quad + c \sum_{s\neq r}\sum_{j=1}^{n}\sum_{k=1}^{n} x_i^{(s)} y_j^{(s)} z_k^{(s)} y_j^{(r)'} z_k^{(r)'}
\end{aligned}
\tag{43}
$$

We can assume that $y_i^{(r)'} \neq y_i^{(r)}$ for $i = 1, \cdots, D_1$, $y_i^{(r)} = y_i^{(r)}$ for $i = D_1 + 1, \cdots, n$, $z_i^{(r)'} \neq z_i^{(r)}$ for $i = 1, \cdots, D_2$, $z_i^{(r)} = z_i^{(r)}$ for $i = D_2 + 1, \cdots, n$, without loss of generality. Let g_1 be denoted as the first term of RHS of the Eq.(43).

$$
\begin{aligned}
g_1 &= c x_i^{(r)} \sum_{j=1}^{n}\sum_{k=1}^{n} y_j^{(r)} z_k^{(r)} y_j^{(r)'} z_k^{(r)'} \\
&= c x_i^{(r)} \Big\{ \sum_{j=D_1+1}^{n} (y_j^{(r)})^2 - \sum_{j=1}^{D_1}(y_j^{(r)})^2 \Big\} \Big\{ \sum_{k=D_2+1}^{n} (z_k^{(r)})^2 - \sum_{k=1}^{D_2}(z_k^{(r)})^2 \Big\} \\
&= c x_i^{(r)} (n - 2D_1)(n - 2D_2)
\end{aligned}
\tag{44}
$$

Let $1/c = (n - 2D_1)(n - 2D_2)$. Then, the Eq.(43) is rewritten as follows:

$$
\begin{aligned}
u_i = x_i^{(r)} + &\frac{1}{(n - 2D_1)(n - 2D_2)} \cdot \\
&\sum_{s\neq r} \Big\{ \sum_{j=D_1+1}^{n} \Big\{ \sum_{k=D_2+1}^{n} x_i^{(s)} y_j^{(s)} z_k^{(s)} y_j^{(r)} z_k^{(r)} - \sum_{k=1}^{D_2} x_i^{(s)} y_j^{(s)} z_k^{(s)} y_j^{(r)} z_k^{(r)} \Big\} \\
&- \sum_{j=1}^{D_1} \Big\{ \sum_{k=D_2+1}^{n} x_i^{(s)} y_j^{(s)} z_k^{(s)} y_j^{(r)} z_k^{(r)} - \sum_{k=1}^{D_2} x_i^{(s)} y_j^{(s)} z_k^{(s)} y_j^{(r)} z_k^{(r)} \Big\} \Big\}
\end{aligned}
\tag{45}
$$

The following relation is obtained by the same method as getting the Eq.(37) from the Eq.(33).

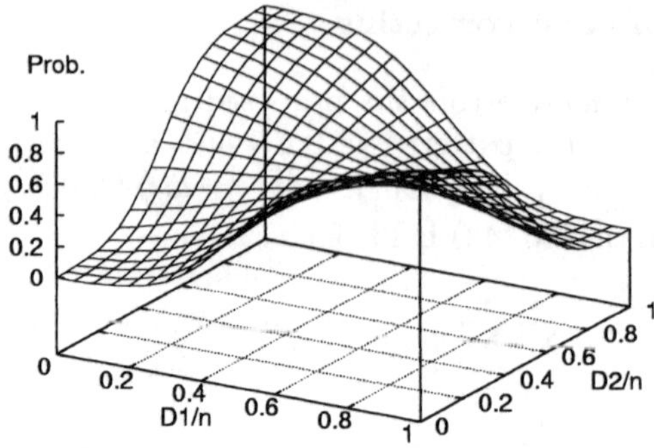

(a) The result from the Eq.(46).

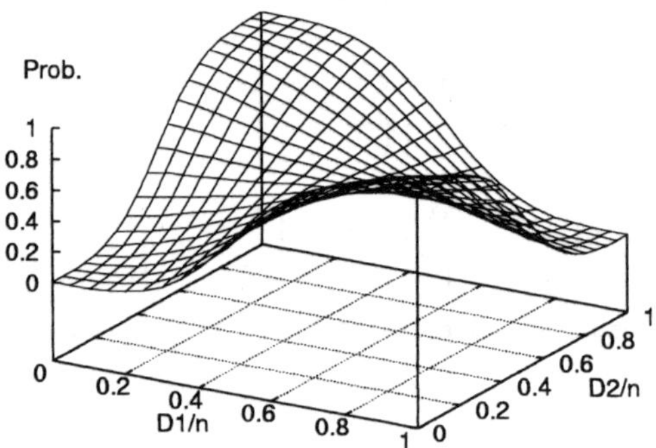

(b) The result of numerical simulation.

Fig. 12. Error recalling ability of HOMAM for $\frac{P}{n^2} = 0.1$. Fig.12.(a) shows the probability of error recalling for $\frac{D_1}{n}$ and $\frac{D_2}{n}$ by using the Eq.(46). Fig.12.(b) shows the graph obtained by numerical simulations, that is, when input patterns y and z with noise rates $\frac{D_1}{n}$ and $\frac{D_2}{n}$ are input to HOMAM, the probability of correct recalling for output pattern x is computed after a transition by the Eq.(24).

$$\mathbf{Prob}(h > -1) = \Phi(\frac{(1 - 2\frac{D_1}{n})(1 - 2\frac{D_2}{n})}{\sqrt{\frac{P}{n^2}}}). \tag{46}$$

Fig.12.(a) shows the results obtained from the Eq.(46) for $\frac{P}{n^2} = 0.1$. The axis of ordinate is $\mathbf{Prob}$(error recalling)$= 1 - \mathbf{Prob}(h > -1)$ and two axes of abscissa are $\frac{D_1}{n}$ and $\frac{D_2}{n}$. Fig.12.(b) shows the result in numerical simulation for $n = 40$ and $\frac{P}{n^2} = 0.1$. The result in numerical simulation is in fairly general agreement with the theoretical ones. Fig.12 also shows that the case with low error rate is the area of $\frac{D_1}{n} \approx \frac{D_2}{n}$. Therefore, let us consider the ability of

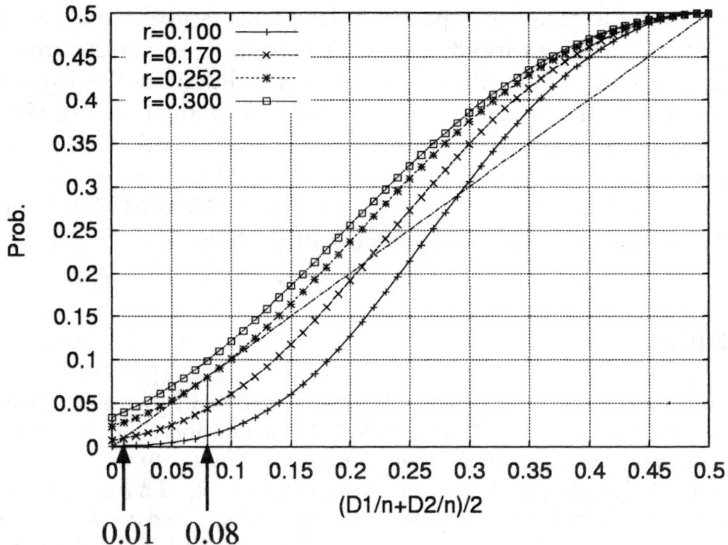

Fig. 13. Error correcting ability for some pattern ratios. The area satisfying the Eq.(47) has the error correcting ability. The critical pattern ratio with the error correcting ability is obtained by finding the curve that is tangent to the dot-dash-line of $\mathbf{Prob}(\cdot) = \frac{1}{2}(\frac{D_1}{n} + \frac{D_2}{n})$. The value is about 0.252. Further, the memory capacity is obtained by finding the intersection among the curve, the dot-dash-line and the line of $\frac{1}{2}(\frac{D_1}{n} + \frac{D_2}{n}) = 0.01$. The value is about 0.17.

error correcting of the case $(\frac{D_1}{n}, \frac{D_2}{n})$ with $\frac{D_1}{n} = \frac{D_2}{n}$.

$$\mathbf{Prob}(\text{error recalling}) < \frac{1}{2}(\frac{D_1}{n} + \frac{D_2}{n}). \tag{47}$$

The area satisfying Eq.(47) has the ability of error correcting, because it means that the rate of error decreases after one transition. Fig.13 is the figure to show the ability of error correcting for $\frac{P}{n^2} = 0.1, 0.17, 0.252, 0.300$, where the axis of ordinate is $\mathbf{Prob}(\text{error recalling})$. Fig.13 shows that the critical case where there exists the area satisfying the Eq.(47) is about $\frac{P}{n^2} = 0.252$ for $\frac{1}{2}(\frac{D_1}{n} + \frac{D_2}{n}) = 0.08$. Further, Fig.13 also shows that the critical pattern ratio satisfying $\mathbf{Prob}(\text{error recalling}) \leq 0.01$, that is $\mathbf{Prob}(\text{correct recalling}) > 0.99$, is about 0.17. The value is near one obtained in Sect. 4.1.

5 Conclusions

In this paper, we proposed higher order multidirectional associative memory (HOMAM) with decreasing energy function and demonstrated the high ability of it by performing numerical simulation and theoretical analysis. Specifically, we showed that the memory capacity of HOMAM is about $0.17{\sim}0.184P/n^2$ and the critical memory ratio of the case where there exists the area of error correcting is about 0.252. Although this paper describes the three layered model with $k = 2$, the result is easily generalized to any case. As the special case, the one with two layered model with $k = 1$ is BAM. It means that the proposed model is a generalized model of BAM in natural style. We will continue to study the relation among BAM, MAM, SOBAM, SOMAM, k-AM and HOMAM and find a stable bidirectional HONNs.

References

1. D. Rumelhart and J. McClelland, Parallel Distributed Processing: Explorations in the Microstructure of Cognition; Vol. 1: Foundations: Vol. 2: Psychological and Biological Models, MIT Press, Cambridge, Mass., 1986.
2. J. Hertz, A. Krogh and R. Palmer, Introduction to the Theory of Neural Computation, Addison-Wesley Publishing Company, Inc., Redwood City, CA, 1991.
3. C. Lin and C. Lee, Neural Fuzzy System, Prentice Hall PTR., 1996.
4. M. Okada, "Notions of associative memory and sparse coding", Neural Networks, Vol.9, pp.1429–1458, 1996.
5. S. Amari and K. Maginu, "Statistical neurodynamics of associative memory", Neural Networks, Vol.1, pp.63–73, 1988.
6. S. Chan and J. Michael, "Memory capacity of artificial neural networks with high order node connections", Proceedings of ICNN, pp.207–216, 1988.
7. D. Psaltis, C. Park and J. Hong, "Higher order associative memories", Neural Networks, Vol.1, pp.149–163, 1988.
8. S. Yatsuki and H. Miyajima, "Associative Ability of Higher Order Neural Networks", in Proc. ICNN'97, Vol.2, pp.1299-1304, 1997.
9. S. Yatsuki, and H. Miyajima, "Statistical Dynamics of Associative Memory for Higher Order Neural Networks", IEEE Proc. of ISCAS 2000, Vol.3, pp.670–673, 2000.
10. L.F. Abbott, and Y. Arian, "Storage Capacity of Generalized Networks", Physical Review A, Vol.36, No.10. pp.5091–5094, 1987.
11. J.J. Hopfield, "Neural networks and physical systems with emergent collective computational abilities", Proceedings of the National Academy of Sciences, Vol.79, pp.2554–2558, 1982.
12. E. Goles, F. Fogelman, and D. Pellegrin, "Decreasing Energy Functions as a Tool for studying threshold networks", Disc. Appl. Math, Vol.12, pp.261–277, 1985.
13. S. Yoshizawa, M. Morita, and S. Amari, "Capacity of Associative Memory using a Nonmonotonic Neuron Model", *Neural Networks*, Vol.6, pp.167–176, 1993.
14. B. Kosko, "Bidirectional associative memory", IEEE Transactions on Systems, Man, and Cybernetics, Vol.18, pp.49–60, 1988.
15. M. Hattori, M. Hasegawa, "Multimodule associative memory for many-to-many associations", Neurocomputing, Vol.19, pp.99–119, 1998.
16. P.K. Simpson, "Higher-ordered and intraconnected bidirectional associative memories", IEEE Transactions on Systems, Man, and Cybernetics, Vol.20, pp.637–653, 1990.

17. C.S. Leung, L.W. Chan and E.M.K. Lai, "Stability and statistical properties of second-order bidirectional memories", IEEE Transactions on Neural Networks, Vol.8, pp.267–277, 1997.

18. Y. Uesaka and K. Ozeki, "Some properties of associative type memories", IEICE of Japan, Vol.55-D, pp.323–330, 1972.

19. Y. Hamakawa, H. Miyajima, N. Shigei and T. Tsuruta, "On Some Properties of Higher Order Correlation Associative Memory of Sequential Patterns", submitted to RISP.

20. R. Meir and E. Domany, "Exact Solution of a Layered Neural Network Memory", Physical Rev. Letter, 59, pp.359–362, 1987.

21. S. Amari, "Statistical Neurodynamics of Various Versions of Correlation Associative Memory", Proceedings of IEEE Conference on Neural Networks, Vol.I, pp.633–640, 1988.

Fast Indexing of Codebook Vectors Using Dynamic Binary Search Trees With Fat Decision Hyperplanes

Frederic Maire[1], Sebastian Bader[2], Frank Wathne[1]

[1] Smart Devices Laboratory, School of SEDC, IT Faculty, Queensland University of Technology, 2 George Street, GPO Box 2434, Brisbane Q 4001, Australia. `f.maire@qut.edu.au` , `frankwathne@hotmail.com`

[2] TU-Dresden, Fakultät Informatik, Institut für künstliche Intelligenz, 01062 Dresden, Germany. `s.bader@gmx.net`

Summary. We describe a new indexing tree system for high dimensional codebook vectors. This indexing system uses a dynamic binary search tree with fat decision hyperplanes. The system is generic, adaptive and can be used as a software component in any vector quantization system. The cost of this higher speed (compared to tabular indexing) is a negligible degradation of the distortion error. Nevertheless, a parameter allows the user to tradeoff speed for a lower distortion error. A distinctive and attractive feature of this tree indexing system is that it can follow non-stationary codebooks by performing local repairs to its indexing tree. Experimental results show that this indexing system is very fast; it outperforms similar tree indexing systems like TSVQ and K-trees.

Key words: tree indexing, dynamic indexing, vector quantization.

1.1 Introduction

Vector Quantization (VQ) is a powerful computational procedure encountered in a wide range of applications including density estimation, data compression, pattern recognition, clustering, function approximation and time series prediction.

The principle of VQ can be formulated as follows. Given an integer m and a pattern space $\mathcal{X}$ (typically a vector space of finite dimension) equipped with a probability distribution $p(x)$, we wish to find a codebook (finite subset of $\mathcal{X}$) such the *distortion error* is minimum. Formally, we wish to find $C = \{x_1, x_2, \ldots, x_m\} \subseteq \mathcal{X}$ that minimizes

$$\int_{\mathcal{X}} \left\| x - x_{c(x)} \right\|^2 p(x)dx$$

where $c(x)$ denotes the index of the closest codebook vector to x in C.

One of the major drawbacks of array based vector quantization is its time complexity; the computational cost of finding the nearest neighbours imposes practical limits on the codebook size and the rate at which the application can operate.

To address this problem, a number of tree-structured indexing systems have been proposed (e.g. R-Trees [9], R^+-Trees [15], K-Trees [8], HiGS [16]). They provide retrieval operators with a running-time which is logarithmic in the size of the codebook (set of reference vectors), by splitting the complete search space recursively into smaller nested regions. In other words, a tree structure guides the search at each level into a smaller region containing the input vector. The new tree indexing system called *Fast Indexing Tree* (FIT) that we introduce in Section 1.5 distinguishes itself by its adaptiveness; instead of regular rebuilding of the tree, local repairs are used for following non-stationary input vector distribution. Moreover, a parameter allows the user to tradeoff speed for a lower distortion.

A fast indexing system should not only provide faster access to the vectors of the codebook compared to a tabular search, but the indexing system should also be independent from the algorithms it serves (k-means, self-organizing maps, neural gas, etc...). From this point of view, FIT is a software component whose interface is composed of a *search* function and three mutator functions. The search function is supposed to retrieve the first closest and second closest codebook vector to a new input vector. The mutator functions are *insert* (inserts a new codebook vector), *delete* (deletes an existing codebook vector) and *update* (assigns a new location to an existing codebook vector). To provide a minimal expected running time the tree of the indexing system should be as balanced as possible. That is, the subtrees of any node should have about the same height.

In Section 1.2, we review some applications of tree indexing systems for higher dimensional vector spaces. In Section 1.3, we sketch the main ideas of other tree indexing systems that are related to FIT. In Section 1.4, we introduce an algorithm to build an exact tree indexing system. The tree indexing system is exact in the sense that it agrees with the Voronoi diagram of the codebook. Section 1.5, introduces FIT and explains how the design objectives of FIT are met, how the tree is built and how the tree is kept balanced. Section 1.6 presents some experimental results.

1.2 Applications

The problem of proximity searching is encountered in a wide range of fields [3]. Some examples are non-traditional databases like repositories of fingerprints, images, audio or video clips, where the concept of searching with a key is generalized to searching with an arbitrary subset of the record being searched.

The queries by content in these multimedia databases are based on the definition of a similarity function among the multimedia objects. Although not considered a multimedia data type, unstructured text retrieval poses similar problems to those of multimedia object retrieval.

Multimedia objects are usually mapped to feature vectors in some high-dimensional space and queries are processed against a database of those features. For example, audio fingerprinting [11] seeks to recognize a song by extracting a compact representation from an arbitrary, say 5 second interval and comparing this fingerprint to a database of known fingerprints. For such a pattern recognition system to be of practical use, it should be scalable to at least one million songs or about 10 billions fingerprints (assuming about 10,000 fingerprints are allocated per song). This example clearly illustrate the need for scalable, similarity based, indexing system.

Audio and video compression algorithms like MPEG rely on searching a subframe buffer. Each time a new frame arrives for transmission, the buffer is searched for a similar subframe (within a tolerance). If a similar subframe has already been sent, then only the index of the similar subframe found is sent. Otherwise, the entire frame is added to the buffer.

Astronomy is another producer of huge databases. A KD-tree variation [12] was introduced to support range queries and solve nearest neighbour search for objects in a 19 millions entries star catalog.

Bioinformatics [17] requires also the manipulation of an astronomical number of data. DNA and protein sequences are the basic objects of computational biology. A fundamental problem in this domain is to find a given sequence of characters in a longer sequence. However, an exact match is unlikey to occur (because of minor differences in the DNA sequences due to typical variations). Computational biologists are therefore interested in finding parts of a sequence that are similar to a given short sequence (an edit distance can be used). Other related problems to tree indexing systems in bioinformatics are the building of phylogenetic trees, summarizing the evolution paths of species with hierarchical vector quantization systems.

A nearest neighbour classification scheme [4] is a simple and robust classifier that assigns to a new pattern p whose class is unknown, the class label of the nearest neighbour of p in a training set (set of class labelled patterns). As the training set may be large, fast seach algorithms are desirable (and necessary for real time systems).

1.3 Previous Work

There is a large literature on Nearest Neighbour Search. In this section, we describe some of the earlier contributions that are related to the tree indexing system that will be introduced in Section 1.5.

In [7], Fukanaga employs standard clustering techniques to build a space decomposition, but branch-and-bound the resulting data structure by exploiting the triangle inequality. The triangle inequality implies that

$$\text{if } |d(x, y) - d(z, y)| > r \text{ then } d(x, z) > r$$

In particular, if x is far from y and z is close to y, then x cannot be close to z. This simple observation eliminates distance computations by pruning regions of the input space represented with pivot points.

The KD-tree [5] has proved to be a useful tool in Euclidian spaces of moderate dimension. A KD-tree is built recursively by partitioning the input space with axis parallel hyperplanes. For a given coordinate, the input space is split at the median of the distribution generated by projecting the codebook (or training set) on that coordinate axis. The coordinate whose distribution exhibits the most spread is selected for the cut.

Like a KD-tree, a KDB-tree [14], partitions the input space with axis parallel hyperplanes, but a KDB-tree uses a height balanced tree similar to a B^+-tree. The distinctive feature of KDB-tree is the disjointness among subregions on a same tree level. However, this disjointness comes at a cost [10]; many leaf nodes are nearly empty. This reduces the performance of KDB-trees for nearest neighbour queries.

The R^*-tree [1] like its predecessor the R-tree [9] is another height balanced tree corresponding to a hierarchy of nested rectangles. In the R-tree and the R^*-tree, the regions associated with the nodes are determined by bounding rectangles that are allowed to overlap, whereas in the KDB-tree the regions are disjoint and defined by axis parallel hyperplanes. The X-tree [2] is an improvement of the R^*-tree that minimizes the overlap between bounding rectangles.

Like a KD-tree, a VP-tree [18] (Vantage Point tree) builds a binary tree recursively. The construction of a VP-tree starts by taking any element p as the root and determining the radius r of a sphere centered at p that partitions the training set evenly; about half of the training points should be inside the sphere and the rest outside the sphere. In other words, r is the median of the set of all distances of the training points to p. The training points that are inside the sphere are inserted in the left subtree of the root, whereas the training points that are outside the sphere are inserted in the right subtree of the root.

Unfortunately, none of these tree indexing systems are suitable for non-stationary codebooks.

1.4 Exact Binary Search Trees

In this section, we build a balanced search tree agreeing with the Voronoi diagram of a given codebook. That is, this search tree does return the nearest codebook vector to a new input vector. Before describing the construction of this search tree, we recall some basic properties of Voronoi diagrams.

The Voronoi diagram of a codebook divides the space into polyhedral cells centered on the codebook vectors. Let $X = \{x_1, x_2, \ldots, x_m\}$ be a codebook. The separating hyperplane $H_{i,j}$ that separates x_i from x_j with the maximal margin is characterized by the following equation

$$(x_j - x_i)^T (x - x_i) = \frac{1}{2} \|x_j - x_i\|$$

The hyperplane $H_{i,j}$ splits the space into two half-spaces

$$H_{i,j}^- = \left\{ x \in R^n \mid (x_j - x_i)^T (x - x_i) \leq \frac{1}{2} \|x_j - x_i\| \right\}$$

and

$$H_{i,j}^+ = \left\{ x \in R^n \mid (x_j - x_i)^T (x - x_i) > \frac{1}{2} \|x_j - x_i\| \right\}$$

We say that x_j lies in the positive half-space $H_{i,j}^+$ of $H_{i,j}$, and that x_i lies in the negative half-space $H_{i,j}^-$ of $H_{i,j}$. The Voronoi cell of x_i is the polyhedron (intersection of a finite number of half-spaces)

$$V_i = \bigcap_j H_{i,j}^-$$

The exact search tree is built recursively from the root. The root contains the hyperplane H that most evenly split the codebook X. Each internal node of the search tree determines on which side of its hyperplane is the input vector. Although the hyperplane H is chosen among the facets of the Voronoi cells, H most likely will cut some Voronoi cells. We define X_{pos} as the subset of vectors of the codebook X whose Voronoi cells intersect the positive half-space of H. Similarly, X_{neg} will denote the subset of vectors of X whose Voronoi cells intersect the negative half-space of H. In particular, if H cuts the Voronoi cell of a codebook vector x_k, then x_k belongs to both X_{pos} and X_{neg}.

The tree is recursively built on X_{neg} and X_{pos}. The recursive construction is stopped when the cardinal of X becomes small enough that a linear search is less expensive than a tree search.

The Matlab code for this algorithm is available at
`http://www.fit.qut.edu.au/ maire/FIT.html`

The expensive step of this algorithm is the determination of whether or not a hyperplane cuts a polyhedron. This test is equivalent to solving a linear programming (LP) problem. Because there are m^2 separating hyperplanes and m Voronoi cells, the number of LP problems solved is in $O(m^3)$. The time complexity of the building of the tree prevents the utilization of this algorithm on very large codebooks. However, the search time is linear in the depth of the tree. Another drawback of this algorithm is that it is not adaptive. If some vectors of the codebook are changed, the whole search tree has to be recomputed. This issue is addressed with our *Fast Indexing Tree* presented in next section.

Algorithm 1 T=BuildExactSearchTree(X)

Require: codebook $X = \{x_1, x_2, \ldots, x_m\}$

Ensure: T is a balanced search tree whose induced partition agrees with the Voronoi diagram of X

1: **if** $|X|$ is small **then**
2:     return X {A linear search is more efficient than a tree search for small codebooks}
3: **end if**
4: Determine all separating hyperplane $H_{i,j}$ between all pair of points (x_i, x_j) {This is done only once}
5: Find the hyperplane H, among all separating hyperplanes $H_{i,j}$, that most evenly splits the codebook X {That is $\left|H_{i,j}^-\right| / \left|H_{i,j}^+\right|$ closest to $1/2$}
6: Set the hyperplane H as the test of the root of T
7: Compute X_{neg} and X_{pos} given H
8: T_{neg}=BuildExactSearchTree(X_{neg})
9: T_{pos}=BuildExactSearchTree(X_{pos})
10: Attach T_{neg} and T_{pos} to the root of T

1.5 Fast Indexing Tree (FIT)

Binary Space Partition (BSP) trees are popular in computer graphics to perform hidden surface removal and shadow generation [6]. But, it is straightforward to generalize the idea to higher dimensional spaces. KD-trees and quad trees are in fact special cases of generalized BSP-trees. FIT, the tree indexing system presented in this section is related to BSP trees too, as its internal nodes contain hyperplane splits. In subsection 1.5.1, we describe the overall structure of FIT. In subsection 1.5.2, we show how to reduce the distortion error using a no-man's-land around the decision hyperplane. In subsection 1.5.3, we explain how the tree is kept balanced when codebook vectors are moved in the input space (non stationary input distribution).

1.5.1 Structure of the decision tree

Once abstracted into an interface, an indexing system can be viewed as a black box containing a set of vectors (the codebook) and providing access to them through its functional interface (*search*, *insert*, *delete* and *update*). Classical data structures like B-trees or AVL-trees implement such an interface for totally order sets. Although AVL-trees cannot be used to index data in R^n, as there does not exist a total order in R^n compatible with the arithmetic operators, AVL-trees have inspired the design of FIT with respect to the rebalancing of the tree. Two types of nodes are used in FIT. The internal nodes of the tree are *decision nodes* that contain a linear discriminant function and two region centres. The leaf nodes are *cluster nodes* that contains a small set of codebook vectors (this set is represented with a table, as below a critical number of elements a tree-search is more expensive than a tabular search).

The tree is built by splitting the input-space recursively into half-spaces. The linear discriminant function is chosen so that it distributes evenly the codebook vectors into the two half-spaces (maximizes the entropy). To find a good splitting hyperplane the principal component of the codebook (viewed as a set of vectors) is used to determine the weight vector of the linear discriminant function. Once the direction for the separating hyperplane is found, the hyperplane is shifted in such a way as to distribute evenly the codebook vectors on each side of the decision surface. The calculation of the principal component could be done by a PCA. But in FIT, a 2-means is performed, and the difference vector of the two means is used as a substitute to the principal component (2-means is computationally cheaper than a PCA). This process generates a shallow and balanced tree, guaranteeing an optimal average access time. Both codebook vector subsets created by this split operation are processed recursively until each subset becomes small enough to be placed in a cluster node. The structure of the tree is sketched in Figure 1.1.

Another way to construct the indexing-system is to build it incrementally. By starting with one empty leaf-node, and inserting the codebook vectors one at a time.

1.5.2 Fat Decision Hyperplane

Because the linear discriminant functions of FIT do not correspond to facets of the cells of the Voronoi diagram of the codebook, a simple binary search along the decision nodes may fail to identify the closest codebook vector. This is due to the fact that when reaching a leaf, the search for the closest codebook vectors is restricted to the cell of the partition containing the input vector (the partition of the vector space into polyhedral cells is induced by the hyperplanes of the decision nodes). As shown in Figure 1.1, the closest codebook vector might be contained in a neighbouring cell. To address this *near miss* problem, a *no-man's-land* slice around the separating hyperplane triggers a search in both half-spaces whenever the input vector falls into the no-man's-land. Otherwise, the search in only continued into the half-space that contains the input vector. Hyperplanes that have a non zero width no-man's-land are called *fat*.

The width of the no-man's-land slice around the separating hyperplane is defined relatively to the distance between the means of the codebook vectors stored in the two subtrees of the decision node. The ratio of the width of the no-man's-land over the distance between the means is a parameter of the indexing system. This user-defined parameter controls the trade-off between the accuracy and the speed of the system. The system is fastest when the no-man's-land has a width of 0 (no fat), but with this setting the system is more likely to return a codebook vector that is not the closest codebook vector to the input vector. At the other end of the spectrum, an infinitely wide no-man's-land forces the system to consider all codebook vectors. Therefore, the

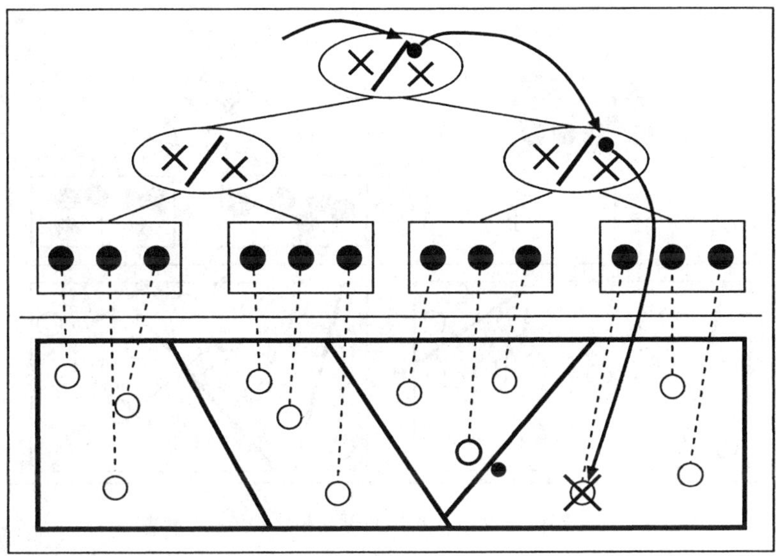

Fig. 1.1. Near miss in a search. The closest codebook vector is marked with a thick circle and the returned vector with a cross.

response in this case is much slower, but the system is then guaranteed to always return the true closest codebook vector. See Section 1.6 for experimental results and recommended magic numbers.

1.5.3 Re-balancing a FIT

To achieve near optimal results it is important to keep the tree as shallow and balanced as possible. This section describes scenarios in which the tree becomes unbalanced and shows ways to rebalance tree. First, we will define and discuss invalid cluster nodes. Then we will discuss the more complex problem of re-balancing whole subtrees.

After a sequence of insert operations, a cluster node (leaf) might become *invalid*, because the number of vectors stored in it exceeds its nominal capacity (*overflow*). Similarly, after a sequence of deletion operations, a leaf can become empty and *invalid* (*underflow*). Figure 1.2 shows an overflow situation and the corresponding repair action.

To solve an overflow problem, the system simply splits the invalid cluster node as described in Section 1.5.1. A leaf underflow problem is solved by replacing the redundant decision node with its non-empty child node (see Figure 1.3).

A decision-node is called *invalid* if the total number of codebook vectors stored in the two subtrees together is so small that all the vectors could be stored in a single leaf-node (*underflow* of the decision node). This problem

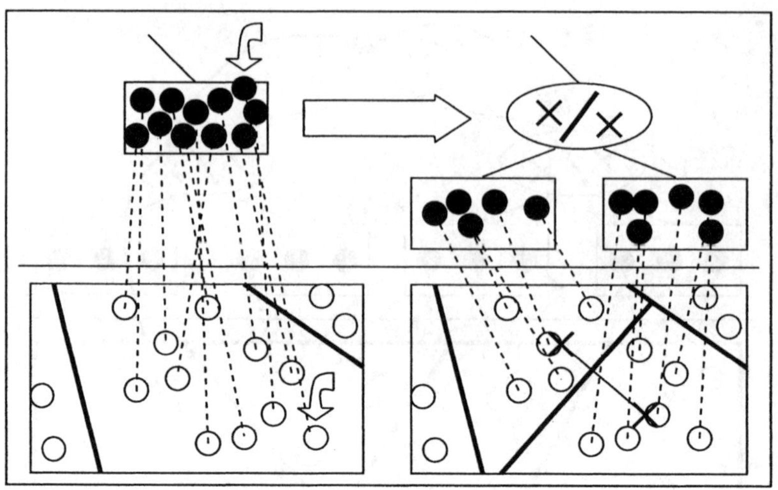

Fig. 1.2. Overflow of a leaf node.

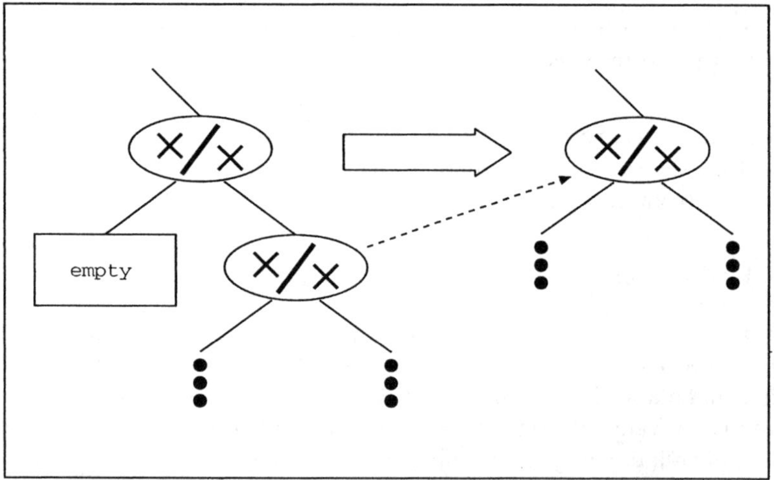

Fig. 1.3. Underflow of a decision node.

can be solved by replacing the decision node with a single cluster node as shown in Figure 1.4.

A decision-node is called *unbalanced* if the difference in height of its two subtrees exceeds one. To solve this re-balancing problem, FIT tries three different strategies (*delegation*, *migration* and *recreation*). These strategies are detailed below.

In some cases, re-balancing an unbalanced decision node N can be achieved by decreasing the height of the deeper subtree. Therefore, the system first

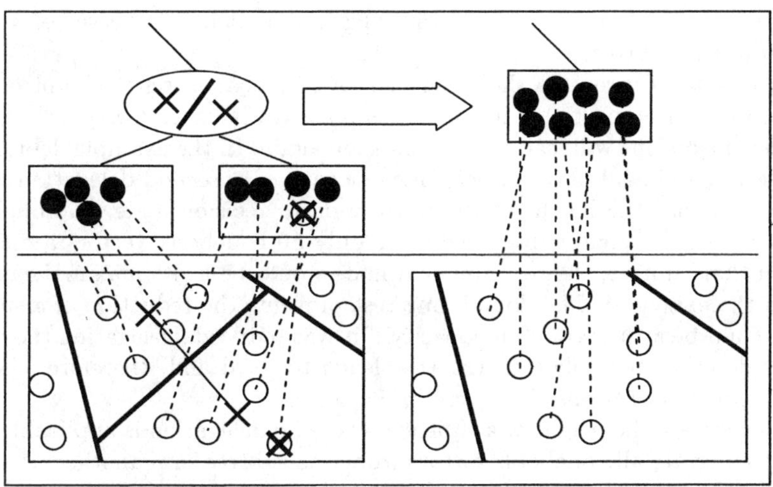

Fig. 1.4. Solving an underflow of a decision node.

attempts to delegate the reduction problem to the deeper subtree (see Figure 1.5).

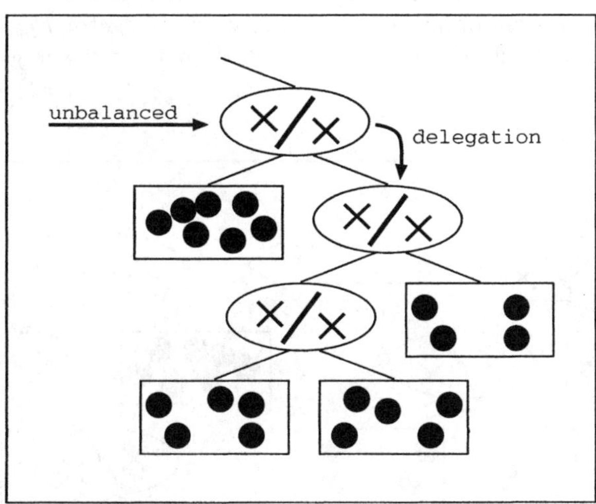

Fig. 1.5. Delegation of the rebalancing. The topmost node is unbalanced. The rebalancing is delegated to its right subtree, because $2^{2-1} \cdot 10 \cdot 0.8 > 14$. It is not delegated further, because $2^{1-1} \cdot 10 \cdot 0.8 \not> 10$.

The system first determines whether or not it is worth trying to decrease the height of a subtree. It is worthwhile, if $2^{d-1} \cdot C \cdot F_D > |V|$, where d denotes the depth of a subtree (a leaf's depth $= 0$), C the nominal capacity of a leaf-

159

node, F_D a factor preventing floundering and V denotes the set of vectors stored in the subtree.

The factor F prevents the delegation of the rebalancing to a subtree that is almost saturated. If this inequality is not satisfied, it is very likely that a few new insertions will lead to an expansion again. In the example depicted in Figure 1.5, FIT will perform only one delegation. A second delegation would try to decrease the height of the bottom-most decision node. Although the reduction of the node is possible, since only 10 codebook vectors are stored within that subtree, the resulting leaf node would be at its maximal capacity (10 in the example). The floundering test prevents the reduction of a subtree that would be at (almost) full capacity if it was reduced. Delegating the repair as far down as possible into the tree leads to local and, therefore, cheaper maintenance operations.

To decrease the height of a subtree, where no delegation is applicable, FIT tries to migrate all codebook vectors from one subtree into another. If all the vectors of one subtree can be added to its sibling-tree, the parent decision node of the subtrees is redundant and can be removed. This process is shown in Figure 1.6.

To determine whether the process is likely to succeed, the system compares the number of codebook vectors to migrate and the available capacity of the other subtree. If the capacity is larger, it tries the migration. Unfortunately, it is not guaranteed that the process does not lead to a split in the target subtree, which could lead to an infinite loop of migrations. A factor F_M, similar to the factor F_D, prevents floundering, i.e. the system tries the migration, only if $2^{d-1} \cdot C \cdot F_M > |V|$.

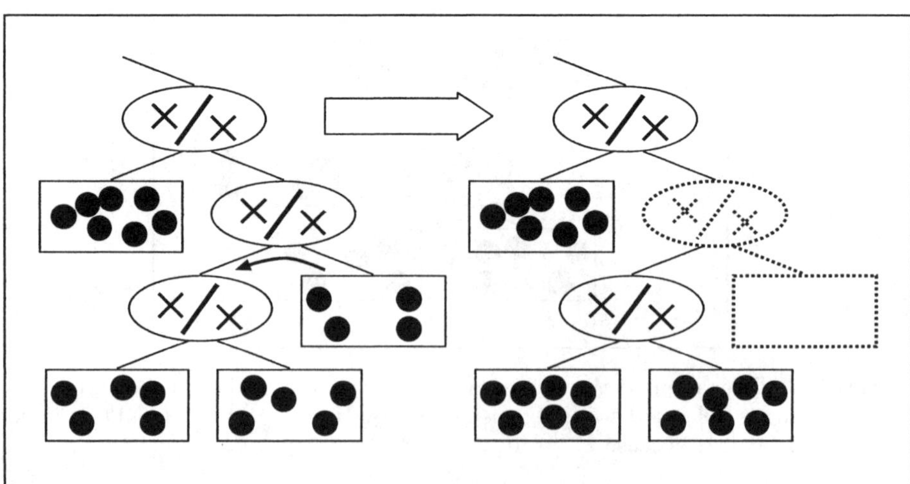

Fig. 1.6. The migration of the codebook vectors from the 2nd level node leads to an empty leaf and therefore a redundant decision-node, which can be removed.

If neither delegation nor migration operations are applicable, the FIT system re-creates a whole subtree from scratch. This is done as described in Section 1.5.1. The recreation is not performed with the subtree which is unbalanced, but with the tree subject of the repair delegation. This strategy favours local repairs as much as possible and keeps the maintenance costs as low as possible.

1.6 Experimental Results

This section presents experimental results that demonstrate the usefulness of FIT. The experiments were done using an implementation of FIT in Java. All the experiments were repeated at least 20 times to obtain statistically significant results.

There are two different ways to assess the performance of an indexing system like FIT. One can evaluate the speed of the codebook by itself (search running time). One can also compare the running time and distortion error of a client algorithm using an array indexing system against a client algorithm using FIT.

1.6.1 Benchmarking experiments

We compared FIT with TSVQ, K-tree and k-means by running the same experiments as those described in [8].

A data set of 3924 vectors of dimension 20 were quantized using the different quantizers. This data set comes from a speech spectrum benchmark problem [13]. Figure 4 shows the distortion errors of the different systems. FIT outperforms all the algorithms except the array implementation of k-means (which is optimal).

1.6.2 Time Complexity

We tested how the speed of the search operator scales with the codebook size. As expected, a logarithmic time complexity was observed for FIT (thanks to the tree structure), and a linear complexity was observed for the array indexing system. The results of these experiments are shown in figure 1.8.

We tested also the incidence of FIT on the speed of the client algorithm running time. LBG was used as the client algorithm. Figure 1.9 shows the results of these experiments.

1.6.3 Incidence of the No-Man's-Land Width Ratio on the Distortion Error

As mentioned in Section 1.5, the distortion error of the client algorithm is affected by the indexing algorithm used. Indeed, it might be the case that the

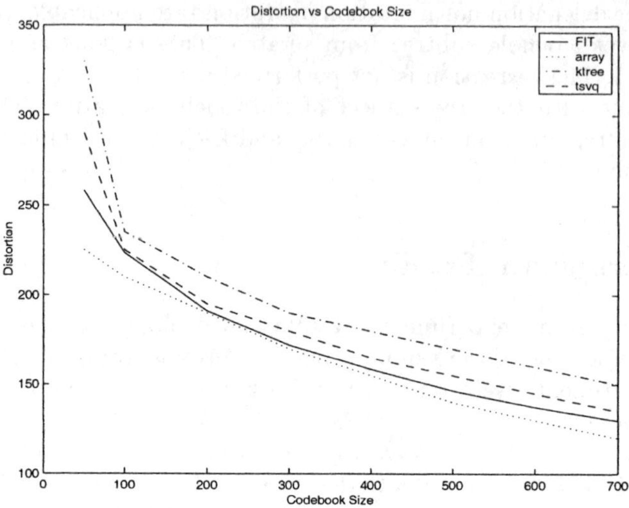

Fig. 1.7. Comparison of distortion errors

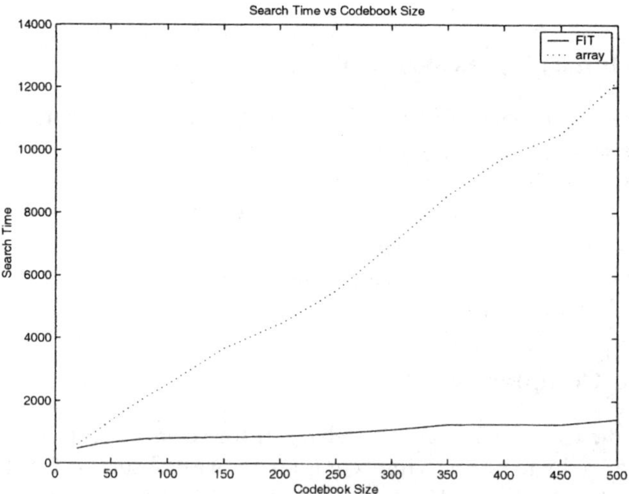

Fig. 1.8. Speed of a nearest neighbour search versus codebook size.

indexing system does not return the codebook vectors which are the closest to a given input (the indexing system might return codebook vectors that are relatively close, but not the closest). The theoretically best result that can be achieved is obtained with a full tabular search.

In the case of FIT, the negative effect of the tree indexing can be controlled via the No-Man's-Land Width Ratio.

162

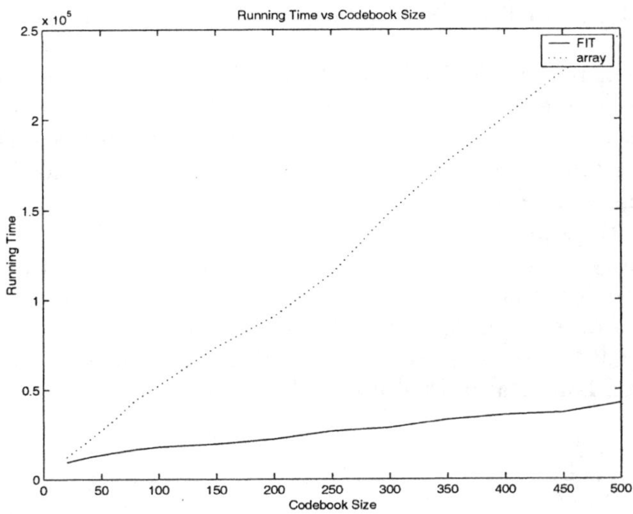

Fig. 1.9. Running time of a client-algorithm versus codebook size.

To measure the incidence of No-Man's-Land Width Ratio on the distortion error, we varied the No-Man's-Land Width Ration from 0.0 to 1.0. The experiment showed that the error distortion initially decreases rapidly with No-Man's-Land Width Ratio. But the gain becomes much smaller if this parameter is set to a value greater than 0.4. The results can be seen in Figure 1.10.

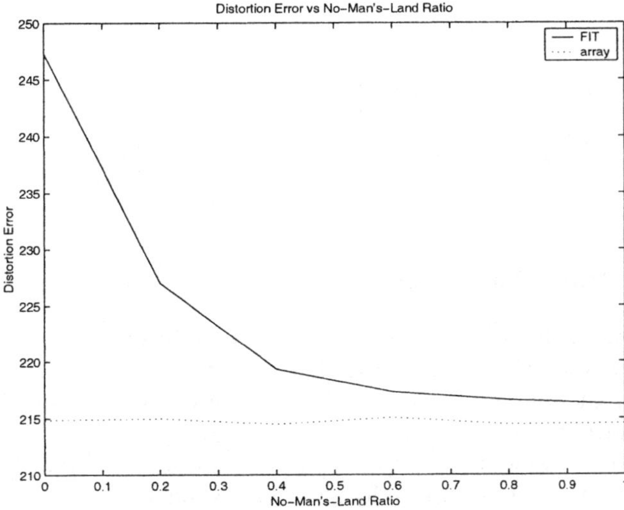

Fig. 1.10. Distortion error versus the No-Man's-Land Width Ratio.

1.6.4 Incidence of the No-Man's-Land Ratio on the Running Time

The cost of increasing the No-Man's-Land Width Ratio is a deterioration in performance with respect to the running time. The user has to find a tradeoff between speed and accuracy. The previous experiment showed that 0.4 was a critical value (almost a turning point) for the No-Man's-Land Width Ratio, as above this value the relative improvement in the distortion will be more expensive with respect to running time.

To quantify the incidence of the No-Man's-Land Width Ratio on the running time, we performed some experiments using a fixed number of codebook vectors and a fixed set of input vectors, and varied the No-Man's-Land Width Ratio from 0.0 to 1.0. We observe that the running time grows linearly with the No-Man's-Land Ratio. Figure 1.11 shows the results of this experiment.

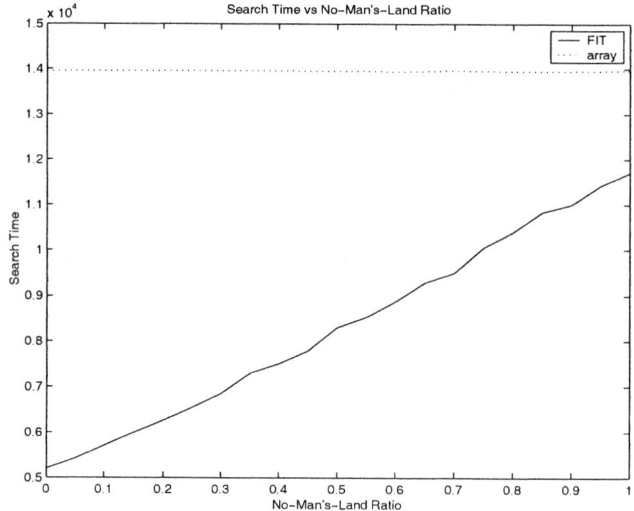

Fig. 1.11. The running time depends linearly on the No-Man's-Land Ratio.

1.6.5 The Other Parameters

The experiments described above were done with fixed values for the parameters used for the rebalancing algorithms. That is the parameters which are used to prevent floundering during deletion, insertion and rebalancing were kept constant. They were all set to a value of 0.8.

During other exploratory experiments, it was observed that these parameters do not have a significant influence. Only when set to extreme values like 0 or 1, do the results differ significantly.

1.7 Conclusions

As shown in Section 1.6, FIT is a fast indexing system. Moreover, FIT is adaptive and can be made as accurate as needed. It is its adaptiveness that makes FIT particularly interesting for the machine learning community.

From a software engineering point of view, FIT was designed as a component that can be used as a plug-in for any algorithm working with codebooks.

The five parameters of the FIT system allow the user to fine-tune the system. However, the default values (that were used in the experiments reported here) can be considered as universal constants, as the performance of the system is not overly sensitive to these parameters.

The development of more sophisticated rebalancing algorithms that avoid subtree re-creation would further improve the system.

A Java implementation of FIT is available at
`http://www.louise15.de/borstel/ents.html`

References

1. N. Beckmann, H. Kriegel, R. Schneider, and B. Seeger. R*-tree: an efficient and robust access methods for points and rectangles. In *Proc. ACM SIGMOD*, pages 322–331, 1990.
2. S. Berchtold, D. Keim, and H. Kriegel. X-tree: an index structure for high-dimensional data. In *Proc. of the 22nd VLDB Conference*, Mumbai (Bombay), India, 1996.
3. E. Cahvez, G. Navarro, R. Baeza-yates, and J. Marroquin. Searching in metric spaces. *ACM Computing Surveys*, 33(3):273–321, 2001.
4. R. Duda, P. Hart, and G. Stork. *Pattern Classification (2nd Edition)*. Wiley-Interscience, London, 2000. ISBN: 0471056693.
5. J. Friedman, F. Baskett, and L. Shusteck. An algorithm to find nearest neighbours. *IEEE Transactions on computers*, 1975.
6. H. Fuchs, Z. Kedem, and B. Naylor. on visible surface generation by a priori tree structures. In *Proc. of SIGGRAPH'80*, pages 124–133, 1980.
7. K. Fukanaga. A branch and bound algorithm for computing k-nearest neighbours. *IEEE Transactions on computers*, 1975.
8. S. Geva. K-tree: A height balanced tree structured vector quantizer. In *IEEE Neural Network for Signal Processing Workshop, Sydney, Australia*, 2000.
9. A. Guttman. R-trees: A dynamic index structure for spatial searching, 1984.
10. N. Katayama and S. Satoh. The sr-tree: An index structure for high-dimensional nearest neighbor queries. In *Proc. ACM SIGMOD*, pages 13–15, 1997.
11. M. Miller, M. Rodriguez, and J. Cox. Audio fingerprinting: Nearest neigbhor search in high dimensional binary space. In *IEEE Multimedia Signal Processing Workshop, St. Thomas, US Virgin Islands*, 2002.
12. D. Nguyen, K. DuPrie, and P. Zographou. A multidimensional binary search tree for star catalog correslations. In *Astronomical Data Analysis software and Systems VII ASP Conference Series, Vol 145*, 1998.
13. Helsinki University of Technology. Lvq_pak. `http://www.cis.hut.fi/research/som-research/nnrc-programs.shtml`.

14. J. Robinson. The k-d-b-tree: a search structure for large multidimensional dynamic indexes. In *Proc. ACM SIGMOD*, pages 10–18, 1981.

15. T. Sellis, N. Roussopoulos, and C. Faloutsos. Tree: A dynamic index for multidimensional objects, 1988.

16. C. Mohan V. Burzevski. Hierarchical growing cell structures. Technical report, Syracuse University, 1996. `ftp://www.cis.syr.edu/users/mohan/papers/higs-tr.ps`.

17. M. Waterman. *Introduction to Computational Biology*. Chapman and Hall, London, 1995.

18. P. Yianilos. Data structures and algorithms for nearest neighbor search in general metric spaces. In *Proc. of the Fourth ACM-SIAM Symposium on dDiscrete Algorithms (SODA'93)*, pages 311–321, 1993.

On Some External Characteristics of Brain-like Learning and Some Logical Flaws of Connectionism

Asim Roy

School of Information Systems, Arizona State University, Tempe, AZ 85287-4606, USA

Abstract. This paper attempts to define some common, externally observed, properties of human learning, properties that are common to all types of human learning. It is expected that any theory/conjecture about the internal learning mechanisms of the brain should account for these common external properties. Characterization of an autonomous learning system such as the brain has been one of the "large" missing pieces in connectionism and other brain-related sciences. The external characteristics of learning algorithms have never been defined in these fields. They largely pursued algorithm development from an "internal mechanisms" point of view. This paper is an attempt to rectify that situation. This paper also argues that some of the ideas of connectionism are not only logical flawed, but also are inconsistent with some commonly observed human learning behavior. The paper does not present any new learning algorithms, but it is about learning algorithms and what properties they should exhibit.

Keywords. Connectionism, artificial neural networks, brain-like learning algorithms, properties of brain-like learning.

1   Introduction

One of the "large" missing pieces in the existing theories of connectionism and artificial neural networks is the definition or characterization of an autonomous learning system such as the brain. Although Hebb, Rumelhart and others (Amari [1988], Churchland and Sejnowski [1992], Fahlman and Hinton [1987], Feldman and Ballard [1982], Grossberg [1982, 1988], Hebb [1949], Kohonen [1993], Rumelhart and McClelland [1986], Smolensky [1989]) have proposed various theories about the "internal mechanisms" of the brain, there has been no corresponding attempt by connectionists to define the external behavioral characteristics that these "internal" mechanisms are supposed to exhibit. As a result, these fields have largely pursued algorithm development from an "internal mechanisms" point of view (e.g. local, autonomous learning by individual neurons in a network of neurons; memoryless, instantaneous learning by such a network) rather than from the point of view of "external behavioral characteristics" of human learning. This paper is an attempt to remedy that flaw and define some common external characteristics of human learning that all learning algorithms must demonstrate during learning. It tries to define external learning characteristics that are "independent of": (1) any conjectures about the "internal" mechanisms of the brain, and (2) the specific learning problem (function approximation, pattern classification, object recognition, learning of grammar, learning to reason and so on).

The main motivation behind proposing a set of external learning properties is to facilitate the development of future learning algorithms, so that future robots (hardware, software) can learn in a manner similar to humans. In addition, these external properties would also allow the field to test the validity of alternative conjectures/ideas/theories about the internal mechanisms of the brain. If a certain conjecture about the internal mechanisms of the brain cannot demonstrate these external properties, then that conjecture cannot be valid. So defining the external characteristics of brain-like learning mechanisms might be quite helpful in searching for the proper internal mechanisms of the brain. This paper does not present any new learning algorithms, but it is about learning algorithms and how they should be constructed and what properties they should exhibit. This paper also points out some major logical flaws of connectionism.

In developing brain-like learning algorithms, a comparison with system identification in science and engineering may also be in order. In system identification, the basic idea is to construct an equivalent system (model) that can produce "behavior" that is similar to the actual system. So the key idea is to produce "matching external behavior." The equivalent system may or may not necessarily match the internal details of the system to be identified. So one's constructed model of the system is allowed to deviate from the internals of the system as long as it matches its external properties. And the external properties to match may be "many." This is not to say that one should not take advantage of any information about the internals of the system.

The paper is organized as follows. Section 2 reviews the basic connectionist framework as a starting point for discussion on connectionism and artificial neural networks. Section 3 identifies some common external learning characteristics of the brain that any equivalent learning system should match. Section 4 discusses some major flaws in the connectionist framework. This is followed by a conclusion in section 5.

2   The Connectionist Framework

This section provides a brief overview of some of the basic ideas of connectionism. The following connectionist framework (from "The Architecture of Mind: A Connectionist Approach" by David Rumelhart in John Haugeland's (ed.) MIND DESIGN II [1997]), as defined by the elements below, will be used in this paper for further discussions on connectionism: 1) a set of processing units; 2) a state of activation for each unit; 3) an output function for each unit; 4) a pattern of connectivity among units; 5) an activation rule for combining the inputs impinging on a unit; 6) a learning rule whereby patterns of connectivity are modified by experience; 7) an environment within which the system must operate. As is well-known, the following additional or related notions are implied here: (a) memoryless learning, (b) instantaneous learning, (c) autonomous local learning by each processing unit, (d) predesigned nets, and (e) no controllers in the system (that is, there is no subsystem controlling other subsystems in this system). Here is a John Haugeland characterization of connectionist networks (from "What is Mind Design" by John Haugeland in John Haugeland's (ed) MIND DESIGN II [1997]): "There is no central processor or controller, and also no separate memory or storage mechanism." Connectionism and artificial neural networks are based on this conjecture about the internal mechanisms of the brain, as defined by the elements above.

In this framework, the brain is assumed to be a massively parallel computing system. Each computing element (a neuron or a brain cell) in this massively parallel system is envisioned to perform a very simple computation, such as $y_i = f(z_i)$, where z_i is assumed to be a real valued input to the ith neuron, y_i is either a binary or a real valued output of the ith neuron, and f a nonlinear function. The nonlinear function f, also called a node function, takes different forms in different models of the neuron; a typical choice for the node function is a step function or a sigmoid function. The neurons get their input signals from other neurons or from external sources such as various organs of the body. The output signal from a neuron may be sent to other neurons or to another organ of the body.

Let $\mathbf{x}_i = (x_{i1}, x_{i2}, \ldots , x_{in})$ be the vector of input signals to the ith neuron, the inputs signals being from other neurons in the network or from external sources. Many neural network models assume that each input signal x_{ij} to the ith neuron is "weighted" by the strength of the ith neuron's connection to the jth source, w_{ij}. The weighted inputs, $w_{ij} x_{ij}$, are then summed in these models to form the actual input z_i to the node function f at the ith neuron: $z_i = \Sigma\ w_{ij} x_{ij} + \theta_i$, where θ_i is a constant, called the threshold value.

2.1   Designing and training networks

A network of neurons is made to perform a certain task by designing and training an appropriate network through the process of learning. The design of a network generally involves determining: (a) the number of layers in the network, (b) the number of neurons in each layer, (c) the connectivity pattern between the layers and the neurons, (d) the node function to use at each neuron, and (e) the mode of operation of the network (e.g. feedback vs. feedforward). The training of a network involves determining the connection weights $[w_{ij}]$ and the threshold values [θ_i] from a set of training examples. For some learning algorithms like back-propagation (Rumelhart et. al. [1986]) and self-organizing maps (Kohonen [1993]), the design of the network is provided by the user or by some other external source. For other algorithms like Adaptive Resonance Theory (Grossberg [1982, 1988]), reduced coulomb energy (Reilly et. al. [1982]), and radial basis function networks (Moody and Darken [1989]), the design of the network is accomplished by the algorithm itself. However, other parameter values have to be externally supplied to these latter algorithms on a trial and error basis in order for them to perform the design task satisfactorily.

A network is trained to perform a certain task by adjusting the connection weights [w_{ij}] by means of a local learning rule or law. A local learning law is a means of gradually changing the connection weights [w_{ij}] by an amount Δw_{ij} after observing each training example. A learning law is generally based on the idea that a network is supposed to perform a certain task and that the weights have to be set such that the error in the performance of that task is minimized. A learning law is local because it is conceived that the individual neurons in a network are the ones making the changes to their connection weights or connection strengths, based on the error in their performance. Local learning laws are a direct descendent of the idea that the cells or neurons in a brain are autonomous learners. The idea of "autonomous learners" is derived, in turn, from the notion that there is no homunculus or "a little man" inside the brain that "guides and controls" the behavior of different cells in the brain (Kenny [1971]). This "no homunculus" argument says that there couldn't be a distinct and separate physical entity in the brain that governs the behavior of other cells in the brain. In other words, as the argument goes, there are no "ghosts" in the brain. So any notion of "extracellular control" of synaptic modification (connection weight changes) is not acceptable to this framework.

So, under the connectionist theory of learning, the connection weight $w_{ij}(t)$, after observing the tth training example, is given by: $w_{ij}(t) = w_{ij}(t-1) + \Delta w_{ij}(t)$, where $\Delta w_{ij}(t)$ is the weight adjustment after the tth example is observed and the adjustment is determined by the local learning law being used. Much of the current research on artificial neural networks is on developing new or modifying old local learning laws (training algorithms). There are now hundreds of local learning laws (training algorithms), but the most well-known among them are back-propagation (Rumelhart et al. [1986]), ART (Grossberg [1982, 1988]), SOM (Kohonen [1993]), and RBF networks (Moody and Darken [1989]) and their different variations. To give an example, the back propagation learning law is as follows: $\Delta w_{ij}(t) = - \eta(\partial E/\partial\, w_{ij}(t)) + \alpha\Delta w_{ij}(t-1)$. Here η is the learning rate (step size) for the weight update at step t (after observing the tth training example) and α is a momentum gain term. E is the mean-square error of the whole network based on some desired outputs, in a supervised mode of learning, where a teacher is present to indicate to the network what the correct output should be for any given input.

3 On some external properties of the brain as a learning system

Any learning system that claims a similarity to the brain has to account for and exhibit a broad set of external properties that characterizes human learning. Understanding and characterizing the phenomenon to be modeled and explained is clearly the first step towards developing a theory for it. If that is not done, it is very likely that wrong theories will be proposed, since it is not known exactly what the theory should account for.

The attempt here is to state some general properties of the brain as a learning system. That is, properties that are independent of a specific learning situation like learning a language, mathematics or a motor skill. There has been no prior attempt to do so in the connectionist literature (Amari [1988], Churchland and Sejnowski [1992], Fahlman and Hinton [1987], Feldman and Ballard [1982], Grossberg [1982, 1988], Hebb [1949], Kohonen [1993], Rumelhart and McClelland [1986], Smolensky [1989]). These properties are classified as external behavioral properties because they can be verified from external observations of humans as a learning system.

3.1 On the recall property of the brain

First, it is a well-observed fact that humans collect and store information in order to learn from it. It is part of everyone's learning experience. That does not mean that humans store any and all information provided to them; they are definitely selective and parsimonious in the choice of information/facts to collect and store. But a very important characteristic of the brain is the ability to recall and present information that has been collected for the purpose of learning. So any brain-like learning system should exhibit this recall property of the brain if it is to be consistent with human learning behavior. As is argued in section 3, connectionist learning systems are incapable of demonstrating this recall property.

3.2 On the ability to generalize from examples

Second, learning of rules from examples involves generalization. Generalization implies the ability to derive a succinct description of a phenomenon, using a simple set of rules or statements, from a set of observations of the phenomenon. So, in this sense, the simpler the derived description of the phenomenon, the better is the generalization. For example, Einstein's $E = MC^2$ is a superbly succinct generalization of a natural phenomenon. And this is the essence of learning from examples. So any brain-like learning system must exhibit this property of the brain - the ability to generalize. That is, it must demonstrate through its equivalent computational model or algorithm that it makes an explicit attempt to generalize and learn. In order to generalize, the learning system must have the ability to design a network on its own. As noted in section 3, many connectionist learning systems, however, depend on external sources to provide the network design to them (Amari [1988], Churchland and Sejnowski [1992], Fahlman and Hinton [1987], Feldman and Ballard [1982], Kohonen [1993], Rumelhart and McClelland [1986], Smolensky [1989]); hence they are inherently incapable of generalizing "without external assistance." This implies again that connectionist learning is not brain-like at all.

3.3   On the ability to learn quickly from a few examples

Third, learning from examples involves collecting and storing examples to learn from. Humans exhibit a wide range of behavior on this aspect of learning. Some can learn very quickly from only a few examples. Others need far too many examples to achieve the same level of learning. But "learning quickly from only a few examples" is definitely the desired characteristic to emulate in any brain-like learning system. So any such system should demonstrate the ability to learn quickly from only a few examples. The so-called "memoryless learning" of connectionism, where no storage of facts and information is allowed, has been shown by (Roy, Govil & Miranda [1995, 1997]) to be extremely slow and time-consuming. This is because it requires many more training examples (by several orders of magnitude more) compared to methods that use memory to store the training examples.

3.4   On the ability to construct and test new solutions without total re-supply of previously provided information

Fourth, normal human learning includes processes such as (1) collection and storage of information about a problem, (2) examination of the information at hand to determine the complexity of the problem, (3) development of trial solutions to the problem, (4) testing of trial solutions, (5) discarding such trial solutions if they are not good enough, and (6) repetition of these processes until an acceptable solution is found. Note that these learning processes can exist only if there is storage of information about the problem. Although these processes are internal to the brain, humans can readily externalize the results of these processes as and when required. Hence these learning processes are not mere conjectures; they are easily verifiable through externalization by humans. Hence, any brain-like learning system should also demonstrate these abilities. It is essentially the ability to construct and verify "internally" a solution to the problem from the available information. It is also the property whereby the learning system can start from scratch to build a new solution without requiring a total re-supply of previously provided information, although new information can be provided if needed. As discussed in section 3, connectionism cannot demonstrate any of these learning characteristics; "memoryless learning" requires a complete re-supply of previously provided information every time a new solution has to be constructed.

3.5 Restrictions on the information that can be supplied to the system

Fifth, humans, in general, are able to acquire a great deal of information about a learning problem from external sources. The information usually acquired by them includes examples or cases of the problem, cues about what features of the problem are important, relationship to other knowledge already acquired and so on. But, on the other hand, humans have no external control of the learning processes inside the brain. For example, one cannot provide a network design to the brain. Nor can one set the parameters of its "learning algorithm." Hence any brain-like learning system should also be restricted to accepting only the kinds of information that are normally supplied externally to a human. Thus it should not obtain any information that pertains to the internal control of the learning algorithm, since humans cannot externally control the learning processes inside the brain. This restriction, by the way, is quite severe; perhaps none of the connectionist learning algorithms can satisfy this requirement because they obtain quite a bit of algorithmic control information from external sources. So, in summary, this requirement states that a brain-like learning system should not receive any information that is not provided to the human brain from its external environment.

4 On some logical flaws of connectionism

This section looks at some of the basic ideas of connectionism and shows that they are logically flawed. Some of the ideas are also inconsistent with other ideas of connectionism and, therefore, violate their own basic principles.

4. 1 Autonomous Local Learning

First, the notion that each neuron or cell in the brain is an "autonomous/independent learner" is one of the fundamental notions of connectionism. Under this notion, it is construed that only the individual neurons or cells themselves can "decide" how to modify their synaptic strengths (connection weights), based on error signals or any other information provided to them from external sources within the brain. In other words, in the connectionist framework, this "adjustment decision" cannot be conveyed to the neuron from outside, by an outside source, although "any other" type of information can be supplied to it from outside. What all this means is that there is some restriction on the kind of information that can be supplied to a neuron or cell; in particular, the cell's operating properties cannot be adjusted by means of signals coming from outside the cell. In other words, this implies that no other physical entity external to a neuron is allowed to "signal" it directly to adjust its synaptic strengths, although other kinds of signals can be sent to it. All of the well-known local learning laws (learning algorithms) (Amari [1988], Churchland and Sejnowski [1992], Fahlman and Hinton [1987], Feldman and Ballard [1982], Grossberg [1982, 1988], Kohonen [1993], Moody and Darken [1989], Reilly, Cooper and Elbaum, [1982], Rumelhart and McClelland [1986], Smolensky [1989]) developed to date quite faithfully adhere to this notion, although this notion is logically quite problematic. The references here are only a sample of the large body of learning algorithms in this field. No attempt has been made to exhaustively list all such algorithms or their variations, since they all are based on the same core notion of autonomous/independent learners.

Here is the problem with this notion. Strict autonomous local learning implies "pre-definition" of a network "by the learning system" without having seen a single training example and without having any knowledge at all of the complexity of the problem. There is no system, biological or otherwise, that can do that in a meaningful way; it is not a "feasible idea" for any system. There is no way any biological or man-made system can magically design a network and start learning in that network without knowing anything about what is there to learn. The other fallacy of the autonomous local learning idea is that it acknowledges the existence of a "master system" that provides the network design and adjusts the learning parameters so that autonomous learners can learn. So connectionism's autonomous learners, in the end, are directed and controlled by other sources. In summary, the idea of autonomous local learning in connectionism is logically flawed and incompatible with other ideas of connectionism.

The notion of "autonomous/independent learners" is not only problematic from a logical point of view, but is also inconsistent with some recent findings in neuroscience; there is actually no neurobiological evidence to support this notion. Connectionism had pursued this idea of autonomous/independent learners (local learning rules that is) for a number of reasons. One such reason might have been that there was no clear evidence for the sources of and the pathways by which additional signals could influence a synapse. But now there is clear evidence from neuroscience for the "different pathways" by which "additional signals" could influence synaptic adjustments directly. There is a growing body of evidence that shows that the neuromodulator/neurotransmitter system of the brain controls synaptic adjustments within the brain (Hasselmo [1995], Kandel et al. [1993], Hestenes [1998] and others). It shows that there are many different neuromodulators and neurotransmitters and many different cellular pathways for them to affect cellular changes. Cellular mechanisms

174

within the cell are used to convert these "extracellular" signals into long-lasting changes in cellular properties. Thus the connectionist notion of "autonomous/independent learners" (local decision-making embodied in local learning rules of the neurons) is not consistent with these recent findings in neuroscience. So the connectionist conjecture that no other physical entity in the brain can directly signal "changes" to a cell's behavior is a major misconception about the brain. And this conjecture is also one of the backbones behind the connectionist notion of "no controllers" in the system.

4.2   Memoryless learning

Second, under connectionism, brain-like learning systems cannot record and store any training examples explicitly in its memory - in some kind of working memory of the learning system, that is, so that it can readily access those examples in the future, if needed, in order to learn. The learning mechanism can use any particular training example presented to it to adjust whatever network it is learning in, but must forget that example before examining others. That's how all connectionist learning rules operate – they are designed to adjust a network from a single instance. The learning rules are not designed to examine more than one example at a time. This is the so-called "memoryless learning" of connectionism, where no storage of facts/information is allowed. The idea is to obviate the need for large amounts of memory to store a large number of training examples or other information. Although this process of learning is very memory efficient, it can be very slow and time-consuming, requiring lots of training examples to learn from, as shown in (Roy, Govil & Miranda [1995, 1997]).

However, the major problem with this notion of memoryless learning is that it is completely inconsistent with the way humans learn; it completely violates very basic behavioral facts about human learning. Remembering relevant facts and examples is very much a part of the human learning process; it facilitates mental examination of facts and information that is the basis for all human learning. And in order to examine facts and information and learn from them, humans remember things.

There is extensive evidence in the experimental psychology literature on the use of memory in learning. In many psychological experiments, memorization actually precedes learning of rules. In fact, there is so much evidence for instance memorization in experimental psychology that it produced the instance theory of learning. From Shanks [1995], p. 81: "On the basis of the overwhelming evidence that instance memorization plays a role in category learning, Medin and Schaffer (1978) proposed that a significant component of the mental representation of a category is simply a set of stored exemplars or instances. The mental representation of a category such as *bird* includes representations of the specific instances belonging to that category, each presumably connected to the label *bird*. In a concept learning experiment, the training instances are encoded along with their category assignment." So even though collecting and storing relevant information is an important part of the human learning process, there is no provision for such a learning process in the learning theories of connectionism and artificial neural networks.

There are other logical problems with the idea of memoryless learning. First, one cannot learn (generalize, that is) unless one knows what is there to learn (generalize). And one can know what is there to learn "only by" collecting and storing some information about the problem at hand. In other words, no system, biological or otherwise, can "prepare" itself to learn without having some information about what is there to learn (generalize). And in order to generalize well, one has to look at a whole body of information relevant to the problem, not just disconnected bits and pieces of information (presented one at a time), as postulated in memoryless learning. Moreover, memoryless connectionist systems would indeed be magicians if they can perform the following tasks without knowing anything about the problem and without having seen a "single" training example before the start of learning: (1) determine how many inputs and outputs there will be in the network, (2) determine what the network design should be (how many layers, how many nodes per layer, their connectivity and so on), and (3) set the learning parameters for the learning to take place. Again, there is no system, biological or otherwise, that can do that. So the notion of "memoryless learning" is a very serious misconception in these fields - it is not only inconsistent with the way humans learn, but is also illogical.

Connectionist and artificial neural network learning systems themselves provide the best evidence that memory is indeed used in learning. Except for some simple cases of learning, these learning systems rely on a trial-and-error process where either a human or a computer program changes the learning parameters in order for the system to learn properly. Such a trial-and-error process implies that the training examples are available to the learning system in some kind of memory store for reuse. If it is to be believed that such trial-and-error learning systems are actually used by the brain, then the brain must also have a corresponding memory store for the training examples.

4.3   The networks are predesigned and externally supplied to the system; and so are the learning parameters

A third major flaw in connectionism is the requirement that network designs and other algorithmic information often have to be externally supplied to some of their learning systems, whereas no such information is observed to be an external input to the human brain. The well-known back-propagation algorithm of Rumelhart et al. [1986] is a case in point. In fact, often many different network designs and other parameter values have to be supplied to these learning systems on a trial and error basis in order for them to learn (Fahlman and Hinton [1987], Feldman and Ballard [1982], Grossberg [1982, 1988], Kohonen [1993], Moody and Darken [1989], Reilly, Cooper and Elbaum, [1982], Rumelhart and McClelland [1986], Smolensky [1989]). However, as far as is known, no one has been able to externally supply any network designs or learning parameter information to a human brain. Plus, the idea of "instantaneous and memoryless learning" is completely inconsistent with these trial and error learning processes; there is supposed to be no storage of learning examples in these systems for such a "trial and error process" to take place. In other words, no such trial and error process can take place unless there is memory in the system, which connectionism disallows. So connectionism violates its own basic principles in the actual operation of its learning systems.

Furthermore, in order for humans to generalize well in a learning situation, the brain has to be able to design different networks for different problems - different number of layers, number of neurons per layer, connection weights and so on - and adjust its own learning parameters. The networks required for different problems are different; it is not a "same size fits all" type of situation. So the networks cannot come "pre-designed" in the brain; they cannot be inherited for every possible "unknown" learning problem faced by the brain on a regular basis. So, in general, for previously unknown problems, the networks could not feasibly come pre-designed to us. Since no information about the design of the network is ever supplied to the brain externally, it therefore implies that the brain performs network design internally. Thus, it is expected that any brain-like learning system must also demonstrate the same ability to design networks and adjust its own learning parameters without any outside assistance. But most of the connectionist learning systems can't demonstrate this capability and that is a problem with their systems.

5 Conclusions

Definition or characterization of an autonomous learning system such as the brain has been one of the "large" missing pieces in connectionism and artificial neural networks. The external behavioral characteristics of learning algorithms have never been defined in these fields. The fields largely pursued algorithm development from an "internal mechanisms" point of view rather than from the point of view of "external behavior or characteristics" of the resulting algorithms. This paper is an attempt to rectify that situation. This paper has suggested some common external characteristics of human learning that all brain-like learning systems should exhibit.

This paper has also tried to point out some problems with some of the basic ideas of connectionism and artificial neural networks. Some of the connectionist ideas have been shown to have logical flaws in them, while others are inconsistent with some commonly observed human learning processes and are even in conflict with other connectionist ideas. The notions of external signal and control inherent in neuromodulation and neurotransmission should allow the field of artificial neural networks to freely explore other means of adjusting and setting connection weights in a network than through local learning laws. Using such alternative means of training, Roy et al. (Roy, Govil & Miranda [1995, 1997]) have developed robust and reliable learning algorithms that have polynomial time computational complexity in both the design and training of networks. So the exploration of other means of learning should be of substantial benefit to the field.

The main motivation for proposing a set of external learning characteristics is to facilitate the development of future learning algorithms in this field, so that future robots (hardware, software) can learn in a manner similar to humans. In addition, these external properties would allow the field to test the validity of alternative theories about the internal mechanisms of the brain. This paper does not present any new learning algorithms, but it is about learning algorithms and how they should be constructed and what properties they should exhibit.

References

[1]   Amari, S. I. (1988), "Mathematical theory of self-organization in neural nets," in *Organization of neural networks*, eds. W. von Seelen, G. Show and U. M. Leinhos, VCH Weinheim, FRG, pp.399-413.

[2]   Churchland, P. and Sejnowski, T. (1992), *The Computational Brain*, MIT Press, Cambridge, MA.

[3]   Fahlman, S. E. and Hinton, G. E. (1987), "Connectionists Architectures for Artificial Intelligence," *Computer*, Vol. 20, pp.100-109.

[4]   Feldman, J. A. and Ballard, D. A. (1982), "Connectionists Models and Their Properties," *Cognitive Science*, Vol. 6, pp. 205-254.

[5]   Grossberg, S. (1982), *Studies of Mind and Brain: Neural Principles of Learning   Perception, Development, Cognition, and Motor Control*, Reidell Press, Boston.

[6]   Grossberg, S. (1988), "Nonlinear neural networks: principles, mechanisms, and architectures," *Neural Networks*, Vol.1, pp. 17-61.

[7]   Hasselmo, M. (1995), "Neuromodulation and cortical function: Modeling the physiological basis of behavior," *Behavioral and Brain Research*, Vol. 67(1), pp. 1-27.

[8]   Haugeland, J. (1996), "What is Mind Design," Chapter 1 in Haugeland, J. (ed), *Mind Design II*, 1997, MIT Press, pp. 1-28.

[9]   Hebb, D. O. (1949), *The Organization of Behavior, a Neuropsychological Theory*, John Wiley, New York.

[10]   Hestenes, D. O. (1998), "Modulatory mechanisms in mental disorders," in D. Stein (Ed.), *Neural Networks and Psychopathology*, Cambridge University Press, Cambridge, UK.

[11]   Kandel E.R., Schwartz J.H. & Jessel T.M. (1993), *Principles of Neural Science,* 3rd ed., Elsevier, New York.

[12]   Kenny, A. (1971), "The homunculus fallacy," in Grene, M. (ed.), *Interpretations of life and mind*, London.

[13]   Kohonen, T. (1993), "Physiological interpretation of the self-organizing map algorithm," *Neural Networks*, Vol. 6, pp. 895-905.

[14]   Medin, D. L. & Schaffer, M. M. (1978), "Context theory of classification learning," *Psychological Review*, Vol. 85, pp. 207-38.

[15]   Moody, J. & Darken, C. (1989), "Fast Learning in Networks of Locally-Tuned Processing Units," *Neural Computation*, Vol. 1(2), pp. 281-294.

[16]   Reilly, D.L., Cooper, L.N. and Elbaum, C. (1982), "A Neural Model for Category Learning," *Biological Cybernetics*, Vol. 45, pp. 35-41.

[17]   Roy, A., Govil, S. & Miranda, R. (1995), "An Algorithm to Generate Radial Basis Function (RBF)-like Nets for Classification Problems," *Neural Networks*,   Vol. 8, No. 2, pp. 179-202.

[18]   Roy, A., Govil, S. & Miranda, R. (1997), "A Neural Network Learning Theory and a Polynomial Time RBF Algorithm," *IEEE Transactions on Neural Networks,*   Vol. 8, No. 6, pp. 1301-1313.

[19]   Rumelhart, D.E., and McClelland, J.L. (eds.) (1986), *Parallel Distributed Processing: Explorations in Microstructure of Cognition, Vol. 1: Foundations,* MIT Press, Cambridge, MA., pp. 318-362.

[20]   Shanks, D. (1995), *The Psychology of Associative Learning*, Cambridge University Press, Cambridge, England.

[21]   Smolensky, P. (1989), "Connectionist Modeling: Neural Computation/Mental Connections," Chapter 9 in Haugeland, J. (ed), *Mind Design II*, 1997, MIT Press, pp. 233-250.

Superlinear Learning Algorithm Design

Peter GÉCZY, Shiro USUI

RIKEN Brain Science Institute
2-1 Hirosawa, Wako-shi
Saitama 351-0198
Japan

Abstract. Superlinear algorithms are highly regarded for the speed-complexity ratio. With superlinear convergence rates and linear computational complexity they are the primary choice for large scale tasks. However, varying performance on different tasks rises the question of relationship between an algorithm and a task it is applied to. To approach the issue we establish a classification framework for both algorithms and tasks. The proposed classification framework permits independent specification of functions and optimization techniques. Within this framework the task of training MLP neural networks is classified. The presented theoretical material allows design of superlinear first order algorithms tailored to particular task. We introduce two such techniques with a line search subproblem simplified to a single step calculation of the appropriate values of step length and/or momentum term. It remarkably simplifies the implementation and computational complexity of the line search subproblem and yet does not harm the stability of the methods. The algorithms are theoretically proved convergent. Performance of the algorithms is extensively evaluated on five data sets and compared to the relevant first order optimization techniques.

Key words: first order optimization, superlinear convergence raters, steepest descent, conjugate gradient, line search, classification framework, neural networks

1 Introduction

Development of superlinear methods has been largely motivated by practical factors [1]. Although first order methods provide first order convergence rates, they are substantially computationally less expensive than second order methods with quadratic convergence rates. The computational expensiveness of the second order methods originates in their second order model that leads to the necessity of obtaining or approximating the second order information about the objective function — Hessian matrix. Calculation of the Hessian matrix is impractical in tasks with large number of parameters and/or data

[2]—[4]. Practically, the only suitable techniques are based on the first order approaches.

Over the past few decades, several modifications of first order line search techniques have been proposed [5]—[10]. The original line search subproblem has been simplified to a one-step-finding of training parameters. This, on one hand, leads to computationally less expensive optimization procedures, however, on the other hand, it may harm the stability of methods.

Early strategies in first order optimization were aimed at choosing the step length $\alpha^{(k)}$ close to the values given by the exact line search. This trend was mainly motivated by theoretical results stating that the steepest descent method with exact line search is globally convergent to a stationary point. Because of the computational excess of accurate line searches the researchers weakened the exact line search subproblem just to the decrease (for minimization case) of the objective function E at each iteration, $E^{(k)} > E^{(k+1)}$. However, the descent property itself is unsatisfactory since it allows negligible reductions of E. Negligible reductions can occur when the learning rate parameter $\alpha^{(k)}$ approaches zero values, $\alpha^{(k)} \to 0$, or when the search direction $s^{(k)}$ is almost perpendicular to the gradient vector $\nabla E^{(k)}$. To avoid these occurrences, two conditions on $\alpha^{(k)}$ have been proposed in the early optimization literature [11];

$$E(u^{(k)} + \alpha^{(k)} s^{(k)}) \leq E(u^{(k)}) + \alpha^{(k)} \sigma \nabla E(u^{(k)}) , \tag{1}$$

$$E(u^{(k)} + \alpha^{(k)} s^{(k)}) \geq E(u^{(k)}) + \alpha^{(k)} (1 - \sigma) \nabla E(u^{(k)}) , \tag{2}$$

where $\sigma \in (0, 0.5)$ is a fixed parameter. The conditions (1) and (2) are plausible when optimizing quadratic functions. In the case of non-quadratic E, the condition (2) may exclude the minimizing point of $E(u^{(k)} + \alpha^{(k)} s^{(k)})$. For this reason the condition (2) may be replaced by the following (see [12]);

$$E(u^{(k)} + \alpha^{(k)} s^{(k)}) \geq \delta \nabla E(u^{(k)}) , \tag{3}$$

where $\delta \in (\sigma, 1)$ is a fixed parameter. This results in more complicated convergence theorems [13]. For practical purposes a stringent condition instead of (3) has also been considered (see [14]);

$$\left| E(u^{(k)} + \alpha^{(k)} s^{(k)}) \right| \leq -\delta \nabla E(u^{(k)}) . \tag{4}$$

In [14] it is also demonstrated that the line search subproblem in nonlinear least squares gains in efficiency by making polynomial interpolations to the individual residuals rather than to the overall objective function. This basically suggests multiple variable step length (or learning rate) parameters. The idea has been extensively used on neural networks. In a neural network literature the line search subproblem has been heuristically simplified to a single step calculation of the learning rate and/or momentum term [15]. In

[16] heuristics are used for adapting a learning rate for a given weight u_i as $c\alpha_i^{(k)}$, $c > 1$, if $sgn(\frac{\partial E(u^{(k)})}{u_i}) = sgn(\frac{\partial E(u^{(k-1)})}{u_i})$; and $b\alpha_i^{(k)}$, $b \in (0,1)$, if $sgn(\frac{\partial E(u^{(k)})}{u_i}) \neq sgn(\frac{\partial E(u^{(k-1)})}{u_i})$. A similar, however theoretically justified method for modifying the learning rate is proposed in [17]: set $\alpha^{(k)} = \alpha^{(k-1)}$ if $sgn(\nabla E(u^{(k)})) = sgn(\nabla E(u^{(k-1)}))$ and $\alpha^{(k)} = 0.5\alpha^{(k-1)}$ otherwise. Another heuristic method for adapting the learning rate utilizes exponential increases/decreases rather than linear and prohibits taking 'steps' when oscillations occur [18].

In stochastic gradient descent algorithms [19], [20], for asymptotic convergence from stochastic approximation theory implies the learning rate adaptation rule [21]: $\alpha^{(k)} = 1/k$, or by [22],[23] as: $\alpha^{(k)} = \alpha^{(0)}/(1 + k)$. These methods lead to very slow convergence when k grows large. Indirect proportionality of learning rate to a number of iterations results in $\alpha^{(k)} \to 0$. Hence it allows negligible reductions of the objective function that result in slow convergence when the number of iterations increases. Increasing $\alpha^{(0)}$ in this case causes instability for small k. To overcome this difficulty a variation has been proposed [24], [25]: $\alpha^{(k)} = \alpha^{(0)}/[1 + (k/\tau)]$.

Taking into account the previous search direction leads to the conjugate gradient methods. Several previous search directions can also be utilized [26], however, this approach increases memory requirements in implementing the method. Modifications of the momentum term for the conjugate gradient methods in early optimization literature [27], [28] were also suggested [29]—[31]. Regarding neural network literature, a heuristic approaches to modifying the momentum term have been proposed [32]—[35].

2 Classification of Functions and Algorithms

In the theoretical context of general optimization there are two essential elements: algorithm and task. Traditionally, they are considered to be fairly independent of each other. The objective function describing particular task does not depend on the algorithm that is applied to optimize it. Likewise, the algorithm is independent of the task. In practice this independence has exceptions: the degree of smoothness of the objective function may influence the choice of an algorithm, the algorithm design may not accommodate certain stopping criteria, convergence speed and computational complexity of some algorithms can be unfeasible for some tasks. The pair 'task-algorithm' cannot be chosen totally *in abstracto*. Introduction of a theoretical construct that highlights and at least partially overbridges these two poles may be very beneficial in practice.

To present more rigorous specification of algorithms and tasks it is necessary to take into account at least the following three factors: **I.** Convergence of the algorithm plus given convergence rates. **II.** The objective function E

given by the problem. **III.** Applied optimization technique. We introduce a boundary between the mentioned factors and their underlying links.

Theorem 1.
Let E be a function optimized by the first order technique having the convergence rates a. The following holds,

$$a \geq \left| 1 - \limsup \frac{\frac{\|\Delta u^{(k)}\|_2 \cdot \|\nabla E(u^{(k)})\|_2}{|E(u^*) - E(u^{(k)})|}}{\left| 1 - \frac{R_{n \geq 2}}{R_{n \geq 1}} \right|} \right| , \tag{5}$$

where $R_{n \geq 2}$ are second and higher order terms and $R_{n \geq 1}$ are first and higher order terms of Taylor series expansion of E at the optimum point u^*. ■

Corollary 1.
Let E be a function optimized by the first order technique having superlinear convergence rates. The following holds,

$$\limsup \frac{\frac{\|\Delta u^{(k)}\|_2 \cdot \|\nabla E(u^{(k)})\|_2}{|E(u^*) - E(u^{(k)})|}}{\left| 1 - \frac{R_{n \geq 2}}{R_{n \geq 1}} \right|} = 1 , \tag{6}$$

where $R_{n \geq 2}$ are second and higher order terms and $R_{n \geq 1}$ are first and higher order terms of Taylor series expansion of E at the optimum point u^*. ■

Corollary 2.
Let E be a function optimized by the first order technique. If,

$$\limsup \frac{R_{n \geq 2}}{R_{n \geq 1}} = 0 \ \ or \ \ 2 , \tag{7}$$

holds for E, where $R_{n \geq 2}$ are second and higher order terms and $R_{n \geq 1}$ are first and higher order terms of a Taylor series expansion of E at the optimum point u^*, then the first order technique optimizes E with superlinear convergence rates if the following expression is satisfied,

$$\limsup \frac{\|\Delta u^{(k)}\|_2 \cdot \|\nabla E(u^{(k)})\|_2}{|E(u^*) - E(u^{(k)})|} = 1 . \tag{8}$$

■

It is clear that the expression (7) is independent of optimization technique. It describes only the property of the objective function E and serves as its relevant descriptor. On the other hand, the expression (8) is algorithm-dependent. Therefore, it can be utilized as a descriptor of a particular optimization technique. In other words, the expression,

$$\mathcal{AD} = \limsup \frac{\|\Delta u^{(k)}\|_2 \cdot \|\nabla E(u^{(k)})\|_2}{|E(u^*) - E(u^{(k)})|} , \tag{9}$$

is an **algorithm descriptor** ($\mathcal{AD}$), whereas the expression,

$$\mathcal{PD} = \left| 1 - \limsup \frac{R_{n\geq 2}}{R_{n\geq 1}} \right| , \tag{10}$$

is a **problem descriptor** ($\mathcal{PD}$). The underlying relationship between the problem descriptor $\mathcal{PD}$ (10) and the algorithm descriptor $\mathcal{AD}$ (9) is formulated in Theorem 1 for general linear convergence rates, and in Corollary 1 for superlinear convergence rates. This allows specification of functions, or classes of functions, that are optimized by first order techniques with superlinear convergence rates.

Definition 1. *(Classification Framework)*

It is said that function E belongs to the class $\mathcal{PD}^{(i)}$ if for a convergent sequence $\{u^{(k)}\}_k \to u^*$ the following holds,

$$\mathcal{PD} = \left| 1 - \limsup \frac{R_{n\geq 2}}{R_{n\geq 1}} \right| \geq i ,$$

where $R_{n\geq 2}$ are second and higher order terms and $R_{n\geq 1}$ are first and higher order terms of Taylor series expansion of E at the optimum point. Specifically, E belongs to the class $\mathcal{PD}^{(1,0)}$ if it belongs to the class $\mathcal{PD}^{(1)}$ and furthermore the following holds,

$$\limsup \frac{R_{n\geq 2}}{R_{n\geq 1}} = 0 .$$

Analogously, it is said that a given algorithm (or optimization technique) belongs to the class $\mathcal{AD}^{(i)}$ if for a convergent sequence $\{u^{(k)}\}_k \to u^*$ holds,

$$\mathcal{AD} = \limsup \frac{||\Delta u^{(k)}||_2 \cdot ||\nabla E(u^{(k)})||_2}{|E(u^*) - E(u^{(k)})|} \geq i .$$

$\blacksquare$

Remark 1. The subclass $\mathcal{PD}^{(1,0)}$ of the class $\mathcal{PD}^{(1)}$ plays an important role in machine learning and neural network field. Thus it is essential to observe analytically the properties of this class. Another important question is: What functions belong to the class $\mathcal{PD}^{(1,0)}$? It was partially answered in [36]. $\square$

Theorem 2. *(Analytic Description of class $\mathcal{PD}^{(1,0)}$)*

Let E be a function $E : \Re^{N_I} \to \Re$, $N_I \in \mathcal{N}$, such that,

$$\text{I.} \quad \forall n \geq 2 \left(\frac{\partial^n E}{\partial u^n} \neq \overline{\infty} , \; n \in \mathcal{N} \right) , \tag{11}$$

and

$$\text{II.} \quad \frac{\partial E}{\partial u} \neq \overline{0} \Rightarrow \exists n \geq 2 \left(\frac{\partial^n E}{\partial u^n} \neq \overline{0} , \; n \in \mathcal{N} \right) , \tag{12}$$

or

$$\frac{\partial^2 E}{\partial u^2} \neq \overline{0} , \tag{13}$$

where $u \in \Re^{N_F}$ denotes an N_F-dimensional real vector of variables of E, $\overline{0}$ is the 0-vector, and $\overline{\infty} = (\psi_1, \ldots, \psi_{N_F})$, $\psi_i = +\infty$ or $\psi_i = -\infty$. A function E satisfying the above two conditions belongs to the class $\mathcal{PD}^{(1,0)}$, $E \in \mathcal{PD}^{(1,0)}$.

$\blacksquare$

Implying from Theorem 2, any function in the class $\mathcal{PD}^{(1,0)}$ must have bounded second and higher order partial derivatives and must have either nonzero curvature or from nonzero gradient must imply at least one nonzero higher order partial derivative. Note that these conditions do not necessarily have to hold on the whole range of definition of E. It is satisfactory if they hold at the limit point u^*.

2.1 Mapping and Training in MLP Networks

Three-layer MLP networks have been proved to have universal approximation capabilities [37]. This means that an arbitrary functional dependency can be approximated to an arbitrary level of accuracy by three-layer artificial neural network with an appropriate number of nonlinear hidden units. Nonlinearity of hidden elements is crucial for universal approximation. Particularly popular is a sigmoidal type of nonlinearity [38]. The mapping of a three-layer MLP network under consideration is defined as follows.

Definition 2. *(Mapping of a Three-Layer MLP Network)*
A mapping $\mathcal{F}$ is said to be a mapping of a three-layer MLP network defined as follows.

$$\mathcal{F} = \mathcal{F}_{HO} \circ \mathcal{F}_{IH} \quad (\mathcal{F}: \Re^{N_I} \to \Re^{N_O}),$$

where N_I is the dimensionality of the input space and N_O is the dimensionality of the output space. $\mathcal{F}_{HO}$ is an affine mapping from N_H-dimensional subspace $\mathcal{D}^{N_H}$ of $\Re^{N_H}$ to $\Re^{N_O}$.

$$\mathcal{F}_{HO} = (\mathcal{F}_{HO_1}, \ldots, \mathcal{F}_{HO_{N_O}}) \quad ; \quad \mathcal{F}_{HO}: \mathcal{D}^{N_H} \to \Re^{N_O}$$

$$\mathcal{F}_{HO_k}^{(p)} = \sum_{j=1}^{N_H} w_{jk} \mathcal{F}_{IH_j}^{(p)} - \theta_{o_k}$$

where $\mathcal{F}_{IH_k}^{(p)}$ is the output of the k-th hidden unit for the p-th training pattern, θ_{o_k} is the threshold value ($\theta_{o_k} \in \mathcal{R}$) for the k-th output unit, w_{jk} is the real valued weight connection connecting the j-th hidden unit with the k-th output unit. $\mathcal{F}_{IH}$ is nonlinear multidimensional mapping

$$\mathcal{F}_{IH} = \overline{f} \circ \mathcal{A}_{IH} \quad (\mathcal{F}_{IH}: \Re^{N_I} \to \mathcal{D}^{N_H}),$$

$$\mathcal{F}_{IH_j}^{(p)} = f \left(\sum_{i=1}^{N_I} v_{ij} x_i^{(p)} - \theta_{h_j} \right)$$

where θ_{h_j} is the threshold value $(\theta_{h_j} \in \mathcal{R})$ for the j-th hidden unit, v_{ij} is the real valued weight connection connecting the i-th input unit with the j-th hidden unit, $x_i^{(p)}$ is the i-th coordinate of the p-th input vector $\overline{x}^{(p)}$, $\overline{f}$ stands for a multidimensional nonlinear sigmoidal transformation in which each dimension of its N_H-dimensional domain vector is transformed by a sigmoidal transfer function f, $(\overline{f} : \Re^{N_H} \to \mathcal{D}^{N_H})$, $\mathcal{A}_{IH}$ is an input-to-hidden affine submapping $\mathcal{A}_{IH} : \Re^{N_I} \to \Re^{N_H}$. ∎

Lemma 1.

Let $\mathcal{F}$ be a mapping of a three-layer MLP network. The following holds.

$$\theta_{o_k} : \quad \frac{\partial \mathcal{F}}{\partial \theta_{o_k}} = 1, \quad \forall n > 1 \left(\frac{\partial^n \mathcal{F}}{\partial \theta_{o_k}^n} = 0, \quad n \in \mathcal{N} \right) \tag{14}$$

$$w_{jk} : \quad \frac{\partial \mathcal{F}}{\partial w_{jk}} = \mathcal{F}_{IH_j}, \quad \forall n > 1 \left(\frac{\partial^n \mathcal{F}}{\partial w_{jk}^n} = 0, \quad n \in \mathcal{N} \right) \tag{15}$$

$$\theta_{h_j} : \quad \forall n \geq 1 \left(\frac{\partial^n \mathcal{F}}{\partial \theta_{h_j}^n} = \sum_{q=1}^{N_O} w_{jq} \frac{\partial^n \mathcal{F}_{IH_j}}{\partial \theta_{h_j}^n}, \quad n \in \mathcal{N} \right) \tag{16}$$

$$v_{ij} : \quad \forall n \geq 1 \left(\frac{\partial^n \mathcal{F}}{\partial v_{ij}} = \sum_{q=1}^{N_O} w_{jq} \frac{\partial^n \mathcal{F}_{IH_j}}{\partial \mathcal{A}_{IH_i}^{\,n}} \cdot x_i^n \quad n \in \mathcal{N} \right) \tag{17}$$

∎

Remark 2. Proof of Lemma 1 is obtained directly by differentiating $\mathcal{F}$ with respect to its parameters ($\mathcal{F}$ is a mapping of three-layer MLP networks (Definition 2)). The obtained expressions are well-known to the researchers in neural network field. Therefore, the proof is not presented here. □

Definition 3. *(Training in MLP Networks)*

Let T be a training set with cardinality N_P,

$$T = \{ [\overline{x}, \overline{y}] | \overline{x} \in \Re^{N_I} \wedge \overline{y} \in \Re^{N_O} \} \quad , \quad |T| = N_P \ ,$$

where each pair $[\overline{x}, \overline{y}]$ contains the input pattern $\overline{x}$ of the dimensionality N_I, and the expected output pattern $\overline{y}$ of the dimensionality N_O. Let u denote a set of free system parameters of a network and the objective function E be defined as follows,

$$E(u, \overline{x}) = C \cdot \sum_{p=1}^{N_P} \sum_{k=1}^{N_O} (\mathcal{F}_k(u, \overline{x}^{(p)}) - y_k^{(p)})^2, \quad \text{where } C \text{ is a constant.} \tag{18}$$

Training in MLP networks is a process of minimization the objective function E,

$$\arg \min_u E(u, \overline{x}) \ , \tag{19}$$

given a finite number of samples $[\overline{x}, \overline{y}] \in T$ drawn from an arbitrary sample distribution. ∎

Lemma 2.
Let E be an objective function, $E = \frac{1}{2}\|\mathcal{F}(u) - \mathcal{F}^*\|_2^2$ of the general least square problem and u be a set of parameters of function $\mathcal{F}$. The following holds.

$$\text{I.} \quad \frac{\partial^2 E}{\partial u^2} = \underbrace{\frac{\partial^2 E}{\partial \mathcal{F}^2} \cdot \frac{\partial \mathcal{F}}{\partial u}}_{(a)} + \underbrace{\frac{\partial E}{\partial \mathcal{F}} \cdot \frac{\partial^2 \mathcal{F}}{\partial u^2}}_{(b)} \tag{20}$$

$$\text{II.} \quad \forall n \geq 2 \left(\frac{\partial^n E}{\partial u^n} = \sum_{i=0}^{n-i} \binom{n-1}{i} \frac{\partial^{(n-1)} E}{\partial \mathcal{F}^{(n-i)}} \cdot \frac{\partial^{(i+1)} \mathcal{F}}{\partial u^{(i+1)}}, \quad n \in \mathcal{N} \right) \tag{21}$$

$$\text{III.} \quad \frac{\partial E}{\partial \mathcal{F}} = \|\mathcal{F}(u) - \mathcal{F}^*\|_1 \tag{22}$$

$$\text{IV.} \quad \frac{\partial^2 E}{\partial \mathcal{F}^2} = 1 \tag{23}$$

$$\text{V.} \quad \forall n > 2 \left(\frac{\partial^n E}{\partial \mathcal{F}^n} = 0, \quad n \in \mathcal{N} \right) \tag{24}$$

$\blacksquare$

Again, proof of Lemma 2 is obtained directly by differentiating E, therefore it is not shown here. Given the definition of mapping of MLP networks and the definition of training in MLP networks it is interesting to observe the following results.

Theorem 3. *(Training Problem Classification)*
Let $\mathcal{F}(u)$ be a mapping of a three-layer artificial neural network with finite nonzero real valued variables, $u_l \in u$; $u_l \neq 0 \;\wedge\; u_l \neq \pm\infty$, and nonlinear transfer functions f_j $(j = 1, \ldots, N_H)$ satisfying the conditions,

$$\text{I.} \quad \forall \psi \in \Re \quad (f_j(\psi) \neq \pm\infty) , \tag{25}$$

$$\text{II.} \quad \forall n \geq 1 \left(\frac{\partial^n f_j}{\partial u_l^n} \neq \pm\infty , \; n \in \mathcal{N} \right) . \tag{26}$$

The problem (19) of minimizing (or maximizing) the least square error(s) (18), given a finite number of samples $[\overline{x}^{(i)}, \overline{y}^{(i)}] \in T, i = 1, \ldots, N_P$, is $\mathcal{PD}^{(1,0)}$ problem. $\blacksquare$

Remark 3. It is important to note that conditions (25) and (26) do not impose strong restrictions on transfer functions. As can easily be verified, the most used nonlinear transfer functions satisfy conditions (25) and (26). For example, for the sigmoidal transfer function $f(\psi) = \frac{1}{1+e^{-\psi}}$ we have,

$$\frac{\partial^n f}{\partial \psi^n} = (-1)^n f(\psi) + (-1)^{n+1} f^2(\psi) .$$

Hence the conditions (25) and (26) are obviously satisfied. $\square$

Recall the relationship of $\mathcal{PD}$ and $\mathcal{AD}$ expressed in Corollary 1, for superlinear convergence rates, $\frac{\mathcal{AD}}{\mathcal{PD}} = 1$. If the problem class $\mathcal{PD}$ is known, the appropriate first order optimization technique (for achieving superlinear convergence rates) should belong to the same class as the optimized problem, that is $\mathcal{AD}^{(a)} = \mathcal{PD}^{(a)}$. Theorem 3 states that the problem of training MLP networks is $\mathcal{PD}^{(1,0)}$ problem. Then a superlinear first order line search training technique for MLP networks must belong to the algorithm class $\mathcal{AD}^{(1)}$ (according to the results of Theorem 1, Corollary 1, Corollary 2, Definition 1, and Theorem 3). How to derive such algorithms is shown in the next section.

3 Design of Superlinear First Order Algorithms

Optimization algorithms optimize the objective function in an iterative manner. Starting from the initial point they generate a sequence of points $\{u^{(k)}\}_k$ in a parameter space of E that should converge to some point in a solution set. General optimization scheme thus includes at least the following elements:

- **Initialization:** determines the starting point of optimization and sets parameters specific to the used algorithm.
- **Progression:** relates to the rules of generation the points of sequence $\{u^{(k)}\}_k$.
- **Termination:** specifies when the algorithm stops — stopping criterion.

The set of rules to construct the iterative sequence $\{u^{(k)}\}_k$ is the main part of the algorithm design. We focus on the design of superlinear first order line search methods. These methods iteratively move in the parameter space of the objective function along the determined direction vector. Once the search direction is decided, the algorithm has to chose the length of the iterative progress along it. This is done by scaling the direction vector(s) by scalars: step length and/or momentum. Of particular interest are the methods that can determine the values of step length and/or momentum term in a single calculation step. This considerably relaxes computational complexity of the line search subproblem.

Superlinear convergence rates of the sequence $\{u^{(k)}\}_k$ to the solution point u^* are measured with respect to the following standard expression:

$$\limsup \frac{||u^{(k+1)} - u^*||_2}{||u^{(k)} - u^*||_2} = 0.$$

Various approaches can be used for deriving superlinear techniques [39], [40]. Straightforward way is to utilize algorithm descriptor expression. Substituting for $\Delta u^{(k)}$ from the parameter update of a given optimization technique the expressions for iterative updates of step length $\alpha^{(k)}$ and/or momentum $\beta^{(k)}$ directly imply. This approach to the line search subproblem results in completely novel first order methods: ALGORITHM 1 and ALGORITHM 2.

ALGORITHM 1 *(Steepest Descent)*

1. Set the initial parameters: $u^{(0)}$, a, and $E(u^*)$.

2. Calculate $E(u^{(k)})$, evaluate the stopping criteria and then either terminate or proceed to the next step.

3. Calculate the gradient $\nabla E(u^{(k)})$.

4. Calculate $\alpha^{(k)}$,

$$|\alpha^{(k)}| = a \cdot \frac{|E(u^*) - E(u^{(k)})|}{\|\nabla E(u^{(k)})\|_2^2} \ . \tag{27}$$

5. Update the system parameters as follows.

$$u^{(k+1)} = u^{(k)} - \alpha^{(k)} \cdot \nabla E(u^{(k)})$$

6. Set $k = k + 1$ and proceed to Step 2. ∎

ALGORITHM 2 *(Conjugate Gradient)*

1. Set the initial parameters: $u^{(0)}$, a, and $E(u^*)$.

2. Calculate $E(u^{(k)})$, evaluate the stopping criteria and then either terminate or proceed to the next step.

3. Calculate the gradient $\nabla E(u^{(k)})$.

4. Calculate $\alpha^{(k)}$,

$$|\alpha^{(k)}| = a \cdot \frac{|E(u^{(k)})|}{\|\nabla E(u^{(k)})\|_2^2} \ , \tag{28}$$

and $\beta^{(k)}$,

$$|\beta^{(k)}| = \frac{a \cdot E(u^*)}{\|s^{(k-1)}\|_2 \cdot \|\nabla E(u^{(k)})\|_2} \ . \tag{29}$$

5. Update the system parameters as follows.

$$u^{(k+1)} = u^{(k)} - \alpha^{(k)} \cdot \nabla E(u^{(k)}) + \beta^{(k)} \cdot s^{(k-1)} \ .$$

6. Set $k = k + 1$ and proceed to Step 2. ∎

ALGORITHM 1 and ALGORITHM 2 have linear computational complexity $O(N_F)$, where N_F is a number of free parameters. ALGORITHM 1 is memoryless. The necessity of storing the previous search direction $s^{(k-1)}$ in conjugate gradient techniques leads to the linear memory requirements $O(N_F)$ of ALGORITHM 2. Despite the simplicity of the line search subproblem, both ALGORITHM 1 and ALGORITHM 2 are convergent.

Theorem 4. *(Convergence Theorem)*
Let E be an objective function defined on the attractor basin $\mathcal{L}$ of the point u^* with continuous first derivatives. A sequence of points $\{u^{(k)}\}_k$ generated by ALGORITHM 1 or ALGORITHM 2 from the initial point $u^{(0)} \in \mathcal{L}$ converges to the terminal attractor u^* (with respect to the negligible residual $R_{n \geq 2}$), $\{u^{(k)}\}_k \to u^*$. ∎

Although ALGORITHM 1 is a slightly modified steepest descent method, it is substantially more powerful, in terms of convergence speed, than the standard steepest descent method used in BP training techniques for MLP neural networks. Figure 1 shows that ALGORITHM 1 not only minimizes oscillations, but progresses considerably faster to the optimum point.

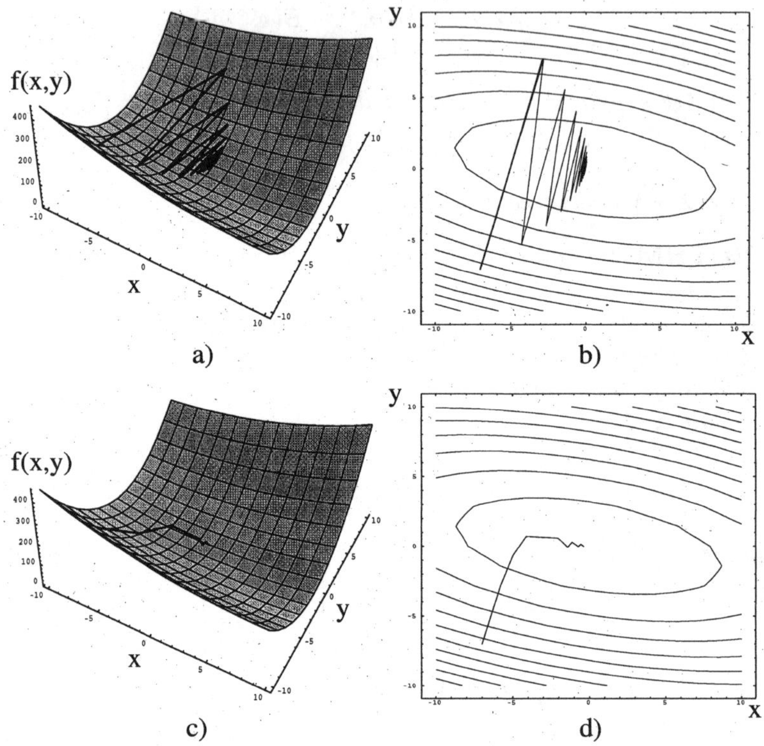

Fig. 1. Comparison of optimization progress between ALGORITHM 1 (charts **c)** and **d)**) and standard BP (steepest descent) (charts **a)** and **b)**) on quadratic function $f(x, y) = 0.5x^2 + 3y^2 + xy$ from the starting point [-7,-7]. BP used step length $\alpha = 0.3$. ALGORITHM 1 had setting: $E(u^*) = 0, a = 1$. Stopping criterion was the value $f(x, y) \leq 0.1$. It is evident that the progress of ALGORITHM 1 is much more smooth and faster than the standard BP.

ALGORITHM 2 is demonstrated in Figure 2. ALGORITHM 2 (charts **c)** and **d)**) clearly converges substantially more smoothly to the optimum point than the conventional conjugate gradient method (charts **a)** and **b)**). Flexibility of the momentum term helps to determine the search direction of the algorithm more appropriately. In comparison to ALGORITHM 1, ALGO-RITHM 2 can be regarded as slightly smoother.

190

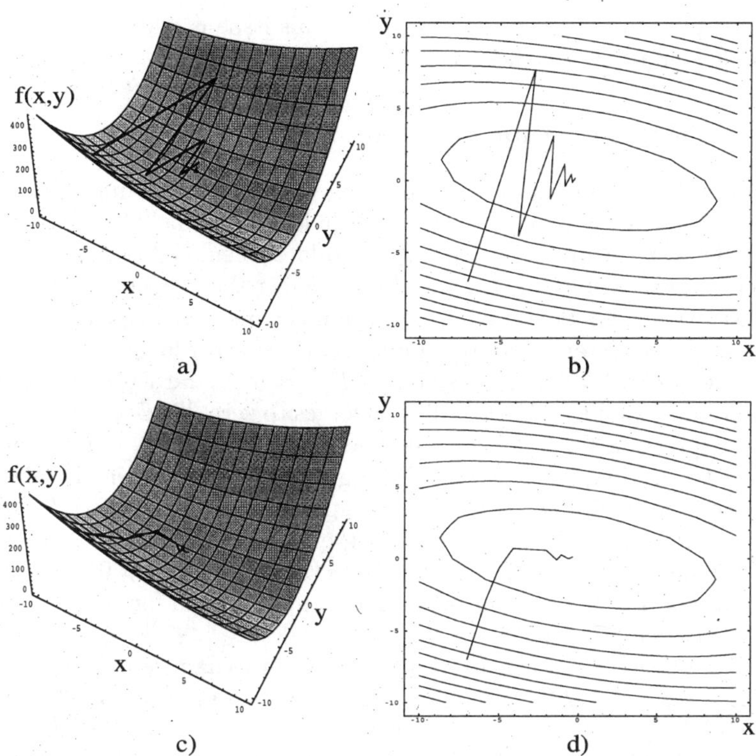

Fig. 2. Comparison between standard BP (charts **a)** and **b)**), with the step length $\alpha = 0.3$ and the momentum $\beta = 0.1$, and ALGORITHM 2 (charts **c)** and **d)**) with setting: $E(u^*) = 0, a = 1$. Objective function was a quadratic function $f(x,y) = 0.5x^2 + 3y^2 + xy$. Optimization procedure started from the initial point [-7,-7] and was terminated when $f(x,y) \leq 0.1$. ALGORITHM 2 displays almost non-oscillatory behavior and smooth and fast progress to the optimum point, unlike the standard BP.

Theorem 5. *(Classification of ALGORITHM 1 and ALGORITHM 2)*
ALGORITHM 1 and ALGORITHM 2 belong to the class $\mathcal{AD}^{(a)}$ where a is the parameter in step length expression (27), (28), and in the expression for adjustable momentum term (29). ∎

Theorem 5 implies that both ALGORITHM 1 and ALGORITHM 2 are capable of optimizing an arbitrary objective function E with superlinear convergence rates. If the optimization task of the objective function E belongs to the class $\mathcal{PD}^{(1)}$ (e.g. training MLP networks) it is required to adjust the parameter $a = 1$. For other function classes it is necessary to set the parameter a appropriately according to the problem descriptor $\mathcal{PD}$. Having a choice of a and not having information about the problem class $\mathcal{PD}$ causes a slight difficulty. However, it is possible to determine the parameter a approximately by a short pre-optimization of the system. This issue is addressed in detail in the 'Implementation Specifics' section.

3.1 Local Minima Escape Capabilities

Due to the dynamic update of step length and/or momentum term ALGORITHM 1 and 2 are capable of escaping from local minima. The escape mechanism is based on the dramatic parameter update when $||\nabla E(u^{(k)})||_2$ is close to zero and objective function E is not approaching the expected value. Details and illustrative example are offered in the following paragraphs.

First order necessary condition for extreme, whether local or global, is zero gradient vector $\nabla E = \bar{0}$, and thus also $||\nabla E||_2 = 0$. As the algorithm converges to the optimum point, $||\nabla E(u^{(k)})||_2$ converges to zero, $||\nabla E(u^{(k)})||_2 \to 0$. Recall the expression for adaptable step length $\alpha^{(k)}$ (27). It is inversely proportional to the squared l_2 norm of the gradient. Thus, when the algorithm approaches the minimum, $||\nabla E(u^{(k)})||_2^2$ approaches zero, $||\nabla E(u^{(k)})||_2^2 \to 0$. If error value $E(u^{(k)})$ is not approaching expected value $E(u^*)$, the expression in the numerator of (27) is nonzero, $|E(u^*) - E(u^{(k)})| \neq 0$. This leads to large values of $\alpha^{(k)}$ and dramatic parameter update even for small coordinate values of gradient $\nabla E(u^{(k)})$.

Similar phenomenon can be observed also in dynamics of adaptable momentum term $\beta^{(k)}$ (29), providing $E(u^*) \neq 0$ and $a \neq 0$. Convergence of $||\nabla E(u^{(k)})||_2$ to zero during the convergence to the local minimum, and nonzero numerator, $a \cdot E(u^*) \neq 0$, results in large values of $\beta^{(k)}$. Then the nonzero coordinates of the vector of the previous search direction, $s^{(k-1)}$, multiplied by large $\beta^{(k)}$, contribute to the dramatic parameter update.

The evidence of the local minima escape capabilities is illustratively demonstrated in Figure 3. ALGORITHM 1 was applied to minimizing function $F(x,y) = x^4 - x^3 + 17x^2 + 9y^2 + y + 102$. The function is quadratic with respect to 'y' and of the 4th order with respect to 'x'. 3D plot of the function is shown in chart **a)** of Figure 3. Function $F(x,y)$ has two minima; one global and one local. Chart **b)** of Figure 3 shows the position of local and global minima at the cutting plane $y = 0$. The starting point of the optimization was $[x,y] = [-5,-5]$ and parameter 'a' was set to 1, $a = 1$. In the first case, the expected stopping function value was set to 55, $F(u^*) = 55$. Progress of ALGORITHM 1, given the above setting, is shown in contour plot **c)** of Figure 3. Since the expected functional value was the stopping criterion, the algorithm stopped after 5 iterations reaching the value $F(u^*) \leq 55$ in the local minimum. In the second case, the expected functional value was set to 6. The optimization progress of the algorithm is shown in contour plot **d)** of Figure 3. Starting from the point $[-5,-5]$, the algorithm initially converged to the local minimum. Small values of $||\nabla F(u^{(k)})||_2^2$ in the area around the local minimum, and discrepancy $|F(u^{(k)}) - F(u^*)|$, led to large values of $\alpha^{(k)}$ and dramatic parameter update. This caused the algorithm to jump out of the local minimum and finally (after escaping from the local minimum) in a

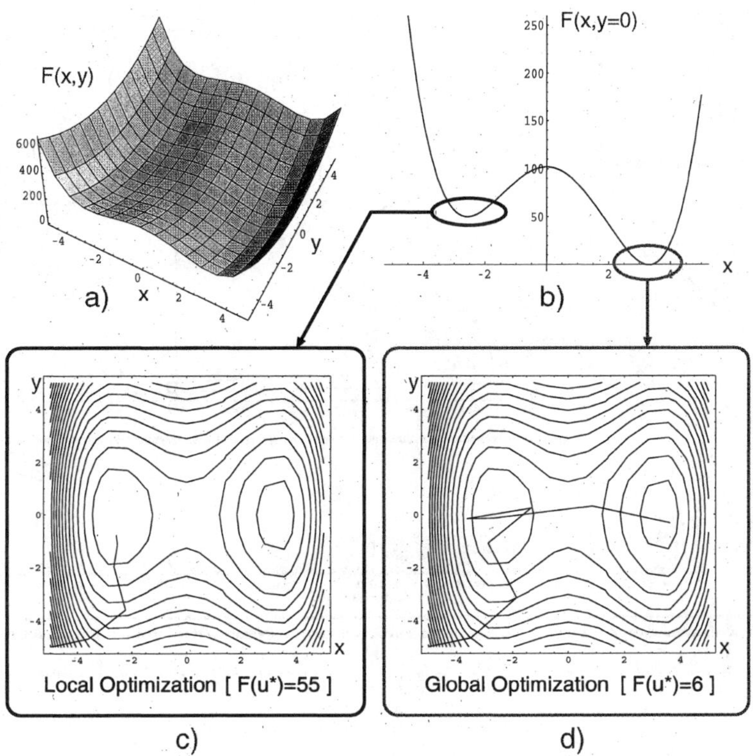

Fig. 3. Demonstration of local minima escape capabilities.

single step to reach the appropriate value of F in the neighborhood of the global minimum.

3.2 Dynamics of Adaptable Step Length and Momentum Term

Typical dynamics of adaptable step length $\alpha^{(k)}$ and momentum $\beta^{(k)}$ are demonstrated on the task of optimizing artificial neural network parameters. A neural network had the configuration 4–3–1 with sigmoidal nonlinear elements and was trained on the Lenses data set [41]. Batch training was terminated when the mean square error (18) (the objective function) had value less than or equal to $5 \cdot 10^{-2}$.

3.2.1 Dynamics of Adaptable Step Length

Typical dynamics of automatically adaptable step length $\alpha^{(k)}$ is depicted in Figure 4. After the initial progress (approximately 5 cycles) the network located the flat area of the error surface as indicated by the almost constant curve of the error E. Details of this phase (120 cycles) are displayed in chart

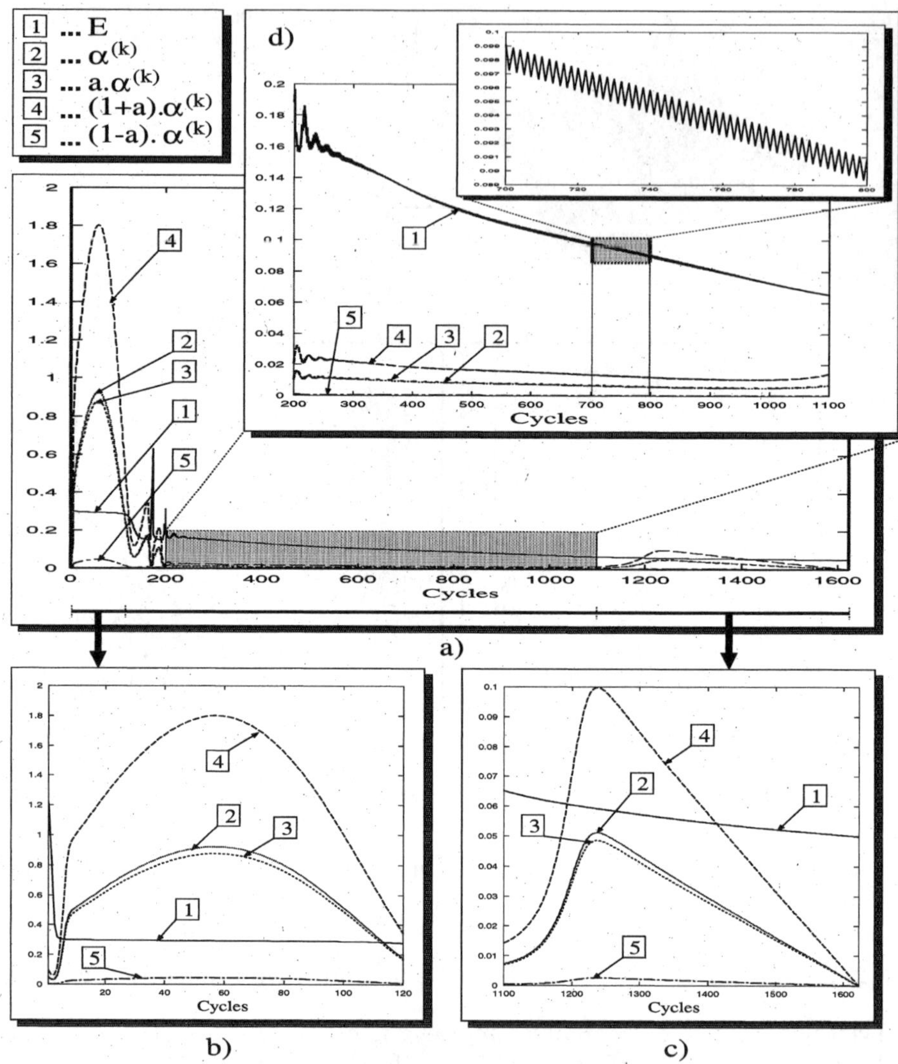

Fig. 4. Typical dynamics of the adaptable step length term $\alpha^{(k)}$.

b). The flatness of the surface results in very low values of the first derivatives and hence the values of the gradient, ∇E. Since the search direction $s^{(k)}$, in steepest descent, is determined by the gradient $\nabla E(u^{(k)})$, very slow progress occurs. The natural action in this situation is to speed up the progress by increasing step length $\alpha^{(k)}$. As it can be seen from curves 2 , 3 , 4 , and 5 , the dynamically adaptable step length $\alpha^{(k)}$ automatically follows this rule.

After approximately 150 cycles the network reached a strongly eccentric region of the error surface. Details of this phase, starting from cycle 200, are shown in chart **d)**. The slight oscillations are enlarged in the upper right chart of chart **d)**. Eccentricity of the surface, on which the network was progressing, caused high values of derivatives with respect to the parameters in which the error surface sharply decreased and low values of derivatives with respect to the parameters where the error surface was relatively flat. The result was that the gradient pointed to the other side of the multidimensional ravine rather than to its minimum. In such situations it is essential for a steepest descent technique to lower the values of step length $\alpha^{(k)}$ so as to reach the bottom of the ravine in the fewest number of cycles. Again, the theoretically derived expressions for dynamically adaptable step length $\alpha^{(k)}$, depicted by curves $\boxed{2}$, $\boxed{3}$, $\boxed{4}$, $\boxed{5}$, clearly follow this necessity.

In the final phase of a network's training, starting from cycle 1100, the eccentricity of the attractor basin around the terminal attractor point u^* slightly relaxed. This led to a relative balance of first partial derivatives with respect to the free parameters. Gradient ∇E pointed almost directly to the optimum point u^*. To reach the terminal attractor faster, the optimization procedure automatically increased the step length $\alpha^{(k)}$. This behavior is depicted in chart **c)**.

3.2.2 Behavior of Adaptable Momentum Term

Figure 5 displays the dynamics of the momentum term $\beta^{(k)}$. First, during approximately 120 cycles, the network progressed on the flat area of the error surface (as indicated by the flat mean square error E). The value of momentum term increased. It reflected the convergence-speed-increase effect of the momentum term. Flatness of the error surface caused low values of the gradient vector ∇E. Small gradient values resulted in a slow progress of an algorithm. However, the algorithm progressed in almost constant direction. Faster progress was initiated by increased value of the momentum term.

The oscillatory phase of the training was between cycles 170 and 1100. The network, oscillating in the ravine-like area of the error surface, updated its parameters by inappropriately high $\Delta u^{(k)}$, so it temporarily over-jumped lower positioned parts of the error surface. Oscillations decreased when the update term $\Delta u^{(k)}$ was lowered. Then the algorithm made smaller steps and reached the bottom of the ravine-like area of the error surface faster than jumping from one side to the other. This stabilizing effect of the momentum term is depicted in the upper chart of Figure 5.

In the final phase of training, (from cycle 1100), the modifiable momentum term helped the network to progress faster to the optimum. It had convergence-speed-increase effect, thus it raised in value.

The modifiable momentum term $\beta^{(k)}$ automatically indicates the necessities of an algorithm for faster progress. Adaptation dynamics of adjustable momentum term $\beta^{(k)}$ helps the algorithm to achieve faster convergence to the optimum point while it keeps its stabilizing effect.

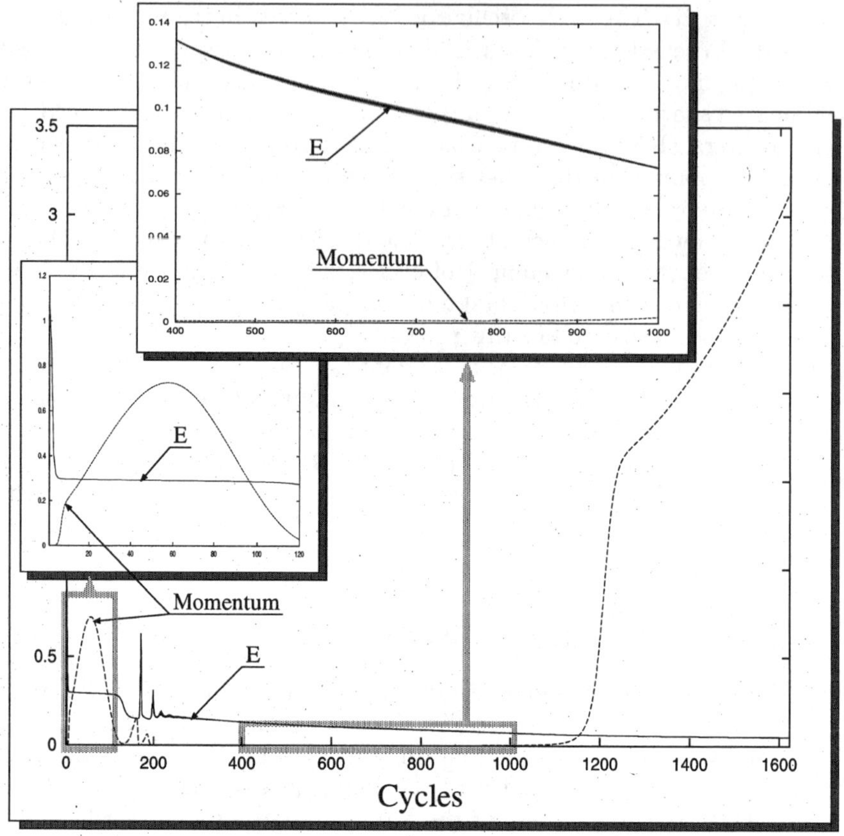

Fig. 5. Typical behavior of adjustable momentum term $\beta^{(k)}$.

4 Implementation Specifics

The presented algorithms incorporate the formulas for automatic adjustments of step length $\alpha^{(k)}$ and/or momentum term $\beta^{(k)}$ containing the choice of a. The parameter a directly determines the algorithm class $\mathcal{AD}$. Knowledge of the function class (or problem) $\mathcal{PD}$ to be optimized allows automatic setting of a for achieving superlinear convergence rates of the algorithms. Although the value of problem class descriptor $\mathcal{PD}$ may be unknown, it is possible to determine a approximately. We determine a approximately for ALGORITHM 1.

Since ALGORITHM 1 is a modification of steepest descent technique it is reasonable to assume comparison of ALGORITHM 1 to the steepest descent algorithm in order to determine a. Without neglecting the second and higher order terms of Taylor expansion, $R_{n\geq2}$, it follows for the standard steepest descent technique with constant α,

$$\left| 1 - \limsup |\alpha| \frac{||\nabla E(u^{(k)})||_2^2}{|E(u^*) - E(u^{(k)}) + R_{n \geq 2}|} \right| . \tag{30}$$

Then for ALGORITHM 1 the following is implied,

$$\left| 1 - \limsup \frac{a \cdot |E(u^*) - E(u^{(k)})|}{|E(u^*) - E(u^{(k)}) + R_{n \geq 2}|} \right| . \tag{31}$$

ALGORITHM 1 has better convergence rates than the standard steepest descent technique with constant step length α if $(31) < (30)$, that is,

$$\limsup \frac{a \cdot |E(u^*) - E(u^{(k)})|}{|E(u^*) - E(u^{(k)}) + R_{n \geq 2}|} > \limsup |\alpha| \cdot \frac{||\nabla E(u^{(k)})||_2^2}{|E(u^*) - E(u^{(k)}) + R_{n \geq 2}|}$$

$$\frac{a \cdot |E(u^*) - E(u^{(k)})|}{|E(u^*) - E(u^{(k)}) + R_{n \geq 2}|} > |\alpha| \cdot \frac{||\nabla E(u^{(k)})||_2^2}{|E(u^*) - E(u^{(k)}) + R_{n \geq 2}|}$$

$$a \cdot |E(u^*) - E(u^{(k)})| > |\alpha| \cdot ||\nabla E(u^{(k)})||_2^2$$

$$a > |\alpha| \cdot \frac{||\nabla E(u^{(k)})||_2^2}{|E(u^*) - E(u^{(k)})|} . \tag{32}$$

The expression (32) is the necessary condition for a when ALGORITHM 1 is used. Then from a short pre-training of the standard steepest descent with the constant step length α the parameter a can be determined as a value satisfying (32) by simply taking the maximum of the pre-training sequence.

$$\max_{k=1,\ldots,C_p} \left\{ |\alpha| \cdot \frac{||\nabla E(u^{(k)})||_2^2}{|E(u^*) - E(u^{(k)})|} \right\}_{k=1}^{C_p} \tag{33}$$

Parameter C_p stands for a number of allowed pre-training cycles. Determination of the parameter a by (33) is naturally more precise when C_p grows large. Moving average, instead of the maximum, in (33) can also be used. Analogously, it is possible to monitor a for various values of constant step length.

Another pertinent issue related to the presented algorithms is a choice of the value $E(u^*)$. Seemingly obvious solution would be to assign $E(u^*)$ the value of the expected error. However, when the algorithm's stopping condition is a value of the expected error, then $\alpha^{(k)}$ converges to 0 as the algorithm approaches the expected error value. This results in very slow convergence in the final phase of optimization. The simple solution avoiding this difficulty is setting $E(u^*)$ slightly lower than the value of the expected error. For certain classes of problems $E(u^*)$ can be determined on different grounds. For example, in the least square problems (whether linear or nonlinear) the value of $E(u^*)$ can be determined from the boundness of the problem. Another solution is to implement adaptable strategy for $E(u^*)$.

The universality and superlinear convergence of the proposed algorithms relies on the appropriate setting of the parameters $E(u^*)$ and a. The exact values of the parameters in practice may be unknown. Though the above mentioned procedures can be applied to approximate the parameters, it should be noted that the theoretically obtained superlinear convergence no longer holds. However, even in such situation the algorithms should perform substantially well.

5 Simulations

The introduced algorithms were compared to the relevant techniques within the same class, that is, the first order methods and to the pseudo-second order method called Kick-Out [26]. Since both ALGORITHM 1 and ALGORITHM 2 are first order methods, it would be unreasonable to compare them to the second order ones. Kick-Out, however, utilizes the approximation of the second order information. It has been observed that Kick-Out outperforms conventional learning algorithms and their variations.

The effectiveness of the algorithms is practically demonstrated on five tasks represented by the following data sets: Lenses [41], Glass, Monks 1 [42], Monks 2 [42], and Monks 3 [42]. The presented algorithms were applied to training various MLP networks to perform tasks given by five, the above mentioned, data sets. Neural networks' performance was optimized according to the mean square error. Stopping criterion was the value of the expected error.

In the case of the Lenses data set [41], a neural network had configuration 4–3–1 with sigmoidal hidden units. Expected error was set to $5 \cdot 10^{-2}$. In the Glass problem, a network was configured as: 9–5–1 (sigmoidal hidden units) and the expected error was equal to 0.35. Finally, for Monks 1, 2, and 3 problems [42] a neural network structure was set as: 6–3–1 (sigmoidal hidden units), and the expected error was equal to 0.103. Network's weights were initialized randomly in the interval $< -0.1, 0.1 >$, which corresponded to the steepest region of the sigmoidal transfer function of the hidden units. ALGORITHM 1 and 2 used the value of $E(u^*) = 0$ (implying from the lower bound of the mean square error function E). The parameter a was equal to 1. In case network's error did not converge to the value less than or equal to the expected error within 20000 cycles, the training process was terminated. It is interesting to note that additional stopping condition of maximum 20000 cycles was practically applied only to the BP employing standard first order techniques. ALGORITHM 1 and 2 always converged.

Outline of the experiments follows. First, the standard BP with the constant step length term α ranging from 0.1 to 0.9 in 0.1 increments and Kick-Out algorithm were compared to ALGORITHM 1 (see Table 1). Remaining experiments were performed with the value of the step length (learning rate) corresponding to the best results of BP and Kick-Out against ALGORITHM 1 (in Monks 1 case $\alpha = 0.8$, and for all other data sets $\alpha = 0.9$). The momentum term ranging from 0.1 to 0.7 in 0.1 increments was then added. Kick-Out

198

Learning Rate α		0.1	0.2	0.3	0.4	0.5	0.6	0.7	0.8	0.9	AVERAGE Task	Total
Lenses	BP	11.1	5.58	3.68	3.58	4.05	5.28	3.99	3.96	1.79	**4.78**	
	KO	3.21	2.36	2.01	1.73	2.21	2.56	1.78	1.67	1.21	**2.08**	
Glass	BP	6.41	9.08	6.42	4.88	3.93	3.27	2.81	2.47	2.23	**4.61**	BP: 4.12
	KO	2.35	3.78	3.21	2.1	1.58	1.32	1.22	1.19	1.11	**1.98**	
Monks 1	BP	7.69	7.51	5.7	4.49	3.67	3.08	2.49	2.22	2.61	**4.38**	
	KO	3.15	3.76	2.63	2.22	1.91	1.49	1.22	1.15	1.21	**2.08**	
Monks 2	BP	9.54	4.8	3.04	2.34	1.88	1.38	1.2	1.23	1.04	**2.93**	KO: 2.08
	KO	4.1	3.12	2.72	2.52	1.23	1.19	0.98	1.1	1.12	**2.01**	
Monks 3	BP	7.2	6.58	4.37	3.2	2.66	2.78	2.9	3.71	1.64	**3.89**	
	KO	3.55	3.28	2.66	1.77	1.35	1.41	1.52	1.83	1.22	**2.07**	

Table 1. Comparison between ALGORITHM 1, standard BP and Kick-Out (KO) on several data sets. Elements of the table indicate how many times was ALGORITHM 1 faster than the standard BP and Kick-Out.

Momentum Term β		0.1	0.2	0.3	0.4	0.5	0.6	0.7	AVERAGE Task	Total
Lenses	BPM	2.05	1.26	1.11	1.88	14.66	20.62	34.84	**10.92**	
($\alpha = 0.9$)	KO	1.5	1.11	1.03	1.25	5.31	5.72	6.11	**3.15**	
Glass	BPM	2.24	2.14	1.96	1.85	1.70	1.61	1.43	**1.85**	
($\alpha = 0.9$)	KO	1.31	1.3	1.24	1.36	1.28	1.13	1.01	**1.23**	
Monks 1	BPM	2.46	2.14	1.76	1.71	2.16	8.19	14.81	**4.75**	BPM: **7.38**
($\alpha = 0.8$)	KO	1.32	1.26	1.21	1.17	1.33	2.21	3.15	**1.66**	KO: **2.18**
Monks 2	BPM	1.01	0.96	1.17	2.55	8.31	37.17	48.31	**14.21**	
($\alpha = 0.9$)	KO	0.97	0.96	1.11	1.35	3.24	5.89	6.73	**2.89**	
Monks 3	BPM	1.73	1.68	1.56	2.50	6.01	9.93	12.87	**5.18**	
($\alpha = 0.9$)	KO	1.25	1.21	1.18	1.77	2.31	2.66	3.24	**1.95**	

Table 2. Comparison of ALGORITHM 2, BPM and Kick-Out (KO) with learning rate setting corresponding to the best obtained former results against ALGORITHM 1.

and BP with the momentum term and the best value of step length (denoted in further text as BPM) were compared to ALGORITHM 2 (see results in Table 2). Kick-Out's additional parameters were set as follows: $\kappa = 0.0001$, $\phi = 0.9$ and $\tau = 0.7$. For a given setting of learning rate and momentum term the simulations were run 10 times for different randomly initialized weights in the interval $< -0.1, 0.1 >$. The values in Table 1 and 2 represent ten-run-averages. Convergence speed increase values in the tables indicate how many times the proposed algorithms converged faster than BP, BPM and Kick-Out techniques. Criterion for comparison of the convergence speed was the number of cycles required to decrease the mean square error E of a neural network below the value of the expected error.

It is clear, from Table 1 and 2 that the proposed algorithms converged
substantially faster. ALGORITHM 1 was on the average over all five tasks
approximately 4 times faster than BP and 2 times faster than Kick-Out. Performance of ALGORITHM 2 indicated approximately 7 times faster convergence than BPM and 2 times faster than Kick-Out. As previously mentioned,
ALGORITHM 1 and 2 converged each time, whereas BP and BPM for some
initial setting of weights and parameters α, β could not achieve convergence
even after 20000 cycles.

6 Conclusions

A novel classification framework for first order optimization was introduced. It
allows classification of functions as well as optimization techniques. Moreover,
the essential links between the problem classes and the optimization algorithm
classes were established. Application of this framework to the neural network
field determines the task of training MLP networks as $\mathcal{PD}^{(1,0)}$ problem.

Based on the theoretical grounds of the proposed classification framework,
superlinear algorithms with automatically adjustable step length $\alpha^{(k)}$ and/or
momentum term $\beta^{(k)}$ were introduced. ALGORITHM 1 features adjustments
of only step length $\alpha^{(k)}$ at each iteration of optimization procedure. It is memoryless with linear computational complexity $O(N_F)$, where N_F is a number
of free parameters in a system. ALGORITHM 2 dynamically adjusts step
length $\alpha^{(k)}$ and momentum term $\beta^{(k)}$. Computational complexity and memory requirements of ALGORITHM 2 are linear, $O(N_F)$. The proposed algorithms are capable of achieving the superlinear convergence rates on an
arbitrary problem. They are convergent, computationally inexpensive, easily implementable, and in practice very suitable for large scale optimization,
whether in terms of number of parameters or extensiveness of data. In cases
where amount of available memory plays inevitable role, ALGORITHM 1 is
advantageous (it is memoryless). The highest flexibility of step length $\alpha^{(k)}$ and
momentum term $\beta^{(k)}$, featured in ALGORITHM 2, however may in practice
be the best choice.

Practical validation of the presented algorithms was performed on five data
sets: Lenses, Glass, Monks 1, 2 and 3. The algorithms were compared to the
relevant first order line search optimization techniques: the steepest descent,
the conjugate gradient and Kick-Out. Simulation results show satisfactory
performance.

Acknowledgment

The authors would like to thank Dr. Shun-ichi Amari of RIKEN and Prof.
Naohiro Toda of Aichi Prefectural University for their useful comments.

Nomenclature

$\mathcal{N}$ set of integers
N_I number of input units
N_H number of hidden units
N_0 number of output units
N_P cardinality of the training set
E objective function for a neural network
J Jacobean matrix
α constant step length
$\alpha^{(k)}$ value of modifiable step length at the k-th iteration
β constant momentum term
$\beta^{(k)}$ value of modifiable momentum term at the k-th iteration
$\mathcal{AD}$ algorithm class
$\mathcal{PD}$ problem class (or function class)
$\mathcal{L}$ attractor basin of the terminal attractor u^*
$\mathcal{F}^*$ true real valued mapping
$\mathcal{F}$ mapping of a three-layer MLP network
$\mathcal{F}_{HO}$ hidden-to-output submapping
$\mathcal{F}_{IH}$ input-to-hidden mapping
$\overline{x}^{(p)}$ the p-th input vector
$\overline{y}^{(p)}$ the p-th output vector
$\Re$ real space
$\Re^{N_I}$ N_I – dimensional real input space
$\Re^{N_H}$ N_H – dimensional real hidden space
$\Re^{N_O}$ N_O – dimensional real output space
T training set
u set of free parameters of a neural network
$u^{(k)}$ set of free parameters of a neural network at the k-th iteration
u^* terminal attractor point

References

1. P. E. Gill, W. Murray, and M. H. Wright. *Practical Optimization.* Academic Press, London, 1982.
2. Leon S. Lasdon. *Optimization Theory for Large Systems.* Dover, New York, 2002.
3. J. Hiriart-Urruty and C. Lemarechal. *Convex Analysis and Minimization Algorithms I and II.* Springer-Verlag (Second Corrected Printing), Heidelberg, 1996.
4. H. Frenk, K. Roos, T. Terlaky, and S. Zhang (Editors). *High Performance Optimization.* Kluwer Academic Publishers, Dordrecht, 1999.
5. E. K. P. Chong and S. H. Zak. *An Introduction to Optimization, 2nd Edition.* John Wiley & Sons, New York, 2001.
6. C. T. Kelley. *Iterative Methods for Optimization.* SIAM, Philadelphia, 1999.
7. Ronald E. Miller. *Optimization: Foundations and Application.* John Wiley & Sons, Essex, 1999.
8. Cornelius T. Leondes. *Optimization Techniques.* Academic Press, London, 1998.
9. R. K. Sundaram. *A First Course in Optimization Theory.* Cambridge University Press, Cambridge, 1996.
10. Donald A. Pierre. *Optimization Theory with Applications.* Dover, New York, 1987.
11. A. A. Goldstein. On steepest descent. *SIAM Journal of Control,* 3:147–151, 1965.
12. P. Wolfe. Convergent conditions for ascent methods. *SIAM Review,* 11:226–235, 1969.
13. M. J. D. Powell. A view of unconstrained optimization. In L. C. W. Dixon, editor, *Optimization in Action,* London, 1976. Academic Press.
14. M. Al-Baali and R. Fletcher. An efficient line search for nonlinear least squares. *Journal of Optimization Theory and Application,* 48(3):359–377, 1986.
15. R. A. Jacobs. Increasing rates of convergence through learning rate adaptation. *Neural Networks,* 1:295–307, 1988.
16. T. P. Vogl, J. K. Manglis, A. K. Rigler, T. W. Zink, and D. L. Alkon. Accelerating the convergence of the back-propagation method. *Biological Cybernetics,* 59:257–263, 1988.
17. Ch. G. Pflug. Non-asymptotic confidence bounds for stochastic approximation algorithms. *Mathematic,* 110:297–314, 1990.
18. T. Tollenaere. SuperSAB: Fast adaptive back propagation with good scaling properties. *Neural Networks,* 3:561–573, 1990.
19. J. C. Spall. *Introduction to Stochastic Search and Optimization.* John Wiley & Sons, Essex, 2003.
20. H. J. Kushner and G. G. Jin. *Stochastic Approximation Algorithms and Applications.* Springer-Verlag, New York, 1997.
21. S. Amari. Theory of adaptive pattern classifiers. *IEEE Transactions,* EC–16(3):299–307, 1967.
22. L. Ljung. Analysis of recursive stochastic algorithms. *IEEE Transactions on Control,* AC–22(3):551–575, 1997.
23. L. Ljung. Strong convergence of stochastic approximation algorithm. *Annals of Statistics,* 6(3):680–696, 1978.

24. C. Darken and J. Moody. Note on learning rate schedules for stochastic optimization. In R. P. Lippman, J. E. Moody, and D. S. Touretzky, editors, *Proceedings of the Neural Information Processing Systems 3 (Denver)*, pp. 832–838, San Mateo, 1991. Morgan Kaufmann.

25. C. Darken and J. Moody. Towards faster stochastic gradient search. In J. E. Moody, S. J. Hason, and R. P. Lipmann, editors, *Proceedings of the Neural Information Processing Systems 4 (Denver)*, pp. 1009–1016, San Mateo, 1992. Morgan Kaufmann.

26. K. Ochiai, N. Toda, and S. Usui. Kick-Out learning algorithm to reduce the oscillation of weights. *Neural Networks*, 7(5):797–807, 1994.

27. R. Fletcher and M. J. D. Powell. A rapidly convergent descent method for minimization. *Comput. Journal*, 6:163–168, 1963.

28. R. Fletcher and C. M. Reeves. Function minimization by conjugate gradients. *Comput. Journal*, 7:149–154, 1964.

29. J. W. Daniel. Convergence of the conjugate gradient method with computationally convenient modifications. *Numerical Mathematics*, 10:125–131, 1967.

30. B. T. Polyak. The conjugate-gradient method. In *Proceedings of The Second Winter School on Mathematical Programming and Related Questions*, volume I, pp. 152–202, Moscow, 1969.

31. D. F. Shanno. Conjugate gradient methods with inexact searches. *Mathematics of Operations Research*, 3(3):244–256, 1978.

32. S. E. Fahlman. Fast learning variations on back-propagation: An empirical study. In D. Touretzky, G. Hinton, and T. Sejnowski, editors, *Proceedings of The 1988 Connectionist Models Summer School (Pittsburgh)*, pp. 38–51, San Mateo, 1989. Morgan Kaufmann.

33. S. J. Perantonis and D. A. Karras. An efficient constrained learning algorithm with momentum acceleration. *Neural Networks*, 8(2):237–249, 1995.

34. X. Yu, N. K. Loh, and W. C. Miller. A new acceleration technique for the back-propagation algorithm. In *Proceedings of The IEEE International Conference on Neural Networks*, pp. 1157–1161, San Francisco, 1993.

35. X. Yu, G. Chen, and S. Cheng. Dynamic learning rate optimization of the backpropagation algorithm. *IEEE Transactions on Neural Networks*, 6(3):669–677, 1995.

36. P. Géczy and S. Usui. Novel first order optimization classification framework. *IEICE Transactions on Fundamentals*, E83-A(11):2312–2319, 2000.

37. K. Hornik. Multilayer feedforward networks are universal approximators. *Neural Networks*, 2:359–366, 1989.

38. A. Menon, K. Mehrotra, C.K. Mohan, and S. Ranka. Characterization of a class of sigmoid functions with application to neural networks. *Neural Networks*, 6(5):819–835, 1996.

39. P. Géczy and S. Usui. Novel concept for first order learning aglorithm design. In *Proceedings of IJCNN 2001*, pp. 382–387, Washington D.C., 2001.

40. P. Géczy, S. Amari, and S. Usui. Superconvergence concept in machine learning. In P. Sinčák, J. Vaščák, V. Kvasnička, and J. Pospichal, editors, *Intelligent Technologies - Theory and Applications*, pp. 3–9, IOS Press, Amsterdam, 2002.

41. J. Cendrowska. Prism: An algorithm for inducing modular rules. *International Journal of Man-Machine Studies*, 27:349–370, 1987.

42. J. Wnek and R. S. Michalski. Comparing symbolic and subsymbolic learning: Three studies. In R. S. Michalski and G. Tecuci, editors, *Machine Learning: A Multistrategy Approach*, volume 4, San Mateo, 1993. Morgan Kaufmann.

Appendix

Proof of Theorem 1

From linear convergence rates of first order line search optimization techniques, by definition, holds,

$$a = \limsup \frac{||u^{(k+1)} - u^*||_2}{||u^{(k)} - u^*||_2} \ . \tag{34}$$

The update rule is given as follows,

$$u^{(k+1)} = u^{(k)} - \Delta u^{(k)} \ . \tag{35}$$

Substituting (35) into (34) it is obtained,

$$a = \limsup \frac{||u^{(k)} - \Delta u^{(k)} - u^*||_2}{||u^{(k)} - u^*||_2}$$

$$a = \limsup \frac{||(u^{(k)} - u^*) - \Delta u^{(k)}||_2}{||u^{(k)} - u^*||_2} \ . \tag{36}$$

Applying triangle inequality to the numerator of the expression (36) it follows,

$$a = \limsup \frac{||(u^{(k)} - u^*) - \Delta u^{(k)}||_2}{||u^{(k)} - u^*||_2} \geq \limsup \left(\frac{|\,||u^{(k)} - u^*||_2 - ||\Delta u^{(k)}||_2\,|}{||u^{(k)} - u^*||_2} \right)$$

$$a \geq \limsup \left(\left| 1 - \frac{||\Delta u^{(k)}||_2}{||u^{(k)} - u^*||_2} \right| \right) \ . \tag{37}$$

From the Taylor expansion of the objective function E around the optimum point u^*, we have,

$$E(u^*) = E(u^{(k)}) + \nabla E(u^{(k)}) \cdot (u^{(k)} - u^*)^T + R_{n \geq 2}$$

$$E(u^*) - E(u^{(k)}) - R_{n \geq 2} = \nabla E(u^{(k)}) \cdot (u^{(k)} - u^*)^T \ . \tag{38}$$

Putting both sides of (38) to absolute values and applying Hölder's inequality leads to the expression,

$$|E(u^*) - E(u^{(k)}) - R_{n \geq 2}| \leq ||\nabla E(u^{(k)})||_2 \cdot ||u^{(k)} - u^*||_2$$

$$\frac{|E(u^*) - E(u^{(k)}) - R_{n \geq 2}|}{||\nabla E(u^{(k)})||_2} \leq ||u^{(k)} - u^*||_2 \ . \tag{39}$$

Substituting (39) into (37) it is obtained,

$$a \geq \limsup \left(\left| 1 - \frac{||\Delta u^{(k)}||_2 \cdot ||\nabla E(u^{(k)})||_2}{|E(u^*) - E(u^{(k)}) - R_{n \geq 2}|} \right| \right) \ . \tag{40}$$

Multiplication of both numerator and denominator of (40) by the term

$$\frac{1}{E(u^*) - E(u^{(k)})}$$

leads to the following,

$$a \geq \limsup \left(\left| 1 - \frac{\frac{\|\Delta u^{(k)}\|_2 \cdot \|\nabla E(u^{(k)})\|_2}{|E(u^*)-E(u^{(k)})|}}{\left| 1 - \frac{R_{n>2}}{E(u^*)-E(u^{(k)})} \right|} \right| \right) . \tag{41}$$

From Taylor expansion of E around the optimum point u^* we have,

$$E(u^*) - E(u^{(k)}) = R_{n \geq 1} . \tag{42}$$

Then substitution of (42) into (41) finally leads to the expression,

$$a \geq \left| 1 - \limsup \frac{\frac{\|\Delta u^{(k)}\|_2 \cdot \|\nabla E(u^{(k)})\|_2}{|E(u^*)-E(u^{(k)})|}}{\left| 1 - \frac{R_{n>2}}{R_{n \geq 1}} \right|} \right| .$$

■

Proof of Corollary 1

From assumption of superlinear convergence rates $(a = 0)$ implies,

$$0 \geq \left| 1 - \limsup \frac{\frac{\|\Delta u^{(k)}\|_2 \cdot \|\nabla E(u^{(k)})\|_2}{|E(u^*)-E(u^{(k)})|}}{\left| 1 - \frac{R_{n>2}}{R_{n \geq 1}} \right|} \right| . \tag{43}$$

Thus it immediately follows from (43) for any convergent sequence $\{u^{(k)}\}_k \to u^*$ the following,

$$\limsup \frac{\frac{\|\Delta u^{(k)}\|_2 \cdot \|\nabla E(u^{(k)})\|_2}{|E(u^*)-E(u^{(k)})|}}{\left| 1 - \frac{R_{n>2}}{R_{n \geq 1}} \right|} = 1 .$$

■

Proof of Corollary 2

Implying from (6) it is obtained,

$$\limsup \frac{\frac{\|\Delta u^{(k)}\|_2 \cdot \|\nabla E(u^{(k)})\|_2}{|E(u^*)-E(u^{(k)})|}}{\left| 1 - \frac{R_{n>2}}{R_{n \geq 1}} \right|} = 1$$

$$1 = \frac{\limsup \frac{\|\Delta u^{(k)}\|_2 \cdot \|\nabla E(u^{(k)})\|_2}{|E(u^*)-E(u^{(k)})|}}{\limsup \left| 1 - \frac{R_{n>2}}{R_{n \geq 1}} \right|} = \frac{\limsup \frac{\|\Delta u^{(k)}\|_2 \cdot \|\nabla E(u^{(k)})\|_2}{|E(u^*)-E(u^{(k)})|}}{\left| 1 - \limsup \frac{R_{n>2}}{R_{n \geq 1}} \right|} . \tag{44}$$

Then from assumption (7) and (44) it is immediately obtained,

$$\limsup \frac{\|\Delta u^{(k)}\|_2 \cdot \|\nabla E(u^{(k)})\|_2}{|E(u^*) - E(u^{(k)})|} = 1 .$$

■

Proof of Theorem 2

The proof is shown for an arbitrary parameter u_l of E, $u_l \in u$. For functions in the class $\mathcal{PD}^{(1,0)}$ holds: $\limsup \frac{R_{n\geq2}(E)}{R_{n\geq1}(E)} = 0$. Then,

$$\limsup \frac{R_{n\geq2}(E)}{R_{n\geq1}(E)} = \limsup \frac{R_{n\geq2}(E)}{\frac{\partial E}{\partial u_l}(u_l^* - u_l) + R_{n\geq2}(E)} =$$

$$= \limsup \frac{1}{\frac{\frac{\partial E}{\partial u_l}(u_l^* - u_l)}{R_{n\geq2}(E)} + 1} = \frac{1}{\underbrace{\limsup \frac{\frac{\partial E}{\partial u_l}(u_l^* - u_l)}{R_{n\geq2}(E)}}_{(a)} + 1} \quad (45)$$

In the further part of this proof we analyze (a) in (45).

$$\limsup \frac{\frac{\partial E}{\partial u_l}(u_l^* - u_l)}{R_{n\geq2}(E)} = \limsup \frac{\frac{\partial E}{\partial u_l}(u_l^* - u_l)}{\sum_{n=2} \frac{1}{n!} \frac{\partial^n E}{\partial u_l^n}(u_l^* - u_l)^n} =$$

$$= \limsup \frac{\frac{\partial E}{\partial u_l}}{\sum_{n=2} \frac{1}{n!} \frac{\partial^n E}{\partial u_l^n}(u_l^* - u_l)^{n-1}} \; . \quad (46)$$

Implying from (46), if the sequence $\{u_l^{(k)}\} \to u_l^*$ does not converge to the stationary point, that is for $u_l^* \Rightarrow \frac{\partial E}{\partial u_l}(u_l^*) \neq 0$, under assumption of boundness of higher order derivatives $\frac{\partial^n E}{\partial u_l^n}$,

$$\limsup \frac{\frac{\partial E}{\partial u_l}}{\sum_{n=2} \frac{1}{n!} \frac{\partial^n E}{\partial u_l^n}(u_l^* - u_l)^{n-1}} = \infty \quad (47)$$

holds. Thus from (47) and (45) we have,

$$\limsup \frac{R_{n\geq2}(E)}{R_{n\geq1}(E)} = 0 \quad \Rightarrow \quad E \in \mathcal{PD}^{(1,0)} \; .$$

If u_l^* is a stationary point (to which the sequence $\{u_l^{(k)}\}$ converges), from (46) by L'Hospital rule follows,

$$\limsup \frac{\frac{\partial^2 E}{\partial u_l^2}}{\sum_{n=2} \frac{1}{n!} \frac{\partial^{(n+1)} E}{\partial u_l^{(n+1)}}(u_l^* - u_l)^{n-1} + \sum_{n=2} \frac{1}{n!} \frac{\partial^n E}{\partial u_l^n}(n-1)(u_l^* - u_l)^{n-2}} = \infty$$

for arbitrary convergent sequence $\{u_l^{(k)}\} \to u_l^*$ under the assumption,

$$\forall n \geq 2 \left(\frac{\partial^n E}{\partial u_l^n} \pm \infty \quad , \quad n \in N \right) \quad \wedge \quad \frac{\partial^2 E}{\partial u_l^2} \neq 0 \; .$$

$\blacksquare$

Proof of Theorem 3

*Implying from Theorem 2, a function E in the class $\mathcal{PD}^{(1,0)}$ must satisfy (11) and (12) or (13). Thus to satisfy **II.** of Theorem 2 it is sufficient to satisfy (13). For satisfaction of (13) from Lemma 2: (20(a)), (23) implies,*

$$\exists u_l \in u \quad \left(\frac{\partial \mathcal{F}}{\partial u_l} \neq 0 \right) . \tag{48}$$

The satisfaction of (48) for the least square training problem of a three-layer MLP network implies directly from (14) of Lemma 1 for $u_l \equiv \theta_{o_k}$: $\frac{\partial \mathcal{F}}{\partial \theta_{o_k}} = 1$.

*In order to satisfy the condition **I.** of Theorem 2, that is (11), for E from (21), (22), and (23) must hold,*

$$\frac{\partial E}{\partial \mathcal{F}} = ||\mathcal{F}(u) - \mathcal{F}^*||_1 \neq \pm\infty , \tag{49}$$

and for $\mathcal{F}$ from (21) must hold,

$$\forall u_l \in u \quad \left(\frac{\partial^n \mathcal{F}}{\partial u_l^n} \neq \pm\infty \right) . \tag{50}$$

Expression (49) implies that E and $\mathcal{F}$ must be bounded. As for (50), for each $u_l \in u$ we have,

θ_{o_k} : (50) *obviously satisfied.*

w_{jk} : $n = 1$: $\dfrac{\partial \mathcal{F}}{\partial w_{jk}} = \mathcal{F}_{IH_j}, \quad \Rightarrow \ \forall j = 1, \ldots, N_H$ (f_j *must be bounded*)

$\qquad\quad n \geq 2$: (50) *obviously satisfied.*

θ_{h_j} : $n \geq 1$: $\dfrac{\partial^n \mathcal{F}}{\partial \theta_{h_j}^n} = \displaystyle\sum_{q=1}^{N_O} w_{jq} \dfrac{\partial^n \mathcal{F}_{IH_j}}{\partial \theta_{h_j}^n}$

$$\Rightarrow \ w_{jq} \neq \pm\infty \ \wedge \ \forall n \geq 1 \left(\frac{\partial^n f_q}{\partial \theta_{h_j}^n} \neq \pm\infty, \ n \in \mathcal{N} \right)$$

v_{ij} : $n \geq 1$: $\dfrac{\partial^n \mathcal{F}}{\partial v_{ij}} = \displaystyle\sum_{q=1}^{N_O} w_{jq} \dfrac{\partial^n \mathcal{F}_{IH_j}}{\partial \mathcal{A}_{IH_i}^{\ n}} \cdot x_i^n$

$$\Rightarrow \ w_{jq} \neq \pm\infty \ \wedge \ x_i \neq \pm\infty \ \wedge \ \forall n \geq 1 \left(\frac{\partial^n f_q}{\partial \mathcal{A}_{IH_i}^{\ n}} \neq \pm\infty, \ n \in \mathcal{N} \right)$$

Hence, given the assumptions of Theorem 3 and the above results the proof immediately follows. ∎

Proof of Theorem 4

ALGORITHM 1: *First, it is shown that ALGORITHM 1 decreases the objective function E. Taylor expansion of E around the point $u^{(k+1)}$ is as follows,*

$$E(u^{(k+1)}) = E(u^{(k)}) + \nabla E(u^{(k)}) \cdot (u^{(k+1)} - u^{(k)})^T + R_{n \geq 2} \; . \qquad (51)$$

From the step 5 of ALGORITHM 1 we have,

$$u^{(k+1)} = u^{(k)} - \alpha^{(k)} \nabla E(u^{(k)})$$
$$u^{(k+1)} - u^{(k)} = -\alpha^{(k)} \nabla E(u^{(k)}) \; . \qquad (52)$$

Hence from (51) and (52) follows,

$$E(u^{(k+1)}) - E(u^{(k)}) = \nabla E(u^{(k)}) \cdot (u^{(k+1)} - u^{(k)})^T + R_{n \geq 2}$$
$$E(u^{(k)}) - E(u^{(k+1)}) = \alpha^{(k)} ||\nabla E(u^{(k)})||_2^2 - R_{n \geq 2} \; . \qquad (53)$$

Neglecting second and higher order terms $R_{n \geq 2}$ of Taylor expansion in (53) it is obtained,

$$E(u^{(k)}) - E(u^{(k+1)}) \approx \alpha^{(k)} ||\nabla E(u^{(k)})||_2^2 \; , \qquad (54)$$

and thus,

$$\frac{E(u^{(k)}) - E(u^{(k+1)})}{||\nabla E(u^{(k)})||_2^2} \approx \alpha^{(k)} \; . \qquad (55)$$

From boundness of the objective function E, and from the expressions (53) and (54) implies that the objective function E decreases from the step k to the step $k+1$ by the factor: $\alpha^{(k)} ||\nabla E(u^{(k)})||_2^2 - R_{n \geq 2}$, or approximately by the factor: $\alpha^{(k)} ||\nabla E(u^{(k)})||_2^2$.

In the next part of the proof is shown convergence to the terminal attractor point u^*. If $E(u^{(k+1)})$ is not the stationary point then there exists $E(u_A)$ such that at any iteration k the following holds,

$$\exists u_A \in \mathcal{L} \; (E(u^{(k)}) - E(u^{(k+1)}) \leq E(u^{(k)}) - E(u_A)) \; . \qquad (56)$$

Since the objective function E is bounded on $\mathcal{L}$ with the terminal attractor u^*, the point $u_A \equiv u^*$. Hence substituting for $\alpha^{(k)}$ to (55) leads to the following inequality,

$$\forall k = 1, 2, \ldots \; \left(\frac{E(u^{(k)}) - E(u^{(k+1)})}{||\nabla E(u^{(k)})||_2^2} \leq a \cdot \frac{E(u^{(k)}) - E(u^*)}{||\nabla E(u^{(k)})||_2^2} \right) \; . \qquad (57)$$

Expression (54) together with (57) implies that ALGORITHM 1 with the modifiable step length $\alpha^{(k)}$ decreases the objective function E at each iteration for both $E(u^*) \neq 0$ and $E(u^*) = 0$ if $a \leq 1$. Since the objective E has minimum $E(u^*)$, ALGORITHM 1 decreases E to the minimum $E(u^*)$, $E(u^{(k)}) \to E(u^*)$. Hence the sequence $\{u^{(k)}\}_k$ generated by ALGORITHM 1 must converge to the terminal attractor u^*, $\{u^{(k)}\}_k \to u^*$.

ALGORITHM 2: *Again, first it is shown that ALGORITHM 2 iteratively decreases the objective function E. Taylor expansion of the objective function E around the point $u^{(k+1)}$ has the form,*

$$E(u^{(k+1)}) = E(u^{(k)}) + \nabla E(u^{(k)}) \cdot (u^{(k+1)} - u^{(k)})^T + R_{n \geq 2}$$
$$E(u^{(k+1)}) - E(u^{(k)}) = \nabla E(u^{(k)}) \cdot (u^{(k+1)} - u^{(k)})^T + R_{n \geq 2}$$
$$E(u^{(k+1)}) - E(u^{(k)}) \geq \nabla E(u^{(k)}) \cdot (u^{(k+1)} - u^{(k)})^T \ . \tag{58}$$

Following the parameter update formula of ALGORITHM 2 (step 5) we have,

$$u^{(k+1)} = u^{(k)} - \alpha^{(k)} \nabla E(u^{(k)}) + \beta^{(k)} s^{(k-1)}$$
$$u^{(k+1)} - u^{(k)} = -\alpha^{(k)} \nabla E(u^{(k)}) + \beta^{(k)} s^{(k-1)} \ . \tag{59}$$

Substituting (59) into (58) it is obtained,

$$E(u^{(k+1)}) - E(u^{(k)}) \geq \nabla E(u^{(k)}) \cdot (-\alpha^{(k)} \nabla E(u^{(k)}) + \beta^{(k)} s^{(k-1)})^T$$
$$E(u^{(k+1)}) - E(u^{(k)}) \geq -\alpha^{(k)} ||\nabla E(u^{(k)})||_2^2 + \beta^{(k)} \nabla E(u^{(k)}) \cdot s^{(k-1)^T} \ . \tag{60}$$

By Hölder's inequality from (60) follows,

$$E(u^{(k)}) - E(u^{(k+1)}) \leq$$
$$\leq |\alpha^{(k)}| \cdot ||\nabla E(u^{(k)})||_2^2 - |\beta^{(k)}| \cdot ||\nabla E(u^{(k)})||_2 \cdot ||s^{(k-1)}||_2 \ . \tag{61}$$

To satisfy (61) so as to iteratively decrease the objective function E the following must hold,

$$\alpha^{(k)} ||\nabla E(u^{(k)})||_2^2 - |\beta^{(k)}| \cdot ||\nabla E(u^{(k)})||_2 \cdot ||s^{(k-1)}||_2 \geq 0$$
$$\alpha^{(k)} ||\nabla E(u^{(k)})||_2^2 \geq |\beta^{(k)}| \cdot ||\nabla E(u^{(k)})||_2 \cdot ||s^{(k-1)}||_2 \ . \tag{62}$$

Substituting for $\alpha^{(k)}$ and $\beta^{(k)}$ into (62) we have,

$$a \cdot E(u^{(k)}) \geq \frac{a \cdot E(u^*)}{||s^{(k-1)}||_2 \cdot ||\nabla E(u^{(k)})||_2} \cdot ||s^{(k-1)}||_2 \cdot ||\nabla E(u^{(k)})||_2$$
$$E(u^{(k)}) \geq E(u^*) \ . \tag{63}$$

Since E has minimum $E(u^*)$ at the terminal attractor point $u^* \in \mathcal{L}$, the expression (63) is obviously satisfied. Hence ALGORITHM 2 decreases the objective function E at each iteration. Thus from boundness of the objective function E implies that ALGORITHM 2 decreases the the objective function E to its minimum $E(u^*)$, and thus the sequence of points $\{u^{(k)}\}_k$ generated by ALGORITHM 2 from the initial point $u^{(0)} \in \mathcal{L}$ converges to the terminal attractor point u^*, $\{u^{(k)}\}_k \to u^*$. ∎

Proof of Theorem 5
ALGORITHM 1: *According to the parameter update of ALGORITHM 1 (step 5) and dynamic step length adjustments of $\alpha^{(k)}$, it is obtained,*

$$||\Delta u^{(k)}||_2 = |\alpha^{(k)}| \cdot ||\nabla E(u^{(k)})||_2$$
$$||\Delta u^{(k)}||_2 = a \cdot \frac{|E(u^*) - E(u^{(k)})|}{||\nabla E(u^{(k)})||_2^2} \cdot ||\nabla E(u^{(k)})||_2$$
$$||\Delta u^{(k)}||_2 = a \cdot \frac{|E(u^*) - E(u^{(k)})|}{||\nabla E(u^{(k)})||_2} \ . \tag{64}$$

Substituting (64) into the expression for $\mathcal{AD}$ (9) it follows,

$$\limsup \frac{a \cdot \frac{|E(u^*) - E(u^{(k)})|}{||\nabla E(u^{(k)})||_2} \cdot ||\nabla E(u^{(k)})||_2}{|E(u^*) - E(u^{(k)})|} = \limsup \frac{a \cdot |E(u^*) - E(u^{(k)})|}{|E(u^*) - E(u^{(k)})|} = a \ .$$

ALGORITHM 2: *Given the parameter update of ALGORITHM 2 (step 5) it follows,*

$$u^{(k+1)} = u^{(k)} - \alpha^{(k)} \nabla E(u^{(k)}) + \beta^{(k)} s^{(k-1)} \ ,$$

where,

$$\Delta u^{(k)} = \alpha^{(k)} \nabla E(u^{(k)}) - \beta^{(k)} s^{(k-1)}$$

$$||\Delta u^{(k)}||_2 = ||\alpha^{(k)} \nabla E(u^{(k)}) - \beta^{(k)} s^{(k-1)}||_2 \ . \tag{65}$$

Applying triangle inequality to (65) we have,

$$||\Delta u^{(k)}||_2 \geq \left| |\alpha^{(k)}| \cdot ||\nabla E(u^{(k)})||_2 - |\beta^{(k)}| \cdot ||s^{(k-1)}||_2 \right| \ . \tag{66}$$

Then substituting (66) into the expression for $\mathcal{AD}$ (9) it is obtained,

$$\limsup \left(\left| |\alpha^{(k)}| \cdot ||\nabla E(u^{(k)})||_2 - |\beta^{(k)}| \cdot ||s^{(k-1)}||_2 \right| \right) \cdot \frac{||\nabla E(u^{(k)})||_2}{|E(u^*) - E(u^{(k)})|} =$$

$$\limsup \left(\left| |\alpha^{(k)}| \cdot \frac{||\nabla E(u^{(k)})||_2^2}{|E(u^*) - E(u^{(k)})|} - |\beta^{(k)}| \cdot \frac{||s^{(k-1)}||_2 \cdot ||\nabla E(u^{(k)})||_2}{|E(u^*) - E(u^{(k)})|} \right| \right) \tag{67}$$

By substituting for $|\alpha^{(k)}|$ and then for $|\beta^{(k)}|$ into (67) it is obtained: $|a|$. ∎

Extension of Binary Neural Networks for Multi-class Output and Finite Automata

Narendra S. Chaudhari[1] and Aruna Tiwari[2]

[1] School of Computer Engineering (SCE), Block N4-02a-32, 50 Nanyang Avenue,
   Nanyang Technological University (NTU), Singapore 639798 SINGAPORE
   Email: asnarendra@ntu.edu.sg; nsc@lycos.com
   and
[2] Department of Computer Engineering, Shri G.S. Inst. of Tech. & Sci. (SGSITS),
   23, Park Road, Indore 452003 (M.P.) INDIA
   Email: aruna_tiwari@rediffmail.com

Abstract.   Neural networks implementing Boolean functions are known as Binary Neural Networks (BNNs). Various methods for construction of BNNs have been introduced in the last decade. Many applications require BNNs to handle multiple classes. In this paper, we first review some basic methods proposed in the last decade for construction of BNNs. We summarize main approach in these methods, by observing that a neuron can be visualized in terms of its equivalent hypersphere. Next, we give some approaches for adopting a BNN construction process for classification problem that needs to classify data into multiple (more than two) classes. We illustrate these approaches using examples. From the theoretical view, the limited applicability of BNNs does not come in the way of expressing a Finite Automaton (FA) in terms of recurrent BNNs. We prove that recurrent BNNs simulate any deterministic as well as non-deterministic finite automaton. The proof is constructive, and the construction process is illustrated by suitable examples.

Keywords: Binary neural network, Recurrent neural network, Hard-limiter neuron, Boolean function, Finite automaton, Nondeterminism.

1   Introduction

Binary Neural Networks (BNNs), or neural networks having Boolean values for their inputs and an output, form a highly restricted model of neural networks. For practical applications, they are generally considered to be of very little use due to their restrictive nature. The classical work by Marvin Minsky and Seymour Papert investigate and highlight the limitations of many such approaches in the context of perceptrons [1]. Seigelmann gives a more recent survey of limitations of many approaches for networks having binary, or integral inputs, outputs, and weights, and develops a framework for handling analog values [2].

211

However, during the last decade, researche.s have explored their use in many fields such as data mining, recognition and classification [3]. Chu and Kim (Oct. 1993) investigate the applications of BNNs for pattern classification of breast cancer data [4]. Further, Kim, Ham, and Park (Oct. 1993) report the applications of BNNs for handwritten digit recognition [5]. In 1997, Windeatt and Tebbs have applied BNN to identify the distorted UV-visible spectra [6]. Non-ordinal variables can be modeled in the framework of BNNs. Vinay Deolalikar mapped a Boolean Function to BNNs with zero threshold and binary $\{1, -1\}$ weights. He substituted a pair of input and output, say X and Y, as a new single normalized variable, say Z_{XY}, which can convert multiple classification problem to two-classification problem [7,8]. Aizenberg *et.al.* introuduce a concept or multivalued and universal binary neurons, discuss learning strategies, and their applications to image processing and pattern recognition [9].

Use of BNNs for many applications need systematic methods to adopt existing construction methods for handling multiple classes. In this paper, we introduce an approach for this problem. To develop this approach for BNNs, in section 2, we first give a brief survey of the construction methods developed in the last decade, and summarize main concepts these approaches. In section 3, we then illustrate our proposed approaches to handle multiple classes and illustrate them using examples.

For highly restrictive models like BNNs, there seems to be a general view that BNN is "highly inadequate" for any practical problem, and hence it is not worthy of consideration for practical problems. We feel that serious discussion for such issues needs careful formulation of a given problem, as well as investigations the computational capabilities of a model. To give a sound basis for investigations about the generality of BNNs for various applications, we focus our attention to the computational power of BNN representation. We now state some of the earlier works in this area. Forcada and Carrasco investigated the relationship and conversion algorithm between finite-state computation and neural network learning [10]. Stephan Mertens and Andreas Engel investigated the *Vapnik-Chervonenkis* (*VC*) dimension of neural networks with binary weights and obtained the lower bounds for large systems through theoretical argument [11]. Kim and Roche discussed and answered [12] two mathematical questions for BNNs: (i) there exists a ρ ($0 < \rho < 1$), such that for all sufficiently large n there is a BNN of n hidden neurons which can separate ρn (unbiased) random patterns with probability close to 1; (ii) it is impossible for a BNN of n hidden neurons to separate $(1 - o(1))n$ random patterns with probability greater than some positive constant.

For such comparisons, the recurrent neural network structure (in which output is used to determine the next input in the recurrent phase) is employed. Recently, a number of papers have explored the ability of recurrent neural networks to learn a symbolic representation of the language [13-26]. Many of the successful approaches mainly concentrate on construction of a deterministic finite state automaton (DFA). For example, one of the important approaches is reported in Omlin and Giles [14, 13,15], where they have presented a second order recurrent neural network that learns DFA. Omlin and Giles have used sigmoid discriminator activation function for the neurons. These proposals include the adaptation of back-propagation technique. These approaches do not use hard-limiting activation functions, and do not consider

BNNs. Hence, from the theoretical as well as practical point of view, it is of interest to investigate the capabilities of BNNs systematically.

In section 4 and 5, we prove that BNNs using hard-limiter neurons are capable of simulating any deterministic as well as non-deterministic automaton. Establishing this link is crucial for the adaptation of "learning" ideas in the field of computational learning in the area of BNN construction. Brief concluding remarks are included in section 6.

2 A Review of Approaches for BNN Construction

Historically, we note that Minsky and Perpet's influential work (1988) proved that neural network models, e.g., Rosenblatt's "Perceptron", is an inadequate tool to represent Boolean logic function XOR. Further they went in depth to prove the inadequacy of neural network approach for many problems [1]. This inadequacy effectively delayed acceptance of neural network models, particularly in United States. However, it did not stop numerous researchers from investigations in neural networks, and their use for machine learning. Systematic understanding of neural network models for implementing any arbitrary Boolean function started emerging only in early 1990's.

Many training algorithms for neural networks have been proposed since 1960's. These training algorithms can be classified into two categories based on their training process. One category fixes the network structure (the number of hidden layers and the number of neurons in each hidden layer); then connection weights and thresholds in parameter space are adjusted by decreasing errors between model outputs and desired outputs. Such methods typically use back-propagation (BP) for modifying the weights. These kinds of algorithms cannot guarantee fast convergence and need more training time. The other category adds hidden layers and hidden neurons in the training process, which are called constructive training algorithms. Examples of such methods are Expand-and-Truncate Learning (ETL) algorithm [35], and Constructive Set Covering Learning Algorithm (CSCLA) [34]. Constructive training algorithms are promising because they guarantee faster convergence and need less training time.

The paradigm of constructive algorithms developed for training BNNs in 1990's have significantly contributed to our understanding of the construction process of BNNs. Donald L. Gray and Anthony N. Michel devised Boolean-Like Training Algorithm (BLTA) for construction of BNNs in 1992 [27]. BLTA does well in memorization and generalization, but many hidden neurons are needed. Jung Kim and Sung Kwon Park proposed Expand-and-Truncate Learning (ETL) algorithm in 1995 [34]. They defined set of included true vertices (SITV) as a set of true vertices, which can be separated from the remaining vertices by a hyperplane. The status of "true" and "false" vertices is swapped, if SITV cannot be expanded further. Such results, like BLTA, and ETL, constructively established that hard-limiter BNN interconnection structures are "powerful enough" to express any Boolean function.

Atsushi Yamamoto and Toshimichi Saito improved ETL (called IETL) by modifying some vertices in SITV as "don't care" [33]. Fewer neurons are needed in IETL.

ETL and IETL begin with selecting a true vertex as the core vertex for SITV. In both of these methods, the number of hidden neurons depends on the choice of the core vertex and an order to examine the status of vertex. Different choice of core vertex and different examining order cause different structure of neural nets. In addition, ETL and IETL need search a lot of training pairs for determining each neuron in the hidden layer. If h hidden neurons are needed for n-dimensional inputs, the number of operations needed are $O(h2^n)$.

Ma Xiaomin introduced the idea of weighted Hamming distance hypersphere in 1999, which improved the representation ability of each hidden neuron, hence improved the learning ability of BNNs [32]. In his later research in 2001, based on the idea of weighted Hamming distance hypersphere he proposed Constructive Set Covering Learning Algorithm (CSCLA) [34]. CSCLA needs an ancillary Neural Network. Hence it results in double work for training space. In his paper, he only considered including vertices with Hamming distance one from the core, not including vertices with Hamming distance more than one in a hidden neuron. So Xiaomin's neural networks have more hidden neurons [34]. Sang-Kyu Sung *et.al.* proposed an optional synthesis method for BNNs using Newly Expanded and Truncated Learning Algorithm (NETLA) to minimize not only the number of connections but also the number of neurons in hidden layer in 2002 [36].

Bernd Sternbach and Roman Kohut discussed how to transfer linearly inseparable mapping to linearly separable mapping by expanding the input dimensions [37]. If we wish to restrict ourselves in lower input dimensions, however, we have to allow a nonlinear hidden neuron to take care of the representation of a given class. A nonlinear hidden neuron has greater representation ability than a linear hidden neuron, but the computation is more complex. Janusz Starzyk and Jing Pang [38] proposed evolvable BNNs for data classification. They introduced the framework for using the ideas in evolutionary computing, for BNN training by generating new features (combination of the input bits), and then selecting some features which make more contribution (activation or inhibition) to linear separability.

Chaudhari and Tiwari proposed combination of BLTA and ETL for adapting BNNs to handle multiple classes, as needed for many classification problems [39]. Wang and Chaudhari introduced an alternative method to train BNNs, in which they begin with several core vertices; this approach is called as Multi-Core Learning (MCL) [40]. They later extended this approach to include expand-and-truncate mechanism as well, and the resulting approach is called as and Multi-Core Expand-and-Truncate Learning (MCETL) [41]. The issue of selection of (multiple) core vertices is left unanswered in their approach. One systematic approach for handling this problem is proposed by Chaudhari and Wang, in which they introduced an idea of multi-level geometrical expansion. In their approach, the learning process is based on the judgment about the region the new vertex belongs to. This algorithm possesses the generalization capability and need no core vertex before training. Further, this method is relatively less sensitive to different input sequences (different input sequences result in similar net structure) [42].

In the remaining part of this section, we review some main concepts in the above constructive learning approaches. Specifically, we give some details of geometric approach for BNNs in section 2.1. Next, in section 2.2, some discussion about the generic approaches for BNN learning is given. In section 2.3, we illustrate how we may visualize the half-space represented by a hard-limiter neuron can be visualized as representing a sphere in n-dimensions.

2.1 A Geometric Approach for BNN Learning: ETL of Kim and Park

Jung Kim and Sung Kwon Park's Expand-and-Truncate Learning (ETL) algorithm is based on geometrical concepts [35], and discussion of some aspects of ETL is useful for appreciating various approaches proposed by various researchers in the last decade. ETL finds a set of required separating hyperplanes and determines the connection weights and thresholds based on geometrical analysis of the given training set. Hence, compared with BP algorithm, the learning speed of ETL is faster. In ETL only integral connection weights and thresholds are used, which greatly facilitate the hardware implication. ETL automatically determines a required number of neurons in the hidden layer.

ETL constructs a neural network having input layer, hidden layer, and an output layer. The structure of such a network is given in Fig. 1.

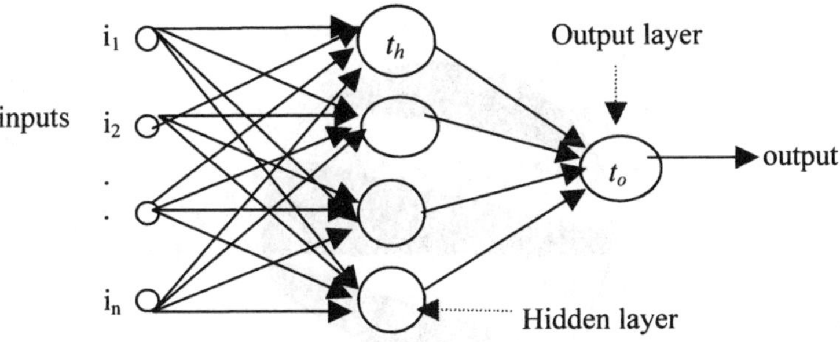

Fig. 1. A neural network structure constructed by ETL.

A set of 2^n binary patterns each with n bits may be considered as forming 2^n vertices of an n-dimensional hypercube. Assume that these patterns can be separated into two sets (true and false) by k $(n\text{-}1)$-dimensional hyperplanes. Kim and Park's ETL algorithm to achieve this separation proceeds as follows. They first define a set of true vertices, which can be separated from the remaining vertices by a hyperplane, as a set of included true vertices (SITV). Using this concept, in their method, they change the status of "true" and "false" vertices, if SITV cannot be expanded further. Thus vertices between two consecutive hyperplanes have the same desired outputs. The two consecutive groups separated by a hyperplane have different desired outputs. So these k $(n\text{-}1)$-dimensional hyperplanes partition all vertices into $k+1$ groups. Vertices in each group have the same desired outputs (either 0 or 1).

ETL begins with selecting one core vertex in SITV. The vertices, which are not included in SITV, are examined one by one. Next, SITV is expanded (using approaches (i), (ii), or (iii) given below in section 2.2) to include as many vertices as it can. If no more vertices can be added to SITV, the first separating hyperplane is found. However, if we have not separated all true vertices from false vertices using this hyperplane, a second separating hyperplane is to be found. To obtain the second hyperplane, false vertices are converted to true vertices, and true vertices which are not in SITV are converted to false vertices, and the second separating hyperplane is obtained. This process goes on until all true vertices are separated from all false vertices.

We illustrate the training process of ETL using analogy shown in Fig. 2. White regions stand for true subsets and black regions stand for false subsets. The number in each region stands for the order of generating hyperplanes (hidden neurons). Based on the selected true core vertex (in region 1), ETL begins to extend SITV to cover as many true vertices as possible. When SITV covers region 1 (reaches the boundary of region 1 and region 2), it meets a false vertex, which prevents SITV from further expansion. Then false vertices out of region 1 are converted to true vertices and true vertices out of region 1 are converted to false vertices. Hence the false vertex which blocks the expansion will not block it now. SITV then expands to include false vertices in region 2 until it reaches region 3. This process goes on until it covers all true vertices or all the remaining false vertices.

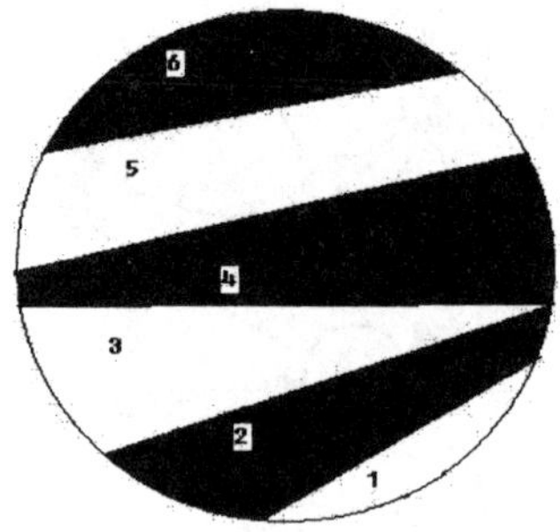

Fig. 2. Training process of ETL.

If only one (n-1)-dimensional hyperplane (one hidden neuron) is needed, this function is linearly separable. Otherwise, the inseparable function should be decomposed into multiple linearly separable functions. It is easy to show that any binary-to-binary mapping can be decomposed into a series of linearly separable functions. ETL can decompose an arbitrary linearly inseparable function into multiple linearly separable functions, each of which is realized by a neuron in the hidden layer. The number of hidden neurons needed equals to the number of separating hyperplanes.

The jth hyperplane can be expressed by a hidden neuron j:

$$\sum_{i=1}^{n} w_{ij}^{h} x_i - T_j = 0, \tag{1}$$

where x_i is the ith bit of the input, w_{ij}^{h} is the connection weight of the jth hidden neuron for x_i, and T_j is the threshold for the jth hidden neuron.

ETL is based on the result that any binary-to-binary mapping can be decomposed into a series of linearly separable functions:

$$y(x_1, x_2, \cdots, x_n) = x_1 \theta(x_2 \theta(\cdots(x_{n-1} \theta x_n))\cdots), \tag{2}$$

where x_i is the ith bit of the input and operator θ is either logical *AND* or logical *OR*. The hard-limiter activation function for the jth hidden neuron can be expressed by

$$y_j = \begin{cases} 1 & if \quad \Sigma_{i=1}^{n} w_{ij}^{h} x_i - T_j \geq 0 \\ 0 & otherwise \end{cases}, \tag{3}$$

where y_j is the output of the jth hidden neuron.

2.2 Some Approches for BNN Construction

In their ETL framework, to construct hidden neurons, Kim and Park proposed three approaches based on different situations, subject to satisfying certain conditions [35]:

<u>Approach (i):</u>

$$w_{ij}^{h} = 1, \text{ if } f(x_i) = 1 \text{ and } x_c^{i} = 1, \tag{4}$$
$$w_{ij}^{h} = -1, \text{ if } f(x_i) = 1 \text{ and } x_c^{i} = 0, \tag{5}$$
$$w_{ij}^{h} = 2, \text{ if } f(x_i) = 0 \text{ and } x_c^{i} = 1, \tag{6}$$
$$w_{ij}^{h} = -2, \text{ if } f(x_i) = 0 \text{ and } x_c^{i} = 0, \tag{7}$$

$$T_j = \sum_{i=1}^{n} w_{ij}^{h} x_i - 1, \tag{8}$$

where x_c^{i} is the ith bit of the core, x_i is the ith bit of the input, w_{ij}^{h} is the connection weight of the jth hidden neuron for x_i, and T_j is the threshold for the jth hidden neuron. Ma Xiaomin's CSCLA makes use of the above approach, and extends it [34].

<u>Approach (ii):</u>

$$w_{ij}^{h} = 1, \text{ if } v_c^{i} = 1, \tag{9}$$
$$w_{ij}^{h} = -1, \text{ if } v_c^{i} = 0, \tag{10}$$

$$T_j = \sum_{i=1}^{n} w_{ij}^{h} x_c^{i} - (d-1), \tag{11}$$

where x_c^{i} is the ith bit of the core, x_i is the ith bit of the input, w_{ij}^{h} is the connection weight of the jth hidden neuron for x_i, and T_j is the threshold for the jth hidden neuron. Gazula and Kabuka's Nearest to an Exampler (NTE) approach generalizes the above construction [29].

Approach (iii):

$$w_{ij}^{\;h} = 2c_i - 1, \qquad\qquad (12)$$
$$w_{ij} = 2C_i - C_0, \qquad\qquad (13)$$
$$C_i = \sum_{k=1}^{C_0} x_k^{\;i}, \qquad\qquad (14)$$

where C_0 is the number of vertices included in the jth hidden neuron, $x_k^{\;i}$ is the ith bit of the kth input, and $w_{ij}^{\;h}$ is the connection weight of ith bit of the input to the jth hidden neuron. Next, compute f_{max} and t_{min} according to the formulation given below:

$$f_{max} = \max_{f(X)=0} \sum_{i=1}^{n} w_i x_i, \qquad\qquad (15)$$

$$t_{min} = \min_{f(X)=1} \sum_{i=1}^{n} w_i x_i. \qquad\qquad (16)$$

If $f_{max} < t_{min}$, then these two sets can be linearly separated, and $T = \left\lceil \dfrac{t_{min} + f_{max}}{2} \right\rceil$;

otherwise if $f_{max} < t_{min}$, these two sets can not be linearly separated, where T_j is the threshold for the jth hidden neuron.

In Approach (i) and (ii), the connection weights are restricted to simple values like 1, -1, 2, and -2. Approach (iii) is more general, and it is possible to extend it in a variety of ways. In section 2.3, we give a framework, which is useful for variety of such extensions.

After all the required hyperplanes (hidden neurons) are found, one output neuron is needed to combine the outputs of hidden neurons. A hidden neuron is defined as a converted hidden neuron, if the neuron was determined based on converted true vertices which are originally given as false vertices and converted false vertices which are originally given as true vertices. So every even-number hidden neuron obtained by above method is a converted hidden neuron. So formulas to determine the connection eights and thresholds is defined as follows:

$$w_j^{\;o} = \begin{cases} 1 & \text{if } j \text{ is odd,} \\ -1 & \text{if } j \text{ is even,} \end{cases} \qquad\qquad (17)$$

$$T^o = \begin{cases} 1 & \text{if the number of hidden neurons is odd,} \\ 0 & \text{if the number of hidden neurons is even,} \end{cases} \qquad (18)$$

where $w_j^{\;o}$ is the connection weight from the jth hidden neuron to the output neuron, and T^o is the threshold of the output neuron.

ETL algorithm has been used in many implications, such as handwritten digit recognition based on simulated light sensitive model and shown a remarkable result in both training speed and number of hidden neurons than other techniques. Also ETL is used in pattern classification of breast cancer. ETL has been used for benchmarking performance for other machine learning methods.

However, the selection of core vertex remains an important open issue for the construction of the binary neural network in ETL.

218

2.3 A Framework – Neuron visualized in terms of a Hypersphere

A set of 2^n binary patterns $(0,1)^n$ can be considered as an n-dimensional unit hypercube. Each pattern is located on one vertex of this hypercube, hence on the surface of the following reference hypersphere (RHP):

$$(x_1 - \frac{1}{2})^2 + (x_2 - \frac{1}{2})^2 + \cdots + (x_n - \frac{1}{2})^2 = \frac{n}{4}.$$
(19)

A Boolean function (of n variables) classifies 2^n vertices into two classes: a set of true vertices, and a set of false vertices. If a Boolean function is linearly separable, then all true vertices lie inside or on some hypersphere (HP), and all false vertices lie outside [2]:

$$\sum_{i=1}^{n} (x_i - c_i)^2 = r^2,$$
(20)

where $(c_1, c_2, \ldots c_n)$ is the center of HP, r is the radius. To obtain the intersection of HP and RHP, we subtract (2) from (1), and get the following separating hyperplane:

$$\sum_{i=1}^{n} (2c_i - 1)x_i = \{\sum_{i=1}^{n} c_i^2\} - r^2.$$
(21)

The true vertices, lie on one side of this hyperplane (having left hand side expression in (21) being $>=$ right hand side).

A typical hyperplane in a BNN has its corresponding hypersphere, with center $\mathbf{c}$, and radius r. While constructing a BNN, suppose that $\{\mathbf{x}^1, \mathbf{x}^2, \ldots, \mathbf{x}^v\}$ are v (true) vertices included in one hypersphere. In terms of these vertices, we define the center $\mathbf{c} = (c_1, c_2, \ldots, c_n)$, as:

$$c_i = \sum_{k=1}^{v} x_i^k / v.$$
(22)

To restrict our discussion for integer valued weights, we multiply both sides of (21) by v.

$$\sum_{i=1}^{n} (\{2\sum_{k=1}^{v} x_i^k\} - v)x_i = v[\{\sum_{i=1}^{n} c_i^2\} - r^2].$$
(23)

We represent (23) using a hard-limiter model for a neuron, which has output zero when its input is less than the threshold, and has output one for values greater than or equal to the threshold.

The radius r denotes the minimum value such that all the v vertices (in "covered region" in Fig. 3) are exactly in or on the hypersphere. Thus, we have:

$$r^2 = \max_{k=1}^{v} \sum_{i=1}^{n} (x_i^k - c_i)^2,$$
(24)

From (23), the (integer valued) weights for neuron are:

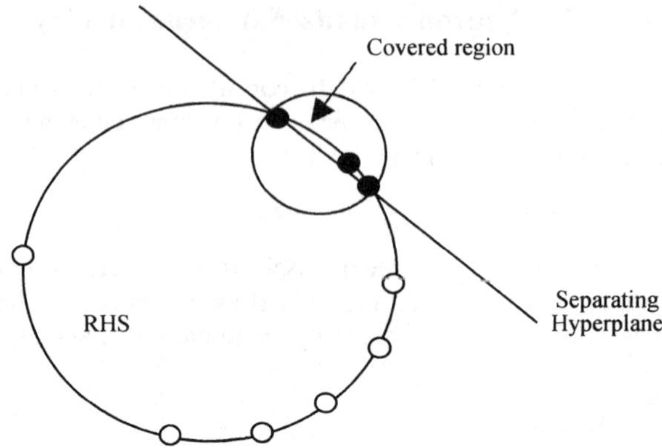

Fig. 3. Visualization (in 2-D) of Reference Hypersphere (RHP), and separating Hypersphere (HP) for a typical hidden Neuron.

and, the threshold t (corresponding to r, given in (24) above) is:

$$t = v\,(\sum_{i=1}^{n} c_i^2 - r^2) = \min_{k=1}^{v}\{\sum_{i=1}^{n}(v(2c_i x_i^k - (x_i^k)^2))\}\,, \tag{26}$$

where x_i^k is $\{0, 1\}$, so $x_i^k = (x_i^k)^2$. Hence,

$$t = \min_{k=1}^{v}\{\sum_{i=1}^{n}(v(2c_i - 1)x_i^k)\} = \min_{k=1}^{v}\sum_{i=1}^{n} w_i x_i^k\,. \tag{27}$$

If a Boolean function is not linearly separable, it can be represented by some m n-dimensional hyperspheres (and hence, hyperplanes). Each hidden neuron in BNN represents one (of these m) hyperplane. Different approaches for construction of BNNs use different method for determining these hyperplanes.

3   Extension for Multi-class Problems

To classify input vertices into more than two classes [3], to represent each of these classes, we use one output neuron. In Example 1, we wish to have three classes as output. We represent the output by three binary neurons $o_1 o_2 o_3$. We use the BNN construction process to construct these three separate neurons. After the output neuron o_1 is obtained, however, we allow the vertices in this group to be "don't care" for the remaining groups. This results in simplified weights for construction of these neurons. However, we need additional "composition" layer to produce correct output, by avoiding two or more outputs to be one at the same time. This is the additional cost that we need to pay for the simplification of weights using "don't care".

Example 1: An illustrative example, which has three classes as output, given in Table 1.

Table 1. Multiclass Example 1.

Data set (Input Vertices)	Output $o_1o_2o_3$
0100	100
0011	100
1100	100
1110	100
1000	010
1001	010
1111	010
0110	010
1101	001
1010	001
0001	001
0101	001
0111	001
0010	001

We partition input vertices in groups of output. This results in Table 2 below.

Table 2. Classes explicitly shown for Example in Table 1.

Data set (Input Vertices)	Classes	Output $o_1o_2o_3$
0100 0011 1100 1110	G_1	100
1000 1001 1111 0110	G_2	010
1101 1010 0001 0101 0111 0010	G_3	001

Let us use hard-limiter neuron with zero-one output (as in (23) of section 2). We first apply ETL construction for the input vertices in G_1, treating these vertices as true vertices and vertices in the rest group as $\{G_2,G_3\}$ as false vertices. Thus,

For G_1:
True Vertices: {0100,0011,1100,1110}
False Vertices: {1000,1001,1111,0110,1101,1010,0001,0101,0111,0010}

We now apply ETL construction for these vertices:

First Expanded Hyperplane equation is:

$$x_1 + 3*x_2 - x_3 - 3*x_4 - 3 = 0,$$ which separates {0100,1100,1110} vertices.

The remaining true vertex {0011} is added into the SITV only after conversion from true vertices to the false and false into the true vertex. Therefore after conversion, the second Hyperplane equation is:

$$x_1 + 3*x_2 - x_3 - x_4 + 1 = 0.$$

Keeping track of this conversion of vertices in the ETL construction (section 2), the weights and threshold values of the output neuron are obtained. This approach results in a neural network structure in Fig. 4.

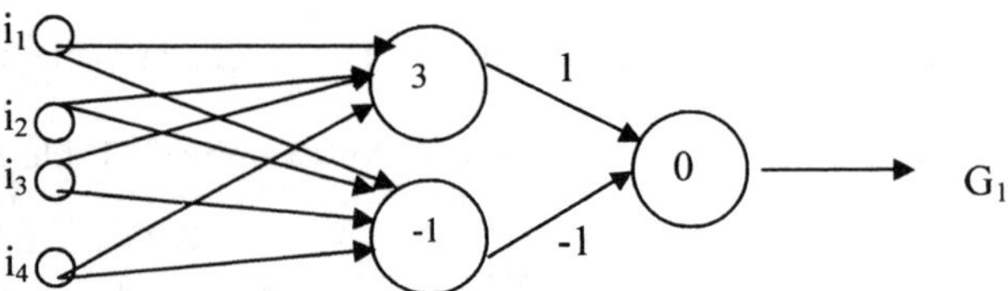

Fig. 4. Neural network for group G_1 of Example in Table 2.

Now, the input vertices in G_1 are regarded as "don't care" vertices for the training of the rest group. Thus for G_2 group the problem is realized as follows:

For G_2:
True Vertices: {1000,1001,1111,0110}
False Vertices: {1101,1010,0001,0101,0111,0010}

After applying ETL algorithm, the hyperplane equations are :

$$x_1 - 2*x_2 - 2*x_3 - x_4 = 0, \qquad 2*x_1 - 4*x_2 - 2*x_3 + 3 = 0$$
$$3*x_1 - 3*x_2 - x_3 + x_4 + 1 = 0, \qquad x_1 - x_2 - x_3 + 3*x_4 + 1 = 0.$$

Network corresponding to the group G_2 is given in Fig. 5.

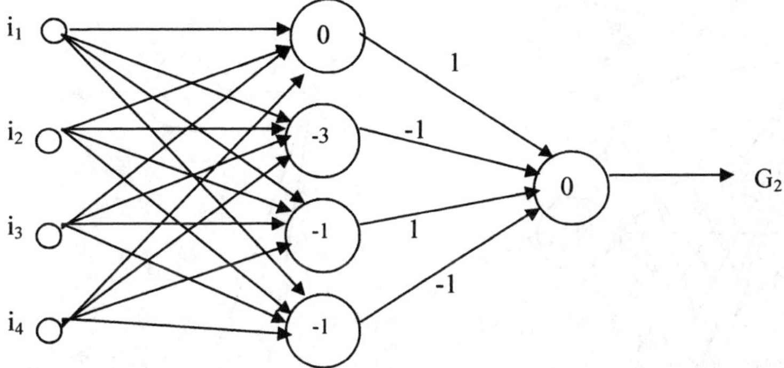

Fig. 5.   Neural network for group G_2 of Example in Table 2.

After training G_2 group, the input vertices in G_1 and G_2 are regarded as "don't care" vertices for the training of the rest group. Thus, for G_3:
True Vertices: {1101,1010,0001,0101,0111,0010}; False Vertices: Nil.

After applying ETL algorithm, the hyperplane equations are :

$$- 3*x_1 + x_2 - x_3 + 3*x_4 + 2 = 0, \qquad - 2*x_1 - 2*x_3 + 2*x_4 + 3 = 0.$$

The neural network structure for G_3 is given in Fig. 6.

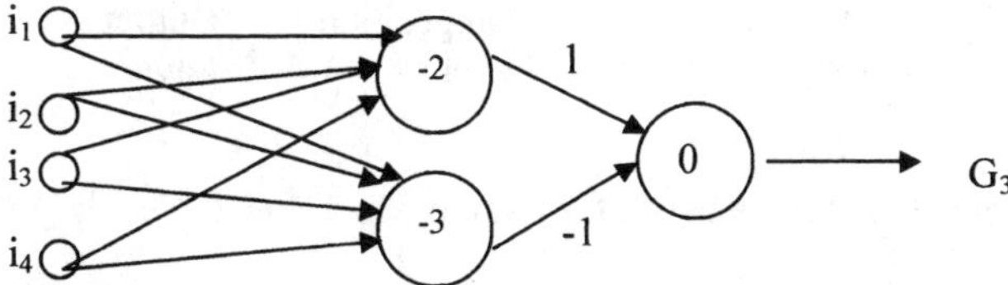

Fig. 6.   Neural network for group G_3 of Example in Table 2.

Thus, the hyperplane equations for hidden layer neurons (for the complete neural net for the example in Table 2) are:

For G1: $x_1 + 3*x_2 - x_3 - 3*x_4 - 3 = 0,$ $\qquad$ $x_1 + 3*x_2 - x_3 - x_4 + 1 = 0.$

For G2: $x_1 - 2*x_2 - 2*x_3 - x_4 = 0,$ $\qquad$ $2*x_1 - 4*x_2 - 2*x_3 + 3 = 0,$
$\qquad$ $3*x_1 - 3*x_2 - x_3 + x_4 + 1 = 0,$ $\qquad$ $x_1 - x_2 - x_3 + 3*x_4 + 1 = 0.$

For G3: $- 3*x_1 + x_2 - x_3 + 3*x_4 + 2 = 0,$ $\quad$ $- 2*x_1 - 2*x_3 + 2*x_4 + 3 = 0.$

To combine the networks (Fig. 4-6) we note that additional "composition" layer is needed due to our approach of using vertices in other groups as "don't care" vertices.

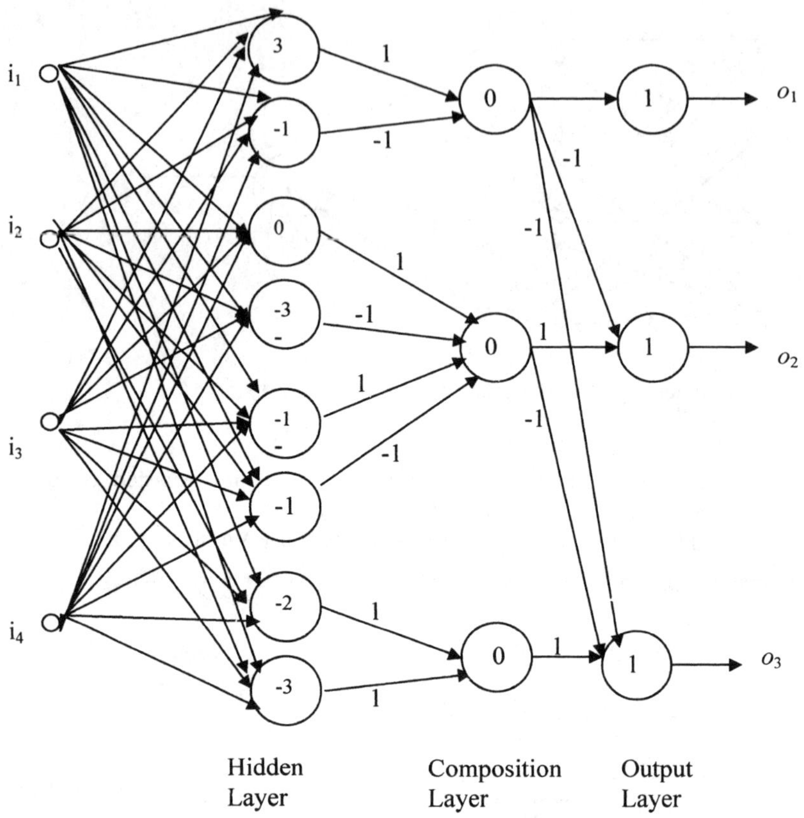

Fig. 7. Complete neural network for Example in Table 2.

Example 2. *Serial Binary Adder*: As a second example, we consider the example of serial binary adder. We have two input bits (to be added), and additionally, a carry, which results from the previous set of inputs. Fig. 8 gives the block diagram of serial binary adder, and corresponding truth table in given in Table 3.

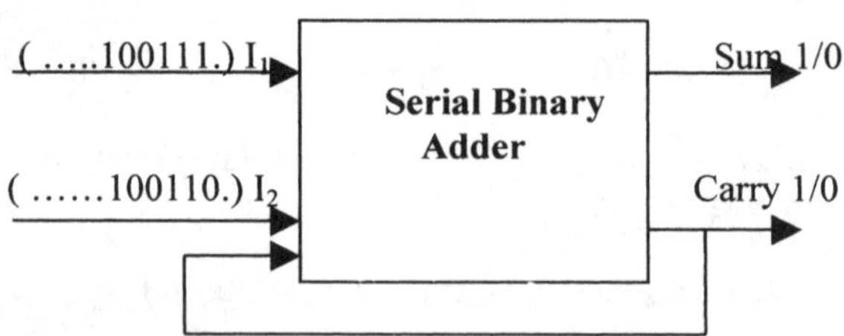

Fig. 8. Block diagram for Example 2: Serial Binary adder.

224

Table 3. The truth table for Example 2: Serial Binary Adder.

C	N_1	N_2	Sum(S)	Carry(C)
0	0	0	0	0
0	0	1	1	0
0	1	0	1	0
0	1	1	0	1
1	0	0	1	0
1	0	1	0	1
1	1	0	0	1
1	1	1	1	1

We formulate this problem as a multi-class problem in the framework of example 1; hence truth table is given in Table 4.

Table 4. Classes identified for Example 2: Serial Binary Adder.

C	N_1	N_2	Sum(S)	Carry(C)	Classes
0	0	0	0	0	G_1
0	0	1	1	0	G_2
0	1	0	1	0	G_2
0	1	1	0	1	G_3
1	0	0	1	0	G_2
1	0	1	0	1	G_3
1	1	0	0	1	G_3
1	1	1	1	1	G_4

The weights and thresholds of hidden layer neurons are obtained using the method similar to Example 1, and are given in the form of equations given below:

For G1: $-2*x_1-2*x_2-2*x_3 + 1=0.$

For G2: From first core={001} $-x_1-2*x_2+x_3 =0,$ $-2*x_2+2*x_3=0.$
         From second core={010} $-x_1+x_2-x_3 =0,$ $4*x_2-2=0.$

For G3: $x_1-x_2+x_3 =0,$ $2*x_1+2*x_3 -3=0.$
For G4: $x_1+x_2+2*x_3 -3=0.$

Using these hyperplane equations, we directly give neural net structure obtained for realizing serial adder in Example 2 in Fig. 9. In Fig. 9, we have chosen to denote the output neurons with labels G_1, G_2, G_3, and G_4 (instead of o_1, o_2, o_3, o_4).

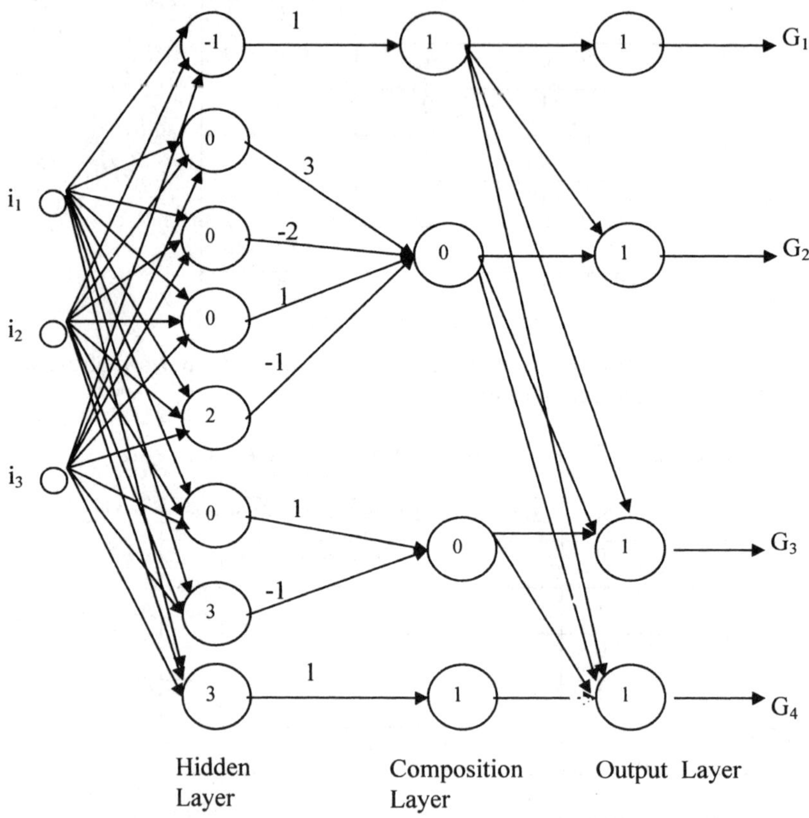

Fig. 9. ETL-based neural network for Example 2: Serial Binary adder (recurrent feedback connections from G_3, G_4 to i_3 are not shown).

In terms of $o_1o_2o_3o_4$, the sum $S = o_2+o_4$, and carry $C = o_3+o_4$ ("+" here denotes Boolean *or* operation). Carry C needs to be fed back as input i_3 to act as carry of next stage. We have not shown the implementations of these Boolean operations in Fig. 9; however, these implementations are straightforward and can directly be done in the framework of ETL.

Due to our approach of "don't care" vertices, the additional "Composition layer" is required in Fig. 9 as well. The weights of the output layer neurons ($o_1o_2o_3o_4$, corresponding to groups $G_1G_2G_3G_4$) are given in Table 5.

Table 5. Weights of output layer neurons in Fig. 9 (Serial Binary Adder).

No.	Output neurons of various classes	Weights
1	G_1	1000
2	G_2	-1100
3	G_3	-1-110
4	G_4	-1-1-11

If we do not want to use the additional composition layer, while constructing neural network for each group of vertices, we should treat the true vertices in other groups as "false" vertices. For the remaining part of this paper (in sections 4 and 5), we shall closely stick to this approach.

4   Finite Automata (FA) and their implementation using BNNs

Implementation of a finite automaton using neural networks is one of the well-studied problems by many researchers. For example, Marvin Minsky, in his book on "Computation: Finite and Infinite Machines", implements state transition function in terms of a neural network model using what he calls as "G-function connection box" [44]. More recent approaches are reported in [13-22, 26]. All these approaches, however, do not use the framework we have introduced in section 2, and 3. Hence, we investigate this problem from different perspective.

Finite Automaton (FA) uses the current state and the next input (symbol being scanned /consumed /applied) to determine the next state [43]. The next state depends upon the previous state and the input (bit) applied. (For the case of nondeterministic finite automaton (NFA), we have a set of possible states; this aspect has been taken care of in our construction given in section 4.2 below). Schematically, the input (bit) applied is taken as external input in Fig. 10. Corresponding to each state in the FA, we prepare a neural network by first identifying suitable Boolean network (BN), and implement it using ETL [35]. The output of each of the BN block shown in Fig. 10 is true (or, 1) whenever we should "go to" the corresponding state (or, a set of states, if automaton is nondeterministic). In order to define each of the BN block (Fig. 10), we need to find out the true vertices and false vertices for each state of FA. After this, by applying ETL separately for each state of the FA, we obtain the entire neural network that is trained for a given FA.

4.1   Neural Network Structure

We use is a three-layered network [35] with the modification of having feedback connections, resulting in a recurrent neural network structure (as is common convention, the feedback lines include unit delay element, which is not explicitly shown in the following figures). In Fig. 10, the input layer is a set of transparent nodes used to distribute the inputs coming from previously generated outputs.

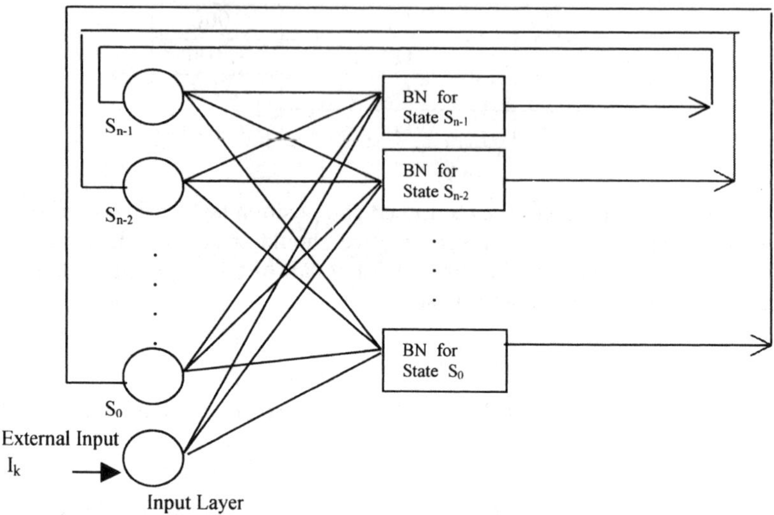

Fig. 10. General network architecture used to construct a FA.

4.2 Algorithm for Neural network Construction for a FA

Step 1 Corresponding to each state of FA, we need a separate neural net; these nets are referred as "BN for S_0", "BN for S_1"...... "BNfor S_{n-1}" in Fig. 10.

Step 2. Defining input output relationship for each BN net.

The next step is to determine the number of inputs and the output for each of this BN. As is shown in Fig. 10, the number of inputs for each of this BN is (n+1), where n is the number of states. We use one bit value to correspond to each of the states. We represent the bit-values for these inputs in the sequence S_{n-1}, S_n,....,S_0, and the last bit corresponds to the present value of input applied.   We refer to input applied by I_k. Thus, we have (n+1) inputs, and we consider them to be in the sequence S_{n-1}, S_n,...,S_0, I_k. Each of the n+1 values in the sequence is either 0, or 1, and hence our input (consisting of n+1 bits) is binary sequence. We also refer to this input as "vertex". Boolean networks (BNs) are defined in terms of "input-output" relationships. The output of each of such BN networks is binary (say, true, or false).

Sub-step 2.1: *For DFA*:
A combination of input $S_{n-1}S_{n-2}....S_0\ I_k$, results in at most one DFA state, say S_r. Hence, $S_{n-1}S_{n-2}....S_0\ I_k$ is a true vertex for BN corresponding to S_r (Fig.

10). It may, however, be noted that there may be more than one such "true" vertex corresponding to a given state S_r. Identification of all such true vertices for each state in DFA is an important step in our construction. After all true vertices for all states are identified, we proceed to identify the "false" vertices for the states.

Since we have already identified all true vertices for each of the states first, it is possible for us to identify the set of vertices which does not correspond to a given state, say S_i. In general, due to the deterministic property of a DFA, it follows that the set of false vertices for a state S_i, denoted by FVS_i, includes the set of true vertices for other states in DFA. Thus,

$$FVS_i \supseteq TVS_{n-1} \cup .... \cup TVS_{i+1} \cup TVS_{i-1} \cup .... \cup TVS_0, \tag{28}$$

where, FVS_i is the set of false vertices for i^{th} state and,
TVS_k is the set of true vertices for the kth state.

Sub-step 2.2: *For NFA*:

A combination of input $S_{n-1}S_{n-2}....S_0\ I_k$, may now result in more than one NFA state (there may be more than one states will be activated at one time, which corresponds to multiple paths from one state, after applying a input). Therefore, we need to check 2^{n+1} combinations of bits to decide true vertices for a given state. Thus, we introduce "extended transition function", δ', (similar constructions are well-known, e.g. [43, pp. 61]) for each combination of the bits applied. The δ' is a function that gives the set of states or empty set for any combination of bits applied ($S_{n-1}\ S_{n-2}\ ...\ S_0\ I_k$). In the construction of network, we make use of δ' to obtain the set of true vertices for a given state.

Formally, δ' is given by

$$\delta'(S_{n-1}\ S_{n-2}\ ...\ S_0\ 1) = \delta(S_{n-1},1) \cup \delta(S_{n-2},1) \cup ..... \cup \delta(S_0,1) \tag{29}$$

$$\delta'(S_{n-1}\ S_{n-2}\ ...\ S_0\ 0) = \delta(S_{n-1},0) \cup \delta(S_{n-2},0) \cup ..... \cup \delta(S_0,0) \tag{30}$$

(where δ is a transition function given for a state.)

Thus, δ' is to be checked for 2^{n+1} combinations of ($S_{n-1}\ S_{n-2}\ ...\ S_0\ I_k$).

If δ' is an empty set then that bit combination neither be taken as true vertex nor false vertex, we say that this bit-combination (vertex) is "don't care".

Using the above information about the transition relation, as in the case of DFA (sub-step 2.1 above), it is first convenient to determine the set of true vertices for all states. Next, in terms of these sets of true vertices, it is possible to identify the set of vertices that do not correspond to a given state. However, due to nondeterminism, unlike (28), we cannot include all true vertices for the other states into this false vertex set. Hence, false vertices for state S_i are decided as follows:

$$FVS_i \supseteq \{ ((TVS_{n-1} - (TVS_{n-1} \cap TVS_i)) \cup ....\cup$$

$$((TVS_0 - (TVS_0 \cap TVS_i)) \} \tag{31}$$

Step 3. Apply the ETL construction for input/output mapping corresponding to each state $S_{n-1}, S_{n-2},....,S_0$ separately thus develop the BN nets "BN for S_0", "BN for S_1"...... "BN for S_{n-1}" (shown in Fig.10).

Step 4. Next we need to introduce the feedback (recurrence) of each state's output to it's own input node, so that state's results depends on the previous state.

4.3 Acceptance/rejection of input string in a network

For DFA:
After applying the last symbol of a string, if network activates a state, which is an accepting state, then the string to be tested is accepted; otherwise it is rejected.
For NFA:
After the last string symbol, if network activates the set of states in which at least one state is accepting state, then the string is accepted; otherwise it is rejected.

5   Illustration of FA Construction Using Examples

Example 3. DFA : Let us consider an example DFA given in Fig. 11.

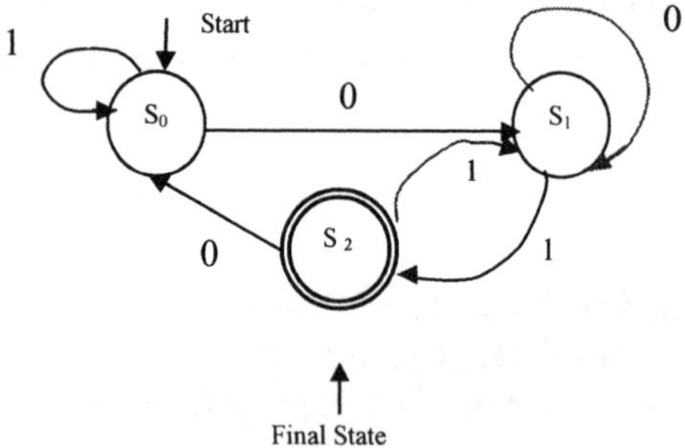

Fig. 11. An example DFA.

In this example, we have taken our language having only two alphabet symbols, which we called as 0, and 1. However, our proposed construction can easily be generalized for languages having more than two alphabet symbols.

Using step 2 of algorithm in Section 4.2 for the DFA in Fig. 11, we first determine the set of true vertices for each of the states using the information of transition function. Thus, we get:

For S_0: True Vertices=\{0011,1000\},
For S_1: True Vertices=\{0010,0100,1001\},
For S_2: True Vertices=\{0101\}.

After having determined the set of true vertices for each of the states as above, we now use the construction in (28) to determine the set of false vertices for each of these states. Thus, we get:

For S_0: False Vertices=\{0010,0100,0101,1001\} (obtained as $= TVS_1 \cup TVS_2$),
For S_1: False vertices=\{0011,1000,0101\} (obtained as $= TVS_0 \cup TVS_2$),
For S_2: False vertices=\{0011,0010,0100,1001,1000\} (obtained as $= TVS_0 \cup TVS_1$).

Using this information, and using the construction in section 2 (ETL construction), we obtain the weights for the hidden layer neurons (hyperplane equations) as follows:

For S_0: $-x_1-x_2+x_3+2*x_4-2=0$; $-3*x_1-x_2-x_3+x_4+2=0$;
For S_1: $-2*x_1-2*x_4+1=0$; $-3*x_1-x_2-x_3-x_4+3=0$;
For S_2: $-x_1+x_2-x_3+2*x_4-2=0$.

Next, we apply step 4 of algorithm, which introduces feedback. The resulting neural net is given in Fig. 9. This neural net accepts a given string whenever original FA is in a final state. For our example, this condition arises whenever "ETL output for S_2" is 1.

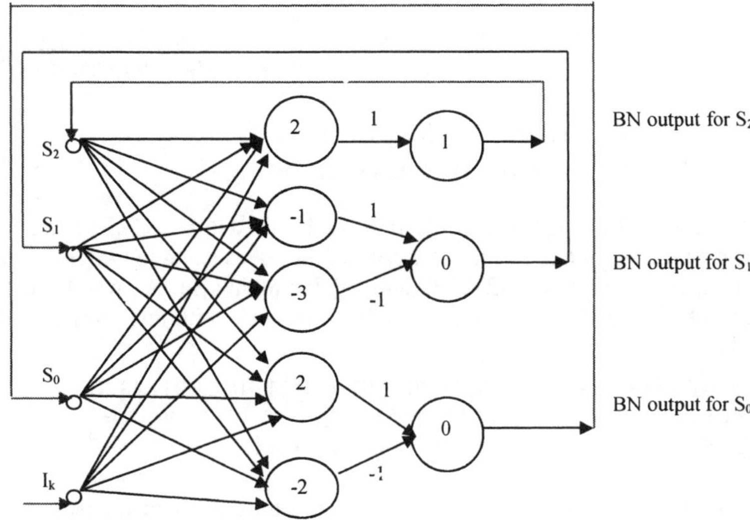

Fig. 12. ETL based final solution for DFA of Fig. 11.

Now let us illustrate the construction of neural net for general (nondeterministic) finite automaton, using Example 4 .

Example 4: Let us take an example of Nondeterministic Finite Automata (NFA).

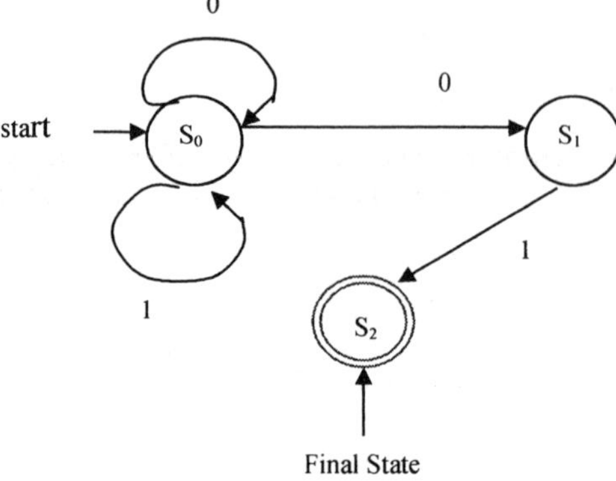

Fig. 13. An Example NFA.

Apply step 2.2 of algorithm (section 4.2) and evaluate δ' as follows:

$\delta'(0110) = \delta(S_1,0) \cup \delta(S_0,0) = \{S_0,S_1\}$, thus 0110 is a true vertex for S_0 and for S_1. (32)
$\delta'(0111) = \delta(S_1,1) \cup \delta(S_0,1) = \{S_2,S_1\}$, thus 0111 is a true vertex for S_2 and for S_1. (33)

$\delta'(1010) = \delta(S_2,0) \cup \delta(S_0,0) = \{S_0,S_1\}$ (34)
$\delta'(1011) = \delta(S_2,1) \cup \delta(S_0,1) = \{S_0\}$ (35)

$\delta'(1100) = \delta(S_2,0) \cup \delta(S_1,0) = \{\phi\}$ ⟵ (ϕ:Null set) thus 1100 is neither a (36)
$\delta'(1101) = \delta(S_2,1) \cup \delta(S_1,1) = \{S_2\}$ true nor a false vertex. ("Don't care") (37)

$\delta'(1110) = \delta(S_2,0) \cup \delta(S_1,0) \cup \delta(S_0,0) = \{S_0,S_1\}$ (38)
$\delta'(1111) = \delta(S_2,1) \cup \delta(S_1,1) \cup \delta(S_0,1) = \{S_0,S_2\}$ (39)

Using step 2 of algorithm in Section 4.2 for the NFA in Fig. 11, we first need to determine the set of true vertices for each of the states using the information of transition relation. From steps (32)-(39) (and continuing in a similar way for the remaining 8 vertices), we get the following sets of true vertices, with their corresponding states, as follows:

For S_0: True Vertices = {0011,0010,0110,0111,1010.1011,1110,1111}, (40)
For S_1: True Vertices ={0010,0110,1010,1110}, (41)
For S_2: True Vertices ={0101,0111,1101, 1111}. (42)

Instead of the above steps (32)-(39), to illustrate this construction in an alternate way, we re-write the transition relation for an NFA in the form of Table 6. We note that the last bit of our vertex represents the alphabet symbol (the header row in table

232

6). The first n bits represent whether the corresponding state can make a transition to the given state (for which we are constructing true vertices) on the alphabet symbol in the last position. The union operation in the above steps is equivalent to having a "don't care" condition in the following sequences for the corresponding states. This helps us in visualizing the set of true vertices for states in quick way, either from the diagrammatic representation of an NFA, or from the state transition relation (expressed in a tabular way).

Table 6. State Transition Relation of the NFA.

State ↘	Alphabet Symbol → 0	1
S_0	$\{S_0, S_1\}$	$\{S_0\}$
S_1	$\emptyset$	$\{S_2\}$
S_2	$\emptyset$	$\emptyset$

Consider the entry S_0 in the above table. We can reach state S_0 either from being in state S_0 and having an input alphabet symbol either 0, or 1. This corresponds to vertices "xx10" (we use first "x" to represent that state S_2 does not have a role as previous state for reaching to state S_0, and we use second "x" to denote that the state S_1 does not have any role as a previous state to reach to state S_0), and "xx11" being in the set of true vertices for state S_0. We observe that this short representation in fact represents all the eight true vertices in (40).

Similarly, we can reach the state S_1 only from being in state S_0 and having an input alphabet symbol 0; hence the set of true vertices for S_1 can be represented by "xx10". We observe that this short representation denotes all the four true vertices in (41).

For the remaining state S_2, we can reach S_2 only from being in state S_1 and having an input alphabet symbol 1; hence the set of true vertices for S_2 can be represented by "x1x1". We observe that this short representation denotes all the four true vertices in (42).

After having determined the set of true vertices for each of the states as in (40)-(42) above, we now use the construction in (?.1) to determine the set of false vertices for each of these states. Thus, we get:

For S_0: False Vertices $=\{0101,1101\}$,
For S_1: False vertices $=\{0011,0111,1011,1111,0101,1101\}$,
For S_2: False vertices $=\{0011,0010,0110,1010,1011,1110\}$.

Using the above, we first obtain the hyperplane equations from the ETL construction (of section 2). The hyperplane equations are:

For S_0 : $8*x_3 - 4 = 0$;

For S_1 : $4*x_3 - 4*x_4 - 2 = 0$;
For S_2 : $4*x_2 + 4*x_4 - 6 = 0$.

The resulting neural network is given in Fig. 14.

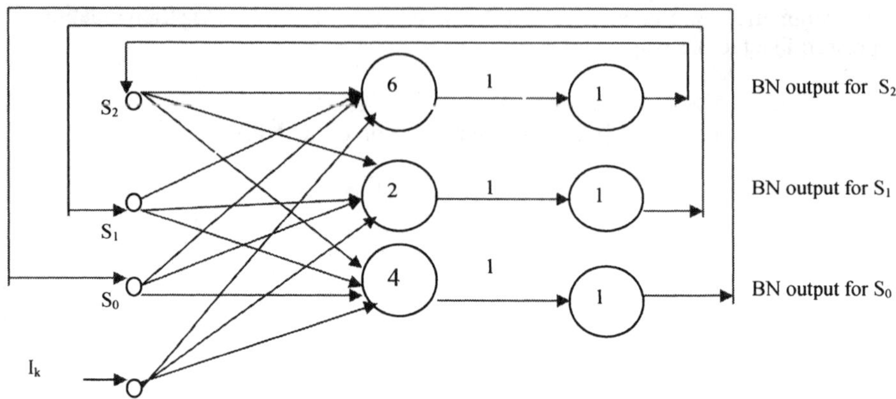

Fig. 14. Final ETL Solution for the different states of the NFA.

6 Concluding Remarks

From the known methods for constructing a neural network for a Boolean function, we have given a methodology for design of multi-class neural network. Next, we have given an approach for expressing a finite state automaton in terms of a recurrent neural network. To illustrate our method, we mainly adopt the methodology similar to ETL (Kim and Park [35]); however, any other methodology can be used for illustrating our approach. Most of the earlier approaches for learning of DFA [13-26] use second order recurrent network with neurons based on sigmoid activation function (or, other type of non-linearity).

Our approach uses hard-limiting activation function, and all the weights are guaranteed to be of integral value. Hence, this method can be used for developing systematic VLSI implementation(s) for a given finite automaton. The number output neurons required in the proposed approach are the number of classes to be learned. Though it's simple, however this can be further enhanced by encoding the number of output neurons in $O(\log(n))$, where 'n' is the number of classes to be learned.

Our approach guarantees integer weights, and is based on BNN construction algorithms; hence it is fast. The neural network model which uses analog (continuous) values for input variables, and use some continuous nonlinear function as an activation function of the neuron, is however, more powerful, and is generally employed for

234

many practical classification problems. For developing a general constructive framework for neural networks having analog values, additional investigations are needed.

References

1. Marvin L. Minsky, and Seymour A. Papert, *Perceptrons*, (Expanded Edition, 1988; first Edition 1969), Cambridge, MA: The MIT Press (1988).

2. Hava T. Seigelmann, *Neural Networks and Analog Computation: Beyond the Turing Limit*, Boston: Birkhauser (1999).

3. Terry Windeatt, Reza Ghaderi, "Binary labeling and Decision-level fusion", *Information Fusion*, Vol. 2, pp. 103-112 (2001).

4. C. H. Chu and J. H. Kim, "Pattern Classification by Geometrical Learning of Binary Neural Networks", *Proceedings of International Joint Conference on Neural Networks*, Nagoya, Japan, (Oct. 1993).

5. Jung H. Kim, Byungwoon Ham, and Sung-Kwon Park, "The Learning of Multi-output Binary Neural Networks for Handwriting Digit Reorganization," *Proceedings of International Joint Conference of Neural Networks (IJCNN)*, Vol.1, pp. 605-508, (Oct. 1993).

6. T. Windeatt, and R. Tebbs, "Spectral technique for hidden layer neural network learning," *Pattern Recognition Letters*, Vol. 18, No. 8, pp. 723-731, (1997).

7. Vinay Deolalikar, "Mapping Boolean functions with neural networks with binary weights and zero thresholds," *IEEE Transactions on Neural Networks*, Vol. 12, No. 1, pp. 1-8, (July 2001).

8. Vinay Deolalikar, "A two-layer paradigm capable of forming arbitrary decision regions in input space," *IEEE Transactions on Neural Networks*, Vol. 13, No. 1, pp. 15-21, (Jan. 2002).

9. Igor N. Aizenberg, Naum N. Aizenberg, and Georgy A. Krivosheev, "Multilayered and universal binary neurons: Learning algorithms, applications to image processing and recognition," In, Lecture Notes in Artificial Intelligence, Berlin: Springer-Verlag, Vol. 1715: *Machine Learning an Data Mining in Pattern Recognition* –Proceedings of the First International Workshop *MLDM'99*, Leipzig, Germany, (Sept. 1999).

10. Milel Forcada, Rafael C Carrasco, "Finite-state computation in analog neural networks: steps towards biologically plausible models?" In, Lecture Notes in Artificial Intelligence, Berlin: Springer-Verlag, Vol. 2036: *Emergent Neural Computational Models Based on Neuroscience*, pp. 482-486, (2001).

11. Stephan Mertens, Andreas Engel, " Vapnic—Chervonenkis dimension of neural networks with binary weights", *Physical Review E*, Vol. 55, No. 4, (April 1997).

12. Jeong Han Kim, James R.Roche, "Covering Cubes by Random Half Cubes, with Applications to Binary Neural Networks", *Journal of Computer and System Science (JCSS)* Vol. 56, pp. 223-252, (1998).

13. C.L. Giles, C.B.Miller, D.Chen, H.H.Chen, G.Z.Sun, and Y.C.Lee, "Learning and extracting finite state automata with second order recurrent networks", *Neural Computation*, Vol. 2:331-402, (1992).

14. C. Omlin and C.L. Giles, "Constructing Deterministic Finite State Automata in recurrent neural networks", *Journal of the Association of Computing Machinery (JACM)*, Vol. 45, No. 6, pp. 937-972, (1996).

15. C.L. Giles, C.B.Miller, D.Chen, H.H.Chen, G.Z.Sun, and Y.C.Lee, "Extracting and learning an unknown grammar with recurrent neural networks", *Advances in Neural Information Processing Systems*, Vol. 4, pp. 317-324, (1992).

16. M. L. Forcada and R. C. Carrasco, " Learning the initial state of a second order recurrent neural network during regular language inference", *Neural Computation* Vol. 7, pp. 1075-1082 (1995).

17. N. Alon, A. Dewdney, and T. Ott, "Efficient simulation of finite automata by neural nets", *Journal of the Association for Computing Machinery*, Vol. 38, no.2, pp.495-514 (April 1991).

18. P. Frasconi, M. Gori, M. Maggini, and G. Soda, "Unified integration of explicity and learning by example in recurrent networks," *IEEE Transactions on Knowledge and Data Engineering(TKDE)*, Vol.7, no. 2, pp. 340-346, (1995).

19. P. Frasconi, M. Gori and G. Soda, "Injecting nondeterministic finite state automata into recurrent networks", Tech. Rep., Dipartimentc di Sistemi e Informatica, University di Firenze, Italy, Florence, Italy, (1993).

20. J. Pollack, "The induction of dynamical recognizers", *Machine Learning*, Vol. 7, pp.227-252, (1991).

21. R. Watrous and G. Kuhn, "Induction of finite state languages using second order recurrent networks", *Neural Computation*, Vol. 4, no. 3, p.406, (1992).

22. Z. Zeng, R. Goodman, and P.Smyth, "Learning finite state machines with self-clustering recurrent networks", *Neural Computation*, Vol.5, No.6, pp. 976-990, (1993).

23. J. Elman ,"Finding structure in time", *Cognitive Science*, Vol.14, pp. 179-211, (1990).

24. C. Giles and C. Omlin, "Rule refinement with recurrent neural networks", In Proceedings IEEE International Conference on Neural Networks (*ICNN'93*), Vol.II, pp.801-806,1993.

25. Mark Steijvers and Peter Grunwald, "A recurrent network that performs a context-sensitive prediction task", *Technical Report* in ESPRIT working group (*NeuroCOLT*), (1996).

26. Rafael C. Carrasco and Mikel L. Forcada, "Second order Recurrent Neural Networks can learn Regular Grammars from Noisy strings", In, *Proceedings IWANN* – International Workshop On Artificial Neural Networks, pp. 605 –610 (1995).

27. D.L. Gray, and, A.N. Michel, "A training algorithm for binary feedforward neural networks,", *IEEE Trans. Nerual Networks*, Vol. 3, No. 2, IEEE, USA, pp. 176-194 (Mar. 1992).

28. N.N. Biswas, and R. Kumar, "A new algorithm for learning representations in Boolean neural networks," *Current Science*, Vol. 59, No. 12, pp. 595-600, (June, 1990).

29. S. Gazula, and M. R. Kabuka, "Design of supervised classifiers using Boolean neural networks," *IEEE Trans. Pattern Analysis and Machine Intelligence*, Vol. 17, No. 12, IEEE, USA, pp. 1239-1246, (Dec. 1995).

30. M. R. Kabuka, "Comments on "Design of supervised classifiers using Boolean neural networks", " *IEEE Trans. Pattern Analysis and Machine Intelligence*, Vol. 21, No. 9, IEEE, USA, pp. 957-958, (Sept. 1999).

31. N. S. V. Rao, E.M. Oblow, C.W. Glover, "Learning separations by Boolean combinations of half-spaces," *IEEE Trans. Pattern Analysis and Machine Intelligence*, Vol. 16, No. 7, IEEE, USA, pp. 765-768, (July 1994).

32. M. Xiaomin, YangYixian, and Z. Zhang, "Research on the learning algorithm of binary neural network," *Chinese Journal of Computers* (China), Vol. 22, No. 9, pp. 931-935, (Sept. 1999).

33. Atsushi Yamamoto, Toshimichi Saito, "An improved Expand-and-Truncate Learning," *Proc. of IEEE International Conference on Neural Networks (ICNN)*, Vol. 2, pp. 1111-1116, (June, 1997).

34. Ma Xiaomin, Yang Yixian, Zhang Zhaozhi, "Constructive Learning of Binary Neural Networks and Its Application to Nonlinear Register Synthesis", *Proc. of International Conference on Neural Information Processing (ICONIP) '01*, Vol. 1, pp. 90-95, Shanghai (China), (Nov. 14 -18, 2001).

35. J..H. Kim, and S-K. Park, "The geometric learning of binary neural networks," *IEEE Trans. Neural Networks*, Vol. 6, No. 1, pp. 237-247, (January, 1995).

36. Sang-Kyu Sung, Jong Won Jung, Joon-Tark Lee and Woo-Jin Choi, "Opitonal Synthedid Method for Binary Neural Network Using NETLA", *Lecture Notes in Artificial Intelligence (LNAI)*, Vol. 2275, pp. 236-244, 2002.

37. Bernd Steinbach and Roman Kohut, "Neural Networks-a A model of Boolean Functions", *Proceeding of 5th International Workshops on Boolean Problems* (Freiberg, Germany), 2002.

38. J. A. Starzyk and J. Pang, "Evolvable Binary Artificial Neural Network for Data Classification", Proceedings of The 2000 International Conference on *Parallel and Distributed Processing Techniques and Applications* (*PDPTA'2000*), Monte Carlo Resort, Las Vegas, Nevada (USA), (June 26 - 29, 2000).

39. Narendra S. Chaudhari, Aruna Tiwari, "Extending ETL for multi-class output," *International Conference on Neural Information Processing, 2002 (ICONIP'02), In, Proceedings: Computational Intelligence for E-Age, Asia Pacific Neural Network Association (APNNA)*, pp. 1777-1780, Singapore, (18-22 Nov, 2002).

40. Di Wang, Narendra S. Chaudhari, "A Multi-Core Learning Algorithm for Binary Neural Networks", In *Proceedings of the International Joint Conference on Neural Networks (IJCNN '03)* Vol. 1 pp. 450-455, Portland, USA, (20-24 July 2003).

41. Di Wang and Narendra S. Chaudhari, "Binary Neural Network Training Algorithms Based On Linear Sequential Learning," *International Journal of Neural Systems (IJNS)*, **13**(5) pp. 1-19, (Oct. 2003).

42. Narendra S. Chaudhari, and Di Wang, "A Novel Boolean Self-Organization Mapping Based on Fuzzy Geometrical Expansion", In, Proceedings, *Fourth International Conference on Information, Communication and Signal Processing* & Fourth IEEE Pacific Rim Conference on Multimedia (*ICICS-PCM-03*), Singapore, (16-18 Dec. 2003).

43. John E.. Hopcraft, Rajeev Motwani, and Jeffrey D. Ullman, "Introduction to Automata Theory, Languages and Computation", (Second Edition) Addison-Wesley Longman Inc, (2001).

44. Marvin L. Minsky, "Computation: Finite and Infinite Machines", Englewood Cliffs, NJ: Prentice Hall, Inc. Chapter 3, pp. 32-68, (1967).

A Memory-Based Reinforcement Learning Algorithm to Prevent Unlearning in Neural Networks

Seiichi Ozawa and Shigeo Abe

Graduate School of Science and Technology, Kobe University, Kobe, Japan
{ozawa,abe}@eedept.kobe-u.ac.jp

Summary. In reinforcement learning tasks, neural networks have often been used as function approximator for an action-value function that gives an expected total rewards for an agent's action. Since a reward is given from the environment only after an agent takes an action, the learning of the agent's action-value function is inevitably done in an incremental fashion. It is well known that this type of incremental learning can cause "catastrophic interference" that leads to unlearning of previously acquired knowledge in neural networks. To solve this problem, in this chapter, we introduce a memory mechanism into an extended model of Radial Basis Function (RBF) networks, in which hidden units are adaptively allocated for incoming inputs. In this model, several representative input-output pairs are extracted from the trained action-value function, and they are stored into an external memory. When the agent improves its policy through trial and error, not only a usual temporal difference error but also network errors for some of the extracted pairs are reduced by a gradient-based learning algorithm. The reduction of both errors allows the proposed model to learn a desired action-value function stably. In order to evaluate the incremental learning ability, the proposed model is applied to two standard problems: Random Walk Task and Mountain-Car Task. In these tasks, the working space of an agent is extended as the learning proceeds. In the simulations, we verify that the proposed model can acquire proper action-values with small memory resources as compared with some conventional models.

Key words: reinforcement learning, memory-based learning, incremental learning, unlearning, catastrophic interference, radial-basis function networks

1 Introduction

The purpose of an agent in many reinforcement learning tasks is to find an optimal policy that maximizes the total amount of rewards under unknown environments [1, 2]. In general, the agent policy is represented by a control function that associates the agent's states with their desired actions. However, in many cases, this function is difficult to learn directly, hence the values of actions in each state* called "action-values" (i.e., values of state-action pairs) are introduced into an agent model. An action-value is defined as the expected total reward that the agent will earn after taking an action in a state, and it is estimated through many trials and errors. Once the action-values are accurately estimated, the agent can infer what actions are more desirable in the current state. This means that the agent has an optimal policy because it can select an action that has the largest action-value.

If the number of states and actions is limited to a finite number, it would be possible to save all the relations between state-action pairs and their values in a look-up table. In many cases of practical interest, however, there are far more state-action pairs that could be entries in such a look-up table. Moreover, the regularity in the action-values can be often assumed: that is, similar state-action pairs are associated with similar values. Therefore, in a large and smooth state-action space, it is more effective to approximate these action-values as a function called "action-value function".

Many function approximation schemes have been proposed so far, which can be classified into two groups: piece-wise linear function approximation and nonlinear function approximation. In the former, the state-action space is divided into a certain number of subspaces, and the value for each subspace is approximated by a linear function that is parameterized by a feature vector. Since the approximation accuracy depends on the number of subspaces (so-called granularity), the linear function approximation schemes generally need large memory resources to store the feature vectors especially when action-values should be approximated by a high-dimensional complex-shaped function. This suggests that the linear scheme tends to be suffered from the curse of dimensionality.

To solve the problem, nonlinear function approximators like feedforward neural networks are often used. An advantage of this approach is that they can handle a high-dimensional state-action space more easily and there is no need to divide the state-action space in advance. However it is well known that the learning of neural networks becomes difficult when the distribution of given training data is temporally varied and training data are incrementally given [3, 4, 5, 6, 7]. In the commonly-used reinforcement learning algorithms, network parameters are modified based on the temporal difference (TD) errors between estimated action-values and immediate rewards given incrementally. Therefore, one can say that reinforcement learning essentially has the nature of incremental learning; thus this might result in unlearning of action-values that had been already acquired. This disruption in neural networks is called "catastrophic interference" caused by the excessive adaptation to incoming rewards. Therefore, suppressing the interference is a crucial issue in reinforcement learning tasks when neural networks are used as function approximator.

---

* Instead of action-values, we can introduce the values of states called "state-values" in our proposed algorithm.

To prevent unlearning in neural networks, we have proposed a memory-based learning approach [6] in which representative data are extracted from an approximated function and some of them are learned with incoming training data. This approach has been devised for supervised learning problems so far; hence, we need several modifications in the architecture and learning algorithm when applying to reinforcement learning tasks. In this chapter, these modifications are specified through some considerations, and then we propose a new memory-based reinforcement model. In Section 2, we present several approaches to suppressing the catastrophic interference in neural networks. In Section 3, we explain how to approximate the agent's action-value function using neural networks. Then, we propose our new memory-based approach in Section 4. In Section 5, the proposed model is applied to two standard reinforcement learning problems, and the approximation accuracy of action-values, the training time, and the needed memory resources are evaluated through some comparisons with the conventional approaches. Finally, we state conclusions and our further work in Section 6.

2 Approaches to Preventing Unlearning in Neural Networks

There have been proposed several approaches to the prevention or alleviation of the catastrophic interference in neural networks [3, 8, 9, 10]. They can be roughly categorized into three approaches.

In the first approach, the connection weights trained formerly are not modified by new training data as much as possible: that is, connection weights adapted for new training data are separated from those for old data. Although this approach is easily implemented by scaling up networks (i.e., adding extra hidden units or module networks) [11], the problem is that the scale of networks tends to be large with the increase of training data and an arbitration mechanism for hidden (or module) outputs is often needed.

In the second approach, neural networks with spatially localized basis functions are adopted to alleviate the interference. As such localized basis functions, tile-like receptive fields [1, 12] and radial-basis functions [9, 13, 20, 14, 15] are used. Since the response fields are spatially localized, the interference can also be limited to a local region. In this sense, this approach can alleviate the catastrophic interference, but it is not completely prevented in principle.

In the third approach, some old training samples as well as a new one are simultaneously trained in neural networks to suppress the interference. That is to say, some (or all) of training samples are accumulated in memory, and they are utilized for learning at every step. This approach is called "memory-based learning", and this is one of the most promising strategy for incremental learning. Locally Weighted Regression (LWR) [16] is a successful example of the memory-based learning approaches. In LWR, linear regression is conducted

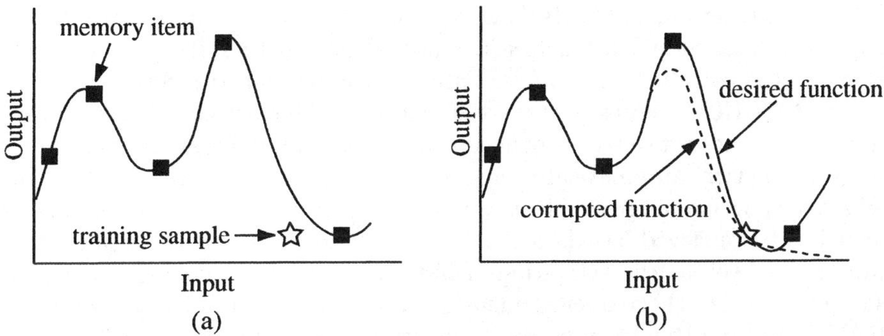

Fig. 1. A schematic explanation on how the interference can occur in neural networks. (a) A network input-output function at time t and a new training sample given at time $t + 1$. (b) Two network input-output relations after the incremental learning of the new sample. The solid and dotted lines mean a desired function of network outputs and a network function corrupted by the interference, respectively. In RAN-LTM, memory items denoted by black squares are extracted from input-output relations during the learning. If some of them are trained with a new sample, RAN-LTM will surely be able to approximate the desired function; this means that the learning of memory items helps prevent the interference.

by using a part of stored training data so that a mean squared error weighted by the distance from a query point (i.e., a new training sample) is minimized. Since LWR is based on linear regression using only local samples around a query point, the training speed is quite fast. However, the regression might be inaccurate if a suitable weight function is not selected, and large memory capacity is often needed to store training samples. Hence, some other learning algorithms have also been proposed in which only representative training samples are extracted from sequentially given training samples and some of them are trained with a current training sample. Yamakawa et al. have proposed Active Data Selection (ADS) in which the contribution of additional data to the approximation accuracy of neural networks are evaluated each time of the learning, and significant data are left in a storage buffer (short-term memory) [17]. Yamauchi et al. have proposed a different type of incremental learning system in which storage data in a buffer are dynamically produced based on the estimation of interference caused by given training data [18]. In these methods, all storage data in a buffer are trained with new data; thus the computation costs of learning tend to be higher.

We have proposed another memory-based learning model called Resource Allocating Network with Long-Term Memory (RAN-LTM). Figure 1 shows a schematic explanation about how the network output function can be corrupted by the interference and it can be prevented in RAN-LTM. When a new training sample is given to a neural network, the connection weights are modified so as to fit the network output to the incoming sample. In many cases, however, the network outputs that have been acquired so far can also be corrupted by this modification (see the dotted line in Fig. 1(b)). This serious side-effect is so-called catastrophic interference mentioned before.

In RAN-LTM, representative data called "memory items" (black squares in Fig. 1(a)) are extracted from accurately approximated regions of the input-output function automatically, and they are stored in long-term memory. When a new training sample is given, some of the memory items are selected to be retrieved based on the curvature information of the input-output function as well as the activation of hidden units. Then, not only a training sample but also the retrieved memory items are simultaneously trained in RAN-LTM, and this learning results in suppressing serious interference.

The above RAN-LTM has been devised for supervised learning problems so far. Therefore, we cannot apply it to reinforcement learning problems without some modifications [19]. In the next two sections, we shall describe how it should be modified.

3 Approximation of Action-Value Functions Using Neural Networks

In reinforcement learning tasks, an agent has to learn a right policy through their own experiences. This means that reinforcement learning inherently has the following nature:

1. A reward is given only when an agent takes an action. Therefore, the learning of the agent's policy is inevitably done in an incremental fashion.
2. Agent's states depend on the initial states and actions taken afterward. Therefore, the occurrence probabilities of the agent's states might be biased depending on the policy whether the learning is conducted based on the on-policy or off-policy control. Moreover, in more practical situations, the occurrence probabilities of many states can be almost zero. That is, the work space for the agent is limited to local regions at the early stage of learning, then the work space would be gradually enlarged in an incremental fashion.
3. As the learning proceeds, it is expected that the agent's policy is improved. As a result, the occurrence probabilities would be changed as well. Hence, one can say that the distribution of training samples is always fluctuated for the agent in general.

Considering the above nature in reinforcement learning tasks, a growing-type (or evolving-type) of neural network is well suited to learn a policy [5]. Resource Allocating Network (RAN) proposed by Platt [20] is one of such neural networks. RAN is an extended model of Radial Basis Function (RBF) networks, in which the hidden units are adaptively allocated for incoming inputs. In the next subsection, let us see how we can approximate action-value functions using RAN.

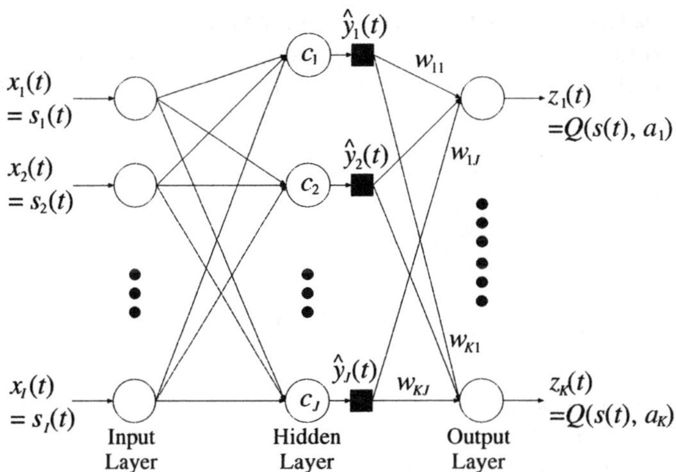

Fig. 2. The structure of RAN with normalized hidden units and the common allocation of the agent's states and action-values to the input and output units in reinforcement learning problems.

3.1 Resource Allocating Network

Let us assume that the state of an agent is represented by an I-dimensional continuous-valued vector $s = \{s_1, \cdots, s_I\}'$, and the agent can select an action from K different actions, $\{a_1, \cdots, a_K\}$. An action-value function defines a mapping from a state-action pair (s, a_k) $(k = 1, \cdots, K)$ to the corresponding value $Q(s, a_k)$. The number of possible state-action pairs is infinite because the states have continuous values. Hence, it is impractical to use all of these state-action pairs as inputs even if a state variable is discretized in a finite number of values.

A practical implementation of action-value functions in neural networks is that each state variable s_i $(i = 1, \cdots, I)$ is set to the ith input unit and the value of each action a_k $(k = 1, \cdots, K)$ is associated with the kth output. In this context, as shown in Fig. 2, we adopt RAN** whose inputs $x(t) = \{x_1(t), \cdots, x_I(t)\}'$ and outputs $z(t) = \{z_1(t), \cdots, z_K(t)\}'$ at time t are given as follows: $x_i = s_i(t)$ and $z_k(t) = Q_t(s(t), a_k)$.

The action-values $Q_t(s(t), a_k)$ are utilized for selecting an agent's action $a(t)$ at time t. In the following, we adopt the following softmax function to select an action based on the outputs $z(t)$ of RAN:

$$P(a(t) = a_k) = \frac{\exp(z_k(t)/T(t))}{\sum_l \exp(z_l(t)/T(t))} \tag{1}$$

** In the original RAN proposed by Platt [20], there is no normalization mechanism in the hidden layer. Hence, the RAN shown in Fig. 2 is a variant of the original RAN.

where $T(t)$ is the temperature at time t. As seen from Eq. (1), the larger value the kth output $z_k(t)$ has, the more frequently the kth action is selected.

In RAN, the outputs $z(t)$ are calculated by the following equations:

$$y_j(t) = \exp(-\frac{\| \boldsymbol{x}(t) - \boldsymbol{c}_j \|^2}{{\sigma_j}^2}) \quad (j = 1, \cdots, J) \tag{2}$$

$$\hat{y}_j(t) = \frac{y_j(t)}{\sum_l y_l(t)} \quad (j = 1, \cdots, J) \tag{3}$$

$$z_k(t) = \sum_{j=1}^{J} w_{kj}\hat{y}_j(t) + \gamma_k \quad (k = 1, \cdots, K) \tag{4}$$

where $\boldsymbol{c}_j = \{c_{j1}, \cdots, c_{jI}\}'$ is the jth RBF center vector, w_{kj} is a connection weight from the jth hidden unit to the kth output unit, γ_k is a bias of the kth output unit, J is the number of hidden units, and σ_j^2 is the variance of the jth radial basis function. Note that outputs of hidden units $y_j(t)$ in Eq. (2) are normalized by their total activations in Eq. (3). Then, the network outputs $z_k(t)$ are obtained from the normalized hidden outputs $\hat{y}_j(t)$.

In general, the RBF network is a powerful function approximator. However, its basis functions have local responses for inputs, hence the outputs might rapidly decrease even when an input is not far from the existing RBF centers. If all the network outputs are almost zero, the action is selected randomly based on Eq. (1). This would be inefficient if the true action-values are smoothly changed in the state space. On the other hand, the normalized RBF network has global responses for the input regions sparsely spanned by radial-basis functions, while it has local responses for the input regions spanned by many radial-basis functions. That is to say, a kind of extrapolation is automatically done for the sparse regions, then we can expect that an agent does not take random actions more often [12, 21]. Hence, it is considered that the normalized RBF networks can learn more efficiently than the conventional ones.

3.2 $Q(\lambda)$ Learning for RAN

When the training gets started in RAN, the number of hidden units is set to one initially. Then, the complexity of approximation is gradually increased by allocating additional hidden units if it is needed to be more accurate.

The learning algorithm of RAN consists of the following two phases:

1. allocation of a new hidden unit,
2. update of connection weights, RBF centers, and biases.

The first phase, the allocation of hidden units, occurs when an agent comes across unknown states. The unknown states can be identified when the activation of all hidden units $y_j(t)$ $(j = 1, \cdots, J)$ is small enough and the following TD error $\delta_k(t)$ is large enough:

$$
\delta_k(t) = \begin{cases} r(t+1) + \gamma \max_a Q_t(s(t+1), a) - Q_t(s(t), a_k)) & (k = k') \\ 0 & \text{(otherwise)} \end{cases} \tag{5}
$$

where k' is the subscript of the action $a(t)$ selected by the agent at time t, and $r(t+1)$ is an immediate reward given by the environment.

On the other hand, in the second phase, the $Q(\lambda)$ learning algorithm is carried out to the agent in which the eligibility trace for all network parameters $\boldsymbol{\theta}$ (i.e., RBF centers, connection weights, and biases) are introduced. The update equation of the eligibility traces $e(t)$ is defined as follows:

$$
e^{NEW} = \gamma \lambda e^{OLD} + \nabla_{\boldsymbol{\theta}} Q(s(t), a(t)) \tag{6}
$$

where $0 \leq \lambda \leq 1$. The second term on the left hand side of Eq. (6) is obtained by $\frac{\partial z_k(t)}{\partial c_{ji}}$, $\frac{\partial z_k(t)}{\partial w_{kj}}$, and $\frac{\partial z_k(t)}{\partial \theta_k}$. Hence, the update equations of the eligibility traces are given as follows:

$$
e_{ji}^{c\,NEW} = e_{ji}^{c\,OLD} + w_{kj} \frac{x_i(t) - c_{ji}}{\sigma_j^2} \hat{y}_j(t)(1 - \hat{y}_j(t)) \tag{7}
$$

$$
e_j^{w\,NEW} = e_j^{w\,OLD} + \hat{y}_j(t) \tag{8}
$$

$$
e_k^{\theta\,NEW} = e_k^{\theta\,OLD} + 1 \tag{9}
$$

where e_{ji}^c, e_{ji}^w, and e_{ji}^θ are the eligibility traces for RBF centers, connection weights, and biases, respectively. The update equations of the network parameters are derived from the squared TD error in Eq. (5) based on the conventional steepest descent method. Then, the $Q(\lambda)$ algorithm for RAN is given as follows:

$$
c_{ji}^{NEW} = c_{ji}^{OLD} + \alpha \delta_k e_{ji}^c \tag{10}
$$

$$
w_{kj}^{NEW} = w_{kj}^{OLD} + \alpha \delta_k e_j^w \tag{11}
$$

$$
\theta_k^{NEW} = \theta_k^{OLD} + \alpha \delta_k e_k^\theta \tag{12}
$$

where α is a positive learning ratio.

4 A Memory-Based Reinforcement Learning Algorithm

4.1 RAN with External Memory

As stated in Section 3, reinforcement learning has some natures of incremental online learning. RAN is a neural network with spatially localized basis functions, hence it is expected that the catastrophic interference is alleviated in RAN. However, since the interference is not completely prevented in principle, the insufficient suppression of the interference might cause serious unlearning in RAN over the long run.

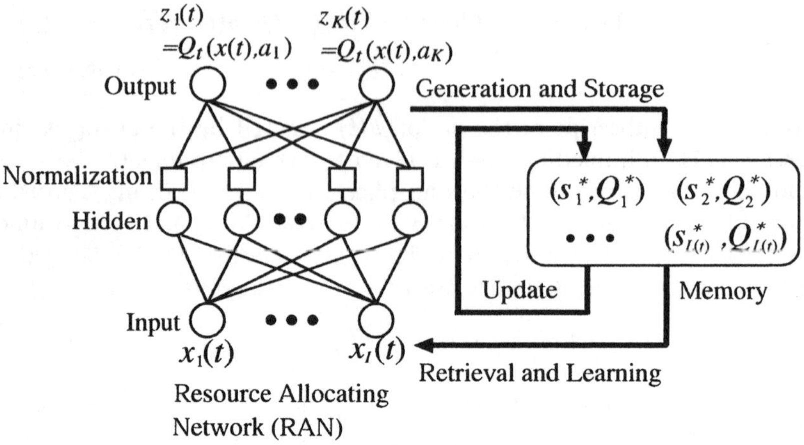

Fig. 3. Structure of RAN-M.

In this section, a variant of Resource Allocating Network with Long-Term Memory (RAN-LTM) is presented, in which its memory items are updated as the need arises rather than kept for a long term. In supervised learning problems, the exact errors between network outputs and their targets are given. In reinforcement learning problems, however, an agent is given only unreliable TD errors which are calculated from currently estimated action-values and immediate rewards [19]. This suggests that memory items do not always hold proper information on action-values until the learning converges. Therefore, we should modify the supervised version of RAN-LTM such that memory items can be updated as the learning proceeds. Here, let us call this model RAN with external Memory (RAN-M).

Figure 3 shows the architecture of RAN-M that consists of two modules: Resource Allocating Network (RAN) and an external memory. The structure of RAN-M is almost the same as that of RAN-LTM. Only a difference between them is that the "update" operation is augmented.

4.2 $Q(\lambda)$ Learning for RAN-M

The proposed $Q(\lambda)$ learning algorithm of RAN-M is similar to that for RAN-LTM discussed in Section 3.2, but there are some differences that are originated from the nature of reinforcement learning. In the following, let us focus on the differences, and then we will present the whole learning algorithm of RAN-M.

Memory items stored in the external memory correspond to the representative input-output pairs that are extracted from accurate regions of the action-value function. In reinforcement learning, only information an agent can estimate the accuracy is the TD error in Eq. (5). Since hidden units have local response for input regions, the accurate regions can be roughly estimated from the RBF centers of highly activated hidden units. Thus, we shall introduce an accuracy index v_j for each hidden unit, and the indexes for only highly activated hidden units are updated as follows:

$$v_j \leftarrow v_j + 2\exp(-\rho|\delta_{k'}(t)|) - 1 \tag{13}$$

where k' is the subscript of the action selected at t and ρ is a positive constant. As you can see from Eq. (13), v_j increases when the TD error is small. If v_j becomes large enough, one can say that a network can learn its action-value function accurately around the region of the jth RBF center. Therefore, an input-output pair $(c_j, z^*(c_j))$ can be extracted as the representative point in this region. This pair is temporarily stored in the external memory in Fig. 3.

To alleviate the forgetting of the previously trained action-value function, not only new information given by an action (i.e., states and rewards) but several retrieved memory items are simultaneously trained in RAN-M. That is to say, the following squared error $E(t)$ between the outputs $z(s_l^*)$ for the lth memory item s_l^* and the corresponding action-value $Q_l^* = \{Q_{1l}^*, \cdots, Q_{Kl}^*\}'$ as well as the squared TD error $\frac{1}{2}\sum_k \delta_k^2$ is simultaneously reduced:

$$E(t) = \frac{1}{2} \sum_{l \in \Omega(t)} \sum_{k=1}^{K} (Q_{kl}^* - z_k(s_l^*))^2 \tag{14}$$

where $\Omega(t)$ is a set of retrieved memory items at time t. Through some simple calculations, we can get the $Q(\lambda)$ learning algorithm for RAN-M as follows:

$$c_{ji}^{NEW} = c_{ji}^{OLD} + \alpha\left[\delta_k e_{ji}^c + \sum_{l \in \Omega(t)} \sum_{k=1}^{K} \Delta_{kl} w_{kj} \frac{s_{li}^* - c_{ji}}{\sigma_j^2} \hat{y}_j(s_l^*)(1 - \hat{y}_j(s_l^*))\right] \tag{15}$$

$$w_{kj}^{NEW} = w_{kj}^{OLD} + \alpha\left[\delta_k e_j^w + \sum_{l \in \Omega(t)} \sum_{k=1}^{K} \Delta_{kl} \hat{y}_j(s_l^*)\right] \tag{16}$$

$$\theta_k^{NEW} = \theta_k^{OLD} + \alpha\left[\delta_k e_k^\theta + \sum_{l \in \Omega(t)} \sum_{k=1}^{K} \Delta_{kl}\right] \tag{17}$$

where $\Delta_{kl} = Q_{kl}^* - z_k(s_l^*)$. The memory items to be recalled are selected based on the distance from the centers of active hidden units: that is, memory items close to the RBF centers are retrieved with high probabilities.

If the forgetting is completely suppressed, the action-value function would be gradually improved by the learning of incrementally given training samples. This is because the target $r(t+1) + \gamma\max_a Q_t(s(t+1), a)$ in Eq. (5) also becomes accurate as the agent's estimation for $Q_t(s, a)$ approaches to a true value. This fact suggests that memory items should be updated with new ones constantly such that they have consistency with the improved action-value function. This operation is easily conducted by recalculating Q_{kl}^* for all memory items using the current action-value function. Unfortunately, however, the adequate frequency of this update is not clear. If the update is frequently carried out, the computation costs could not be neglected. On the other hand, if it is rarely done, memory items will become inconsistent with newly given rewards: that is, the recent experiences of an agent is hardly reflected to the action-value function. Through some preliminary experiments, we determined that this update is carried out every time after each episode is over.

The whole learning algorithm of RAN-M including the procedure of extracting and retrieving memory items is shown below.

[Learning Algorithm]

1. Start an episode. Initialize the agent state as $s(0)$.
2. Set $s(t)$ to RAN's inputs $x(t)$, and calculate the output $z(t)$ from Eqs. (2)-(4).
3. Select an action a_k based on the probability in Eq. (1), and carry it out.
4. After observing the next state $s(t+1)$ and immediate reward $r(t+1)$, calculate the TD error in Eq. (5).
5. If $\delta_k(t) > \eta_1$ and $\|x(t) - c^*\| > \eta_2$, add a hidden unit (i.e., $J \leftarrow J+1$) and initialize connection weights, RBF centers, and biases as follows: $c_{Ji} = x_i(t)$, $w_{kJ} = E_k(t)$, $\sigma_J = \kappa\|x(t) - c^*\|$. Here, c^* is the nearest RBF center to the input $x(t)$ and κ is a positive constant. Otherwise, the following procedure is conducted.
   a) According to the retrieval procedure, recall memory items (s_l^*, Q_l^*) $(l \in \Omega(t))$.
   b) For all retrieved memory items, obtain RAN's outputs $z(s_l^*)$ and calculate the squared error $E(t)$ based on Eq. (14).
   c) To minimize $E(t)$ as well as the squared TD error $1/2 \sum_k \delta_k(t)^2$, update the network parameters of RAN based on Eqs. (15)-(17).
6. Execute the extraction procedure of memory items.
7. If the episode is not over, update all memory items by recalculating $z(s_l^*)$.
8. $t \rightarrow t+1$ and go back to Step 2).

[Retrieval Procedure]

1. Obtain all subscripts j of hidden units whose outputs y_j are larger than η_3, and define a set of these subscripts as $\mathcal{I}_1$.
2. If $\mathcal{I}_1 = \phi$, terminate this procedure.
3. For all hidden units belonging to $\mathcal{I}_1$, obtain memory item s_j^* that has the nearest distance from the center vectors c_j.
4. Add all memory items (s_j^*, Q_j^*) satisfied with the condition $\|c_j - s_j^*\| < \eta_4$ to a set of retrieved memory items $\Omega(t)$.

[Extraction Procedure]

1. Obtain all subscripts j of hidden units whose outputs $\hat{y}_j$ are larger than η_3, and define a set $\mathcal{I}_2$ of these subscripts.
2. If $\mathcal{I}_2 = \phi$, terminate this procedure.
3. Update the value v_j for all $j \in \mathcal{I}_2$ based on Eq. (13).
4. If $v_j < \eta_5$ is satisfied with all $j \in \mathcal{I}_2$, terminate this procedure. Here, η_5 is a positive constant.
5. For all hidden units that satisfy $v_j \geq \eta_5$, carry out the following procedures.
   a) Initialize v_j.
   b) If the corresponding memory item has not been generated yet, increment the number of memory items $L(t)$ (i.e., $L(t) \leftarrow L(t) + 1$). Otherwise, go back to Step (a).
   c) Set the center vector c_j to RAN as its inputs, and calculate the outputs $z^*(c_j)$.
   d) Store $(c_j, z^*(c_j))$ into the external memory as the $L(t)$th memory item.

5 Experiments

To investigate the performance of the proposed memory-based learning approach, let us consider the following situation: there are two adjacent work spaces; an agent moves around in the first region to estimate its action-value function, then the agent learns the other region. Note that, during the learning of the second region, the agent never experiences the same states as those of the first region. In such a situation, one can easily expect that the interference around the border between these two regions would be serious unless the interference is sufficiently suppressed. Once the interference occurs during the learning of the second region, the action-value function acquired in the first region will be lost especially for the states around the border. Therefore, let us check the suppression degree of the interference through the problems in which goal states for the agent is located on the border. To evaluate the proposed $Q(\lambda)$ learning for RAN-M, we adopt two standard reinforcement learning problems here: *Random Walk Task* and *Mountain-Car Task* [1], and they are made some modifications such that the performance of incremental learning is estimated.

In the following experiments, the performance of Tile Coding (TC) [1], Locally Weighted Regression (LWR) [16], and the original RAN are compared with the performance of the proposed model. As stated in Section 1, TC is one of the linear methods in which action-values are approximated by linear functions of feature vectors that roughly code the agent's states. In general, a large number of receptive fields (tiles) are needed for accurate function approximation. It is well known that LWR is a powerful function approximator but they needs a large memory capacity to store training samples. Hence, let us compare our approach with TC and LWR in term of approximation

Fig. 4. Random Walk Task.

Table 1. Theoretical state-values $V^*(s)$ in the Random Walk Task.

State	1	2	3	4	5	6	7	8	9	10
State-value	1/6	2/6	3/6	4/6	5/6	5/6	4/6	3/6	2/6	1/6

accuracy and memory capacity. On the other hand, the original RAN has no external memory. Thus, comparing with RAN, we can see the effectiveness of introducing external memory.

5.1 Random Walk Task

In the original Random Walk Task, an agent has five different States 1 to 5 and each episode starts at State 3 (see Fig. 4). The agent moves either left or right at each step with an equal probability (i.e., move at random) in learning. On the other hand, in the Random Walk Task adopted here, five more States 6 to 10 are added to the original task. The learning of these states is conducted after the learning of States 1 to 5.

In Fig. 4, terminal states are denoted by black squares. The agent should learn action-value functions to select right actions for all states. In this task, each episode starts from State 3 or 8. When an episode is terminated at the central State 5, a reward (+1) is given to the agent; otherwise, the rewards are zero. There are two actions taken by the agent: "move inside" and "move outside". These actions are respectively represented by the following numbers: 1 and 2. For the notational convenience, the left hand side of the region including States 1 to 5 is denoted as R_1, the right hand side of the region including States 6 to 10 is denoted as R_2. The number of input units is 1, and for States 1 to 8 the number is given to the agent as its input.

In the Random Walk Task, the theoretical state-values $V^*(s)$ can be calculated as shown in Table 1. Since the action-values $Q(s, a)$ is obtained by

$$Q(s, a) = r_{ss'}^a + \gamma V(s'), \tag{18}$$

we can easily calculate the theoretical action-values $Q^*(s, a)$ using $V^*(s)$ in Table 1. Here, s' means the next state of the state s and $r_{ss'}^a$ is the immediate reward. To estimate the action-values efficiently, let us assume a random policy for an agent: that is, the agent selects all actions with an equal probability in learning. After the learning completes, the errors between the estimated action-values $z = Q(s, a)$ and the theoretical action-values $Q^*(s, a)$ are evaluated.

Table 2 shows the average errors over the 10 different evaluations. Each error evaluation is conducted after 5,000 episodes are trained by the agent.

Table 2. Average errors between the true and estimated action-values in the Random Walk Task after the learning of R_2 is done.

	TC	LWR	RAN	RAN-M
R_1	0.148	0.111	0.216	0.108
R_2	0.140	0.102	0.169	0.098

As seen from Table 2, the approximation accuracy of TC and RAN is inferior to the accuracy of LWR and RAN-M in both R_1 and R_2. The best accuracy is attained for RAN-M, but it is almost similar to the accuracy of LWR. Hence, we can say that RAN-M is a powerful function approximator as well. As seen from Table 2, the approximation accuracy of RAN in R_1 is seriously degraded as compared with that in R_2. This suggests that the interference was not sufficiently suppressed in RAN during the learning of R_2. On the other hand, we can also see a little increase in errors in the other models, but the increase is about one fifth of that in RAN. From this result, we can conclude that introducing the external memory into RAN contributes to enhancing the approximation accuracy and suppressing the interference effectively.

5.2 Mountain-Car Task

The mountain-car task is a problem in which a car driver (agent) learns an efficient policy to reach a goal as fast as possible. Figure 5 shows the work space in the one-dimensional case. In the original problem, only the left region (R_1) in Fig. 5 is used for learning. Here, to evaluate the suppression degree of the catastrophic interference, the right region (R_2) is also trained after the learning of R_1. Furthermore, we extend this problem to a two-dimensional one where the agent moves around a two-dimensional space spanned by u_1 and u_2. In the one-dimensional problem, when the car agent arrives at the left most and right most places in Fig. 5, the velocity is reset to zero. The goal of the

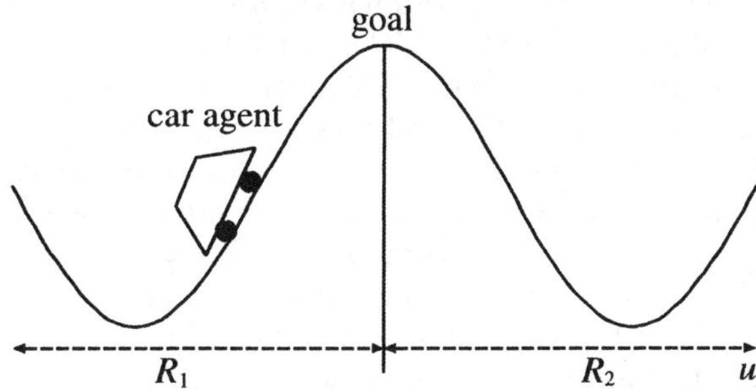

Fig. 5. The work space of the one-dimensional mountain-car task.

car agent is to drive up the steep incline successfully and to reach a goal state at the top of the hill as fast as possible. Hence, the reward in this problem is -1 at all time steps until the car agent reaches the goal. The car agent is initially positioned in either of the two regions: R_1 and R_2. The position $u(t)$ and velocity $\dot{u}(t)$ are updated based on the following dynamics:

$$u(t+1) = B[u(t) + \dot{u}(t)] \tag{19}$$
$$\dot{u}(t+1) = B[\dot{u}(t) + 0.001a(t) - 0.0025\cos(3u(t))] \tag{20}$$

where $B[\cdot]$ is a function to restrict the agent's work space in the following two regions: R_1: $\{u \mid -1.2 \leq u < 0.5\}$ and R_2: $\{u \mid 0.5 < u \leq 2.2\}$. There are three actions to be selected: "full throttle to goal" and "zero throttle" and "full throttle to opposite side of goal". For these actions, $a(t)$ in Eq. (20) is respectively given as follows: $a_1 = 1,\ 0,\ -1$. The goal is located at $u = 0.5$. The inputs of RAN are the position $u(t)$, velocity $\dot{u}(t)$, and previous action $a(t-1)$.

In the two-dimensional problem, the work space is composed of the following two regions: $R_1 : \{(u_1, u_2) \mid -1.2 \leq u_1 < 0.5, -1.2 \leq u_2 < 0.5\}$ and $R_2 : \{(u_1, u_2) \mid 0.5 < u_1 \leq 2.2, -1.2 \leq u_2 < 0.5\}$. The agent's location (u_1, u_2) and velocity $(\dot{u}_1, \dot{u}_2)$ are subject to the following differential equations:

$$u_i(t+1) = B[u_i(t) + \dot{u}_i(t)] \quad (i = 1, 2) \tag{21}$$
$$\dot{u}_i(t+1) = B[\dot{u}_i(t) + 0.001a_i(t) - 0.0025\cos(3u_i(t))] \quad (i = 1, 2) \tag{22}$$

The agent can select the following five actions: *do nothing* and step on the accelerator in four directions (*right, left, up, down*). For these actions, $a_1(t)$ and $a_2(t)$ in Eq. (22) are respectively given as follows: $(a_1, a_2) = (0, 0),\ (1, 0),\ (-1, 0),\ (0, 1),\ (0, -1)$. The other experimental conditions are the same as in the one-dimensional case.

Table 3 shows the experimental results of the one-dimensional and two-dimensional mountain-car tasks. As seen from Table 3, for both problems, RAN needs a considerable number of steps to reach the goal when initial positions are located in R_1. Obviously, serious forgetting of the acquired action-value function is caused by the additional learning of R_2. On the other hand, for the proposed RAN-M as well as TC and LWR, we cannot find distinctive increase in the average steps in R_1. This result suggests that these models can suppress the interference effectively. The average steps in both LWR and RAN-M are almost the same, but TC needs more steps. Since action values are trained separately for every tile in TC, the interference does not occur. However, the continuity of the action values for neighbor tiles is not taken into consideration; hence, this fact might lead to the poor result in TC.

Table 3 also shows the maximum size of shared memory and the average convergence time to learn action-value functions. As we can see, although the fast learning is realized in LWR and TC, they need large memory capacity. On the other hand, RAN and RAN-M need quite small memory capacity;

Table 3. Average numbers of steps to reach the goals in R_1 and R_2, the maximum shared memory, and the convergence time in learning.

(a) One-dimensional problem

	Steps in R_1	Steps in R_2	Memory (KB)	Time (sec.)
TC	377	379	259	73
LWR	196	184	1538	61
RAN	2259	221	15.4	581
RAN-M	185	195	21.4	1211

(b) Two-dimensional problem

	Steps in R_1	Steps in R_2	Memory (KB)	Time (sec.)
TC	517	603	5120	601
LWR	310	281	15396	536
RAN	2011	362	83.7	4121
RAN-M	290	289	118.9	8015

however, the learning of RAN-M and RAN is very slow. This is because the learning is conducted based on the gradient descent algorithm. This problem can be solved by applying the linear method [22] to $Q(\lambda)$ learning for RAN-M.

6 Conclusions

Reinforcement learning inherently has the nature of incremental learning. To learn action-value functions stably, we developed a new version of Resource Allocating Network in which an external memory is introduced. In this model, representative data called "memory items" are automatically extracted from the accurately approximated parts of the agent's action-value function. To prevent unlearning the previously trained action-value function, not only the current experience but several retrieved memory items are simultaneously trained. Moreover, in order to keep more accurate memory items, they are properly updated as the learning proceeds.

To evaluate the incremental learning ability, the proposed model was applied to the following two tasks: Random Walk Task and Mountain-Car Task. We modified these tasks such that there are two adjacent work spaces: that is, an agent moves around in the first region to estimate its action-value function, then the agent learns the other region. From the experiments to estimate the approximation accuracy of action-value functions, we verified that the proposed model outperformed Tile Coding (TC) and the original RAN, and had almost the same performance as that of Locally Weighted Regression (LWR) which is well-known as a powerful function approximator. Moreover, we showed that the shared memory capacity in the proposed model was smaller than that in LWR and TC.

Several problems still remain in our memory-based reinforcement learning
approach. One of these is that the learning algorithm is very slow to converge.
The main reason for this is that the learning algorithm is derived based on
the gradient descent method. However, this problem can be solved by improving our recently proposed supervised incremental learning algorithm [22].
Another problem is that our current algorithm includes many parameters to
be optimized, some of which are very sensitive to the overall performance.
These problems are left as our open questions.

Acknowledgement

The authors would like to thank Mr. Naoto Shiraga for his efforts in developing
programs and conducting computer simulations. This research was partially
supported by the Ministry of Education, Science, Sports and Culture, Grant-
in-Aid for Young Scientists (B).

References

1. Sutton, R. S. and Barto, A. G. (1998). Reinforcement learning - An introduction.
   The MIT Press
2. Kaelbling, L. P., Littman, M. L., and Moore, A. W. (1996). Reinforcement learning: A survey. Journal of Artificial Intelligence Research, 4 : 237–285
3. Carpenter, G. A. and Grossberg, S. (1998). The ART of adaptive pattern recognition by a self-organizing neural network. IEEE Computer, 21 : 77–88
4. Polikar, R., Udpa, L., Udpa, S., and Honavar, V. (2001). Learn++: An incremental learning algorithm for supervised neural networks. IEEE Trans. Systems,
   Man, and Cybernetics – Part C, 31 : 497–508
5. Kasabov, N. (2002). Evolving connectionist systems: Methods and applications
   in bioinformatics, brain study and intelligent machines. Springer-Verlag
6. Kobayashi, M., Zamani, A., Ozawa, S., and Abe, S. (2001). Reducing computations in incremental learning for feedforward neural network with long-term
   memory. Proc. of Int. Joint Conf. on Neural Networks, 1989–1994
7. Shiraga, N., Ozawa, S., and Abe, S. (2002). A reinforcement learning algorithm
   for neural networks with incremental learning ability. Proc. of Int. Conf. on Neural
   Information Processing 2002, 5 : 2566–2570
8. Weaver, S., Baird, L., and Polycarpou, M. (1998). An analytical framework for
   local feedforward networks. IEEE Trans. on Neural Networks, 9 : 473–482
9. Schaal, S. and Atkeson, C. G. (1998). Constructive incremental learning from
   only local information. Neural Computation, 10 : 2047-2084
10. Nakayama, H. and Yoshida, M. (1997). Additional learning and forgetting by
    potential method for pattern classification. Proc. of Int. Conf. on Neural Networks
    97, 1839–1844
11. Kotani, M., Akazawa, K., Ozawa, S., and Matsumoto, H. (2000). Detection of
    leakage sound by using modular neural networks. Proc. of Sixteenth Congress of
    Int. Measurement Confederation, IX : 347–351

12. Kretchmar, R. M. and Anderson, C. W. (1997). Comparison of CMACs and radial basis functions for local function approximators in reinforcement learning. Proc. of Int. Conf. on Neural Networks, 834–837

13. Poggio, T. and Girosi, F. (1990). Networks for approximation and learning. IEEE Trans. on Neural Networks, 78 : 1481–1497

14. Orr, M. J. L. (1996). Introduction to radial basis function networks. Technical Report of Institute for Adaptive and Neural Computation, Division of Informatics, Edinburgh University

15. Haykin, S. (1999). Neural networks – A comprehensive foundation (2nd Ed.). Prentice Hall

16. Atkeson, C. G., Moore, A. W., and Schaal, S. (1997). Locally weighted learning. Artificial Intelligence Review, 11 : 11–73

17. Yamakawa, H., Masumoto, D., Kimoto, T., and Nagata, S. (1993) Active data selection and subsequent revision for sequential learning (in Japanese). Technical Report of IEICE, NC92/99

18. Yamauchi, K., Yamaguchi, and N., Ishii, N. (1999). Incremental learning methods with retrieving of interfered patterns. IEEE Trans. on Neural Networks, 10 : 1351–1365

19. Shiraga, N., Ozawa, S., and Abe, S. (2001). Learning action-value functions using neural networks with incremental learning ability. Proc. of Fifth Int. Conf. on Knowledge-Based Intelligent Information Engineering Systems & Allied Technologies. I : 22–26

20. Platt, J. (1991). A resource allocating network for function interpolation. Neural Computation, 3 : 213-225

21. Morimoto, J. and Doya, K. (1998). Reinforcement learning of dynamic motor sequence: Learning to stand up. Proc. of IEEE/RSJ Int. Conf. on Intelligent Robots and Systems, 3 : 1721–1726

22. Okamoto, K., Ozawa, and S., Abe, S. (2003). A fast incremental learning algorithm of RBF networks with long-term memory. Proc. of Int. Joint Conf. on Neural Networks, 102–107

Structural Optimization of Neural Networks by Genetic Algorithm with Degeneration (GA^d)

Tetsuyuki Takahama[1], Setsuko Sakai[2], and Yoshinori Isomichi[1]

[1] Department of Intelligent Systems, Hiroshima City University, Asaminami-ku, Hiroshima 731-3194 Japan takahama@its.hiroshima-cu.ac.jp
[2] Faculty of Commercial Sciences, Hiroshima Shudo University, Asaminami-ku, Hiroshima 731-3195 Japan setuko@shudo-u.ac.jp

Abstract. There are some difficulties in researches on supervised learning using neural networks: the difficulty of selecting a proper network structure, and the difficulty of interpretation of hidden units. In this work, GA^d (Genetic Algorithm with Degeneration) is proposed to solve the difficulties by optimizing the network structure of neural networks. The GA^d employs real-coded genetic algorithm and introduces the idea of genetic damage. In GA^d, the information of damaged rate is added to every gene. The GA^d inactivates the genes that have lower effectiveness using genetic damage. The performance of GA^d for structural learning is shown by optimizing a simple problem. Also, it is shown that GA^d is an efficient algorithm for the structural learning of layered neural networks by applying GA^d to the learning of a logic function.

Keywords. Structural Learning, Structural Optimization, Degeneration, Genetic Algorithms, Neural Networks, Information Criteria

1 Introduction

Neural networks are applied in various fields. Among those, supervised learning using neural networks is actively studied. However, there are some difficulties in supervised learning using neural networks as follows[9]:

1. It is difficult to select a proper network structure.
   If the network is too large, the generalization ability becomes poor. If the network is too small, the learning ability becomes insufficient. In many cases, the information about the proper network structure is not available. Thus, it is necessary to search the network structure in trial and error, in general.
2. It is difficult to interpret the meanings of hidden units.
   Generally, the sufficient number of hidden units is prepared in order to keep the estimation error small enough. The learned knowledge is distributed to the multiple hidden units. As the result, the meanings of each unit become unclear. The interpretation of the learned knowledge becomes difficult.

256

3. The local minimum problem is inevitable.

   The descent method, which often falls into the local minima, is often adopted as the learning algorithm of neural networks. The problem becomes serious with the increase of the network size.

In order to solve these problems, the researches on *structural learning*, in which not only the parameter values in estimation systems but also the parameter structure of the systems are optimized, are actively carried out. The structural learning of neural networks means not only to optimize the values of weights and thresholds, but also to optimize the network structure, for example, by deleting unnecessary connections and thresholds.

The new methods of structural learning have been proposed based on the idea of degeneration: MGGA (Genetic Algorithm with Mutant Genes)[18, 19] that employs binary-coded Genetic Algorithms (GAs)[5], DGGA (Genetic Algorithm with Damaged Genes)[20, 21] that employs real-coded GAs[11] and GA^d(Genetic Algorithm with Degeneration)[22, 23] that is an extension of DGGA. These algorithms inactivate the genes that less contribute to the survival of the individuals based on genetic damage, and reduce the unnecessary parameters such as weights and thresholds in neural networks, rules in fuzzy rule-based systems, and so on.

The problem 1 can be solved by these algorithms, because they have the ability of optimizing the parameter structure of the estimation systems to be learned. It is expected that problem 2 can be solved, because the obtained parameter structure is the minimum or near minimum and leads the proper interpretation of the systems. It is expected that problem 3 can be solved, because they employ GAs that are comparatively difficult to fall into the local minimum problem[17].

In this work, the performance of GA^d for structural learning is shown by optimizing a simple problem. Also, it is shown that GA^d is an efficient algorithm for the structural learning of layered neural networks by applying GA^d to the learning of a logic function.

2 Structural Learning

2.1 Model Estimation

Model estimation is to identify the input-output relation between variables that consist in training data. Let the explanatory (input) variables be denoted by $x = (x_1, x_2, \cdots, x_n)$, the criterion (output) variable by y, and the model type to express the input-output relation by M and the parameter of the model by P. Then, the relation is represented by the following equation:

$$y = M(x, P) \tag{1}$$

In statistics, a regression model is often used as the model type M. Recently neural networks, RBF(Radial Basis Function) networks and fuzzy inference rules are also used for model estimation. The model parameter P is defined by the structure and values of parameters. The parameter structure is determined by the number of parameters and the meanings of each parameter. For example, in layered neural networks, the parameter structure is defined by the number of layers and the number of neurons in each layer. The parameter values are the values of weights and thresholds. To carry out structural learning and obtain the best model, both of the parameter structure and the parameter values must be optimized.

If the training data $Tr = \{(x^{(k)}, y^{(k)}) \mid k = 1, 2, \cdots, K\}$ are given, then the estimation error (mean square error, MSE) σ^2 of the model M is defined as follows:

$$\sigma^2(M) = \frac{1}{K} \sum_{k=1}^{K} (y^{(k)} - M(x^{(k)}, P))^2 \qquad (2)$$

In general, as the number of parameters increases, the estimation error of training data can be decreased to any extent. If training data are learned by the model with many parameters, the generality of the model will be lost and the model cannot sufficiently cope with untrained data. In this situation, some evaluation criteria such as AIC(Akaike Information Criterion)[1], MDL(Minimum Description Length) Principle[15, 16], and GPE(Generalized Prediction Error)[12] are proposed. These criteria evaluate a model based on not only the estimation error but also the complexity of the parameter structure.

For example, if the estimation errors are statistically independent of each other and follow the normal distribution $N(0, \sigma^2)$, then AIC of the model M is defined as follows:

$$\text{AIC}(M) = K(\log 2\pi\sigma^2 + 1) + 2P_{eff} \qquad (3)$$

where P_{eff} is the number of effective parameters.

2.2 Structural Learning of Neural Networks

There are some researches on the structural learning of neural networks as follows:

- Selective methods: The values of weights and thresholds are optimized under some network structures, and the best structure is selected according to proper evaluation criteria or information criteria[10].
- Destructive methods: Beginning with a large network, a unit or a connection with small contribution to the learning performance for training data will be deleted, usually one by one, from the network while the performance is sufficiently enough[6, 8, 13].
- Constructive methods: Beginning with a small network, a unit or a connection will be added, usually one by one, to the network until the learning performance becomes sufficiently enough[4, 7, 14].
- Reducing methods: Reducing methods are considered as a special case in the destructive methods. Beginning with a large network, the weights of ineffective connections are led to zero while learning the weights. Ishikawa reduced ineffective connections by minimizing the sum of MSE and a penalty criterion for the network size with using a steepest descent method[9]. Takahama et al.

reduced ineffective connections by introducing the idea of degeneration into GAs[18, 19, 20, 21, 22, 23].

In destructive/constructive methods, the process of learning parameter values, or weights, and the process of deleting or adding a unit or a connection are repeated as separate processes. Changing the network structure one by one can be considered as a hill-climbing search in the space of network structures. Thus, it is difficult to search an optimal network structure. Also, usually much computation is needed for re-learning the weights, because there is rather large gap between before and after changing the structures.

In reducing methods, the process of deleting connections is included in the process of learning weights. There is little gap between before and after reduction, because the values of reduced parameters are nearly zero just before they are reduced. The re-leaning does not need. It is expected that the methods based on degeneration can find better network structures, because they uses GAs that will be more robust than steepest descent methods.

In the following, the methods based on degeneration with using GAs are described.

3 Degeneration

In the nature, the phenomenon *degeneration* is well known. Degeneration is the phenomenon that unnecessary organs are lost in the process of evolution. The process of degeneration can be defined from the viewpoint of genetic damage as follows:

1. Some genes are damaged and cannot be repaired.
   DNA (deoxyribonucleic acid), which is the entity of gene, is damaged by the effect of ultraviolet rays, radiation, chemical materials, and so on. In many cases, the damaged DNA will return to its normal state by repair function of DNA. But in some cases, the damaged DNA cannot be repaired and stay in the damaged state. This not-repaired DNA is called as a damaged gene bellow.
2. The individual who has damaged genes loses the function of an organ.
   Genes control protein synthesis. Through the synthesis of protein, some genes control the rate at which new cells are produced and some genes provide the instructions for determining what type of cell is produced and its function - whether it is a skin cell, nerve cell, muscle cell, and so on. If the damaged genes are related to the cells in an organ, the function of the organ will be lost.
3. The individual survives if the lost organ is not important.
   If the lost organ is very important for the individual and he cannot live without the organ, the individual will die. But if the lost organ is not important under current environment, the individual can survive and will leave his descendants.
4. The damaged genes are inherited by the descendants.
   If damaged genes are generated in DNA of somatic cell, the genetic disease in which the organ is lost is not inherited to the descendants. But if damaged genes are generated in DNA of germ cell, the damaged genes and the genetic disease

are inherited by the descendants. The function of the organ, which is related to the damaged genes, of the descendants will be lost, too.

5. All individuals lose the organ.
   If one had such disease, it would mean falling into the disadvantageous situation for survival, and it would be thought that there was no advantage. However, if the lost organ is not related to survival so much, it may work advantageously because his energy will be distributed to the other organs that are more closely related to survival, and he will more adapt himself to the environment. The damaged genes spread over other individuals gradually. All individuals lose the organ. At last, the degeneration of the organ occurs.

The GA^d models these situations. In GA^d, it is assumed that the damaged genes, which are different from the normal genes, are generated by mutation and degeneration is caused by it. If a parameter of estimation systems is treated like the lost organ, the unnecessary parameters can be reduced and the number of parameters can be optimized in the systems described by plural parameters. Thus, GA^d can be applied to the structural learning.

4 Genetic Algorithms with Degeneration (GA^d)

4.1 Genetic Algorithm

Genetic Algorithm (GA) is an optimization algorithm that simulates the heredity and evolution of living things[5, 11]. In GA, a candidate solution is called an individual that is encoded as a list of genes, called chromosome. The GA operates on a population of individuals by applying the principle of survival of the fittest to produce better individuals or candidate solutions. At each generation, a new set of candidate solutions is created by the process of selecting individuals according to their fitness in the problem domain and altering them using operators borrowed from natural genetics such as crossover and mutation operators. This process leads to the evolution of population of individuals that are better suited to their environment than the individuals in earlier generations, just as in natural adaptation. The individual that holds the highest fitness is thought as the best approximate solution in the problem domain. Figure 1 shows the relation among the population, individuals, chromosomes and genes.

The use of a collection of candidate solutions provides a multi-point search. Instead of exploring from one point, GA explores simultaneously from a set of points scattered in the search space. This reduces the possibility of GA getting stuck at local optima, since the chance of all individuals being trapped within a small area is small.

4.2 Damaged Genes

The GA usually does not distinguish a damaged gene with a normal gene. The GA expresses a damaged gene and a normal gene as the same type of a gene. In mutation

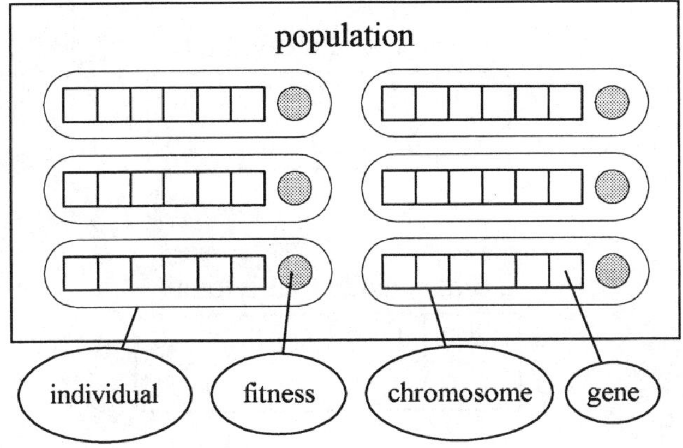

Fig. 1. Population in genetic algorithm

operation, which is very important in GA, all alleles are treated equally. Any gene can be transformed into one of the other alleles reversibly with equal probability.

But in GA^d, a damaged gene is considered as a gene of which state is changed by mutation and of which state is different from that of a normal gene. There are various types of mutation such as substitution, insertion, deletion and so on. It is difficult to assume all states that a damaged gene may have. To settle this situation, a gene is represented by the pair of its *normal value* and *damaged rate* as follows:

- The normal value shows the property or type of gene when the gene is in the normal state. It is assumed that the normal value can change its value among all possible values with equal probability as in GA.
- The damaged rate shows how much the gene is damaged in the interval of $[0, 1]$. The damaged rate of a normal gene is 0. The damaged rate of the gene of which character is completely lost is 1. It is assumed that the damaged rate change its value with biased probability that is newly introduced in GA^d.

4.3 Representation of Individuals

In GA, an individual is usually represented by a chromosome, which holds genetic information. A chromosome is represented by an array of genes. Let the array of genes be denoted by $G = g_1 g_2 \cdots g_L$, the mapping function from genotype to phenotype by h, and the fitness function by f. Then the fitness of the individual is given by $f(h(G))$.

The GA^d is the algorithm of modeling the situation where damaged genes are introduced and the character of individuals or phenotype is affected by damaged genes. An individual in GA^d holds the following information:

- Array of genes (GD)
  $GD = (g_1\, d_1)(g_2\, d_2) \cdots (g_L\, d_L)$. g_i and d_i show the normal value and damaged

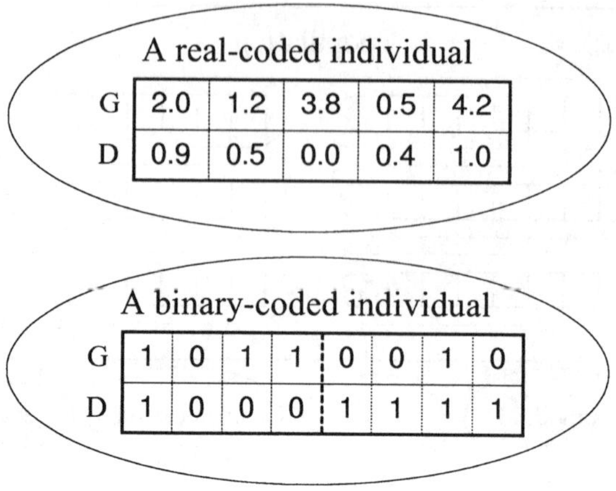

| G | 2.0 | 1.2 | 3.8 | 0.5 | 4.2 |
| D | 0.9 | 0.5 | 0.0 | 0.4 | 1.0 |

| G | 1 | 0 | 1 | 1 | 0 | 0 | 1 | 0 |
| D | 1 | 0 | 0 | 0 | 1 | 1 | 1 | 1 |

Fig. 2. Representation of real-coded and binary-coded individuals

rate of the i-th gene, respectively. L is the chromosome length. It is assumed that a gene is damaged or the damaged rate increases with some probability. Also it is assumed that a gene is repaired or the damaged rate decreases with some probability.

- Fitness value

  The mapping function h^d, which includes the effect of damaged genes, depends on damaged rates $D = d_1 d_2 \cdots d_L$ as well as G. Thus, the fitness value of an individual in GA^d is given by $f(h^d(G, D))$.

4.4 Coding Schema and Mapping Functions

Two types of coding schema, real-coding schema and binary-coding schema, are proposed for GA^d. Figure 2 shows the examples of an individual with real-coding schema in the upper and an individual with binary-coding schema in the lower.

4.4.1 Real-coding schema

In real-coding schema, normal values and damaged rates are represented by real numbers. As a mapping function h^d, a linear function, where the character of individual that are related to damaged genes weakens in proportion to the damaged rate, is proposed. Let assume real value parameters of f be $\boldsymbol{p} = (p_i)$ where p_i is represented by the normal value g_i and the damaged rate d_i of the i-th gene. Then p_i can be defined as follows:

$$p_i = (1 - d_i)g_i \tag{4}$$

For example, in Figure 2, $p_1 = 0.2$ because $g_1 = 2.0$ and $d_1 = 0.9$. Also, $p_5 = 0.0$ and p_5 is reduced because $g_5 = 4.2$ but $d_5 = 1.0$.

4.4.2 Binary-coding schema

In binary-coding schema, normal values and damaged rates are represented by 0/1 bits or bit strings. In this case, a gene is a normal gene if the damaged rate is 0, and a gene is a completely damaged gene if the damaged rate is 1.

As a mapping function h^d, a linear function, where the character of individual that are related to damaged genes weakens in proportion to the number of damaged genes, is proposed. Let real value parameters of f be denoted by $\boldsymbol{p} = (p_i)$, a part of G which represents p_i by G_i, and a part of D which represents p_i by D_i. Then p_i, of which the value is in the interval $[L, H]$, is defined as follows:

$$p_i = (1 - d_i)\{L + (H - L)bin(G_i)/(2^{l_i} - 1)\} \tag{5}$$
$$d_i = one(D_i)/l_i \tag{6}$$

where l_i is the length of G_i and D_i that are represented by two bit strings. The function bin is a function to convert a bit string G_i into an integer in $[0, 2^{l_i} - 1]$. The function one is a function to count the number of "1" in a bit string D_i. d_i indicates the ratio of damaged genes in D_i and corresponds to a damaged rate. For example, in Figure 2, when $L = 0$, $H = 15$ and $l_1 = l_2 = 4$, $p_1 = 33/4$ because $bin(G_1) = 11$, $one(D_1) = 1$ and $d_1 = 1/4$. Also, $p_2 = 0.0$ and p_2 is reduced because $bin(G_2) = 2$ but $one(D_2) = 4$ and $d_2 = 1$.

Of course, the other coding schema, where an individual consists of binary-coded normal values and real-coded damaged rates and vice versa, can be adopted. For example, binary-coded G_i and real-coded D_i is seems to be an interesting idea, but not studied yet.

4.5 Genetic Operations in GA^d

There are some genetic operations in GA^d as well as in GA, such as selection, crossover and mutation. There are two types of mutation in GA^d. One is the mutation for normal values and the other is for damaged rates.

- Selection: The GA^d can adopt various selection schemata such as roulette-wheel selection, ranking selection, tournament selection and so on.
- Crossover: The GA^d can adopt various crossover operations such as one-point crossover, two-point crossover, uniform crossover and so on. In GA^d, the normal values and damaged rates in parents are inherited as a pair by their children. Figure 3 shows the example of one-point crossover.
- Mutation for normal values: The GA^d can adopt various mutation operations for normal values such as uniform mutation, Gaussian mutation and so on. The mutation for normal values is called "reversible mutation", because normal values can be changed into the smaller or larger values reversibly with equal probability.
- Mutation for damaged rates: The GA^d can also adopt various mutation operations for damaged rates. If the damaged rate of a gene is high, it is difficult for the gene to be repaired and the damaged rate tends to increase. If the damaged rate of a gene is small, it is easy for the gene to be repaired, or the damaged rate tends

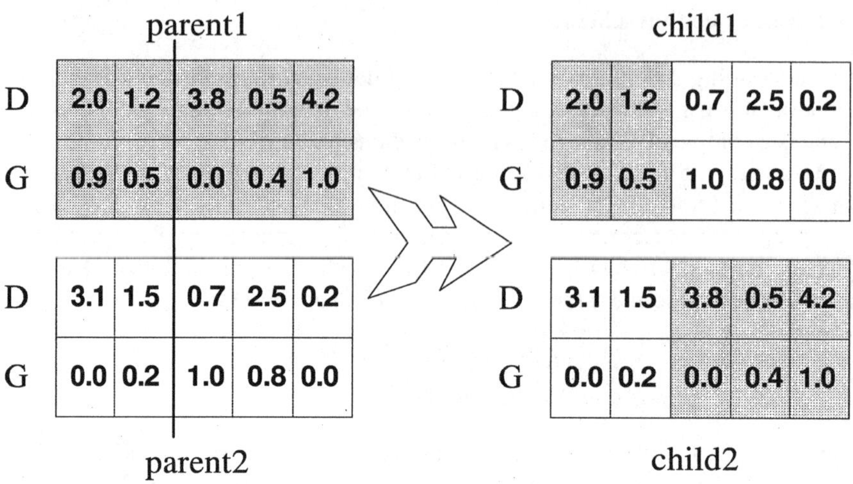

Fig. 3. Crossover operation in GA^d

to decrease. Whether a damaged rate tends to become smaller or larger depends on the damaged rate. Thus, the mutation for damaged rates is called "biased mutation" or "irreversible mutation". In the following, the biased mutation is described in detail.

4.6 Biased Mutation and Damaging Probability Function

In GA^d, the damaged rate is increased or decreased by biased mutation. The probability of increasing the damaged rate is given by a mapping from a current damaged rate $d \in [0, 1]$ to a probability $p \in [0, 1]$, or a damaging probability function P_{dam}:

$$P_{dam} : d \in [0, 1] \rightarrow p \in [0, 1] \tag{7}$$

When biased mutation occurs, the damaged rate increases with the probability $p = P_{dam}(d)$ and decreases with the probability $1 - p$. Generally, it is thought that the smaller the damaged rate is, the easier it is to be repaired. Also, the larger the damaged rate is, the more it is to be damaged. In this work, the following functions are considered as P_{dam}:

$$P_{dam}(d) = \text{constant} \tag{8}$$

$$P_{dam}(d) = d \tag{9}$$

$$P_{dam}(d) = \begin{cases} 2d^2 & (d \leq 0.5) \\ 1 - 2(1 - d)^2 & (d > 0.5) \end{cases} \tag{10}$$

$$P_{dam}(d) = 1 - 0.75(1 - d)^2 \tag{11}$$

The equation (8) is adopted by DGGA. If the constant is 1, or $P_{dam}(d) = 1$, the

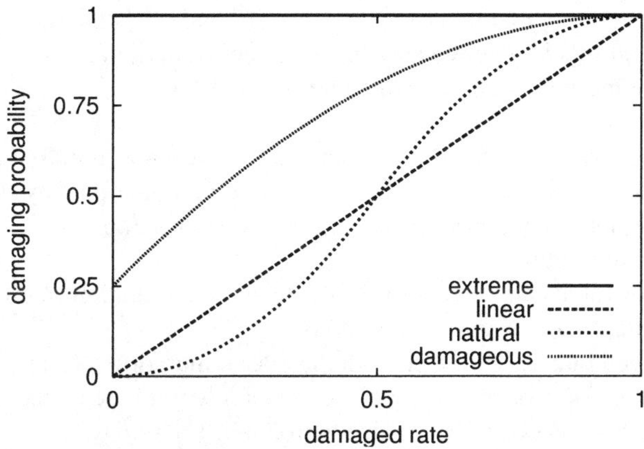

Fig. 4. Damaging probability function P_{dam}

function is the most extreme damaging probability function that a gene is always damaged, and not repaired. The equation (9) and (10) is a linear damaging probability function and a natural damaging probability function, respectively. In the functions, the damaged rate tends to increase when the rate is bigger than 0.5 and the damaged rate tends to decrease when the rate is smaller than 0.5. The equation(11) is the damageous damaging probability function that the damaging probability considerably increases when the damaged rate increases a little. Figure 4 shows these damaging probability functions.

4.7 Algorithm of GA^d

The algorithm of GA^d is as follows:

1. Initialization
   Initial population is created randomly. Usually each damaged rate of the initial individual can be defined as a random number from the interval $[0, 1]$. However all damaged rates can be set all 0, or even all 1 if a proper damaging probability function is selected.

2. Selection
   The parents are selected from the population. Because the evaluation function of the structure will often be minimized in structural learning, the linear ranking selection strategy[2, 3], which is known as a robust strategy, is adopted in this work. All individuals are ranked according to their fitness values. Let the rank of the i-th individual be denoted by r_i ($r_i \in \{1, 2, \cdots, N\}$). The i-th individual's selection probability S_i is defined as follows:

$$S_i = \frac{1}{N}\left(\eta^+ - (\eta^+ - \eta^-)\frac{r_i - 1}{N - 1}\right) \tag{12}$$

where N is the population size, $\eta^- = 2 - \eta^+$ and η^+ is the maximum expected value, which specifies how many times the best individual is selected more than the median individual and is in the interval $[1.0, 2.0]$.

3. Crossover

   The parents are mated with crossover rate P_c and their children are generated. The normal value and damaged rate are inherited as a pair by the children. If parents are not mated, they remain in the next generation.

4. Reversible mutation

   The normal values are mutated with the reversible mutation rate P_{rm}.

5. Biased mutation (irreversible mutation)

   The damaged rates are mutated with the biased mutation rate P_{bm}. When the biased mutation occurs, the damaged rates increase with the damaging probability P_{dam} and decrease with the repairing probability $1 - P_{dam}$.

6. Change of generation

   The current population is replaced by the children. Go back to 2.

This algorithm is coded by C-like language as follows:

```
GA^d ()
{
    t=0;
    Create initial population P(t);
    while(!termination condition) {
        P'=select individuals from P(t);
        for(each pair p,q in P') {
            Crossover p and q with probability P_c;
            for(all locus i in p,q) {
                g_i=g_i + △g_i with prob. P_rm;
                d_i={ d_i + △d_i with prob. P_bm · P_dam
                    { d_i − △d_i with prob. P_bm · (1 − P_dam) ;
            }
        }
        t = t + 1;
        P(t) = P';
    }
}
```

where $\triangle g_i$ is a random number, and $\triangle d_i$ is a positive random number.

5 Properties of GA^d

In this section, the properties of GA^d are examined by optimizing a simple sphere function.

5.1 Test function

The following sphere function F_1 in (13), which includes the variable x_3 that is independent of the value of F_1, is used for the test function. F_1 with the parameters $p = (x_1, x_2, x_3)$ is minimized.

$$F_1(x_1, x_2, x_3) = (x_1 - 1)^2 + (x_2 - 4)^2 \tag{13}$$

The function F_1 has the minimum value at $(x_1, x_2) = (1, 4)$. Since x_1 and x_2 contribute to the value of F_1, it is preferable that they are represented by effective genes. On the contrary, since x_3 is independent of the value of F_1, it is preferable that it is represented by an ineffective or completely damaged gene and is reduced finally.

The condition of the experiment is as follows:

The representation of genes: Every variable is represented by a gene of which the normal value and damaged rate are represented by real numbers.

The generation of initial population: The normal values and damaged rates in each initial individual are generated as random numbers from the interval $[-2, 2]$ and $[0, 1]$, respectively.

The change of normal values and damaged rates:

Each normal value is changed by reversible mutation and the increment/decrement $\triangle g_i$ is given by a random number from $[-0.2, 0.2]$. Each damaged rate is changed by biased mutation and the increment/decrement $\triangle d_i$ is given by a random number from $[0, 0.2]$.

The other conditions: The population size $N = 50$, the maximum expected value in the ranking strategy $\eta^+ = 2.0$, one-point crossover, the crossover rate $P_c = 0.6$, the reversible mutation rate $P_{rm} = 1/2L = 1/6$, and the biased mutation rate $P_{bm} = 1/2L = 1/6$.

To examine the effect of the damaging probability function P_{dam}, $P_{dam} = $ constant (0.5-1.0), linear, natural and damageous functions are tested and the change of F_1's values is observed for each P_{dam}. Every trial in an experiment is continued for 200 generations and the average result of 20 trials is used for evaluation. To compare GA^d with GA of ranking strategy, the same problem is solved using GA^d under the condition that $P_{bm} = 0.0$, $P_{rm} = 1/L = 1/3$ and all damaged rates in initial population are 0.

5.2 Evaluation

Table 1 shows the minimum values of F_1 (Best F_1), the average of the minimum values (F_1), the average of damaged rates of the genes corresponding to x_3 in the best individual (damaged rate), and the number of trials in which the damaged rate of x_3 in the best individual becomes 1 (lost), in 20 trials. In GA, every damaged rate is 0, because every initial damaged rate is 0 and is not damaged.

When P_{dam} was constant, the average of damaged rates of x_3 and the number of trials in which x_3 is reduced tended to increase with the increase of the constant

P_{dam}	Best F_1	F_1	damaged rate	lost
const. 0.5	1.141e-08	4.452e-06	0.7311	11
const. 0.6	1.515e-07	3.060e-06	0.8451	13
const. 0.7	2.152e-07	4.981e-06	0.9652	19
const. 0.8	1.027e-07	2.798e-06	0.9844	18
const. 0.9	1.373e-07	5.504e-06	1.0000	20
const. 1.0	5.033e-08	3.564e-06	1.0000	20
linear	6.786e-10	4.875e-07	0.5500	11
natural	9.305e-11	4.414e-07	0.3500	7
damageous	3.275e-08	9.587e-07	0.9500	19
GA	2.042e-07	2.189e-06	0	0

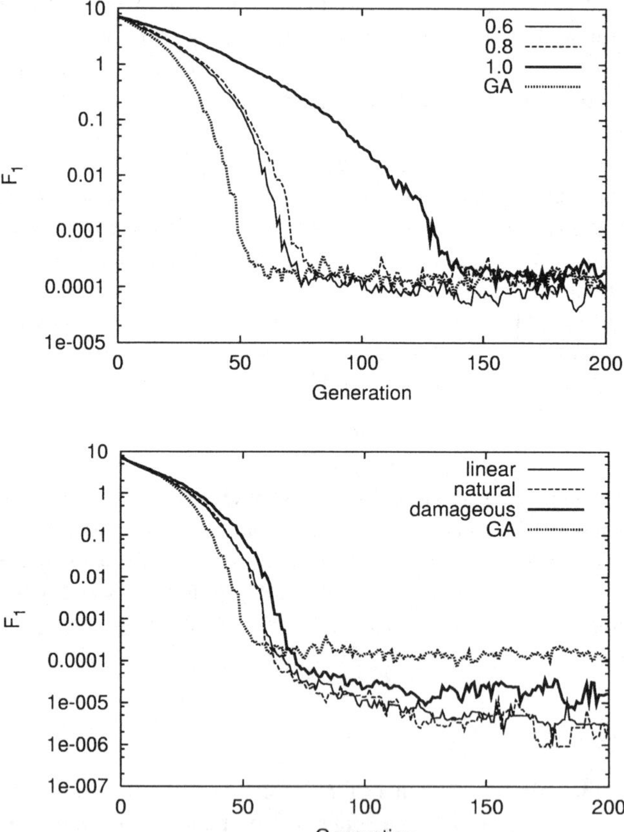

Fig. 5. Change of function values for some P_{dam}s

268

value. When the constant values were 0.7 and 0.8, x_3 was reduced in 19 and 18 trials. When the constant values were 0.9 and 1.0, x_3 was reduced in all trials. Thus, the GA^d with constant P_{dam} has enough ability to reduce unnecessary parameters if a proper constant value is selected. In other P_{dam}, when damaging probability functions are the linear, natural and damageous functions, x_3 was reduced in 11, 7, and 19 trials. Thus, the ability to reduce unnecessary parameters is strong for the damageous function and is weak for the linear and natural functions. The GA^d with constant P_{dam} found worse solutions than GA, but the difference was very small. The GA^d with the linear, natural and damageous functions found better solutions than GA.

Figure 5 shows the graph of the function values over the number of generations. The GA optimized the function faster than GA^d in earlier generations, but GA^d with constant functions found equivalent solutions and GA^d with the linear, natural and damageous functions succeeded to find better solution than GA in later generations. Thus, the optimization ability of GA^d with constant P_{dam} was equivalent to GA. But the optimization ability of GA^d with the linear, natural and damageous functions was better than GA.

Figure 6 shows the graph of the averages of damaged rates of x_1, x_2 and x_3 in the best individual over the number of generations. The GA^d with the linear, natural and damageous functions found that x_1 was an effective parameter, because the damaged rate of x_1 became nearly 0. But GA^d with constant P_{dam} increased the damaged rate of x_1 gradually and it was difficult to find that x_1 was effective. All type of GA^d could find that x_2 was an effective parameter, because the damaged rate of x_2 became almost 0 except that the damaged rate became about 0.2 in GA^d with $P_{dam} = 1.0$. The damaged rate of x_2 was lower than that of x_1, because the optimal value of x_1 ($x_1 = 1$) was in the interval of initial normal values $[-2, 2]$ but the optimal value of x_2 ($x_2 = 4$) was outside the interval $[-2, 2]$. The GA^d with the damageous function and $P_{dam} = 0.8, 0.9, 1.0$ could find that x_3 was an unnecessary parameter, because the damaged rate of x_3 became 1 or nearly 1.

In this section, it was shown that GA^d could reduce unnecessary parameters and was a very effective algorithm for structural learning. Also, it was shown that there were cases where GA^d could overcome simple GA in the optimization ability if a proper damaging probability function was selected, such as the damageous probability function.

6 Structural learning of neural network

In this section, the structural learning of a layered neural network is discussed. The connection weights and thresholds in the neural network are optimized by GA^d. If the damaged rates of genes representing connection weights become 1, then it is considered that the connections are removed. Also, the thresholds are optimized in the similar way.

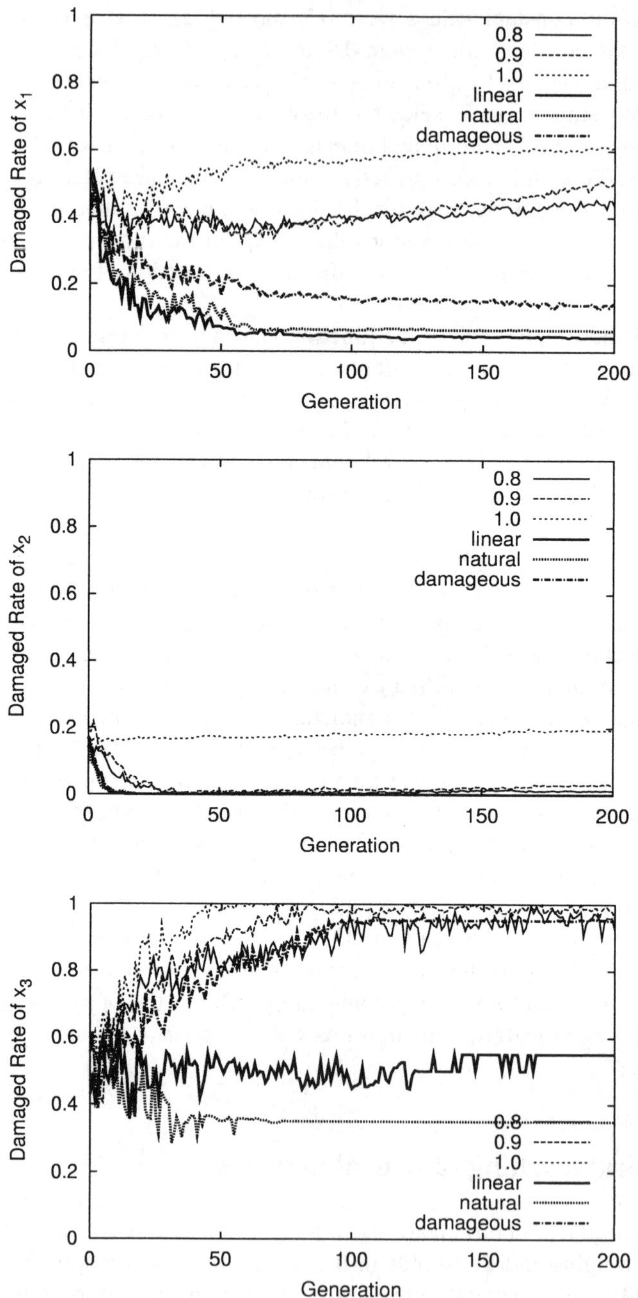

Fig. 6. Change of damaged rates of x_i for some P_{dam}s

6.1 Experiment

6.1.1 Problem

Logical function L in (14) is learned[9]. L includes the variable d which is independent of the function value.

$$L(a, b, c, d, e) = (a \cup b) \cap (c \cup e) \tag{14}$$

The neural network consists of the input layer with 5 neurons, the second layer with 4 neurons, the third layer with 4 neurons and the output layer with 1 neuron. Thus, the maximum number of parameters is 49 ($=6 \times 4 + 5 \times 4 + 5$). Training data are selected randomly from total $2^5 = 32$ data, which are obtained by assigning 0 or 1 to a, b, c, d and e.

6.1.2 Encoding neural networks

In GA^d, each individual or chromosome in population represents one neural network. Figure 7 shows how to encode a neural network into a chromosome. The chromosome consists of three layer parts, from second layer part to output layer part. Each l-th layer part, $l = 2, 3, 4$, consists of the genes corresponding to the parameters of neurons in the l-th layer. The parameters of the i-th neuron in the l-th layer are connection weights w_{ij} to the j-th neuron in the $(l-1)$-th layer and a threshold θ_i. The connection weight w_{ij} is represented by a gene with a normal value w'_{ij} and a damaged rate d_{ij}. The threshold θ_i is represented by a gene with a normal value θ'_i and a damaged rate $d_{\theta i}$.

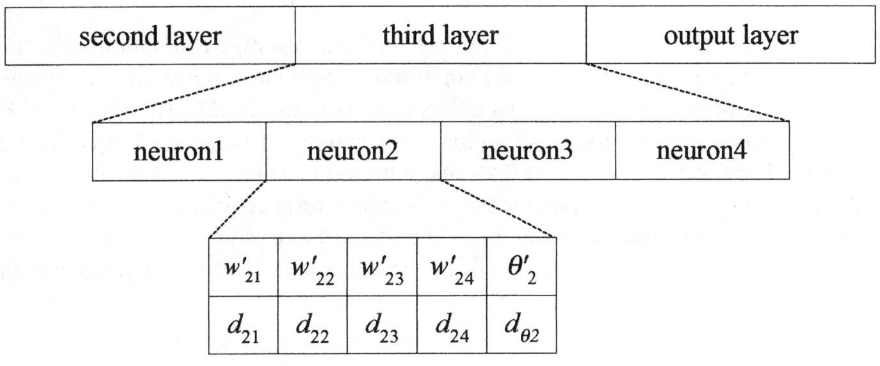

Fig. 7. Encoding of a neural network

The output o_i of the i-th neuron in the l-th layer is given by the following equations:

$$o_i = f\left(\sum_j w_{ij} o_j - \theta_i\right) \qquad (15)$$

$$w_{ij} = (1 - d_{ij})w'_{ij}, \quad \theta_i = (1 - d_{\theta i})\theta'_i \qquad (16)$$

$$f(u) = 1/(1 + \exp(-u)) \qquad (17)$$

where o_j is the output of the j-th neuron in the $(l-1)$-th layer.

The output of the neural network, $M(x, P)$, is o_1 in the output layer. The mean square error of the network is given by eq. (2).

6.1.3 Settings of GA^d

Every gene is represented by a normal value and a damaged rate that are real numbers. The initial values of the normal value and the damaged rate are random numbers from the intervals $[-5, 5]$ and $[0, 1]$, respectively. The fitness value, which is defined by the mean square error σ^2 to the training data, is minimized. The setting for GA^d is as follows: the population size $N = 50$, $\eta^+ = 2$, one-point crossover, P_c=0.6, $P_{rm} = 2/L \approx 0.041$, $P_{im} = 2/L \approx 0.041$. Each normal value g_i is changed by reversible mutation letting $\triangle g_i$ be a random number in $[-1.0, 1.0]$. Each damaged rate d_i is changed by biased mutation letting $\triangle d_i$ be a random number in $[0, 0.2]$. The high constant probability functions (constant 0.98 and 0.99) and the damageous probability function are used as P_{dam}. The number of training data is 24 and the rest of the data is used as test data. Every trial is continued for 2,000 generations, 10 trials are performed in one experiment, and the averages of 10 trials are used for evaluation.

6.2 Experimental Results

Table 2 shows the minimum estimation error σ^2, the average estimation error, the best AIC, the average AIC, the minimum number of effective parameters P_{eff}, and the average number of effective parameters in 10 trials. In all experiments, GA^d found the solution with good estimation error and small number of effective parameters. The GA^d with constant 0.99, the damageous and constant 0.98 functions reduced 36.6, 35.3 and 34.8 parameters on average, respectively. The GA^d with the damageous function found the solution of the best average estimation error. The GA^d with constant 0.99 found the solution of the best average number of effective parameters.

Table 2. Results of structural learning of neural network by GA^d

P_{dam}	min. σ^2	avg. σ^2	min. AIC	avg. AIC	min. P_{eff}	avg. P_{eff}
const. 0.98	5.4423e-57	1.4750e-10	-3015.17	-1376.52	13.00	14.20
const. 0.99	1.9689e-42	3.7343e-13	-2216.64	-1221.31	10.00	12.40
damageous	3.5205e-154	3.0608e-18	-8382.04	-5259.77	10.00	13.70

const. 0.98		const. 0.99		damageous	
min. error	avg. error	min. error	avg. error	min. error	avg. error
3.2443e-57	1.4158e-10	9.8896e-43	3.0746e-13	1.7696e-154	1.9855e-18

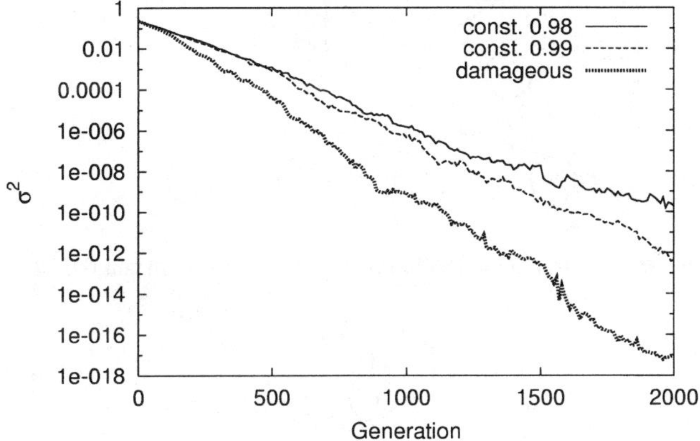

Fig. 8. Change of estimation errors for some P_{dam}s

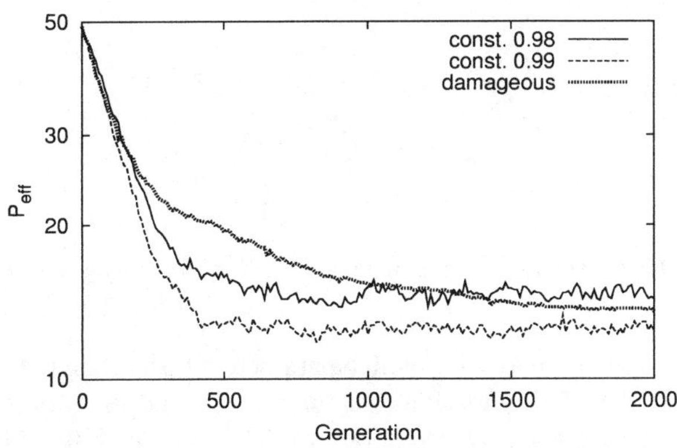

Fig. 9. Change of the number of effective parameters for some P_{dam}s

Table 3 shows the minimum and average mean square error for 8 test data in 10 trials. This shows that the neural networks learned by GA^d can infer the output for the unlearned data.

Figure 8 shows the graph of estimation errors over generations. The GA^d with the damageous function decreased the estimation error faster than other functions and obtained very good solution for the error. Figure 9 shows the graph of the number of effective parameters over generations. The GA^d with constant 0.99 and 0.98

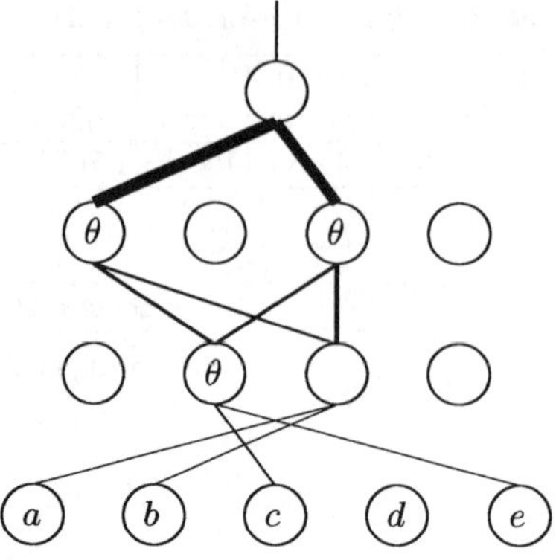

Fig. 10. A neural network obtained by GA^d with constant 0.98 P_{dam}

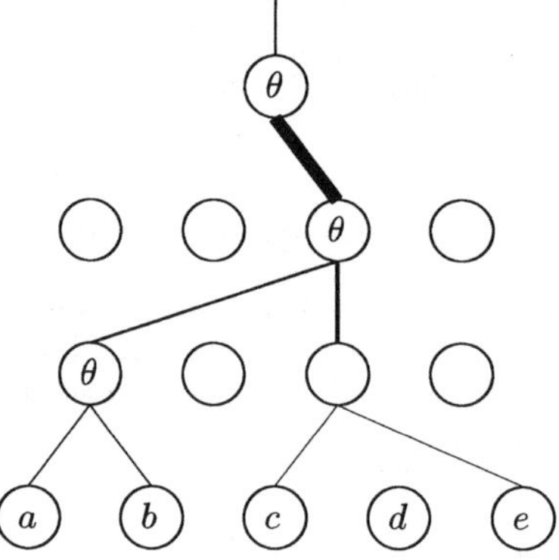

Fig. 11. A neural network obtained by GA^d with constant 0.99 P_{dam}

functions reduced the number of effective parameters faster than GA^d with the damageous function. The GA^d with constant functions found the minimum number of effective parameters at about 800 generations. The GA^d with the damageous function decreased the number of effective parameters slowly but steadily and found the smaller structure than that by GA^d with constant 0.98 finally.

Figures 10, 11 and 12 show the sample of neural networks obtained by GA^d. The thickness of the line represents the relative strength of each weight. θ represents the neuron having the threshold value that is not reduced. From the result, when the number of effective parameters was smallest, the numbers of effective neurons in the second and third layer are 2 and 1, respectively. The structure of the logical function was discovered. The input neuron for independent variable d was removed in almost all trials except that GA^d with the damageous function could not remove d in one trial.

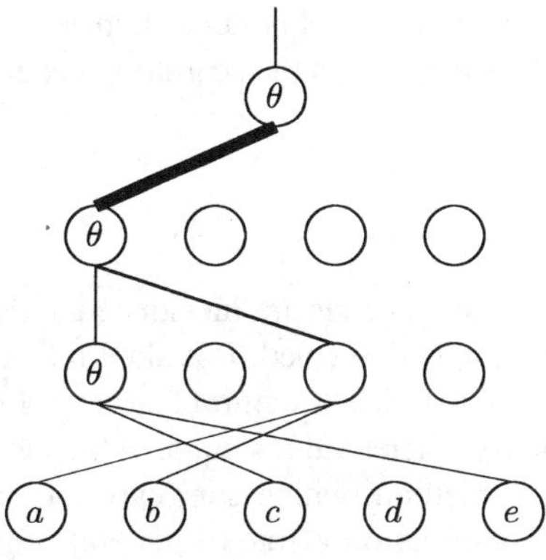

Fig. 12. A neural network obtained by GA^d with damageous P_{dam}

6.3 Discussion

In Section 2.2, the advantage of GA^d was described. In this section, GA^d is compared with other methods in more detail.

Selective methods have the benefit of using information criteria. If proper information criteria are used, the methods can evaluate model structures more precisely than by using estimation errors and can select the model that has high generalization ability. But the methods are not efficient because they need to optimize parameters under different structures. On the contrary, the approach where the information criterion AIC is optimized by GA^d was proposed[24]. The approach is thought to be more efficient because GA^d can optimize many different structures simultaneously.

Destructive/constructive methods are efficient methods of structural learning, because the methods optimize parameters under the structures that are changed step by step and they do not need much time to optimize the structures. But the methods are thought to do hill climbing in structure space and it is difficult to search the optimal structure. It is thought GA^d is less efficient than the methods, but GA^d optimizes parameters under many structures simultaneously and it is possible for GA^d to find better structures. There is a trade-off between the efficiency of destructive/constructive methods and the searching accuracy of GA^d.

There are two type of pruning methods: Ishikawa's method and GA^d. Ishikawa's method needs the differentiability of the evaluation function of models to be learned, because it uses a steepest descent method. Also, it is difficult for the method to optimize information criteria because information criteria usually contain the number of parameters that is not differentiable. On the contrary, GA^d uses only the value of the evaluation function and can be applied to the structural learning of indifferentiable evaluation functions such as information criteria. When the evaluation function is differentiable, it is thought Ishikawa's method is more efficient than GA^d, because the method utilizes the gradient of the function. Since GA^d is an extension of the robust algorithm GA and a multi-point search algorithm, it is thought that GA^d is more robust than the method and can find better structures. There is also a trade-off between the efficiency of the method and the searching accuracy of GA^d.

7 Conclusion

In this work, GA^d, where damaged genes are introduced and they are changed probabilistically by biased mutation, is described. It is shown that GA^d has the ability of structural learning and is suitable to the structural learning of neural networks, which could optimize not only the parameter values but also the network structure.

The GA^d is an excellent method, which can learn the parameter structure naturally with minimizing the estimation error. In general, high P_{bm}, high $\triangle d_i$ and high damaging probability P_{dam} tend to increase the damaged rates and reduce more parameters but tend to increase estimation errors. In the experiments, $P_{bm} \in \{1/2L, 1/L, 2/L\}$, $\triangle d_i \in \{0.1, 0.2\}$ is tested. However, it is necessary to choose proper values for each problem to realize the best structural learning.

The GA^d has various possibilities, such as optimizing information criteria like AIC directly, controlling the damaging probability function through generations, and incorporating the other crossover operation like BLX-α and the other mutation operation like Gaussian mutation, and so on. In future, we'll try to inspect these items.

The GA^d has been applied various models such as regression models, layered neural network models and RBF network models. In future, we plan to apply GA^d to other models such as fuzzy rule bases and apply GA^d to various fields.

References

1. Akaike H (1974) A new look at the statistical model identification. IEEE Trans. Automatic Control AC-19(6):716–723
2. Baker JE (1984) Adaptive selection methods for genetic algorithms. In: Proc. of the First International Conference on Genetic Algorithms and Their Applications. Lawrence Erlbaum Associates, Hillsdale, pp 101–111
3. Bäck T, Hoffmeister F (1991) Extended selection mechanisms in genetic algorithms. In: Proc. of the Fourth International Conference on Genetic Algorithms. Morgan Kaufmann, San Mateo, pp 92–99
4. Campbell C (1997) Constructive learning techniques for designing neural network systems. In: Leondes CT (ed) Neural Network Systems Technologies and Applications. Academic Press, San Diego
5. Goldberg DE (1989) Genetic algorithms in search, optimization and machine learning. Addison Wesley, Reading
6. Le Cun Y, Denker JS, Solla SA (1990) Optimal brain damage. In: Touretzky DS (ed) Advances in Neural Information Processing Systems 2. Morgan Kaufmann, San Mateo, pp 598–605
7. Fahlman SE, Lebiere C (1990) The cascade-correlation learning architecture. In: Touretzky DS (ed) Advances in Neural Information Processing Systems 2. Morgan Kaufmann, San Mateo, pp 524–532

8. Hassibi B, Stork DG (1993) Second order derivatives for network pruning: optimal brain surgeon. In: Hanson SJ, Cowan JD, Giles CL (eds) Advances in Neural Information Processing Systems 5. Morgan Kaufmann, San Mateo, pp 164–171

9. Ishikawa M (1996) Structural learning with forgetting. Neural Networks 9(3):509–521

10. Kurita T (1990) A method to determine the number of hidden units of three layered neural networks by information criteria (In Japanese). IEICE Trans. on Information and Systems J73-D-II(11):1872-1878

11. Michalewicz Z (1996) Genetic algorithm + data structures = evolution programs 3rd ed. Springer-Verlag, Berlin

12. Moody JE (1992) The effective number of parameters: an analysis of generalization and regularization in nonlinear learning systems. In: Moody JE, Hanson SJ, Lippmann RP (eds) Advances in Neural Information Processing Systems 4. Morgan Kaufmann, San Mateo, pp 847–854

13. Mozer MC, Smolensky O (1989) Using relevance to reduce network size automatically. Connection Science 1(1):3–16

14. Parekh R, Yang J, Honavar V (2000) Constructive neural-network learning algorithms for pattern classification. IEEE Trans. on Neural Networks 11(1):436–451

15. Rissanen J (1983) A universal prior for integers and estimation by minimum description length. The Annals of Statistics 11(2):416–431

16. Rissanen J (1986) Stochastic complexity and modeling. The Annals of Statistics 14(3):1080–1100

17. Sexton RS, Dorsey RE, Johnson JD (1998) Toward global optimization of neural networks: a comparison of the genetic algorithm and backpropagation. Decision Support System 22:171–185

18. Takahama T, Sakai S, Isomichi Y (2001) MGGA: genetic algorithm with mutant genes (In Japanese). IEICE Trans. on Information and Systems J84-D-I(9):1297–1306

19. Takahama T, Sakai S, Isomichi Y (2002) MGGA: genetic algorithm with mutant genes. Systems and Computers in Japan 33(14):23–33

20. Takahama T, Sakai S (2002) Structural learning by genetic algorithm with damaged genes. In: Proc. of the IASTED International Conference on Artificial and Computational Intelligence. ACTA Press, Anaheim, pp 161–166

21. Takahama T, Sakai S (2002) Structural optimization of neural network by genetic algorithm with damaged genes. In: Proc. of the 9th International Conference on Neural Information Processing. vol 3, pp 1211–1215

22. Takahama T, Sakai T, Ichimura T, Isomichi Y (2003) Structural optimization by genetic algorithm with degeneration (GA^d) (In Japanese). IEICE Trans. on Information and Systems J86-D-I(3):140–149

23. Takahama T, Sakai S (2003) Learning structure of RBF-fuzzy rule bases by degeneration. In: Proc. of 2003 International Conference on Fuzzy Information Processing. Tsinghua University Press, Beijing, vol 2, pp 611–616

24. Takahama T, Sakai S, Iwane N (2003) Structural learning of RBF-fuzzy rule bases based on information criteria and degeneration. In: Proc. of 2003 IEEE International Conference on Systems, Man, and Cybernetics. IEEE, pp 2581–2586

Adaptive Training for Combining Classifier Ensembles

Nayer M. Wanas and Mohamed S. Kamel

PAMI Lab, University of Waterloo, Canada
{nwanas,mkamel}@uwaterloo.ca

Summary. Classifier ensembles, and multiple classifier systems, have been established in the literature as a means to achieve higher classification accuracy through their combining. This recent interest has been marked by the introduction of a variety of combining methods that improve the overall accuracy. However, it has been noted that for multiple classifier approaches to be useful its members should demonstrate error-independence. Ideally, in terms of ensembles of classifiers, would be a set of classifiers which do not show any coincident errors. That is, each of the classifiers generalized well, and when they did make errors on the test set, these errors were not shared with any other classifier. Generally, training of members of the ensemble to achieve this independence is achieved by training each member independent while manipulating the training data. Although these approaches have been shown to be useful, they might not be sufficient. In this work we present an algorithm to train the members of an ensemble concurrently. The algorithm is applied to an ensemble, based on the weighted average, and the feature based aggregation architecture. An empirical evaluation shows a reduction in the number of training cycles when applying the algorithm on the overall architecture, while maintaining the same or improved performance. The performance of these approaches is also compared to standard approaches proposed in the literature. The results substantiate the use of adaptive training for both the ensemble and the aggregation architecture.

Key words: Multiple Classifier Systems, Decision Fusion, Incremental Training, Neural Networks

1 Introduction

Research in the area of machine intelligence has achieved significant progress in the concept of learning from sampled labelled instances. Although many efficient algorithms have been proposed, they have been limited to simple concepts or problems. Furthermore, numerous results suggest that learning more difficult concepts tends to be extremely difficult. Classifiers are required, based on a limited set of training data, to estimate the target function. Unless the function is simple or the training set is a perfect representative of the data in order to achieve perfect generalization, it is inevitable that the estimate and desired target will differ. This has been the motivation

278

for constructing Multiple classifier systems. Combining classifiers in redundant ensembles improvement the generalization abilities of classifiers. achieved by combining classifiers in redundant ensembles is the main motivation behind using multiple classifiers.

Combining a set of imperfect classifiers is a methodology by which to overcome the limitations of the individual classifiers. Each component classifier is known to make errors; however, the fact that the patterns that are misclassified by the different classifiers are not necessarily the same [1], suggests that the use of multiple classifiers can enhance the decision about the patterns under classification. Combining these classifiers in such a way as to minimize the overall effect of these errors can prove useful. Tumer and Ghosh [2] have shown that the ensemble error ($E_{ensemble}$) decreases with an increase in the number of distinct members of the ensemble (K). The ensemble error improves with the reduction in correlation (ρ). This error is related to that achieved by Bayes rule (E_{bayes}) with the following equation:

$$E_{ensemble} = \frac{1 + \rho(K - 1)}{K} E_{error} + E_{bayes}. \tag{1}$$

Hence, if $\rho = 1$, the error of the ensemble is equal to that of a single classifier. Therefore, it is useless to combine identical classifiers. The individual classifiers must be substantially different. If $\rho = 0$, the error decreases as the number of members of the ensemble increase. However, introducing needless classifiers doesn't necessarily yield improved decisions. Similarly, the agreement of one poor classifier with another doesn't necessarily produce better decisions. Hansen and Salamon [3] have suggested that the ensemble classifiers are most useful when each makes independent errors and the error rate is less than 50% for the ensemble error to decrease monotonically with the number of classifiers.

The error of a classifier can be expressed in terms of the square of the bias and the variance [4, 5]. The bias and variance, can be used to examine the effect of combining redundant ensembles. Based on a training set $(x_1, y_1), \cdots, (x_m, y_m)$, a classifier can be trained to construct a function $f(x)$ for the purpose of approximating y for previously unseen observations of x. Then, the mean squared error of f as a predictor of y may be written

$$E_D\left[(f(X) - E[y \mid x])^2\right] \tag{2}$$

where E_D is the expectation operator with respect to the training set D, and $E[y \mid x]$ is the target function. The mean squared error can be decomposed into

$$\overbrace{(E_D[f(X)] - E[y \mid x])^2}^{bias} + \underbrace{E_D\left[(f(X) - E[f(X)])^2\right]}_{variance} \tag{3}$$

The first term is called the bias and the second is the variance of a classifier. The bias represents a means to measure the ability of a classifier to generalize correctly to a test set after it is trained. While the measure of the level of sensitivity of the output of a classifier to the data on which it was trained is called variance. That is, the extent to which the same results are obtained if we use a different set of training data.

There is a tradeoff between the bias and variance of training classifiers. The best generalization requires a compromise between the conflicting requirements of a small variance and a small bias. Such a tradeoff exists because attempting to decrease the bias by considering more of the data, will likely result in a higher variance. Also, efforts to decrease the variance by paying less attention to the data, usually results in an increased bias. What is required of an estimator is that it generalizes well after training on noisy or unrepresentative data, and to avoid over-fitting.

The improvement in performance from combining classifiers is usually the result of a reduction in variance, rather than a reduction in bias. This occurs because the usual effect of ensemble averaging is to reduce the variance of a set of classifiers, while leaving the bias unaltered. Therefore, an effective approach is to create (or select) a set of classifiers with a high variance, but low bias. The variance component can be removed by combining this set of classifiers. Therefore, it is reasonable to try to reduce the bias, since the resultant increased variance is removed by combining. Combining can thus provide a way of circumventing, or at least reducing, the bias-variance tradeoff.

An ensemble which exhibits a high variance should also show a low correlation of errors. Then, one of the main determinants of the effectiveness of an ensemble is the extent to which the members are error-independent, that is, that they make different errors [1]. Wolpert [6] points out that the more each generalizer has to say, the better the resultant stacked generalization. Jacobs [7] also asserts that the major difficulty with combining expert opinions is that these opinions tend to be correlated or dependent. Ideally, the ensemble is a set of classifiers that do not show any coincident errors. In other words, each of the classifiers generalizes well (a low bias component of error), and when they do make errors on the test set, these errors are not shared with any other classifier (a high variance component of error).

Ensembles of classifiers have been utilized successfully in various real world problems [8, 9, 10, 11]. This is motivated by the fact that using the classification capabilities of multiple experts tends to yield an improved performance over that of a single expert. It also improves the classifiers reliability and generalization ability [12]. With these considerations in mind, there are two key issues in the development of multiple classifier systems, how to create the individual classifiers, and how to perform the combination of these classifiers. In the following we will introduce some of the commonly used approach for both.

2 Combining Classifier Ensembles

Numerous approaches to combining classifier ensembles have been presented in the literature. The most popular of these approaches is the majority vote. Other voting schemes are average, maximum, and the Nash vote [13]. These voting schemes are both static and data independent. The Borda count [14] and other rank based approaches are also data independent.

Weighted average [15], Bayes approach and probabilistic schemes [16], Dempster Shafer theory [17], as well as fuzzy integrals [18] are all trainable approaches for classifier combining. In these approaches a set of weights are assigned to each classifier. These weights are obtained by evaluating the performance of the classifiers on a training set, hence categorizing these approaches as being trained.

The basic categorization of multiple classifier systems has been by the method these classifiers are arranged; parallel or serial suite or a combination of them [19]. Multiple classifiers can also be categorized based on the method of mapping between the input and output of the fusion module. This mapping may be linear or non-linear. Linear mapping being the simplest approaches. The representation methodology; same representation or multi-representation [1] can also be used to categorize the various combining methods. Another categorization of classifier combining methods can be based on whether they encourage specialization in certain areas of the feature space, such are modular approaches [20]. Ensemble of classifiers [21, 22, 23] do not encourage such specialization and hence the classifiers themselves must have different classification powers.

Sharkey [24, 12] presented an account of categorization of the various multiple classifier approaches. The basis of the categorization is the distinction between ensemble and modular approaches, as well as differentiating between cooperative and competitive approaches. Some research has also focused on comparing different types of multiple classifier systems, such as the work of Auda and Kamel [20] which focuses on modular approaches. Kamel and Wanas [25], presented a categorization of multiple classifiers based on the relationship of the final output to the input data. The way the aggregation utilizes the input data can be categorized into data independent, implicitly data dependent or explicitly data dependent. Explicitly data dependent approaches take into account the performance of each classifier as well as the local superiority of certain classifiers into the aggregation. This enhances the performance of explicitly data dependent approaches.

Various architectures to classifier combining that are explicitly data dependent have been introduced. Jordan and Jacobs [26] presented a hierarchical mixture of experts which is based on the divide and conquer approach. Wanas and Kamel [27] presented the feature based approach. Different from the other approaches, this architecture uses both the input features and classifier outputs to train the aggregation to obtain the weights assigned to the different classifiers. The classifier output and the corresponding weights are combined using an aggregation neural network. This architecture showed improvement over various approaches. In the following we present the feature based architecture.

3 Feature Based Decision Aggregation Architecture

This architecture is composed of three different modules. The classifier ensembles, the Detector module and finally the aggregation module. In the following the various components of this architecture will be introduced.

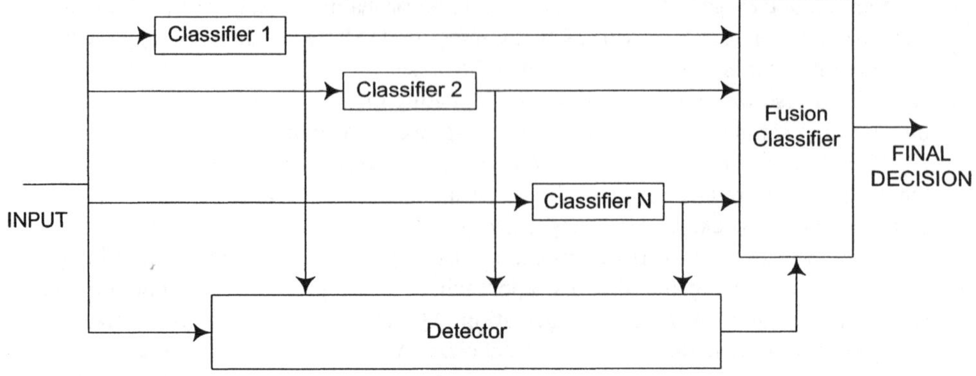

Fig. 1. Feature Based Decision Aggregation Architecture.

3.1 Classifier Ensembles

Each individual classifier, C_i, produces some output representing its interpretation of the input. we are interested in utilize sub-optimal classifiers in the proposed architecture, to make the development overhead of such a system worthwhile. The output of the classifiers are interpreted as a value and a confidence. These two values are used by the detectors and final aggregation schemes.

3.2 Detector Module

The detectors uses the output of the ensemble of classifiers along with the input features to generate weights for the different classifiers. These weights reflect the degree of confidence in each classifier, this confidence is donated by D_i. A neural network approach is used to implement the learning mechanism of this module. The training of this module is performed on the training set after the individual classifiers are trained. These weights are then used in combining the output of the classifier ensemble.

Combining of the classifiers can be done by standard voting approaches. Voting techniques used are the maximum, and average vote. The weights may also be used as an input to the aggregation module.

3.3 The Aggregation Module

The aggregation procedure represents the fusion layer of all the different outputs to generate a more competent output. The aggregation procedure uses the detectors' output to guide the means of combining different classification results. The aggregation scheme can be divided into two phases: a learning phase and a decision making phase. The learning phase assigns a weight factor to each input to support each decision maker. This weighting factor represents a confidence in the output of each classifier. These confidences are then aggregated using standard classifier-combining methods. A neural network approach is selected to perform the weighting of the individual classifier. The neural network would take the inputs from both individual classifiers and the detector and presents a newly modified probability of success of each classifier.

4 Methods for Creating Ensemble Members

Since the main reason for combining classifiers in an ensemble is to improve their performance, there is clearly no advantage to be gained from an ensemble that is composed of a set of identical classifiers; identical that is, in that they generalize in the same way. It is therefore necessary that the individual classifiers are substantially different. Sharkey [12] presents an account of the various training parameters that can be manipulated in order to create the ensemble. Although varying the data is the most common approach, ensembles can be generated by having different initial condition, training data, network topologies or training algorithms.

1. *Varying the initial weights:* A set of networks can be created by varying the initial random weights from which each net starts training.
2. *Varying network topology:* Various network topologies and architectures can be used. These variations can be in the number of hidden nodes or layers.
3. *Varying the training algorithm:* Different training algorithms can be used to train each network. Different classification approaches can also be used.
4. *Varying the training data:* The training data presented to each member of the ensemble can be different. These alterations can be done by, sampling, disjoint training sets, boosting, different features, different data sources or preprocessing. Since this approach is the most frequently used approach we will shed some more light on the various methods of varying the training data.

   - Sampling Data A common approach is training members of an ensemble on a different sub-sample of the training data. Re-sampling methods which have been used for this purpose include cross-validation [28], and bootstrapping [29]. In bagging [29], a training set containing N cases is perturbed by sampling with replacement (bootstrap) from the training set. The perturbed data set may contain repeats. This procedure is repeated several times to create a number of different, although overlapping, data sets. Such statistical re-sampling techniques are particularly useful where there is a shortage of data.
   - Disjoint Training Sets A similar method to data sampling is the use of disjoint, or mutually exclusive training sets, by sampling the data without replacement. There is then no overlap of the data used to train different classifiers. The problem is that, as noted by [30], the size of the training set may be reduced, and this may result in a deteriorated performance.

- Boosting and Adaptive Re-sampling Schapire [31] has demonstrated that a series of weak learners can be converted to strong learners as a result of training the members of an ensemble on patterns that have been filtered by previously trained members of the ensemble. One problem with boosting is that it requires large amounts of data. Freund and Schapire [32] have proposed an algorithm, Adaboost, that largely avoids this problem, although it was developed in the context of boosting. The training sets are adaptively re-sampled, so that the weights in the re-sampling are increased for those cases which are most often misclassified.

- Different Data Sources Using data from different input sources is another method for varying the data on which classifiers are trained. This can be achieved when different sensors, collecting different kinds of information, are used.

- Preprocessing The may vary the data on which classifiers are trained by using different preprocessing methods. This maybe achieved by applying different signal processing methods to the data, or extracting different feature sets. Alternatively, the input data for a set of classifiers can be distorted in different ways, for example, by using different pruning methods [2], by injecting noise [33], or by using non-linear transformations [30, 5].

Generally, the individual classifiers of the ensemble are trained independently and empirically. Auda and Kamel [34] presented the EVOL architecture which tries to achieve autonomous training of an ensemble. In this approach, members of the ensemble utilize voting results to direct the training process. During training, EVOL gates each individual training input to one of the classifiers in the ensemble. This approach reduces crosstalk as well as the correlation amongst the ensemble modules. However, the EVOL approach lacks any guarantee of convergence. In this work we present an adaptive training algorithm that addresses the two observations made concerning classifier ensembles. This algorithm is capable of autonomously directing the training of each classifier in the ensemble by selecting the training records to be used in re-training. This re-training is based on both the performance of the ensemble and the performance of the individual classifier.

5 Adaptive Training for Classifier Ensembles

As mentioned previously, each module in a classifier ensemble is typically trained independently. During the testing phase, each module makes a decision, based on its training, as to the class of the input data. The decisions are fused based on some aggregation scheme and the result of this fusion is the final classification. These different approaches can be viewed as a means by which relative weights are assigned to each classifier. Although the ensemble model has many benefits, there are still some disadvantages to using such a model. For example, in practice, we see that although the individual classification accuracy of some of the modules may be high, the final classification accuracy can be much lower [12]. This is due in part to the fact that the decision fusion mechanism may not have enough information about the accuracy expected from a module during the testing phase. This information could allow the aggregation module to carry out a more informed fusion process. Hence the relative important of each of the modules' accuracy is just as important as its behavior amongst the ensemble. Duin [35] suggests that retraining of the base classifiers after training and evaluating the combining classifier will be useful. This will consequently mean that the design of the combined classifier system becomes an iterative procedure. Another problem shared by all classifier is that of the duration of the training. In other words, how much the individual modules should be trained. Depending on the class groupings, individual modules need to be trained to different levels of generalization or specification as a result of the severity of the overlap of the classes within the training data. The adaptive training algorithm presented attempts to alleviate these problems to some extent in order to create a more robust and systematic training procedure.

The adaptive training attempts to allow the final classification by the aggregation layer to determine whether or not further training should be carried out at the modular level [36]. In addition, a computed confidence factor for each of the modules allows the algorithm to utilize the best weight set available for each module. The idea is that by increasing the quality of training at the modular level, the final aggregation process is expected to be more accurate. The training and testing algorithms are described below. We note here that the data must be divided into training, testing and evaluation sets. The evaluation sets should be distinct from the training and test sets but, like the training set, should include representative vectors for all classes. Let C_i represent module i, CF_i, the confidence factor of the classification from C_i and be a user defined threshold such that $0.0 < \Gamma < 1.0$. We also define Err as the number of evaluation samples in error, $0 < P < 100\%$ as a base percentage and δ to be the modifier for P. K is defined as the number of modules. Let $CF_{i_{best}}$ be the best confidence factor obtained for C_i during training. Figure 2 shows the flowchart of the algorithm.

During testing, we utilize the best weight set obtained for each of the modules. This assures the highest confidence factors for the modules' local decisions. The algorithm utilizes the evaluation set to gauge how well each of the modules is performing not only with respect to itself, but also relative to other modules. The confidence factor acts as a measure of this ability since it is based on the proportion of incorrectly classified records relative to the total number of records in the evaluation set. Referring to the training algorithm, we continue training the network based on the classifications and votes that were in error (from the evaluation set). In order not to destroy the classifications and votes already learnt by a module, the algorithm randomly chooses records from the training set that represents the set of learned classes. The number of records chosen is $\lceil Err \times P \times \delta / 100 \rceil$, in this manner depends on the number of records in error and the user defined constant P. A higher error from the evaluation set results in more records being chosen from the training set that have been classified correctly. The number of records chosen to represent the correctly classified classes is therefore a function of the number of records chosen to represent

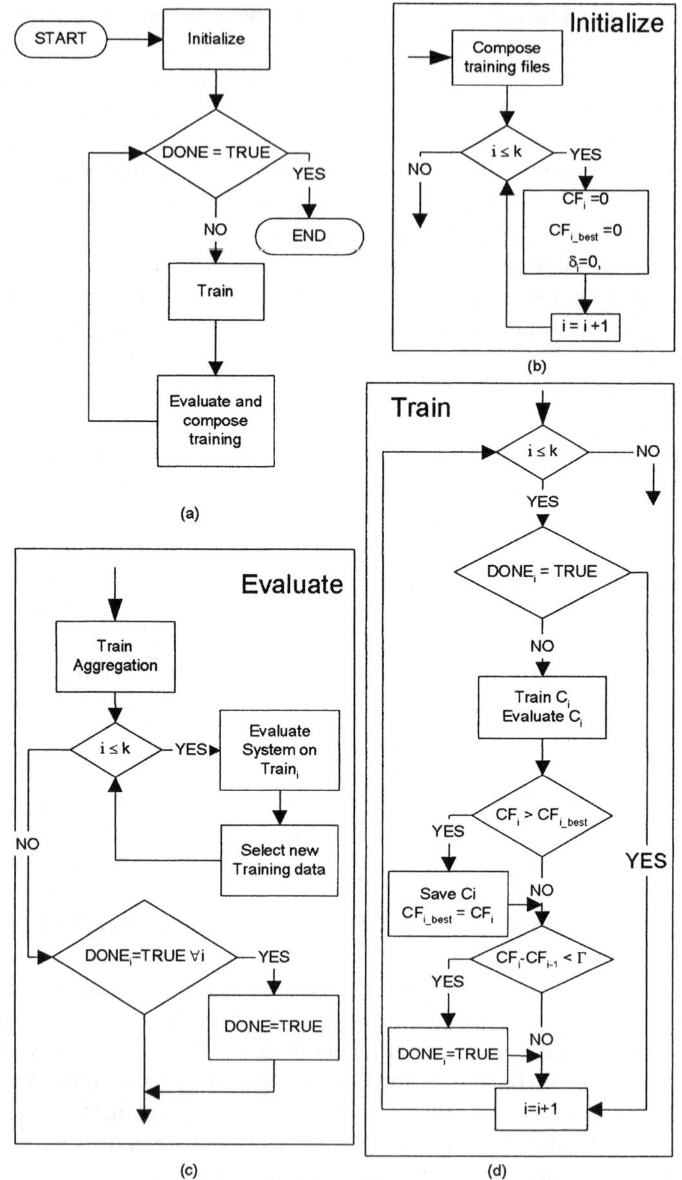

Fig. 2. Adaptive training algorithm, (a) Basic Algorithm which included (b) An Initialization Loop, (c) An Evaluation Algorithm, (d) A Training algorithm

those classes that were incorrectly classified. This limits the amount of user intervention necessary in the retraining of the modules.

This algorithm maybe applied with any ensemble based combining approach. The aggregation procedure maybe trained based on the current ensemble being evaluated. This aggregation procedure will determine the records that will be used for re-training of the ensemble modules. In the following experiments we study the advantage of using the adaptive training algorithm based on the weighted average aggregation and on the feature based architecture. The later will be referred to as Evolving feature based architecture. The objective of our experiments was to compare the classification abilities of the ensemble architecture using a traditional training technique and the adaptive training algorithm presented. We implemented these experiment on the 20 class Gaussian data problem [37], and Satimages database [38]. In all the tests presented in this investigation, we compose an ensemble of five classifiers. Each classifier is a one-hidden layer backpropagation neural network with ten sigmoidal hidden units. These classifiers are trained using the error backpropagation algorithm. In order to decrease the likelihood of external factors, all the models were set up using the same computing platform, language and neural net implementation. All the neural networks utilized the standard backpropagation scheme. The training sets were randomized before being presented to the networks. The training set is divided into five different parts. Each classifier in the ensemble is trained by using four parts of the training set, but a different part is omitted each time, which renders the different training sets partially disjointed. Each classifier is trained for 1000 epochs, and is tested on the evaluation set every 100 epochs. Next, the network instance that provides the best classification is saved. In the case of the Evolving feature based architecture, the detector and aggregation modules are also both modules are one-hidden layer backpropagation neural networks with ten sigmoidal hidden units that are trained separately for each set of ensemble modules. For the adaptive training algorithm, default values for P and Γ of 35% and 0.1 respectively were used. All the networks utilized a 3 layer feed-forward topology. The parameters of all the networks are maintained for all the classifiers that are trained. To reduce any external factors, the models are set up using the same platform, language, and implementation of the neural networks.

In order to investigate the proposed architecture and training algorithm, we implemented various classifier combining approaches. These approaches can be divided into three groups. The *voting approaches*, which include the majority, maximum, average Nash and Borda count, the *trained approaches*, including the weighted average, Bayesian and fuzzy integrals. Finally, the *feature based approach*, which includes the feature based (FB) approach, and voting techniques using the detector output. These voting techniques are the maximum and average vote. Also included are the same approaches using the evolving feature based (EFB) training algorithm.

5.1 Adaptive training based on weighted average aggregation

Table 1 compares the classification errors achieved with standard voting and trained methods on an ensemble that is trained either independently or using the adaptive training algorithm for the 20 Class Gaussian problem. In this case, the adaptive training is modified to compute the combined aggregation by utilizing the weighted average; that is, once the classifiers are trained, the weighting matrix is generated and used, in turn, to evaluate the classification accuracy of each classifier, with respect to its own training patterns. The aim of this change is to factor out the architecture, while evaluating the amount of improvement we can expect from the EVOL algorithm alone. The results indicate a reduction in classification errors ranging from 1.5% to 6%. Although the improvement is consistent with all the aggregation approaches, the majority vote and Borda count experience the greatest improvement. Again, the weighted average and Bayesian approach are among the best aggregation approaches. This improvement is due to the fact that, in their re-training, the classifiers are allowed to focus on improving the overall classification. Consequently, the diversity in the classifiers is enhanced.

Table 1. Comparison of ensemble approaches based on EVOL Algorithm: 20 Class Gaussian Problem

Aggregation method	Normal Ensemble	Adaptive Training
Majority	13.45 ± 0.34	12.67 ± 0.13
Maximum	12.94 ± 0.23	12.74 ± 0.12
Average	12.97 ± 0.12	12.50 ± 0.08
Nash	12.87 ± 0.18	12.49 ± 0.11
Borda	13.44 ± 0.31	12.64 ± 0.14
Weighted Average	12.65 ± 0.10	12.32 ± 0.07
Bayesian	12.73 ± 0.29	12.21 ± 0.13
Choquet Integral	13.00 ± 0.24	12.75 ± 0.12

Table 2 compares the classification error achieved with standard voting and trained methods on an ensemble that is trained either independently or using the adaptive training algorithm on the Satimage data base. Except for the Nash vote, the classification error is reduced by as much as 4%. This increase in performance is also a reflection of the reduction in the average correlation among the members of the ensemble. The correlation reduced from $\rho = 0.9606$ to $\rho = 0.9536$. This slight reduction is responsible for the improved performance and reflected most in the weighted and normal average. Though there is no significant difference in performance amongst the various aggregation methods, the weighted average and Bayesian are amongst the best. The most important improvement maybe reflected in the total number of epochs required to train the ensemble. The number of epochs required to train the ensemble alone is reduced from 2925 epochs to 1775 epochs for the Satimages data set. A similar reduction is also noted for the 20 Class Gaussian problem, where the number of epochs of training the ensemble were reduced to 2275 from the original 3475. This reduction was due to the adaptive learning algorithm.

Table 2. Comparison of ensemble approaches based on EVOL Algorithm: Satimages Data

Aggregation method	Normal Ensemble	Adaptive Training
Majority	13.29 ± 0.15	12.95 ± 0.12
Maximum	13.45 ± 0.16	13.32 ± 0.25
Average	13.21 ± 0.11	12.75 ± 0.09
Nash	14.25 ± 0.31	14.94 ± 0.53
Borda	13.63 ± 0.14	13.33 ± 0.14
Weighted Average	13.11 ± 0.13	12.69 ± 0.11
Bayesian	13.16 ± 0.15	13.15 ± 0.14
Choquet Integral	13.51 ± 0.11	13.33 ± 0.30

5.2 Evolving Feature Based Architecture

In Tables 3 and 4 the performance of the feature based approach on an ensemble created independently, standard feature based, or with the adaptive training algorithm, Evolving feature based, is presented. The feature based approach has improved on the overall classification accuracy beyond other classifier combining methods. The feature based approach based on an ensemble trained using the adaptive training algorithm has also a slight improvement on that of the standard ensemble. The adaptive training algorithm of the feature based approach improved the accuracy compared to the standard training method. The results of the 20 Class problem show that the best performing approach to be the adaptive training method of both the ensemble and the detector and aggregation modules. The satimages database however showed a slight reduction when using adaptive training algorithm for all the components of the architecture. Tables 6, 5 summarizes the total number of epochs; training time required to train the various components of this architecture using the normal training algorithm (each component independently), and the adaptive training algorithm applied to various parts of the architecture separately. Applying the Adaptive training to the ensemble alone reduced the number of epochs by at least 6.5%. While the Adaptive training applied to the Detector and aggregation module alone reduced the number of epochs required for the Satimages data by 27%. A similar large reduction was not seen in the 20Class problem (3% reduction only). Applying the Adaptive training algorithm on the whole architecture produced a reduction of a least 30% from the normal training approach.

The time requirement of the algorithm might be divided into two portions (i) training time, (ii) time used in evaluation of the classifiers. The iterative nature of the algorithm requires more time dedicated to the evaluation of classifiers as compared to independently training the classifiers. This effect can be seen in the amount of time required per-training epoch. There is a considerable increase in the time each epoch of training takes in the adaptive training then normal training. However, the combined effect of the training and evaluation times is favorable since there is a large reduction in the training time required.

Table 3. Performance of the feature based approach using different training: 20 Class Gaussian Problem

	Normal Ensemble	Adaptive Training
Standard FB	8.64 ± 0.60	8.62 ± 0.25
Weighted FB Max	10.99 ± 0.21	11.15 ± 0.23
Weighted FB Avg	11.71 ± 0.23	11.58 ± 0.09
Evolving FB	8.13 ± 0.71	8.01 ± 0.19
Weighted EFB Max	10.94 ± 0.33	10.45 ± 0.29
Weighted EFB Avg	11.65 ± 0.42	10.64 ± 0.48

Table 4. Performance of the feature based approach using different training: Satimage Data

	Normal Ensemble	Adaptive Training
Standard FB	12.48 ± 0.19	12.40 ± 0.12
Weighted FB Max	12.91 ± 0.11	13.50 ± 0.04
Weighted FB Avg	12.67 ± 0.22	12.96 ± 0.02
Evolving FB	12.41 ± 0.09	12.33 ± 0.14
Weighted EFB Max	12.96 ± 0.16	13.74 ± 0.55
Weighted EFB Avg	12.76 ± 0.21	13.07 ± 0.28

Table 5. Training epochs / time (in mins): 20 Class Problem

	Normal Ensemble	Adaptive Training
Standard FB	6727 / 1.11	6290 / 1.11
Evolving FB	6021 / 1.11	4717 / 1.08

Table 6. Training epochs / time (in mins): Satimages Data

	Normal Ensemble	Adaptive Training
Standard FB	7156 / 1.24	7578 / 1.34
Evolving FB	5183 / 1.34	5023 / 0.84

6 Discussions and Conclusion

In this work we have presented an algorithm to train an ensemble of classifiers that addresses some of the disadvantages of manual ad-hoc training of the individual modules. The algorithm presents a method to determine the required training time, increase the useful diversities among the classifiers, and improve the over all classification accuracy. The algorithm is capable of autonomously directing the training of each of the individual modules by selecting the training records used in re-training based on the performance of the ensemble. By evaluating the ensemble with a separate evaluation data set, performance is measured by the degree of generalization illustrated by the system. This evolving algorithm can be applied at various levels (i) on the classifiers only, (ii) on the detector and aggregation module together, or (iii) on the feature based architecture as a whole. Applying the EVOL algorithm on the feature based approach presents the best performance. However, the adaptive algorithm presented, like all iterative algorithms, is more suitable for large data sets. This is especially true when it is used with more sophisticated architectures such as the feature based approach.

We have demonstrated that our method can produce results that are as good as or better than independently training members of the ensemble over a wide variety of aggregation methods. In addition, the automated training method can produce such results with significantly less training effort than the traditional manual approach. Applying this training algorithm on the feature based architecture showed another improvement in classification accuracy. Though this improvement can be considered marginal, the algorithm also managed to reduce the number of training cycles required to achieve this performance. Individually applied to different modules, the adaptive training managed to reduce training times required by 3% to 40%. When this algorithm was applied on the feature based architecture a reduction of up to 50% of the number of epochs was noted in the Satimages data set. This reduction was achieved without the need to compromise on the accuracy achieved.

Acknowledgment

This work was supported by the Natural Science and Engineering Research Council of Canada (NSERC) strategic grant.

References

1. J. Kittler, M. Hatef, R. Duin, and J. Matas, "On combining classifiers," *IEEE Transactions on Pattern Analysis and Machine Intelligence*, vol. 20, no. 3, pp. 226–239, 1998.
2. K. Tumer and J. Ghosh, "Estimating the Bayes Error through classifier combining," in *International Conference on Pattern Recognition*, Vienna, Austria, 1996, pp. 695–699.
3. L. Hansen and P. Salamon, "Neural networks ensembles," *IEEE Transactions on Pattern Analysis and Machine Intelligence*, vol. 12, no. 10, pp. 993–1001, 1990.
4. B. Parmanto and P. Munro, "Reducing variance of committee prediction with resampling techniques," *Connection Science, Special issue on Combining Estimators*, vol. 8, pp. 405–426, 1996.
5. Y. Raviv and N. Intrator, "Bootstrapping with noise: An effective regularization technique," *Connection Science, Special issue on Combining Estimators*, vol. 8, pp. 356–372, 1996.
6. D. Wolpert, "Stacked generalization," *Neural Networks*, vol. 5, pp. 241–259, 1992.
7. R. Jacobs, "Methods of combining experts' probability assessments," *Neural Computation*, vol. 7, pp. 867–888, 1995.
8. J. Kittler and F. Roli, Eds., *Multiple Classifier Systems, First International Workshop, MCS2000, Cagliari, Italy June 2000, Proceedings*, vol. 1857 of *Lecture Notes in Computer Science*, Springer-Verlag Publishers, Berlin, 2000.
9. J. Kittler and F. Roli, Eds., *Second International Workshop, MCS 2001 Cambridge, UK, July 2-4, 2001, Proceedings*, vol. 2096 of *Lecture Notes in Computer Science*, Springer-Verlag Publishers, Berlin, 2001.

10. F. Roli and J. Kittler, Eds., *Third International Workshop, MCS 2002 Cagliari, Italy June 24-26, 2002, Proceedings*, vol. 2364 of *Lecture Notes in Computer Science*, Springer-Verlag Publishers, Berlin, 2002.

11. T. Windeatt and F. Roli, Eds., *Fourth International Workshop, MCS 2003 Guilford, UK, June 11-13, 2003, Proceedings*, vol. 2709 of *Lecture Notes in Computer Science*, Springer-Verlag Publishers, Berlin, 2003.

12. A. Sharkey, "Multi-net systems," in *Combing Artificial Neural Nets*, pp. 1–30. Springer-Verlag Publishers, 1999.

13. L. Xu, A. Krzyżak, and C. Suen, "Methods of combining multiple classifiers and their applications to handwriting recognition," *IEEE Transactions on Systems, Man, and Cybernetics*, vol. 22, no. 3, pp. 418–435, 1992.

14. T. Ho, J. Hull, and S. Srihari, "Decision combination in multiple classifier systems," *IEEE Transactions on Pattern Analysis and Machine Intelligence*, vol. 16, no. 1, pp. 66–75, 1994.

15. S. Hashem, "Algorithms for optimal linear combinations of neural networks," in *International Conference on Neural Networks*, Houston, 1997, vol. 1, pp. 242–247.

16. L. Lam and C. Suen, "Optimal combination of pattern classifiers," *Pattern Recognition Letters*, vol. 16, pp. 945–954, 1995.

17. G. Rogova, "Combining the results of several neural network classifiers," *Neural Networks*, vol. 7, no. 5, pp. 777–781, 1994.

18. P. Gader, M. Mohamed, and J. Keller, "Fusion of handwritten word classifiers," *Pattern Recognition Letters*, vol. 17, no. 6, pp. 577–584, 1996.

19. B. Dasarathy, *Decision Fusion*, IEEE Computer Society Press, 1994.

20. G. Auda and M. Kamel, "Modular neural network classifiers: A comparative study," *Journal of Intelligent and Robotic Systems*, vol. 21, pp. 117–129, 1998.

21. L. Kuncheva, "A theoretical study on six classifier fusion strategies," *IEEE Transactions on Pattern Analysis and Machine Intelligence*, vol. 24, no. 2, pp. 281–286, 2002.

22. A. Verikas, A. Lipnickas, K. Malmqvist, M. Bacauskiene, and A. Gelzinis, "Soft combination of neural classifiers: A comparative study," *Pattern Recognition Letters*, vol. 20, pp. 429–444, 1999.

23. R. Duin and D. Tax, "Experiments with classifier combining rules," in *Multiple Classifier Systems, First International Workshop, MCS2000, Cagliari, Italy June 2000, Proceedings*, J. Kittler and F. Roli, Eds., vol. 1857 of *Lecture Notes in Computer Science*, pp. 16–29. Springer-Verlag Publishers, Berlin, 2000.

24. A. Sharkey, "Types of multinet systems," in *Multiple Classifier Systems, Third International Workshop, MCS2002, Cagliari, Italy June 24-26, 2002, Proceedings*, F. Roli and J. Kittler, Eds., vol. 2364 of *Lecture Notes in Computer Science*, pp. 108–117. Springer-Verlag Publishers, Berlin, 2002.

25. M. Kamel and N. Wanas, "Data dependence in combining classifiers," in *to appear in: Multiple Classifier Systems, Fourth International Workshop, MCS2003, Surrey, England June 2003, Proceedings*. 2003.

26. M. Jordon and R. Jacobs, "Hierarchical mixtures of expert and the em algorithm," *Neural Computing*, pp. 181–214, 1994.

27. N. Wanas and M. Kamel, "Decision fusion in neural network ensembles," in *International Joint Conference on Neural Networks*, Washington, DC, June 2001, vol. 4, pp. 2952–2957.

28. A. Krogh and J. Vedelsby, "Neural network ensembles, cross validation, and active learning," in *Neural Information Processing Systems*, G. Tesauro, D. Touretzky, and T. Leen, Eds., vol. 7, pp. 231–238. MIT Press, Cambridge, Cambridge, 1995.

29. L. Breiman, "Bagging predictors," *Machine Learning*, vol. 26, no. 2, pp. 123–140, 1996.

30. K. Tumer and J. Ghosh, "Error correlation and error reduction in ensemble classifiers," *Connection Science, Special issue on Combining Estimators*, vol. 8, pp. 385–404, 1996.

31. R. Schapire, "The strength of weak learnability," *Machine Learning*, vol. 5, pp. 197–227, 1990.

32. Y. Freund and R. Schapire, "Experiments with a new boosting algorithm," in *Proceedings of the Thirteenth International Conference on Machine Learning*. 1996, pp. 149–156, Morgan Kaufmann.

33. Y. Raviv and N. Intrator, "Variance reduction via noise and bias contraints," in *Combining Artificial Neural Nets*, A. Sharkey, Ed. Springer-Verlag Publishers, 1999.

34. G. Auda and M. Kamel, "EVOL: Ensemble voting on-line," *International Conference on Neural Networks*, pp. 1356–1360, 1998.

35. R. Duin, "The combining classifier: To train or not to train?," in *International Conference on Pattern Recognition*, Quebec City, QC, Canada, June 2002.

36. L. Hodge, G. Auda, and M. Kamel, "Learning decision fusion in cooperative modular neural networks," in *Proceedings of the International Joint Conference on Neural Networks*, Washington, D.C., 1999.

37. G. Auda and M. Kamel, "CMNN: Cooperative modular neural networks for pattern recognition," *Pattern Recognition Letters*, vol. 18, pp. 1391–1398, 1997.

38. C. Blake and C. Merz, "UCI repository of machine learning databases [http://www.ics.uci.edu/~mlearn/MLRepository.html]," 1998.

Combination Strategies for Finding Optimal Neural Network Architecture and Weights

Brijesh Verma[1] and Ranadhir Ghosh[2]

[1]School of Computing and Information Systems,
Central Queensland University, Australia
Email: b.verma@cqu.edu.au
[2]School of Information Technology and Mathematical Sciences,
University of Ballarat, Australia
Email: r.ghosh@ballarat.edu.au

Abstract: The chapter presents a novel neural learning methodology by using different combination strategies for finding architecture and weights. The methodology combines evolutionary algorithms with direct/matrix solution methods such as Gram-Schmidt, singular value decomposition, etc., to achieve optimal weights for hidden and output layers. The proposed method uses evolutionary algorithms in the first layer and the least square method (LS) in the second layer of the ANN. The methodology also finds optimum number of hidden neurons and weights using hierarchical combination strategies. The chapter explores all the different facets of the proposed method in terms of classification accuracy, convergence property, generalization ability, time and memory complexity. The learning methodology has been tested using many benchmark databases such as XOR, 10 bit odd parity, handwriting characters from CEDAR, breast cancer and heart disease from UCI machine learning repository. The experimental results, detailed discussion and analysis are included in the chapter.

Keywords: Neural network architecture, learning algorithms, evolutionary learning algorithms, direct solution methods

1. Introduction

Last few decades have witnessed the use of Artificial Neural Networks (ANNs) in many real-world applications and have offered an attractive paradigm for a broad range of adaptive complex systems. In recent years ANNs have enjoyed a great deal of success and have proven useful in wide variety of pattern recognition problems. However, finding appropriate neural network architecture, uncertain training and long training time are still major problems. Many different kinds of ANN architectures and learning algorithms have been proposed by the researchers to solve the some of the problems mentioned above.

Earlier work by Verma and Ghosh [1-4] proposed and investigated a novel learning methodology, which uses an evolutionary learning [5-19] for the hidden layer weights and least square solution based method for the output layer weights.

The proposed methodology significantly reduced the time complexity of the evolutionary algorithm by combining a fast searching methodology using a least square technique. Also, by using a matrix based solution method, it could avoid further the inclusion of other iterative local search method such as error back propagation (EBP), which had its own limitations. In this chapter we present the proposed learning approach and discuss the possibilities of finding different types of connection topologies in detail with their relative advantages and disadvantages based on experimental results.

1.1 Background

Artificial neural network (ANN), fuzzy system and evolutionary computation are part of the soft computing or computational intelligence discipline [19]. In last few decades several researchers have shown a lot of interests in these areas. Each of these fields or a combination of them has made possible to use for a large number of different application areas. One of the most important constituents of soft computing is artificial neural network (ANN). The basic architecture of an ANN consists of layers of interconnected processing units. These processing units, also known as nodes or processing elements, are the functional counterparts of biological neurons. The number of layers and the degree of interconnection between nodes vary among different ANN designs. The interconnections between the nodes represent the flow of information. Frequently the inner layers of an ANN are referred to as hidden layers since they neither receive input from nor output information outside of the network. The problem domain of the ANN is commonly formulated as a multivariate nonlinear optimization problem over a very high dimensional space of possible weight configurations. The problem of finding a set of parameters for a neural network, which allows it to solve the given problem, can be viewed as a parameter optimization problem. The range of the various parameters such as weights, thresholds and time constants can be bounded between a minimum and maximum value so that the size of the search space is finite. Though having conceded the fact that search space is finite, the optimization of such a problem is a very high order time complexity problem, often known as NP hard problem. The process of finding such optimal parameters is also known as learning algorithm of the ANN [20].

A learning algorithm is at the heart of the neural network based system. Over the past decade, a number of learning algorithms have been developed. However, in most cases learning or training of a neural network is based on a trial and error method. There are many fundamental problems such as a long and uncertain training process, selection of network topology and parameters that still remain unsolved. Learning can be considered as a weight-updating rule of the ANN. Almost all the neural learning methods strictly depend on the architecture of the ANN. The connection rules determine the topological structure of an ANN. Two most important factors in the weight updating rules are the ordering of the nodes and the connection mechanism of the degree of inputs to the ANN. The number of nodes determines the dimensionality of the problem. The response of the ANN can be realized as a function of the weights of an ANN. These surface areas, which can be highly complex, are the areas of concern for a learning algorithm, so that a decision hyper-plane can be constructed by proper combinations of the weight vectors. There are many problems associated in the learning algorithm. For a classification, where the generalization ability of the ANN is important in terms of classification, because of the complexity of the error surface, it may be very difficult for the learning algorithm to achieve this goal, because many of the existing learning algorithms are very much depended upon the error surface information as a priori knowledge. Also, getting trapped in a sub-optimal solution is also a possibility. Another problem can be the step size of a learning process. A very high step size, which means a faster learning, can miss an optimal solution easily where as a very low step size can mean a very high time complexity for the learning process. There is a possibility that the step size can be adapted during the learning process, so that near optimal solution the value becomes less.

One of the learning techniques that attracted the researchers is to apply genetic algorithm based technique. The genetic algorithm is based on the "survival of the fittest" theory of the natural evolution. One of the major problems that can be solved using this method is that because of the stochastic nature of this algorithm the learning process can reach an optimal solution with much higher probability than many standard neural based techniques, which are based on so much on the gradient information of the error surface. One of the problems though, with this global search based technique is the time complexity of the algorithm. For a very large application size, a very powerful computation facility is required to solve the problem. There are other global search methods also such as simulated annealing etc.

GA based learning provides an alternative way to learn for the ANN. The task involves controlling the learning complexity by adjusting the number of weights of the ANN. The use of GA for ANN learning can be viewed as follows
☐ Search for the optimal set of weights
☐ Search over topology space
☐ Search for optimal learning parameters
☐ A combination to search for all the above [5]
The primary feature of the genetic algorithm, which distinguishes it from other evolutionary algorithms, is that it represents the specimen in the population as bit strings. Wright introduced the concept of an adaptive "landscape", which describes the fitness of organisms. Individual genotypes are mapped to respective phenotypes, which are in turn mapped onto the adaptive topography [21]. This is an analogue to the DNA strands used in nature to encode the traits of real organisms. The encoding allows the genetic algorithm to use a set of genetic operators to manipulate the bit strings when creating new specimen. These operators are similar to the types of operations that are naturally encountered by the DNA strands in real organisms during reproduction. The advantages of this approach lie in its generality. The ability of the genetic algorithm to produce progressively better specimen lies in the selective pressure it applies to the population. The selective pressure can be applied in two ways. One way is to create more child specimen then is maintained in the population and selects only

the best ones for the next generation. Even if the parents were picked randomly this method would still continue to produce progressively better specimen due to the selective pressure being applied on the child specimen. The other way to apply selective pressure is to choose better parents when creating the offspring. With this method only as many child specimen as maintained in the population need to be created for the next generation. Because of its nature to find a global solution in the search space, normally GA based learning takes a very long time to train the ANN. So to have a better time complexity, it is a good idea to combine the global search GA with the standard neural learning methods, which perform some local search. Also, many proposals had been given by the researchers to apply such a hybrid algorithm, where the Evolutionary Algorithm (EA) can find a good solution in terms of its weight and architecture of the ANN and then a normal local search procedure can be applied to find the final solution [22] [23] [24].

Another topic of interests for a learning process is the matrix solution based techniques. There are many different matrix solution methods available to solve overly determined equation, based on the iterative nature of the method. One of the major problems in finding the solution is that finding the optimal values for the weights depends on the curve fitting technique. These are also known as least squares methods, because the goal is to reach the minimum error between the fitted curve and the actual target values. Most of the matrix-based solution depends on the inverse property of the matrix. Hence, there is a possibility to find an ill-conditioned matrix. The other challenging point is that the coefficient matrix of the resulting linear system may be full. The major advantage of these methods is the less time complex nature of these methods. Also, these methods can be considered as a local search method, to find optimal values for the weights. Many large-scale optimization problems display sparsity in both second derivative (Hessian) matrices and constraint matrices. Efficient optimization methods must exploit this fact. In recent years researchers have worked on this issue from a variety of directions resulting in several successful and original contributions. These include the development of techniques for the efficient estimation of sparse Jacobian and Hessian matrices (including complete theoretical analysis and high-quality software); an original analysis of the sparse null basis problem (given a sparse rectangular matrix with more columns than rows, determine a sparse (or compact) representation for the null space) and the development of algorithms for finding a sparse null basis, and new direct-factorization methods for solving large and sparse bound-constrained quadratic programming problems.

The computational demands of many large-scale optimization problems can be extremely high; therefore, it is important to explore the practical potential of parallelism in the solution of optimization problems. Researchers have been active in this arena in recent years, having considered several important computational problems related to optimization, including the fast parallel solution of triangular systems of equations; the parallel solution of general systems of non-linear equations; and the parallel solution of non-linear least-squares problems.

In this research, the ideas of exploring these GA based method and the matrix solution methods are considered because of the respective advantages of the two techniques.

2. Proposed learning approach

The proposed novel algorithm consists of mainly two parts – finding the weights and finding the architecture (in terms of finding number of hidden neurons). The two modules are called findArchitecture (this module finds the number of hidden neurons) and findWeight (this modules finds the weight values using the hybrid method) respectively. The two modules are combined using a hierarchical structure, where a feedback mechanism exists for both the modules. In Figure 1, the flowchart for combining these two modules is given.

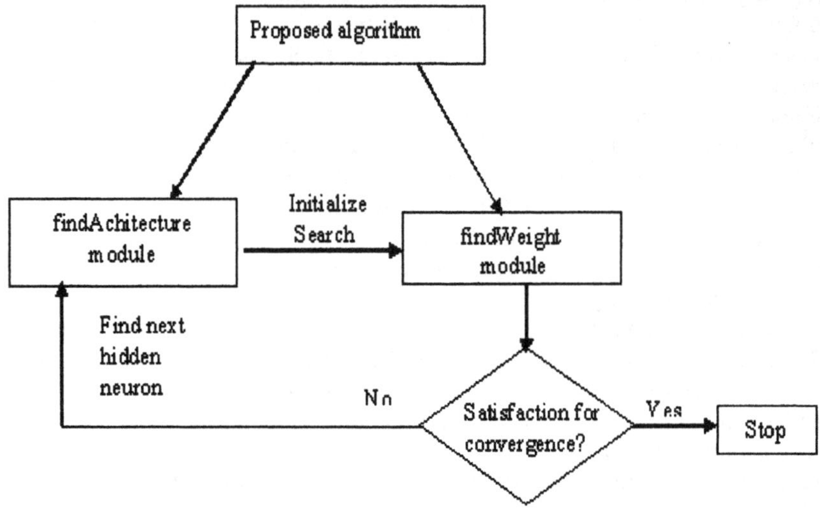

Fig. 1. Combination dynamics for findArchitecture and findWeight modules

2.1 Module details for finding weights

The proposed findWeight module is mainly a hybrid approach, which uses genetic algorithms and the least square method. Hence there are two common issues that need to be discussed, one is the ANN architecture where the proposed algorithm is applied, and because it is a hybrid method that uses two different techniques, the rules joining these two techniques are important. Because of so many possibilities due to the variations that exist in applying such a hybrid method, all possibilities of combinations for variations for the new proposed algorithm are studied. Finally based on the improvement of the results obtained from them, all the varied methods are ranked in a stepwise improvement.

2.2 Common architecture details

A two-layer network architecture is considered. The input nodes for the ANN do the range compression for the input and transmit output to all the nodes in the hidden layer. The activation function used for all the nodes in hidden and output layer is sigmoidal. The first layer is composed of input nodes, which simply range compress the applied input (based on pre-specified range limits) so that it is in the open range interval (0, 1) and fan out the result to all nodes in the second layer. The hidden nodes perform a weighted sum on its input, and then pass that through the sigmoidal activation function before sending the result to the next layer. The output layers also perform the same weighted sum operation on its input and then pass that through the sigmoidal activation function to give the final result.

The weight variables for each layer are found using a hybrid method, which uses the genetic algorithm (GA) and a least square method. The architecture is shown in Figure 2. The genetic algorithm is applied for the first layer weight and the least square method is applied to find the weights for the output layer. We initialize the hidden layer weights with a uniform distribution with closed range interval [-1, +1].

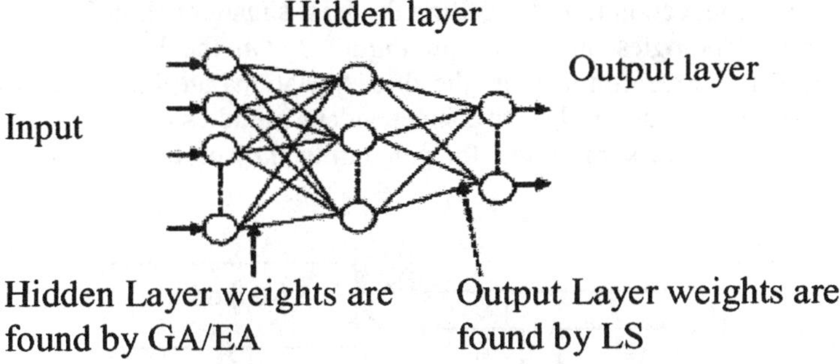

Fig. 2. A two layer ANN architecture for the proposed hybrid learning method

2.3 Combination dynamics

The combination of the genetic algorithm and the least square method provides many possibilities of joining the two different methods. The rules of using the hybrid method can vary not only because there exists different variations of techniques that are suitable in general or may be specific to a particular type of problem, but also because the dynamics of combining the two can vary on the time of joining them. Depending on the different genetic algorithm strategies – two variations of genetic algorithms that can be applied for searching for weights of ANN are investigated. The first strategy is the primitive or straight GA, which is applied to ANN phenotype using a direct encoding scheme. This GA methodology uses the standard two-point crossover or interpolation as recombination operator and a small Gaussian noise addition as mutation operator. A slight variations of the standard GA that is used for testing called evolutionary algorithm, where the only genetic operation to be considered is mutation. Also, mutation is applied using a variance operator. Table 1 shows the two different genetic algorithm strategies that are used in investigation of various combinations.

Table 1. Variations of strategies depending on the GA methods

Genetic Algorithm Strategy	Description
Genetic Algorithm (GA)	Uses the natural GA form using direct encoding schemes. Genetic operators are crossover and mutation with certain probabilities[1].
Evolutionary Algorithm (EA)	Uses the evolutionary form. Uses variance operator as a vector term for performing mutation. No crossover operation is used.

Depending on the connectivity for combining the GA/EA and LS method, three different connection topologies are devised. The variations of the methods are mainly due to the connection time for calling the least square method from the GA / EA. The three topologies are shown in Figure 3. On the basis of the time complexities for the three connections the three variations are known as T1, T2 and T3 on a descending order. The two independent modules of GA/EA and LS are connected together at some point for generating the solution for the weight matrix.

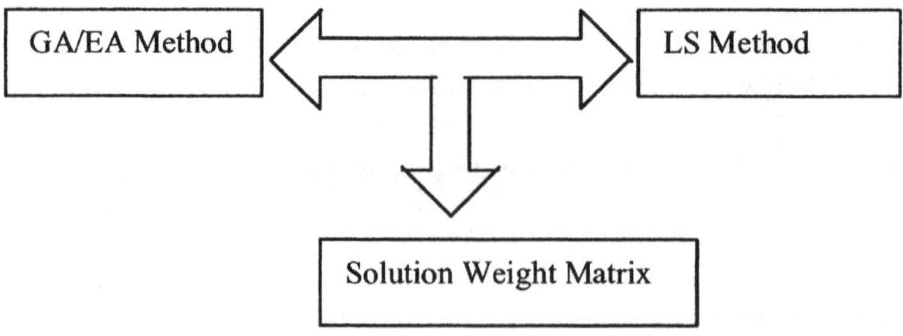

Fig. 3. A general T (1/2/3) connection for GA and LA methods

---

[1] The mutation and crossover probability values will be analyzed in the experimental results chapter.

300

The Table 2 shows a brief description of the three connection strategies before giving their individual details.

Table 2. Three different connection strategies for calling LS from GA/EA

Connection Strategy	Description
T1	The GA/EA and the LS method are called for every generation and every population to find the corresponding fitness value of the population.
T2	The LS method is called after one generation run for GA/EA method. The best fitness population is halved and the upper half is saved as the weights for output layers. Afterwards, the GA/EA is run for the remaining generation.
T3	The LS method is called after the convergence of the GA/EA is over. After certain number of generations for the GA/EA, the best fitness population is halved and the lower half is used as the weights for the hidden layer and those weights are used for the LS method.

The three connection methodologies are explained in simple flowcharts in the following section.

2.3.1 T1 connection

In T1 connection the LS method is called for calculating the fitness for every population and for every generation. The fitness is calculated using the weight represented by the population genome and the LS output of the output layer weights. The flowchart is given in Figure 4.

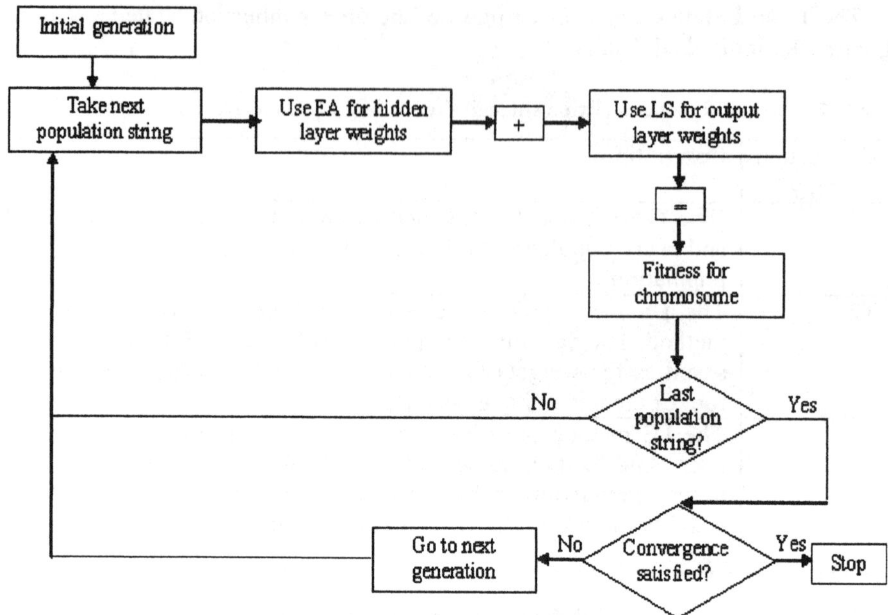

Fig. 4. T1- Connection architecture

2.3.2 T2 connection

In T2 connection the LS method is called for calculating the fitness for every population and only after the first generation. The fitness is calculated using the weight represented by the population genome and the LS output of the output layer weights. The flowchart is given in Figure 5.

302

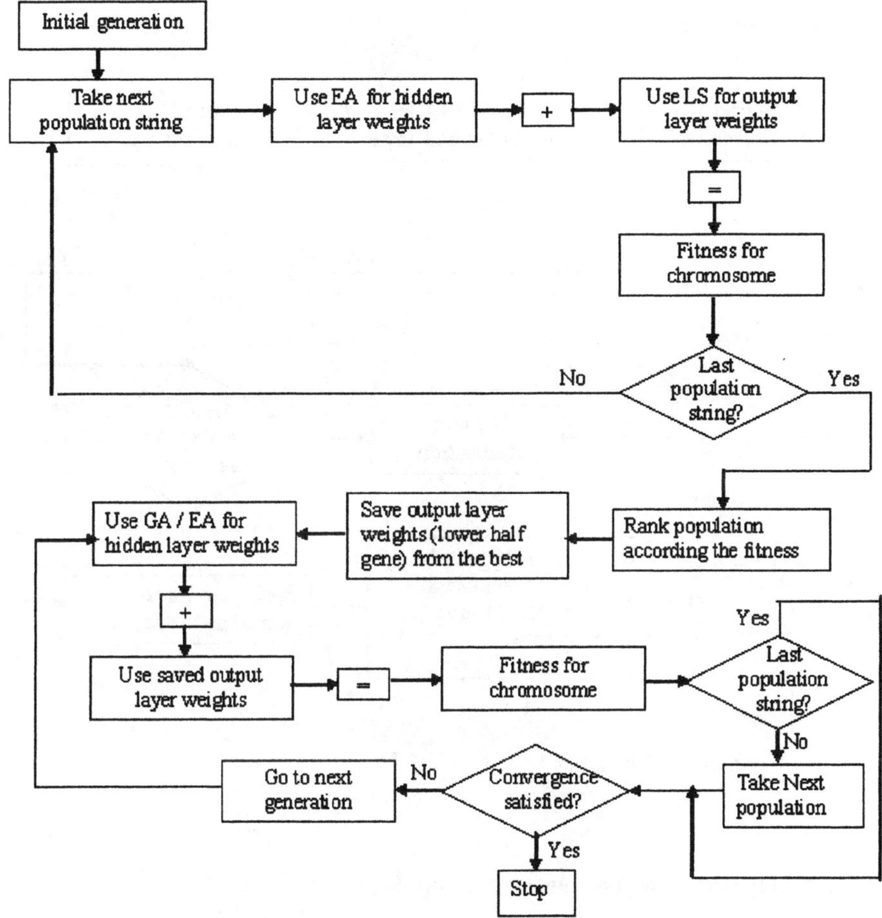

Fig. 5. T2 - Connection architecture

2.3.3 T3 connection

In T3 connection the LS method is called only after the stopping criteria of the GA / EA method is satisfied. The fitness is calculated using the weight represented by breaking the best population genome into half and combining the first half for the hidden layer and the LS output of the output layer weights. The flowchart is given in Figure 6.

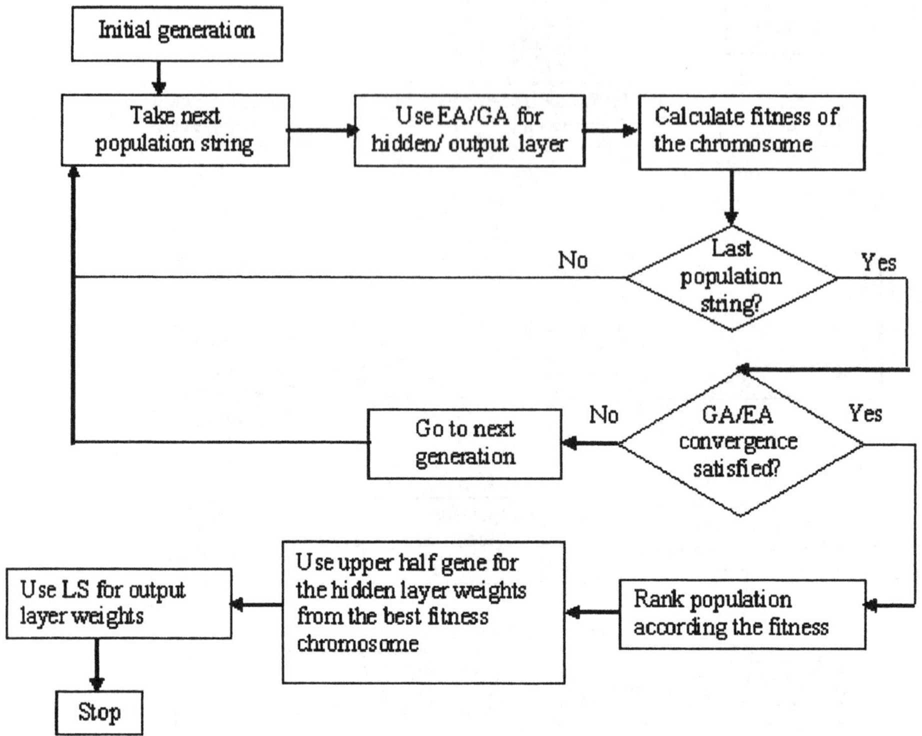

Fig. 6. T3 - Connection architecture

2.4 Efficiency order for the connection topology

Based on the performance of the genetic algorithm strategy the chronological order for all the combination according to their time and memory complexities are GALS (GALS–T1, GALS-T2, GALS-T3) and EALS (EALS-T1, EALS-T2, EALS-T3) respectively.

In the following section, we discuss EALS-T3 method in detail.

Step 1: *Initialize the input range:* All the inputs are mapped into a range of the open interval (0,1). The method of normalization is based on calculating the mean and standard deviation for each element column and uses these to perform additional scaling if required.

Step 2: *Start with some number of hidden neurons:* We start the training process using number of hidden neurons obtained from the **findArchitecture** module.

Step 3: *Initialize all the weights for the hidden layer:* We initialize all the hidden layer weights using a uniform distribution of a closed interval range of [-1, +1]. We also encode the genotype, which represents the weights for the hidden layer with those values for all the population strings. A sample genotype for the lower half gene from the population pool for an n input, h hidden and m output neurons can be written as.

$$\left| w_{11}\mu_{11} w_{12}\mu_{12}..w_{1n}\mu_{1n} w_{21}\mu_{21} w_{22}\mu_{22}..w_{2n}\mu_{2n}..w_{h1}\mu_{h1} w_{h2}\mu_{h2}..w_{hn}\mu_{hn} \right|.$$

Where, range(w) initially is set between the closed interval [-1 +1] μ are the variance vectors, each values of μ is initialized by a Gaussian distribution of mean 0 and standard deviation 1.

Step 4: *Apply Evolutionary algorithm:* We create an intermediate population from the current population using a selection operator. We use roulette wheel selection. The method creates a region of a wheel based on the fitness of a population string. All the population strings occupy the space in the wheel based on their rank in fitness. A uniform random number is then generated within a closed interval of [0,1]. The value of the random number is used as a pointer and the particular population string that occupies the number is selected for the offspring generation. Once the intermediate population strings are generated, we randomly choose two parents from the pool of the intermediate population, and apply the genetic operators (mutation only) to generate the offspring with some predefined probability distribution. We continue this process till such time the number of offspring population becomes same as the parent population. Once the new population has been created, we find the fitness for all the population based on the weights of the hidden layer (obtained from the population string) and the output layer weights (obtained from the least square method). We also normalize the fitness value to force the population string to maintain a pre-selected range of fitness.

Intermediate population generation

$$netOutput = f(hid * weight). \tag{1}$$

where f is the sigmoid function

$$RMSError = \sqrt{\frac{\sum_{i=1}^{n}\left(netOutput_i - net_i\right)^2}{n * p}} \tag{2}$$

$$popRMSError_i = norm(RMSError_i) \tag{3}$$

norm function normalized the fitness of the individual, so the fitness of each individual population is forced to be within certain range.

Step 5: *Offspring generation*
Each individual population vector (w_i, η_i) , $i = 1, 2, \ldots, \mu$ creates a single offspring vector (w_i', η_i')
for $j = 1, 2, \ldots, n$

$$\eta_i'(j) = \eta_i(j)\exp(\tau'N(0,1) + \tau Nj(0,1)) \tag{4}$$

$$w_i'(j) = w_i(j) + \eta_i'(j)\, Nj(0,1) \tag{5}$$

where $w_i(j)$, $w_i'(j)$, $\eta_i(j)$, and $\eta_i'(j)$ denote the jth component of the vectors w_i, w_i', η_i, and η_i', respectively. $N(0,1)$ denotes a normally distributed one-dimensional random number with mean and variance of 0 and 1 respectively. $Nj(0,1)$ denotes that the random number is generated a new for each value of j.

The parameter τ and τ' are set to $\left(\sqrt{2\sqrt{n}}\right)^{-1}$ and $\left(\sqrt{2n}\right)^{-1}$

Step 6: *Check the stopping criteria for EALS:*
If the stopping criteria for EALS is satisfied then
            goto step 7
    else
            goto step 5.

Step 7: *Compute the weights for the output layer using a least square method:*
After applying the evolutionary algorithm for all the weights for hidden and output layers, the best population fitness from the population pool is selected and the lower half gene of the selected population is used to replace its upper half, using least square method. The lower half gene is first used to generate the hidden layer output for all the training pairs data set, generating a matrix *hid*, whose size is P X n (Where P is the number of training pattern). The target output matrix *net* (P X m) is then linearized. As we are using non-linear sigmoidal function, we use the following linearlizing formula

$$netb_i = -\log(\frac{1 - net_i}{net_i}). \tag{6}$$

We then require to solve the over determined system of equations as given below

$$hid * weight = netb. \tag{7}$$

where *hid* is the output matrix from the hidden layer neurons and *weight* is the weight matrix output neurons. We use least square method, which is based on the QR factorization technique to solve the equation for the weight matrix using the *qr* function

$$[Q, R] = qr(hid). \tag{8}$$

The *qr* method we used to decompose the hid is known as householder decomposition method. The solution matrix can be found from the R matrix using one step iterative process as

$$x = \frac{R}{R^T / (hid^T * netb)}. \tag{9}$$

The error e can be calculated as

$$r = netb - hid * x \tag{10}$$

$$e = \frac{R}{R^T / (hid^T * r)}. \tag{11}$$

The final value of solution for weight matrix can be then found as

$$weight = x + e. \tag{12}$$

Step 8: *Check the error goal:* After replacing the upper half gene from the best selected candidate from the evolutionary algorithm using least square method described in step 5, the fitness of the new transformed gene is calculated. If the fitness of the new transformed gene meets the error criteria then stop, else goto findArchitecture module.

2.5 Stopping criteria

Stopping criteria for EALS-T3 is based on few simple rules. The rules were same for other algorithms as well for consistency. All the rules are based on current train and test output and the maximum number of generation for the evolutionary algorithm. Following, we describe the stopping criteria for the convergence of the evolutionary algorithm.

If (best_RMS_error[2] < goal_RMS_error) then
     Stop
Else if (number_of_generation = total_number_of_generation[3]) then
     Stop
Else if (train_classification_error is increased in #m[4] consecutive generation) then
     Stop

---

[2] The best_RMS_error is the best of the RMS error from the population pool
[3] Total number of generations is considered as 30
[4] m is considered as 3

2.6 Finding optimal number of hidden neurons (findArchitecture)

A simple rule base can describe the working of the stopping criteria for the proposed algorithm.

Rule 1: If the current output is satisfactory, then stop the algorithm, else check rule 2.

Rule 2: If the stopping criteria for weight search are met and the search is completely exhausted (in terms of number of iterations) then stop else check rule 3.

Rule 3: If the stopping criteria for weight search are met then go to rule 4 else go to the next generation for the findWeight module.

Rule 4: If the stopping criteria for findWeight module are met then go to rule 1, else initialize for the next number of hidden neurons.

Henceforth EALS-T3 will be referred as new proposed algorithm when only finding weight modules is required to be mentioned without the architecture modules.

2.7 Combining EALS-T3 and findArchitecture modules

Two different types of experiments – Linear incrementing for EALS (LI-EALS-T3) and binary tree search type for EALS (BT-EALS-T3) to find the number of hidden neurons-
1. Starting with a small number, and then incrementing by 1 (LI-EALS-T3)
2. Using a binary tree search type (BT-EALS-T3)

2.8 Experiment A (LI-EALS-T3)

In experiment A, we start with a small number of hidden neurons and then increment the number by one. The stopping criterion for this experiment is as follows:
If (train_classification_error $= 0$) then stop
Else If (the test classification error is high in #n^5 consecutive generation) then
Stop

---

^{5} We use n $= 3$ for our experiments, which is determined by trial and error method

2.9 Experiment B (BT-EALS-T3)

In experiment B, we use a binary tree search type to find the optimal number. A pseudo-code of the algorithm is given below:

Step 1:   Find the % test classification error & train_classification_error(error_min) for #min number of hidden neurons error_min = (train_classification_error (%)+ test_classification_error (%)) / 2.

Step 2: Find the % test classification error & train classification error(error_max) for #max number of hidden neurons
error_max = (train_classification_error (%) + test_classification_error (%)) / 2

Step 3:   Find the % test classification error & train classification error (error_mid) for #mid (mid = (min + max) / 2) number of hidden neurons
error_mid = (train_classification_error (%) + test_classification_error (%)) / 2

Step 4: If (error_mid < error_min) and (error_mid > error_max) then
       min = mid
       mid = (min + max / 2)
  else
       max = mid
       mid = (min + max / 2)
       end if

Step 5:   If (mid > min) and (mid < max)
           Go to Step1
        Else go to Step 6.

Step 6:   #number of hidden neurons = mid.

LI-EALS-T3 and BT-EALS-T3 are given the name GV1 and GV2 methods respectively.

3. Experimental Analysis and Discussion:

3.1 Comparison of results for different T connections

In the following graphs, the analysis for different T connections for EALS algorithm is given. The comparison is based on test classification accuracy, time complexity and memory complexity.

The following figure (Figure 7) shows the comparison for classification accuracy results for different T connections. Only data set, which are run by all the three different T connections, are considered. All algorithms are run for the four data sets and the best results are reported.

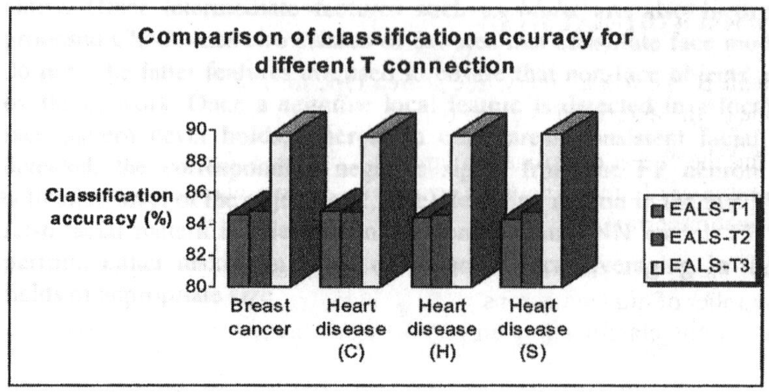

Fig. 7. Comparison of classification accuracy for different T connections

The following figure (Figure 8) shows the time complexity for all the different T connections.

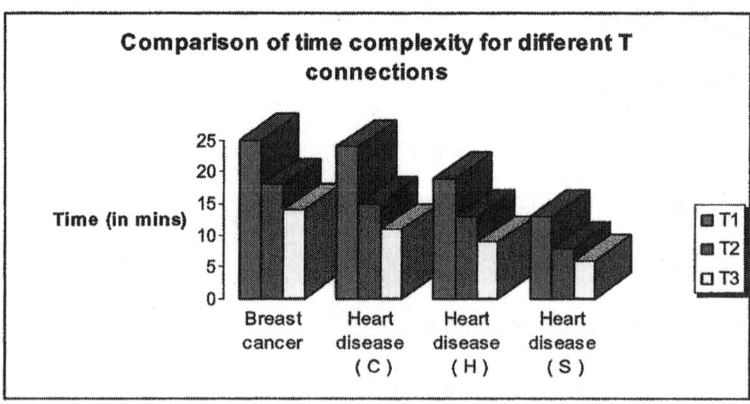

Fig. 8. Comparison of time complexity for different T connections

The comparison of memory complexity graph is not given, as it is already seen that the memory complexity of T1 and T2 are much higher than that of T3. Algorithm based on T1 and T2 are not able to run for all the data sets (for input patterns length > 500 and feature vector of length 100).

From the above two figures (Figure 7 and 8), it can be seen that T3 connection is the best among the three connections in terms of both the classification accuracy and time complexity.

310

3.2 Comparison of results for different GA strategies

The following graphs (Figures 9 and 10) show the result and analysis based on the different GA strategies that are used. The two different strategies GA and EA are compared on the basis of classification accuracy and time complexity. The results are given after the best connection strategy T3 with LS are used for both the strategies.

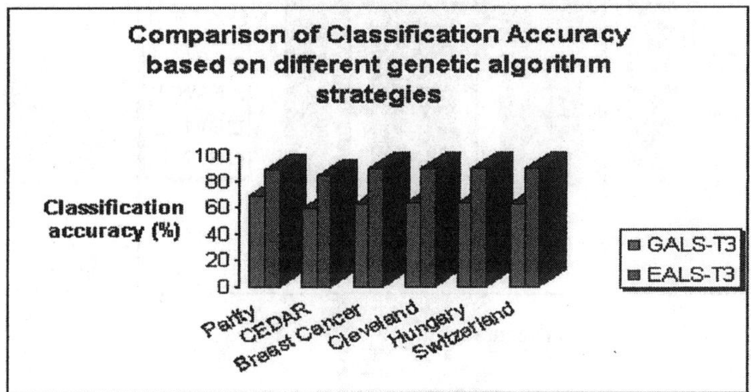

Fig. 9. Comparison of classification accuracy based on different genetic algorithm strategies

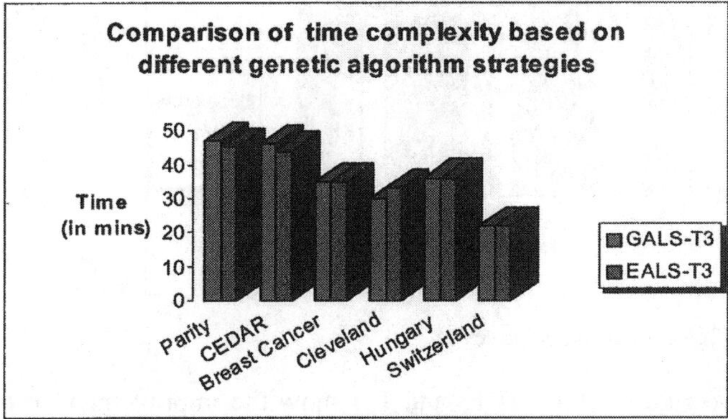

Fig. 10. Comparison of time based on different genetic algorithm strategies

From the figures, it can be seen that based on the classification accuracy EA is better than GA. However the difference in time complexity is very minimal.

3.3 Comparison of classification accuracy

The following two figures (Figures 11 and 12) show the percentage classification accuracy for all the algorithms. To make analysis easier, results from the data sets are divided into two different graphs. Only best classification accuracy results from the rest data sets are considered.

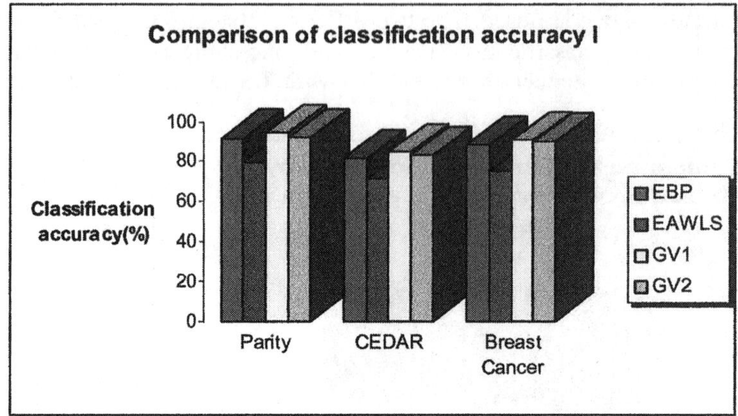

Fig. 11. Comparison of classification accuracy I

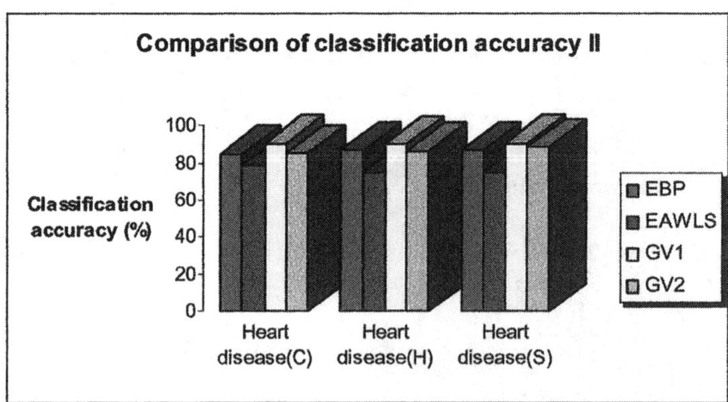

Fig. 12. Comparison of classification accuracy II

The following two figures (Figures 13 and 14) show the improvement of test classification accuracy in percentage over the standard EBP and the EAWLS methods.

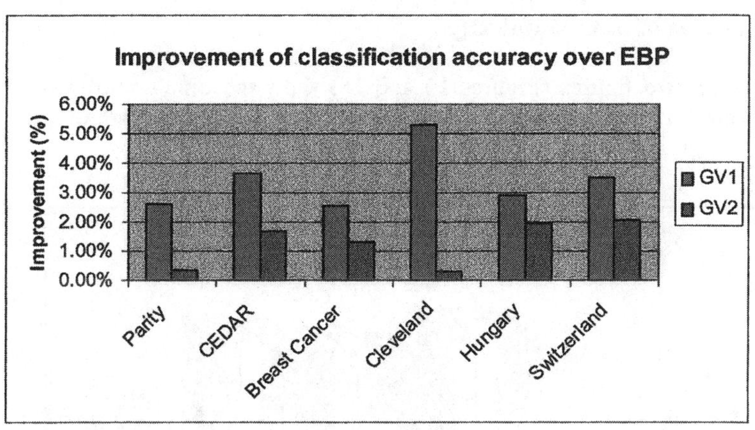

Fig. 13. Improvement of classification accuracy over EBP

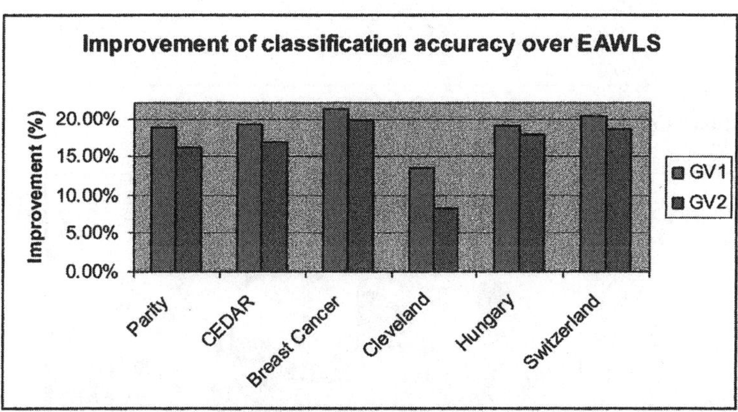

Fig. 14. Improvement of classification accuracy over EAWLS

From the Figures 13 and 14, it shows that in cases for the proposed algorithm the test classification accuracies are higher than the standard EBP and EAWLS methods. In all the cases both the GV1 and GV2 gives better classification accuracy than their counterpart. Whereas in case of EBP the improvement range varies from 0.5% to 6%, the results improve a lot when compared with standard EAWLS method. In latter case, the improvement range varies from 8% to 21%. One interesting thing that is observed here that the improvement of classification accuracy over EBP is very large specially in cases for all the heart disease data set, whereas the improvement is within a limited range for the other three data set (Odd parity, CEDAR and breast cancer data set).

3.4 Comparison of time complexity

The following two figures (Figures 15 and 16) show the time complexity for all the algorithms. To make analysis easier, results from the data set are divided into two different graphs.

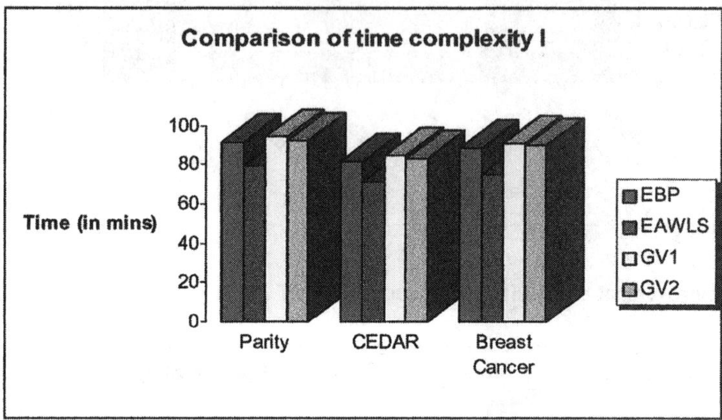

Fig. 15. Comparison of time complexity I

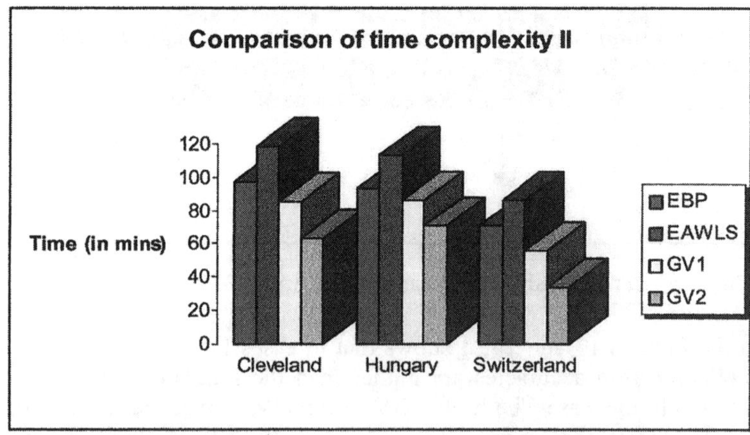

Fig. 16. Comparison of time complexity II

The following two figures (Figures 17 and 18) show the improvement of time complexity of training dataset in percentage over the standard EBP and the EAWLS methods.

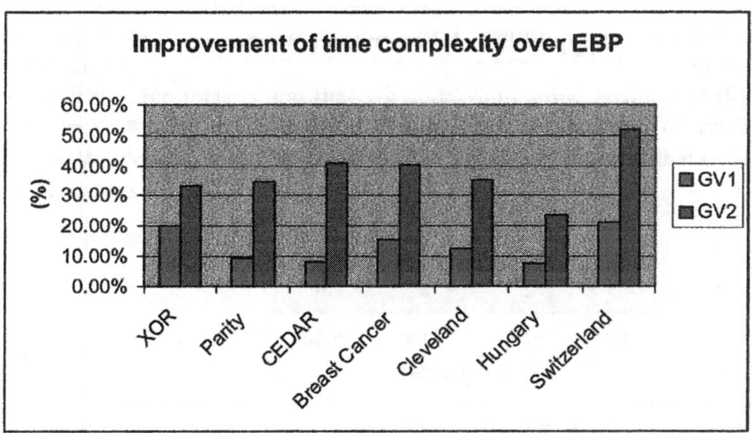

Fig. 17. Improvement of time complexity over EBP

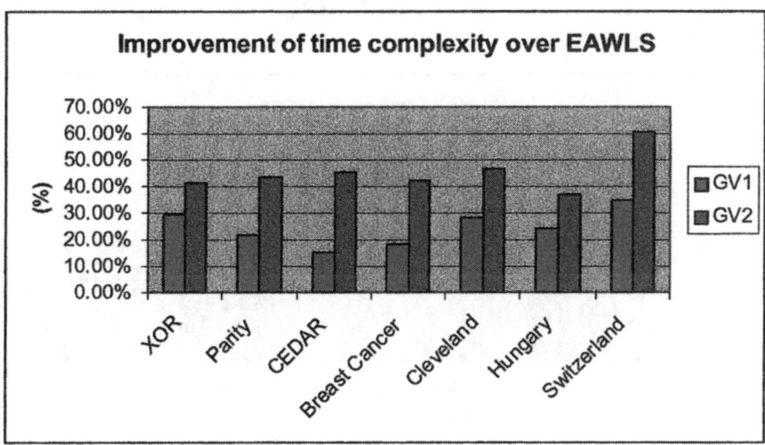

Fig. 18. Improvement of time complexity over EAWLS

3.5 Comparison of memory complexity

The following figure (Figure 19) shows the comparison of memory usage for the EBP, EWALS and the new algorithm[6]. Only the CEDAR dataset for the analysis with different data size is considered.

---

[6] Proposed algorithm refers only to the findWeight module, as the findArchitecture module is not required to add in those cases. Based on the results the proposed algorithm basically refers to the EALS-T3 algorithm.

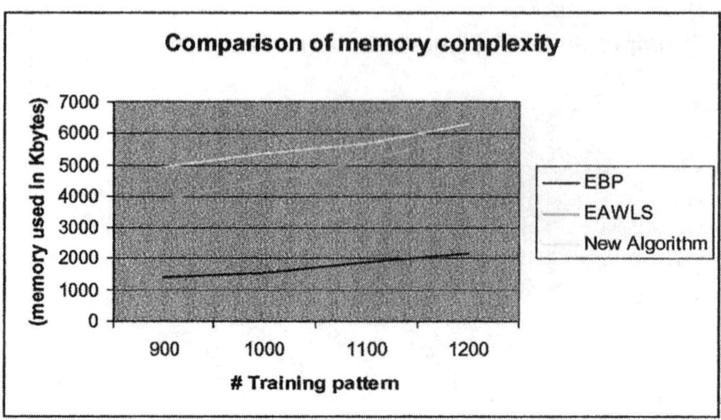

Fig. 19. Comparison of memory complexity

The following figure (Figure 20) shows the increment of memory usage over EBP and EAWLS. Only the CEDAR dataset with variable data length is considered for the analysis.

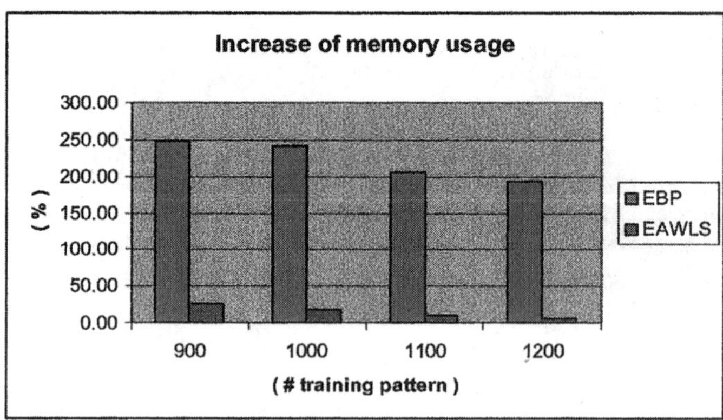

Fig. 20. Increment of memory usage

From the above figure (Figure 20), it shows that, in case of EBP the increment is from 193% to 246%, where as for the EAWLS, the increment is from 5% to 23%. The memory usage is slightly more than the memory usage in EAWLS but much higher than standard EBP (Figure 19).

3.6 Comparison of sensitivity analysis

As initialization is done randomly in many training algorithms, hence it is important to show the behavior of the algorithm with in same environment but for different runs. Hence all the algorithms, which use random initialization for its initial weights are trained with different set of weight matrices keeping all the other variables fixed.

The following figure (Figure 21) shows the comparison of sensitivity analysis for the three algorithms. Only the CEDAR dataset with different data length is considered for the analysis. The algorithms are executed 100 times with same data and the test classification accuracies are stored. Then the minimum, maximum and the standard deviation for all the algorithms are shown.

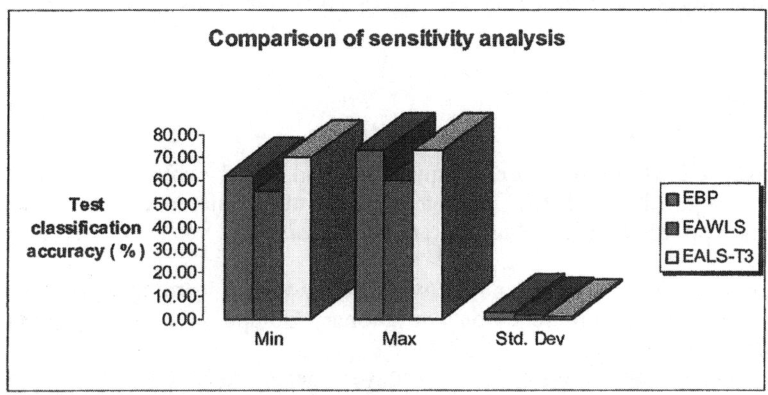

Fig. 21. Comparison of sensitivity analysis

From the above figure (Figure 21) it is clear that the new algorithm is the least sensitive to the initial condition. The variation of EAWLS with respect to the new algorithm is 40% but for the standard EBP with respect to the proposed algorithm it is 73%. The least sensitivity gives the new algorithm to have a better predictable ability with the most confidence

4. Conclusion & Further research

In this chapter various combination strategies for evolutionary learning and least square based learning are explored. The results show that in terms of classification accuracy and time complexity T3 connection is the best. Two different types of GA are considered and it was seen that in terms of the classification accuracy the results based on EA based strategy are much better than those of GA based strategy. Also the time complexity for EA based strategy is slightly less than that of GA.

Only two types of naïve architecture search are combined with the weight search module. It is shown that the architecture search based on the type of binary search technique takes much less time than the linear architecture search counter part. This looks at first quite obvious but considering that the test or train error does not always behave in a consistent way, it is quite interesting to note that the combination of train and test error as is done in the binary search type module, averages out most of the time the inconsistency of the result. This is further proved by the fact that the classification accuracy from this search is only slightly worse than the linear architecture search. Further research should concentrate mainly on two things that are needed to improve the existing algorithm. The first thing is that because the memory complexity is quite high for the proposed hybrid learning, there needs to be further research to improve this shortcoming. Secondly, some kind of clustering method could be used for training of output layer weights, where the training could be based on learning the cluster centroid.

References

[1] R. Ghosh and B. Verma, "Least square method based evolutionary neural learning algorithm," IEEE International Joint Conference on Neural Networks, pp. 2596 -2601, Washington, IEEE Computer Society Press, USA, 2001.

[2] B. Verma and R. Ghosh, "A novel evolutionary neural learning algorithm," IEEE International Conference on Evolutionary Computation 2002, pp1884-89, Honolulu, USA, 2002.

[3] B. Verma, "Fast training of multilayer perceptrons (MLPs)," IEEE Transactions on Neural Networks, vol. 8, no. 6, pp. 1314-1321, 1997.

[4] R. Ghosh and B. Verma, "Finding architecture and weights for ANN using evolutionary based least square algorithm," International Journal on Neural Systems, vol. 13, no. 1, pp. 13-24, 2003.

[5] V. Petridis, S. Kazarlis, A. Papaikonomu and A. Filelis, "A hybrid genetic algorithm for training neural network", Artificial Neural Networks, 2, pp. 953-956, 1992.

[6] A. Likartsis, I. Vlachavas, and L. H. Tsoukalas, "New hybrid neural genetic methodology for improving learning", Proceedings of the 9th IEEE International Conference on Tools with Artificial Intelligence, Piscataway, NJ, USA, pp. 32-36, IEEE Press, 1997.

[7] D. Whitley, T. Starkweather and C. Bogart, "Genetic algorithms and neural networks - optimizing connections and connectivity", Parallel Computing, 14, pp. 347-361, 1990.

[8] M. Koeppen, M. Teunis, and B. Nickolay, "Neural network that uses evolutionary learning", Proceedings of the 1994 IEEE International Conference on Neural Networks. Part 5 (of 7), Piscstaway, NJ, USA, pp. 635-639, IEEE press, 1994.

[9] A.P. Topchy and O.A. Lebedko, "Neural network training by means of cooperative evolutionary search", Nuclear Instruments & Methods in Physics Research, Section A: Accelerators, Spectometers, Detectors and Associated Equipment, vol. 389, no. 1-2, pp. 240-241, 1997.

[10] G. Gutierrez, P. Isasi, J. M. Molina, A. Sanchis and I. M. Galvan, "Evolutionary cellular configurations for designing feedforward neural network architectures, connectionist models of neurons, learning processes and artificial intelligence, Jose Mira et al (Eds), Springer Verlag – Germany, LNCS 2084, pp. 514-521, 2001.

[11] X. Yao and Y. Liu, "Making use of population information in evolutionary artificial neural networks", IEEE Transactions on Systems, Man and Cybernetics, 28(3): pp. 417-425, 1998.

[12] T. Jansen and I. Wegener, "Evolutionary algorithms - how to cope with plateaus of constant fitness and when to reject strings of the same fitness", IEEE Transactions on Evolutionary Computation, vol. 5, no. 6, pp. 589-599, 2001.

[13] H. Jun, X. Jiyou, and X. Yao, "Solving equations by hybrid evolutionary computation techniques", IEEE Transactions on Evolutionary Computation, vol. 4, issue 3, pp. 295-304, 2000.

[14] C. S. Leung, A. C. Tsoi and L. W. Chan, "Two regularizers for recursive least squared algorithms in feedforward multilayered neural networks", IEEE Transactions on Neural Networks, vol. 12, no. 6, pp. 1314-1332, 2001.

[15] O. Stan and E. Kamen, "A local linearized least squares algorithm for training feedforward neural networks", IEEE Transactions on Neural Networks, vol. 11, no. 2, pp. 487-495, 2000.

[16] A. D. Brown and H. C. Card, "Evolutionary artificial neural networks", IEEE 1997 Canadian Conference on Voyage of Discovery, vol. 1, pp. 313-317, 1997.

[17] J. Zhang and A. J. Morris, "A sequential approach for single hidden layer neural networks", Neural Networks, vol. 11, no. 1, pp. 65-80, 1998.

[18] B. A. Charters and J. C. Geuder, "Computable error bounds for direct solution of linear equations", Journal Of The Association For Computing Machinery, vol. 14, no. 1, pp. 63-71, 1967.

[19] L. A. Zadeh, "From computing with numbers to computing with words: From manipulations of measurements to manipulation of perceptions," 3rd International Conference on Application of Fuzzy Logic and Soft Computing, pp. 1-2, Wiesbaden, 1998.

[20] R. J. Williams and D. Zipset, "A learning algorithm for continually running fully recurrent neural networks," Neural Computation, vol. 1, pp. 279-280, 1989.

[21] S. Wright, "The role of mutation, inbreeding, crossbreeding, and selection in evolution," Proceedings of 6th International Congress of Genetics, Ithaca, NY, vol. 1, pp. 356-366, 1932.

[22] D. R. Hush and B. G. Horne, "Progress in supervised neural networks," IEEE Signal Processing Magazine, vol. 10, no. 1, pp. 8-39, 1993.

[23] M. F. Moller, "A scaled conjugate gradient algorithm for fast supervised learning," Neural Networks, vol. 6, pp. 525-523, June 1993.

[24] R. K. Belew, J. McInerney, and N. N. Schraudolph, "Evolving networks: using genetic algorith, with connectionist learning," Technical Report #CS90-174 (Revised), Computer Science & Engineering Department (C-014), University of California at San Diego, La Jolla, CA 92093, USA, 1991.

Biologically inspired recognition system for car detection from real-time video streams

Predrag Neskovic[1], David Schuster[2]*, and Leon N Cooper[1]

[1] Brown University
   Physics Department and
   Institute for Brain and Neural Systems
   Providence, RI 02912, USA
   Predrag_Neskovic@brown.edu, Leon_Cooper@brown.edu
[2] Yale University
   Physics Department
   New Haven, CT 06520, USA
   David.Schuster@yale.edu

Abstract. In this work we present a system for detection of objects from video streams based on properties of human vision such as saccadic eye movements and selective attention. An object, in this application a car, is represented as a collection of features (horizontal and vertical edges) arranged at specific spatial locations with respect to the position of the fixation point (the central edge). The collection of conditional probabilities, that estimate the locations of the car edges given the location of the central edge, are mapped into the weights of the neural network that combines information coming from the edge detectors (bottom-up) with expectations for edge locations (top-down). During the recognition process, the system efficiently searches the space of possible segmentations by investigating the local regions of the image in a way similar to human eye movements, probing and analyzing different locations of the input at different times. In contrast to motion-based models for vehicle detection [7, 8], our approach does not rely on motion information, and the system can detect both still and moving cars in real-time. However, adding motion information should improve the accuracy.

Key words: car detection, video streams, biologically inspired, feature-based, segmentation, saccades, selective attention, bottom-up, top-down.

1 Introduction

Identification of vehicles from video streams is a challenging problem that incorporates several important aspects of vision including: translation and scale

---

*   The author performed the work while at Brown University.

invariant recognition, robustness to noise and occlusions and ability to cope with significant variations in lighting conditions. In addition, the requirement that the system work in real-time often precludes the use of more sophisticated but computationally involved techniques.

In constructing an artificial recognition system for real-time processing of video streams we draw inspiration from the way the human visual system analyzes complex scenes. The properties of the human visual system that we utilize in our system are: selective attention, saccadic eye movements and hierarchical processing of visual information.

Due to the structure of the eyes, the human visual system does not process the whole visual input with the same resolution. The region of the scene that is perceived with the highest quality is the one that projects to the fovea, an area of the retina corresponding to only about the central 2 degrees of the viewed scene. The regions that are further away from the fixation point are perceived with progressively lower resolutions. The visual system overcomes this limitation by making rapid eye movements, called *saccades*. Human recognition is therefore an active process of probing and analyzing different locations of the scene at different times and integrating information from different regions.

Numerous experiments and computational theories [1, 2, 3] advocate the representation of objects in terms of parts or features and relations among these features. In particular, edges have been proposed as the most basic features for object representation. Compared to pixels, edges are much more stable, less susceptible to noise and changes in lighting conditions, which makes them good candidates for image analysis and representation. Since the classical experiments of Hubel and Wiesel [4], in which they discovered neurons in visual cortex that were selective to edges of various orientations and sizes, edges became important ingredients in understanding the biological processing of information.

In this paper, an object (a car) is represented as a collection of horizontal and vertical edges arranged at specific spatial locations with respect to each other. Within a single fixation, the locations of features are always measured with respect to the fixation point - the central edge. During the recognition process, the system efficiently searches the space of possible segmentations by investigating the local regions of the image in a way similar to human eye movements.

This work is an extension of our previous work [5, 6] that was applied to segmentation and recognition of one-dimensional objects, handwritten words. In this work we show that our model can be successfully applied to recognition of two-dimensional objects, such as cars.

The paper is organized as follows: In Section 2 we review the related work and discuss one of the main problems in applications to scene analysis: the segmentation problem. In Sections 3 and 4 we describe our model and the architecture of the neural network. Section 5 illustrates some implementation details and Section 6 describes the recognition process. In Section 7 we summarize the main properties of our method and present experimental results.

2 Background

The problem of vehicle identification from video streams has been widely addressed in computer vision literature [7, 8, 9]. Very often, an underlying assumption is that the vehicles are moving and motion information is used to segment the image into moving regions and a static background. Based on its overall size and shape, a region can then sometimes be recognized as a vehicle even without a detailed description. Furthermore, motion information can reduce the computational complexity since only the regions that contain motion have to be analyzed. However, in many situations, motion information is not available or is insufficient, and other ways of dealing with computational complexity and segmentation problems have to be used. In contrast to motion-based models for vehicle detection [7, 8], our approach does not rely on motion information, and the system can detect both still and moving cars in real-time.

Several approaches use edge information in order to detect vehicles either from still images or video streams. Betke et al. [10, 11] make use of motion and edge information to hypothesise the vehicle locations. However, a vehicle is not represented as a collection of edges. Instead, the edges are used only in order to capture the boundary of a vehicle. Once a region is hypothesized as vehicle, the recognition is performed by matching a template with the potential object marked by the outer horizontal and vertical edges. Edge information is also contained in Haar wavelets that were used as features supplied to Support Vector Machine in [12, 13]. Similarly, Goerick et al. [14] use Local Orientation Coding (LOC) to extract edge information. The histogram of LOC within the area of interest was then fed to a Neural Network (NN) for classification.

Other researchers have also used feature-based (or parts-based) approaches to detect vehicles from images. Wavelets were chosen as features in [12, 13, 15], rectangle features (which are similar to Haar wavelets) were used in [16] while the authors in [17, 18] use the interest operator for automatically selecting features. In [17], a generative probabilistic model is learned over the selected features and, due to a computational complexity, the method they use relies on a very small number of fixed parts.

Biologically-inspired approaches have been much less successful than computer vision or statistical approaches when applied to real-word problems. Biologically-based recognition systems have been proposed for various applications such as face recognition, handwriting recognition and vehicle detection [19]. An approach to object recognition that is based on human saccadic behavior is proposed in [20]. While this model does capture properties of saccadic behavior, it represents an object as a fixed sequence of fixations. In contrast to this approach, our system does not make such an assumption and detects a car regardless of the order in which saccades were perform.

One of the most important problems related to the detection of cars from video streams is the segmentation problem. Given an image, it is not known where a car is or what its size is. Therefore, all the methods that assume a fixed size input vector during the recognition, (e.g. NNs), are not very well suited for this problem. In order to detect a vehicle regardless of its location, the detection system has to be convolved over the whole image, and in order to detect a vehicle at different scales the original image has to be rescaled and the convolution procedure repeated [12, 15]. Since the methods that rely on the exhaustive search and not very efficient they are mostly applied to detection of vehicles from static images.

In case of NNs, the difficulty arises both during the training phase and during the testing phase. One of the important questions related to the preparation of training samples is: how much of the vehicle should be present in the training window? In order for the network to be able to recognize occluded vehicles it would help if it is trained on parts of vehicles. However, not only that it is difficult to define the "minimal size" of the part of the vehicle that should be present in the window but it is much more difficult to train the network on parts of vehicles. During the recognition phase, the network will give multiple detections corresponding to a single vehicle in the region around the vehicle. This problem is not unique to NNs but to all the methods that assume fixed size input window ([12, 13, 15, 16, 18]). Possible remedies are suggested in [16, 18].

One solution to the segmentation problem is to represent a vehicle in terms of its parts or features and to extract them from the whole image [17] as opposed to the window of a specified size as in [12, 13, 15, 16, 18]. However, then the problem is how to group or select only a portion of the feature from the entire image and this can be a serious computational issue, especially if the number of features per object is large [17]. In our previous work [5] we addressed this problem and showed how selective attention and contextual information can be used to search the space of possible segmentations of one dimensional objects - cursive words. In this work, we extend our approach to two dimensional case and show that the same approach can be used for segmentation and detection of cars from video-streams.

3 The model

In our model, an object is represented as a collection of features of specific classes arranged at specific spatial locations with respect to the fixation point. In the current application, car detection from video images, the features are the edges of different orientations and sizes. In contrast to other models that also view an object as a collection of local features [17, 21], the positions of the features in our model are always measured with respect to the location of one feature that we call the central feature (the central edge). During the recognition process the central edge becomes the edge on which the system fixates. We further assume that saccades cannot be made on any point in the image but only on edges, more specifically on their centers.

Once a saccade is made on an edge the question is then how do we know which edge of a car it represents? Obviously, from the strength of the edge itself it is not possible to answer that question since edge strength is obtained using only local and bottom-up information. What is needed is a presence of other (car) edges, to put an edge in context, and we want to know how to measure their influence.

Due to the fact that cars come in different shapes and sizes, the location (and size) of any edge is not fixed with respect to the location of a chosen central edge. So, for a given edge (the central edge), the location of any other edge is not at any specific location but within a region. These uncertainties in edge locations can be calculated from car statistics and can be expressed through conditional probabilities.

Consider two edges of a car, e.g. the bottom edge below the grill (denote it as e_i) and the top edge above the windshield (denote it as e_j). Given the location of the bottom edge, $\mathbf{r}_i$, the position of the top edge, $\mathbf{r}_j$, varies due to variations in car sizes. The collection of all the possible locations of the top edge forms a region (R_j) where the top edge can be found given the location of the bottom edge. If we assume that all the locations of the top edge within the region are equally likely then the probability of finding the top edge at any place within the region R_j is inversely proportional to the size of the region $p(e_j|e_i, \mathbf{r}_{ij}) = cons * 1/S(R_j)$, where $S(R_j)$ denotes the size of the region R_j and $\mathbf{r}_{ij}$ is the location of the center of the jth edge with respect to the location of the ith edge. The edge e_i provides context for the existence of the edge e_j in the region R_j and its influence is expressed through the conditional probability $p(j|i, \mathbf{r}_{ij})$. Similarly, the edge e_j provides context for the edge e_i with strength $p(e_i|e_j, \mathbf{r}_{ji})$, where $\mathbf{r}_{ji} = -\mathbf{r}_{ij}$. It can be easily shown that $p(e_j|e_i, \mathbf{r}_{ij}) = p(e_i|e_j, \mathbf{r}_{ji})$. In the rest of the paper we will use abbreviated notation $p(e_j|e_i, \mathbf{r}_{ij}) = p(e_j|e_i)$ assuming the spatial dependence.

Let us consider the ith edge of a car and assume that it is detected with probability d_i at some location in the image. In order to increase the confidence about the identity of this edge, it would help if every other edge of the vehicle is detected at its expected locations and with high confidence. The support from other edges can be incorporated in a number of ways and one of the simplest is to average their contributions. We define a "context" for the ith edge to be the sum $\sum_{j=1, j\neq i}^{N} p(e_i|e_j)d_j$ divided by the number of contributing edges, where N is equal to the number of edges representing a car.

The collection of conditional probabilities, associated with a given central edge thus constitutes a model of a car from the point of view of the particular central edge. We can think of the collection of local regions where the edges are expected to be found as a template that represents a car from the point of view of the given central edge. Therefore, a car is represented with as many templates as there are edges that can be used as fixation points. The conditional probabilities between the edge pairs are mapped into the weights of a neural network and the task of the network will be to combine the information coming from the edge detectors (bottom-up) with expectations for edge locations (top-down).

4 The architecture of the network

In this section, we describe the architecture of the network that represents a car. We will assume that every edge (from an image) can be used as a fixation point during the recognition process and with each (car) edge we will associate one unit, an object-unit, that will represent an object from the point of view of that edge. We will now focus on one such object-unit and the group of units from which it receives inputs as illustrated in Figure 1.

At the bottom of the hierarchy are the edge detectors whose receptive fields completely cover the input image. An output of an edge detector is the probability, d, of detecting the edge to which it is selective, e.g. an edge of specific orientation and size. The outputs of the edge detectors are supplied to the next layer of units - the *simple units*. Among the simple units, we distinguish a central unit, the one that is positioned above the fixation point, and the surrounding units. The size of the receptive field of the central unit is the smallest, when compared to other simple units and the sizes of the receptive fields of the surrounding simple units increase with their distance from the central unit. The sizes of the receptive fields of the simple units are designed in such a way as to accommodate the uncertainties associated with locations of the edges with respect to the fixation point.

The output of a simple unit, given the location of the ith central unit, is given as

$$s_{ij} = max_{\mathbf{r} \in R_j}(d_j(\mathbf{r})), \tag{1}$$

where $\mathbf{r}$ is the location of the edge detector (selective to the jth edge) with respect to the location of the central edge (that represents the ith edge of a car) and R_j is the receptive field of the jth simple unit. Therefore, a simple unit selects the strongest edge within its receptive field and outputs the probability that this edge represents the jth edge of a car. The next layer of the units, called the *complex units*, incorporates contextual information. The complex unit that receives input from the central simple unit outputs the probability that the region R_i (or the edge it contains) now represents part of the object

$$c_{ii} = d_i \frac{1}{N-1} \sum_{j=1, j \neq i}^{N} p(e_i|e_j)s_{ij}, \tag{2}$$

where N represents the number of edges in the object. This means that the detection of the central edge is now viewed within the *context* of all the other edges of the object. Similarly, the jth complex unit that receives input from the jth simple unit views the jth edge within the context of the central edge

$$c_{ij} = d_j p(e_j|e_i)d_i. \tag{3}$$

According to our model, each local region (an edge) can represent an object with different confidence. The probability that the collection of all the regions

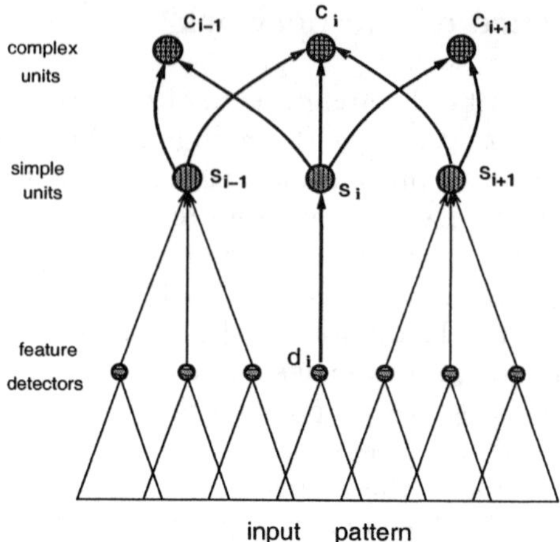

Fig. 1. The network architecture. An ith object unit represents a car from the perspective of an ith edge of a car. There are as many object units as there are possible views of a car which is equal to the number of edges that represent a car. Similarly, for a given fixation point (the central edge), there are as many complex units as there are edges in a car and each complex unit combines bottom-up information and contextual information. The ith complex unit combines bottom-up information coming from the edge detector selective to the ith edge of a car with information coming from all other simple units that represent other edges of the car. In contrast, any other complex units incorporates contextual information coming only from the central edge. Simple units, on the other hand, receive only bottom-up information and find the strongest edges, to which they are selective, within their receptive field. The sizes of the receptive fields of the simple units get progressively larger, further away they are from the location of the central edge (fixation point), thus allowing variations in edge locations (car shapes and sizes).

that contain object edges represents the object from the point of view of the ith edge is captured by the *object unit*

$$o_i(object|fixation\ point\ i) = \frac{1}{N} \sum_{k=1}^{N} c_{ik}, \qquad (4)$$

where the index k goes through all the complex units, the central (i) and surrounding (j) units, of a given view. It is clear that there are as many object units as there are possible views of the object, which in our case, is equivalent to the number of edges in the object.

5 Implementation

Ideally, the system would contain an array of feature detectors that completely cover the input image and process information in parallel. Similarly, the system would benefit from a large number of feature classes, since they would provide richer and more detailed description of objects. However, in order to make a system run in real time on a regular computer and without dedicated preprocessing hardware, we had to make several approximations.

5.1 Feature Selection

In our current implementation, we represent a car as a collection of only horizontal and vertical edges. Since an edge is an extended spatial object, we choose to specify its location in terms of the location of its central point. In this way, a car is modeled as a collection of points, arranged in 3D space, where each point represents an edge of specific size and orientation. Using the statistics for the car sizes and their edges, one can easily calculate the mean size μ_j and the variance ν_j for each edge. Given the location of the fixation point and knowing the variations in size for every edge, it is then straightforward to propagate these uncertainties and calculate the regions where the centers of the edges should be. In order to map this 3D configuration of regions into a 2D image plane we use perspective transformation Eqs. (5) - (7). In this way, for a given location of the fixation point within an image, we associate a group of 2D regions as being allowable locations for the edge centers. Each such region represents a receptive field of one simple unit of the network.

5.2 Perspective Transformations

The perspective transformation equations that we use are described in more detail in [22]. Here we briefly review the main equations. Let us denote with (x_p', z_p') and with $\mathbf{v} = (x, y, z)$ the coordinates of a point in an image plane and in the real-world respectively. The location of the camera (the gimbal center) is at the point $\mathbf{v_o} = (x_o, y_o, z_o)$. The vector $\mathbf{l} = (l_1, l_2 + f, l_3)$ denotes the constant offset between gimbal center and image plane center where f is a constant offset along the optical axis. The world coordinates are transformed into image coordinates using the direct perspective transformation equations:

$$x_p' = f \frac{(x - x_o)cos\theta + (y - y_o)sin\theta - l_1}{-(x - x_o)cos\psi sin\theta + (y - y_o)cos\psi cos\theta + (z - z_o)sin\psi - l_2}, \quad (5)$$

$$z_p' = f \frac{x - x_o)sin\psi sin\theta - (y - y_o)sin\psi cos\theta + (z - z_o)cos\psi - l_3}{-(x - x_o)cos\psi sin\theta + (y - y_o)cos\psi cos\theta + (z - z_o)sin\psi - l_2}. \quad (6)$$

The image coordinates are transformed into world coordinates using the inverse perspective transformation equations:

$$\mathbf{v} = \begin{bmatrix} x \\ y \\ z \end{bmatrix} = \begin{bmatrix} x_o + l_1 cos\theta - l_2 cos\psi sin\theta + l_3 sin\psi sin\theta \\ y_o + l_1 sin\theta + l_2 cos\psi cos\theta - l_3 sin\psi cos\theta \\ z_o + l_2 sin\psi + l_3 cos\psi \end{bmatrix}$$

$$+ \lambda \begin{bmatrix} x_p' cos\theta - f cos\psi sin\theta + z_p' sin\psi sin\theta \\ x_p' sin\theta + f cos\psi cos\theta - z_p' sin\psi cos\theta \\ f sin\psi + z_p' cos\psi \end{bmatrix} , \qquad (7)$$

where the λ is a free non-negative parameter. Knowing the real coordinates of a point (x, y, z) it is easy to calculate the image coordinates (x_p', z_p') using Eqs. (5)-(6). However, in order to go from the image coordinates to the real coordinates one has to provide more information. In most cases it is assumed that the point is on the ground and therefore $z = 0$. Solving Eq. (7) for λ (by setting the third component to zero) and substituting it back in (7) one easily arrives at the values for the (x, y) real-world coordinates.

5.3 Feature Detectors

Another approximation is related to the construction and use of edge detectors. Instead of having an array of edge detectors for detecting the horizontal and vertical edges of all the sizes, the system extracts only the prominent edges (with activations above the predefined threshold) and estimates their sizes. Figure 2 illustrates some of the extracted edges and their estimated sizes. In our current implementation, the edges were extracted from the difference image that is obtained as a difference between the original (gray-scale) image and the background image that contains no vehicles.

Fig. 2. Original image (left) and processed image (right) that illustrates some of the prominent horizontal edges. The extracted edges (right image) are projected on the difference image that is obtained by subtracting the background image (that contains no vehicles) from the original image. This is done only for the illustrative purposes - to make the extracted edges more visible.

The value of the pixel (i, j) of the background image at time $t + 1$ is calculated using the updating rule

$$B_{t+1}(i, j) = B_t(i, j) + \alpha \cdot D_t(i, j) \cdot O(i, j), \qquad (8)$$

where α is an updating constant, $D_t(i, j)$ is the difference between the pixel values at times $t + 1$ and t and $O(i, j)$ is 0 if the pixel (i, j) belongs to an object that has been identified and 1 if it is part of the background. Therefore, the current image is used for updating the background image after the object identification is performed on the current image.

Each edge detector is selective to only an edge of a specific orientation, but can detect edges of various sizes around the preferred size. Since the distribution of sizes for any given car edge is fairly uniform, we use a Gaussian distribution to model the probability of an edge having a specific size. Therefore, an edge detector for an edge of horizontal/vertical orientation is specified with two parameters: the mean length of an edge and its variance. The input to the edge detector (of a given orientation) is an edge of specific size l and the output is the probability that measures how well this edge matches the expected edge size, $d = 1/(\nu\sqrt{2}\pi) * exp(-(\mu - l)^2/2\nu^2))$.

6 Recognition process

The recognition process starts with selection of the most prominent edge in an image, the one with the highest activation. The center of this edge becomes the fixation point from which the locations of other edges are measured. The system now has to determine whether the central edge represents an edge of a car, and if it does which edge it represents.

The first step is to hypothesize what edge of a car the central edge represents. Once the hypothesis is made, the distance of the central edge from the ground is fixed and one can then calculate the constant λ in Eq.(7). The next step is to compute the expected sizes of the central edge (in the image coordinates) and other edges of a car that are to be matched against the edges found in the image. In order to perform the matching the network is first "rescaled". Rescaling the network means a) adjusting the means and the variances of the letter detectors from which the simple units are receiving inputs and b) adjusting the sizes of the receptive fields of the simple units and their centers with respect to the central unit (for every object unit).

Finally, the network is centered over the fixation point, which is the center of the hypothesized edge. This is done by positioning *all* of the object units over the fixation point (central edge) and measuring how much they are activated by the arrangement of the detected surrounding edges. The object unit with the highest activation selects some of the neighboring edges as representing a car and the central edge is given the identity as being a specific edge of a car (e.g. the bottom horizontal edge). In order to associate the group of edges

Fig. 3. The left image illustrates the input to the recognition system while the right image is the result of the recognition process. Once the system detects a car, it draws a white box around the location of the edge on which it fixated. Since the recognition is done without using any motion information or previous history, and as a result of the continuously changing lighting conditions, the location of the central edge (the fixated edge) within the car changes from one frame to another. This selection of different fixation edges at different times within the same car is similar to saccadic eye movements.

(the central edge and the surrounding edges) as a car as opposed to noise, the activation of the object unit has to be above some predefined threshold. Once the group of edges is selected as a representative of a car, their activations are suppressed and the system makes a saccade on another prominent edge and the previous procedure is repeated. The system makes as many saccades as there are prominent edges.

An example of the outcome of the recognition process is illustrated in Fig. 3. Once the system detects a car, it draws a white box around the location of the edge on which it fixated (the central edge). The recognition is done on static images - without using any motion information or previous history. As a result, and due to the variations in lighting conditions, the system does not lock onto one edge of a vehicle but often selects a different edge of a car for each frame in a way similar to saccadic eye movements.

7 Summary and results

In this work we presented a biologically inspired system for car detection from video streams. The architecture of the network reflects the properties of foveal vision through the arrangement and sizes of the simple units. During the recognition process, the system efficiently searches the space of possible segmentations by investigating the local regions of the image in a way similar to human eye movements, probing and analyzing different locations at different times. The computational complexity associated with searching the space of edge activations is greatly reduced using selective attention thus allowing the system to process information in real time. The architecture described in this paper is implemented on a Pentium III, 700MHz processor using an input from a simple web camera.

It is very difficult to compare the performance of our system to other approaches since there has been very little formal characterization of the performance of car detection ([12]) and there is no existing video database for benchmarking detection of cars from video streams. Comparison of our system to detection of vehicles from static images is not appropriate since our system performs detection in real-time whereas approaches for detection from static images do not have a time constraint. As a consequence, the algorithms that have been used for detection of vehicles from video images (e.g. [7, 11]) are simpler compared to algorithms used for vehicle detection in static images (e.g. [12, 15, 17]). Compared to approaches for detection of vehicles from video streams, our approach offers much more detailed representation of a vehicle despite the fact that, in the current implementation, we use only horizontal and vertical edges. On the other hand, many of the algorithms used for detection of vehicles in static images, such as those that require exhaustive scanning of the input image [12, 15] are unlikely to be applicable for real-time vehicle detection.

We tested the performance of the system on several thousand video sequences. Once a system detects a still car it locks onto it (although it might fixate on different edges at different times) and the recognition is nearly 100%. If the cars are moving and are separate from one another, the recognition accuracy is in the neighborhood of 90%. However, when the cars become close to one another the recognition drops to about 70%, depending on how close the cars are and how much they are occluding each other. The system mistakenly recognized a van as a car in about 30% of the time. It never substituted a pedestrian for a vehicle and it incorrectly detected side-of-the-road clutter (producing false positive detections) less than 1% of the time. The system does not make use of any information related to motion that could assist in target detection. However, including motion information should improve the results.

The system's performance regarding the correct identification of cars does not deteriorate if the edges are extracted from an original gray-scale image as opposed to a difference image. However, in that case, the number of false alarms is higher. Most of the false alarms are located on the sides of the road (the regions that contain significant edge-like structures) and can easily be filtered out using the road model.

The fact that we use feature-based object representation allows translation invariant recognition and makes the system very robust to occlusions. Similarly, the system can easily deal with variable lighting conditions since the features are edges and their extraction is not affected with overall change in illumination. One of the consequences of edge-based object representation is that the system can detect both still and moving cars equally well.

There are several limitation of our current implementation that we plan to address in our future work: a) The system currently uses multiple saccades per scene but only one saccade per object. While in some cases that is sufficient, in situations where there are several vehicles close to one another or occluding each other, multiple saccades per object might be necessary in order to make a correct segmentation. b) The richer feature space, compared to only horizontal and vertical edges, would make the system more robust to the absence of some of the features. c) Similarly, the larger number of object classes, compared to using just a model for a car, should increase the recognition accuracy. d) The system currently uses camera calibration information in order to deal with scale invariant recognition and in many cases that information is not available. We believe that the system could be fairly easily modified and improved to overcome all these limitations and that the most difficult problem to deal with, the bottleneck, will be the time constraint - the real-time performance.

Acknowledgments

This work was supported in part by the Army Research Office under Contract DAAD19-01-1-0754.

References

1. N. Logothetis and N. Sheinberg. Visual object recognition, *Annual Review of Neuroscience*, 19:577–621, 1996.
2. E. Wachsmut, M. Oram and D. Perrett. Recognition of objects and their component parts: responses of single units in the temporal cortex of the macaque, *Cerebral Cortex*, 4:509-522 1994.
3. I. Biederman. Recognition by components: a theory of human image understanding, *Psychological Review*, 94:115-147, 1987.
4. D. H. Hubel and T. N. Wiesel. Receptive fields and functional architecture of monkey striate cortex, *J. Physiol.*, 195:215-244, 1968.
5. P. Neskovic and L. Cooper. Neural network-based context driven recognition of on-line cursive script. In *7th International Workshop on Frontiers in Handwriting Recognition*, pp. 352–362, 2000.
6. P. Neskovic, P. Davis, and L. Cooper Interactive parts model: an application to recognition of on-line cursive script. In *Advances in Neural Information Processing Systems*, 2000.
7. D. Koller, J. Weber, J. Malik, Robust multiple car tracking with occlusion reasoning, *Proceedings of the 5th European Conference on Computer Vision*, Springer-Verlag, Berlin, pp. 189-196, 1994.
8. A. Lipton, H. Fujiyoshi, F. Patil, Moving target classification and tracking from real-time video, *IEEE Workshop on Applications of Computer Vision (WACV)*, Princeton NJ, pp. 8-14, 1998.
9. D. Koller, K. Danilidis, H. Nagel, Model-Based Object Tracking in Monocular Image Sequences of Road Traffic Scenes, *International Journal of Computer Vision*, 10-3, pp. 257-281, 1993.

10. M. Betke, E. Haritaoglu and L. Davis, Highway Scene Analysis. *IEEE Conference on Intelligent Transportation Systems*, Boston, 1997.

11. M. Betke, E. Haritaoglu and L. Davis, Real-time multiple vehicle detection and tracking from a moving vehicle, *Machine Vision and Applications*, 12:69-83, 2000.

12. C. Papageorgiou and T. Poggio, A Trainable Object Detection System: Car Detection in Static Images, *A.I. Memo* No. 1673, MIT, 1999.

13. Z. Sun, G. Bebis and R. Miller, Quantized wavelet features and support vector machines for on-road vehicle detection, *IEEE International Conference on Control, Automation, Robotics and Vision*, 2002.

14. C. Goerick, N. Detlev and M. Werner. Artificial neural networks in real-time car detection and tracking applications, *Pattern Recognition Letters*, 17:335-343, 1996.

15. H. Schneiderman and T. Kanade. A statistical method for 3D object detection applied to faces and cars, *IEEE Conference on Computer Vision and Pattern Recognition*, 2000.

16. P. Viola and M. Jones. Rapid object detection using a boosted cascade of simple features, *IEEE Conference on Computer Vision and Pattern Recognition*, 2001.

17. M. Weber, M. Welling and P. Perona. Towards Automatic Discovery of Object Categories, *Computer Vision and Pattern Recognition*, 2000.

18. S. Agarwal and D. Roth. Learning a sparse representation for object detection, *7th European Conference on Computer Vision*, 4:113-130, 2002.

19. S-W. Lee, H. H. Bulthoff and T. Poggio (Eds.), *Biologically motivated computer vision*, Berlin: Springer-Verlag, 2000.

20. J. Keller, S. Rogers, M Kabrisky and M. Oxley, Object Recognition Based on Human Saccadic Behavior, *Pattern Analysis and Applications*, Vol. 2, Springer-Verlag, London, pp. 251-263, 1999.

21. L. Wiskott, J. M. Fellous, N. Kruger, and C. von der Malsburg. Face recognition by elastic bunch graph matching. *IEEE Transactions on Pattern Analysis and Machine Intelligence*, 19(7):775-779, 1997.

22. Rafael Gonzales and Richard Woods, *Digital Image Processing*, Addison-Wesley Publishing Company, 1993.

Financial Time Series Prediction Using Non-fixed and Asymmetrical Margin Setting with Momentum in Support Vector Regression

Haiqin Yang, Irwin King, Laiwan Chan, and Kaizhu Huang

Department of Computer Science and Engineering
The Chinese University of Hong Kong
Shatin, N.T., Hong Kong
{hqyang, king, lwchan, kzhuang}@cse.cuhk.edu.hk

Abstract. Recently, Support Vector Regression (SVR) has been applied to financial time series prediction. The financial time series usually contains the characteristics of small sample size, high noise and non-stationary. Especially the volatility of the time series is time-varying and embeds some valuable information about the series. Previously, we had proposed to use the volatility in the data to adaptively change the width of the margin in SVR. We have noticed that up margin and down margin would not necessary be the same, and also observed that their choice would affect the upside risk, downside risk and as well as the overall prediction performance. In this work, we introduce a novel approach to adopt the momentum in the asymmetrical margins setting. We applied and compared this method to predict the Hang Seng Index and Dow Jones Industrial Average.

Key words: Non-fixed and Asymmetrical Margin, Momentum, Support Vector Regression, Financial Time Series Prediction

1 Introduction

A time series is a collection of observations that measures the status of some activities over time [7, 8]. It is the historical record of some activities, with a consistency in the activity and the method of measurement, where the measurement is taken at equally spaced intervals, e.g., day, week, month, etc. In practice, there are various time series and they are collected in a wide range of disciplines, from engineering to economics. For example, the air temperatures of a certain city measured in successive days or weeks consists of a series; a certain share prices occurred in successive days, months is another series.

Of all the different possible time series, the financial time series is unusual since it contains several specific characteristics: small sample sizes, high noise, non-stationarity, non-linearity, and varying associated risk.

Support Vector Machines (SVMs) are recent generalization models, which find a generalization function through training samples, especially by small samples. It also extends to solve the regression problem by Support Vector Regression (SVR) [31, 26]. Nowadays, SVR has been successfully applied to time series prediction [17, 15] and financial forecasting [29, 27].

Usually, SVR uses the ε-insensitive loss function to measure the empirical risk (training error). This loss function not only measures the training error, but also controls the sparsity of the solution. When the ε-margin value is increased, it tends to reduce the number of support vectors [30]. Extremely, a constant objective function may occur when the width of margin is too wide. Therefore, the ε-margin value setting affects the complexity and the generalization of the objective function indirectly.

Since the ε-margin value setting is very important, researchers proposed various methods to determine it. Usually, there are four kinds of methods to deal with it. First, most practitioners set the ε-margin value to a non-negative constant just for convenience. For example, in [29], they simply set the margin width to 0. This amounts to the least modulus loss function. In other instances the margin width has been set to a very small value [31, 15, 6]. The second method is the cross-validation technique [17]. It is usually too expensive in terms of computation. A more efficient approach is to use another variant called ν-SVR [24, 20], which determines ε by using another parameter ν. It states that ν may be easier to specify than ε. This introduces another parameter setting problem. Another approach by Smola et al [25] is to find the "optimal" choice of ε based on maximizing the statistical efficiency of a location parameter estimator. They showed that the asymptotically optimal ε should scale linearly with the input noise of the training data, and this was verified experimentally, but their predicted value of the optimal ε does not have a close match with their experimental results. In sum, the previous methods tries to use a suitable or an optimal ε-margin value for that particular data set; the ε-margin value is always fixed and symmetrical for that data set. However, in stock market, it is volatile and the associated risk changes with time. A fixed and symmetrical ε-margin setting may lack the ability to capture stock market information promptly and may not be suitable for stock market prediction. Furthermore, our experience showed that ε-margin value is not necessary the same all the time [34].

In [34], we have extended the standard SVR with adaptive margin and classified it into four categories: Fixed and Symmetrical Margin (FASM), Fixed and Asymmetrical Margin (FAAM), Non-fixed and Symmetrical Margin (NASM), and Non-fixed and Asymmetrical Margin (NAAM). Comparing FASM with FAAM, we know that the downside risk can be reduced by employing asymmetrical margins. A theoretical result can be seen in [35]. While comparing FASM, FAAM with NASM, a good predictive result is obtained by exploiting the standard deviation to calculate the margin. However, NAAM requires the adaptation of the margin width and the degree of asymmetry, and no exact algorithm for such margin setting has been introduced.

In [36], we have proposed to use NAAM which combines two characteristics of the margin; non-fixed and asymmetry, to reduce the predictive downside risk while controlling the accuracy of the financial prediction. More specifically, we add a momentum term to achieve this. The width of the margin is determined by the standard deviation [34]. The asymmetry of the margin is controlled by the momentum. This momentum term can trace the up trend and down tendency of the stock prices. Since the financial time series often follows a long term trend but with small short term fluctuations, we exploit a larger up margin and a smaller down margin to under-predict the stock price when the momentum is positive and we use a smaller up margin and a larger down margin to over-predict the stock price while the momentum is negative. A simple illustration is shown in Fig. 1. We will use this downside risk avoiding strategy in the prediction. The work here is a more extensive version of the work in [36]; furthermore, we perform more related work on the experiments, especially normalizing the experimental data.

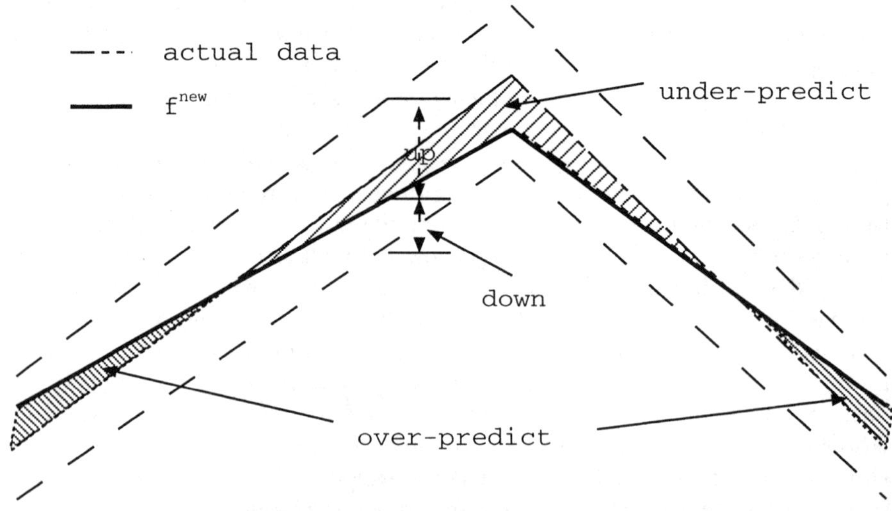

Fig. 1. Margin setting

We organize the paper as follows. First, we give an overview of the time series analysis models in Sect. 2. Next, we introduce the SVR with a general ε-insensitive loss function and the concept of momentum in Sect. 3. The accuracy metrics and experimental results are elucidated in Sect. 4. Finally, we conclude the paper with a brief discussion and final remarks in Sect. 5.

2 Time Series Analysis Models

There are many models for time series analysis. Generally, they are classified into *linear* and *non-linear* models, see Fig. 2.

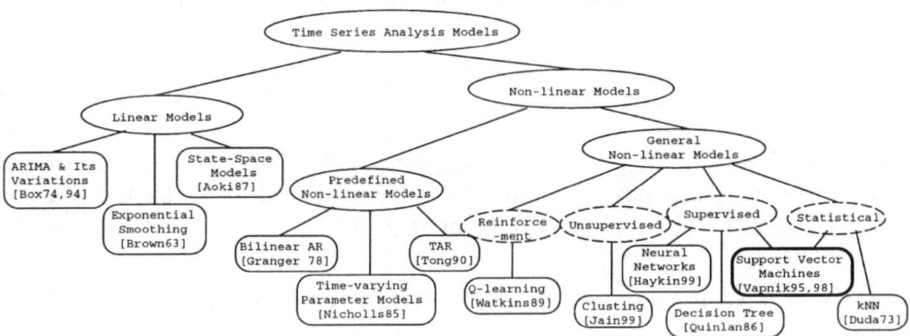

Fig. 2. Time series analysis models

Linear models have the characteristics of simplicity, usefulness and easy application. They work well for linear time series, but may fail otherwise. ARIMA models are typical linear models and used as the benchmark models for time series analysis [4].

Although linear models have both mathematical and practical convenience, there is no reason why real life time series should all be linear, and so the use of non-linear models seems potential promising [8]. In the 1980's, non-linear models were investigated and were proposed by the existing linear models [12, 21]. For example, Bilinear autoregressive or Bilinear AR models [11], time-varying parameter models [23, 19] and threshold autoregressive (TAR) model [28]. These models are agreeable due to the scrutiny given in their development for the standard statistical considerations of model specification, estimation, and diagnosis, but their general parametric nature tends to require significant a prior knowledge of the form of relationship being modeled. Therefore, they are not effective for modeling financial time series because the non-linear functions are hard to choose. Another class of non-linear models are general non-linear models, also called machine learning. These models can learn a model from a given time series without non-linear assumptions. They include reinforcement learning, e.g., Q-learning [32], unsupervised learning, e.g., clustering methods [14], supervised learning, e.g., decision tree [22] and neural network (NN) models [9, 1, 13], and statistical learning, e.g., k-nearest-neighbors(kNN) [10].

SVMs are recently proposed to model the non-linear relationship of the data. They have attracted the interests of researchers due to the following reasons. First, SVMs are grounded on the VC theory, which claims to guarantee the generalization [30]. Second, SVMs were proposed to solve the classification problem in the beginning. The margin maximization has visual geometric interpretation [30, 2]. Third, training SVM leads to solve the Quadratic Programming (QP) problem. For any convex programming problem, every local solution will also be global. Therefore, SVM training always finds a global solution, which is usually a unique solution [5]. Fourth, SVMs can tackle the non-linear cases by introducing the kernel function [16]. Here, our work just concentrate on the regression model, Support Vector Regression.

3 SVR with Momentum

In this section, we will give a brief introduction of Support Vector Regression with a general ε-insensitive loss function. Then we will describe the concept of momentum for the margin setting in Sect. 3.2.

3.1 SVR with a General ε-insensitive Loss Function

Usually, a regression problem is to estimate (learn) a function

$$f(\mathbf{x}, \boldsymbol{\lambda}) : \quad X(\mathbb{R}^d) \to \mathbb{R},$$

where $\boldsymbol{\lambda} \in \Lambda$, Λ is a set of abstract parameters, from a set of independent identically distributed (i.i.d.) samples with size N,

$$(\mathbf{x}_1, y_1), \dots, (\mathbf{x}_N, y_N), \quad \mathbf{x}_i \in X(\mathbb{R}^d), \quad y_i \in \mathbb{R}, \tag{1}$$

where the above samples are drawn from an unknown distribution $P(\mathbf{x}, y)$.

Now the aim is to find a function $f(\mathbf{x}, \boldsymbol{\lambda}^*)$ with the smallest possible value for the *expected risk* (or test error) as

$$R[\boldsymbol{\lambda}] = \int l(y, f(\mathbf{x}, \boldsymbol{\lambda})) P(\mathbf{x}, y) dx dy, \tag{2}$$

where l is a loss function which can be defined as one needs.

However, the probability of distribution $P(\mathbf{x}, y)$ is usually unknown. We are unable to compute, and therefore to minimize the expected risk $R[\boldsymbol{\lambda}]$ in (2), but we may know some information of $P(\mathbf{x}, y)$ from the samples in (1), we can compute a stochastic approximation of $R[\boldsymbol{\lambda}]$ by the so called *empirical risk*:

$$R_{emp}[\boldsymbol{\lambda}] = \frac{1}{N} \sum_{i=1}^{N} l(y_i, f(\mathbf{x}_i, \boldsymbol{\lambda})). \tag{3}$$

This is because of that the law of large numbers guarantees that the empirical risk converges in probability to the expected risk. However, for practical problem, the size of samples is small. Only minimizing the empirical risk may cause problems, such as bad estimation or overfitting, and we cannot obtain good result when new data come in.

338

To solve the small sample problem, the statistical theory, or VC theory, has provided bounds on the deviation of the empirical risk from the expected risk [30]. A typical uniform Vapnik and Chervonenkis bound, which holds with probability $1 - \eta$, has the following form:

$$R[\boldsymbol{\lambda}] \leq R_{emp}[\boldsymbol{\lambda}] + \sqrt{\frac{h(\ln\frac{2N}{h} + 1) - \ln\frac{\eta}{4}}{N}}, \quad \forall \boldsymbol{\lambda} \in \Lambda, \tag{4}$$

where h is the VC-dimension of $f(\mathbf{x}, \boldsymbol{\lambda})$.

From this bound, it is clear that in order to achieve small expected risk, i.e., good generalization performance, both the empirical risk and the ratio between the VC-dimension and the number of samples has to be small. Since the empirical risk is usually a decreasing function of h, it turns out that for a given number of samples, there is an optimal value of the VC-dimension. The choice of an appropriate value of h (which in most techniques is controlled by the number of free parameters of the model) is very important in order to get good performance, especially when the number of samples is small.

Therefore, a different induction principle, *Structural Risk Minimization Principle*, was proposed and developed by Vapnik [30] in the attempt to overcome the problem of choosing an appropriate VC-dimension.

SVMs were developed to implement the SRM principle [30]. They were used in the classification at first [3]; they were also extended to solve the regression problem [30]. When SVMs were used to solve the regression problem, they were usually called Support Vector Regression (SVR). The aim of SVR is to find a function f with parameters $\mathbf{w}$ and b by minimizing the following regression risk:

$$R_{reg}(f) = \frac{1}{2}\langle \mathbf{w}, \mathbf{w} \rangle + C \sum_{i=1}^{N} l(f(\mathbf{x}_i), y_i), \tag{5}$$

where $\langle , \rangle$ denotes the inner product, the first term can be seen as the margin in SVMs and therefore can measure the VC-dimension [30]. A common interpretation is that the Euclidean norm, $\langle \mathbf{w}, \mathbf{w} \rangle$, measures the flatness of the function f, minimizing $\langle \mathbf{w}, \mathbf{w} \rangle$ will make the objective function as flat as possible [26].

The function f is defined as

$$f(\mathbf{x}, \mathbf{w}, b) = \langle \mathbf{w}, \phi(\mathbf{x}) \rangle + b, \tag{6}$$

where $\phi(\mathbf{x}) : \mathbf{x} \to \Omega$, maps $\mathbf{x} \in X(\mathbb{R}^d)$ into a high (possible infinite) dimensional space Ω, and $b \in \mathbb{R}$.

In general, ε-insensitive loss function is used as the loss function [30, 26]. For this function, when the samples are in the range of $\pm\varepsilon$, they do not contribute to the output error. Thus, it leads to the sparseness of the solution. The function is defined as

$$\Gamma(f(\mathbf{x}) - y) = \begin{cases} 0, & \text{if } |y - f(\mathbf{x})| < \varepsilon \\ |y - f(\mathbf{x})| - \varepsilon, & \text{otherwise} \end{cases}. \tag{7}$$

In [34], we have introduced a general ε-insensitive loss function, $\Gamma'(f(\mathbf{x}_i) - y_i), i = 1, \ldots, N$, which is given as

$$\begin{cases} 0, & \text{if } -d(\mathbf{x}_i) < y_i - f(\mathbf{x}_i) < u(\mathbf{x}_i) \\ y_i - f(\mathbf{x}_i) - u(\mathbf{x}_i), & \text{if } y_i - f(\mathbf{x}_i) \geq u(\mathbf{x}_i) \\ f(\mathbf{x}_i) - y_i - d(\mathbf{x}_i), & \text{if } f(\mathbf{x}_i) - y_i \geq d(\mathbf{x}_i) \end{cases}, \tag{8}$$

where $d(\mathbf{x}), u(\mathbf{x}) \geq 0$, are two functions to determine the down margin and up margin respectively.

When $d(\mathbf{x})$ and $u(\mathbf{x})$ are both constant functions and $d(\mathbf{x}) = u(\mathbf{x})$, equation (8) amounts to the ε-insensitive loss function in (7) and we labeled it as FASM (Fixed and Symmetrical Margin). When $d(\mathbf{x})$ and $u(\mathbf{x})$ are both constant functions but $d(\mathbf{x}) \neq u(\mathbf{x})$, this case is labeled as FAAM (Fixed and Asymmetrical Margin). In the case of NASM (Non-fixed and Symmetrical Margin), $d(\mathbf{x}) = u(\mathbf{x})$ and they are varied with the data. The last case is with a non-fixed and asymmetrical margin(NAAM), where $d(\mathbf{x})$ and $u(\mathbf{x})$ are varied with the data and $d(\mathbf{x}) \neq u(\mathbf{x})$.

After using the standard method to find the solution of (5) with the loss function of (8) as [30], we obtain $\mathbf{w} = \sum_{i=1}^{N}(\alpha_i - \alpha_i^*)\phi(\mathbf{x}_i)$, by solving the following Quadratic Programming (QP) problem:

$$\min Q(\alpha, \alpha^*) = \frac{1}{2} \sum_{i=1}^{N} \sum_{j=1}^{N} (\alpha_i - \alpha_i^*)(\alpha_j - \alpha_j^*)\langle \phi(\mathbf{x}_i), \phi(\mathbf{x}_j) \rangle$$
$$+ \sum_{i=1}^{N}(u(\mathbf{x}_i) - y_i)\alpha_i + \sum_{i=1}^{N}(d(\mathbf{x}_i) + y_i)\alpha_i^*,$$

subject to

$$\sum_{i=1}^{N}(\alpha_i - \alpha_i^*) = 0, \alpha_i, \alpha_i^* \in [0, C], \tag{9}$$

where α_i and α_i^* are corresponding Lagrange multipliers used to push and pull $f(\mathbf{x}_i)$ towards the outcome of y_i respectively.

The above QP problem is very similar to the original QP problem in [30], therefore, it is easy to modify the previous algorithm to implement this QP problem. Practically, we implement our QP problem by modifying the libSVM from [6] with adding a new data structure to store both margins: up margin, $u(\mathbf{x})$, and down margin, $d(\mathbf{x})$. Obviously, this will not impact the time complexity of the SVR algorithm; we just need more space, linear to the size of data points, to store the corresponding margins.

Furthermore, using a kernel function, the estimation function in (6) becomes

$$f(\mathbf{x}) = \sum_{i=1}^{N}(\alpha_i - \alpha_i^*)\kappa(\mathbf{x}, \mathbf{x}_i) + b, \tag{10}$$

where the kernel function, $\kappa(\mathbf{x}, \mathbf{x}_i) = \langle \phi(\mathbf{x}), \phi(\mathbf{x}_i) \rangle$, is a symmetric function and satisfies the Mercer's condition. In this work, we select a common kernel function, RBF function, as the kernel function,

$$\kappa(\mathbf{x}, \mathbf{x}_i) = \exp(-\beta \|\mathbf{x} - \mathbf{x}_i\|^2), \tag{11}$$

where β is the kernel parameter.

In the following, we exploit the Karush-Kuhn-Tucker (KKT) conditions to calculate b. Here, they are

$$\alpha_i(u(\mathbf{x}_i) + \xi_i - y_i + \langle \mathbf{w}, \phi(\mathbf{x}_i) \rangle + b) = 0,$$
$$\alpha_i^*(d(\mathbf{x}_i) + \xi_i^* + y_i - \langle \mathbf{w}, \phi(\mathbf{x}_i) \rangle - b) = 0,$$

and

$$(C - \alpha_i)\xi_i = 0,$$
$$(C - \alpha_i^*)\xi_i^* = 0.$$

Therefore, when there exists i, such that $\alpha_i \in (0, C)$ or $\alpha_i^* \in (0, C)$, b can be computed as follows:

$$b = \begin{cases} y_i - \langle \mathbf{w}, \phi(\mathbf{x}_i) \rangle - u(\mathbf{x}_i), & \text{for } \alpha_i \in (0, C) \\ y_i - \langle \mathbf{w}, \phi(\mathbf{x}_i) \rangle + d(\mathbf{x}_i), & \text{for } \alpha_i^* \in (0, C) \end{cases}.$$

When no $\alpha_i^{(*)} \in (0, C)$, the average method [6] is used.

3.2 Momentum

Momentum is a well known term in physics. We borrow this term in the work. The differences are: in physics, momentum is used to measure the change of state of a body by external forces; in our work, the momentum is used to measure the up and down trend of stock market, which is impelled by the investors. The term in both areas reflects the difference of change, but with different kinds of external forces: in our work, the external forces are the investment of investors.

More specifically, we construct a margin setting, which is a linear combination of the standard deviation and the momentum. The up margin and down margin are set in the following forms:

$$u(\mathbf{x}_i) = \lambda_1 \times \sigma(\mathbf{x}_i) + \mu \times \Delta(\mathbf{x}_i), \quad i = 1, \ldots, N,$$
$$d(\mathbf{x}_i) = \lambda_2 \times \sigma(\mathbf{x}_i) - \mu \times \Delta(\mathbf{x}_i), \quad i = 1, \ldots, N, \tag{12}$$

where $\sigma(\mathbf{x}_i)$ is the standard deviation of input $\mathbf{x}_i$, $\Delta(\mathbf{x}_i)$ is the momentum at point $\mathbf{x}_i$, λ_1, λ_2 are both positive constants, called coefficients of the margin width and μ is a non-negative constant, called coefficient of momentum. Using this margin setting (12), the width of margin at point $\mathbf{x}_i$ is determined by $\sigma(\mathbf{x}_i)$ and the sum of λ_1 and λ_2, i.e.,

$$W(\mathbf{x}_i) = (\lambda_1 + \lambda_2) \times \sigma(\mathbf{x}_i).$$

The standard deviation here is used to reflect the change of volatility; therefore, when in a high volatility mode, we use a broad width of margin; when in a low volatility situation, we use a narrow width of margin.

For the setting of momentum, in fact, there are many ways to calculate it. For example, it may be set as a constant. In this work, we exploit the Exponential Moving Average (EMA), which is time-varying and can reflect the up trend and down tendency of the financial data. An n-day's EMA is calculated by

$$EMA_i = EMA_{i-1} \times (1 - r) + y_i \times r,$$

where $r = 2/(1 + n)$ and it begins from the first day, $EMA_1 = y_1$. Here, n is called the length of EMA. The current day's momentum is set as the difference between the current day's EMA and the EMA in the previous k day, i.e.,

$$\Delta(x_i) = EMA_i - EMA_{i-k} \tag{13}$$

where k is called the lag of momentum. Equation (13) actually detects the degree of the change in the stock market.

From above configurations, we know that the margin setting of (12) includes the case of NASM (when $\mu = 0$). When $\mu \neq 0$, it is the case of NAAM. If $\Delta(\mathbf{x}) > 0$, we know that an up trend occurs. Based on our downside risk avoiding predictive strategy, we would use a larger up margin and a smaller down margin to under-predict the stock price. While if $\Delta(\mathbf{x}) < 0$, i.e., in the situation of down trend, we would use a smaller up margin and larger down margin to over-predict the stock price.

In addition, in the margin setting of (12) and momentum setting of (13), we have to specify the concrete setting of parameters. For the coefficients of margin width, λ_1 and λ_2, they are set to $\frac{1}{2}$; therefore, we can make the margin width at day i equal to the standard deviation of input $\mathbf{x}_i$. For the coefficient of momentum, μ, it is equal to 1; the lag of momentum, k, is equal to 1. The setting of these two parameters is coming from our experience in [33]. Actually, the only undetermined parameter is the length of EMA, n. In the following experiments, we use different length of EMA to test their effects and we find that it is related to the volatility of financial data.

4 Experiments

In this section, we first define the performance measurement of our experiments. Then we detail the setup of experiments with their results compared.

4.1 Accuracy Metrics

We use the following statistical metrics to evaluate the prediction performance, including Mean Absolute Error (MAE), Up side Mean Absolute Error (UMAE), and Down side Mean Absolute Error (DMAE). The definitions of these criteria are listed in the Table 1. MAE is the measure of the discrepancy between the actual and predicted values. The smaller the MAE, the closer are the predicted values to the actual values. UMAE is the measure of up side risk. DMAE is the measure of down side risk. The smaller the UMAE and DMAE, the smaller are the corresponding predictive risks.

Table 1. Accuracy metrics

Metrics	Calculation
MAE	$\text{MAE} = \frac{1}{m} \times \sum_{i=1}^{m} |a_i - p_i|$
UMAE	$\text{UMAE} = \frac{1}{m} \times \sum_{i=1, a_i \geq p_i}^{m} (a_i - p_i)$
DMAE	$\text{DMAE} = \frac{1}{m} \times \sum_{i=1, a_i < p_i}^{m} (p_i - a_i)$

a_i and p_i are the actual values and predicted values at day i respectively. m is the number of testing data.

4.2 Experimental Procedure and Results

In this section, we conduct the SVR algorithm with four categories of margin settings, Autoregressive (AR) model with order four and RBF network on two indices respectively and compare their results. The experiments are conducted on Sun Blade 1000, RAM 2GB and Solaris 8.

Two indices are used in the experiments:

1) HSI: daily closing prices of Hong Kong's Hang Seng Index from January 2nd, 1998 to December 29th, 2000.
2) DJIA: daily closing prices of Dow Jones Industrial Average from January 2nd, 1998 to December 29th, 2000.

The ratio of the number of training data and the number of testing data is set to 5:1. Therefore, the corresponding training time periods and testing periods are obtained and listed in Table 2.

Furthermore, we model the system as $p_t = f(\mathbf{x}_t)$, where f is learned by the stated three models: SVR, AR and RBF network, from the training data; $\mathbf{x}_t = (a_{t-4}, a_{t-3}, a_{t-2}, a_{t-1})$, a_t is the daily closing index in day t, an intrinsic assumption here is that there is (non)linear relationship between sequential five days' index values. After finding the function f, we use the testing data to test the predictive performance of the models.

Table 2. Indices, time periods and parameters

Indices	Training time periods	Testing time periods	C	β
HSI	02/01/1998 - 04/07/2000	05/07/2000 - 29/12/2000	2^6	2^{-3}
DJIA	02/01/1998 - 29/06/2000	30/06/2000 - 29/12/2000	2^{-1}	2^1

4.2.1 SVR Algorithm

Before generating the model, we perform a cross-validation on the training data to determine the parameters that are needed in SVR. They are C, the cost of error and β, the parameter of kernel function. The parameters we used are listed in Table 2.

4.2.1.1 NASM and NAAM

The margins setting is based on (12). More specifically, in the case of NASM, we set $\lambda_1 = \lambda_2 = \frac{1}{2}$ and $\mu = 0$, thus the overall margin widths are equal to the standard deviation of input $\mathbf{x}$. In the case of NAAM, we also fix $\lambda_1 = \lambda_2 = \frac{1}{2}$; therefore, we have a fair comparison of NASM case. From our experience [33], $k = 1$ and $\mu = 1$ are suitable for different data sets. The uncertain term for the margin setting is n, the length of EMA. Hence, we use different n, equal to 10, 30, 50, 100 respectively to test the effect of the length of EMA. From the result of Table 3 and Table 4, we can see that the DMAE values in all cases of NAAM are smaller than that in NASM case, thus we have a smaller predictive downside risk in NAAM case. This also meets our assumption, i.e., it is a downside risk avoiding strategy for the prediction. We also see that the MAE gradually decreases with the length of EMA increases and when the length equals 100, the MAE is the smallest in all case of NAAM and is smaller than that of NASM. In Table 4, when the length equals 30, the MAE is the smallest in all cases of NAAM and it is also smaller than that of NASM.

Table 3. Effect of the length of EMA on HSI

type	n	MAE	UMAE	DMAE
NASM		217.18	108.95	108.23
	10	221.01	119.70	101.31
NAAM	30	218.32	123.56	94.76
	50	217.12	120.31	96.81
	100	**216.60**	120.60	96.00

Here, we plot the daily closing prices of HSI with 100-days' EMA and the prices of DJIA with 30-days' EMA in Fig. 3 and Fig. 4 respectively and list the Average Standard Deviations (ASD) of input $\mathbf{x}$ of the training data sets

Table 4. Effect of the length of EMA on DJIA

type	n	MAE	UMAE	DMAE
NASM		87.17	44.17	43.00
	10	86.61	43.79	42.81
NAAM	30	**86.58**	45.10	41.48
	50	87.36	47.02	40.34
	100	87.02	45.67	41.35

of both data sets, the Average of Absolute Momentums (AAM) of input **x** for the best length of both training data sets respectively in Table 5. We can observe that the ASD of HSI is higher than that of DJIA and the ratio of AAM to ASD is smaller for HSI than that for DJIA. This indicates that the data is more volatile in the HSI data; hence we may use a longer length of EMA to represent this volatility for the prediction.

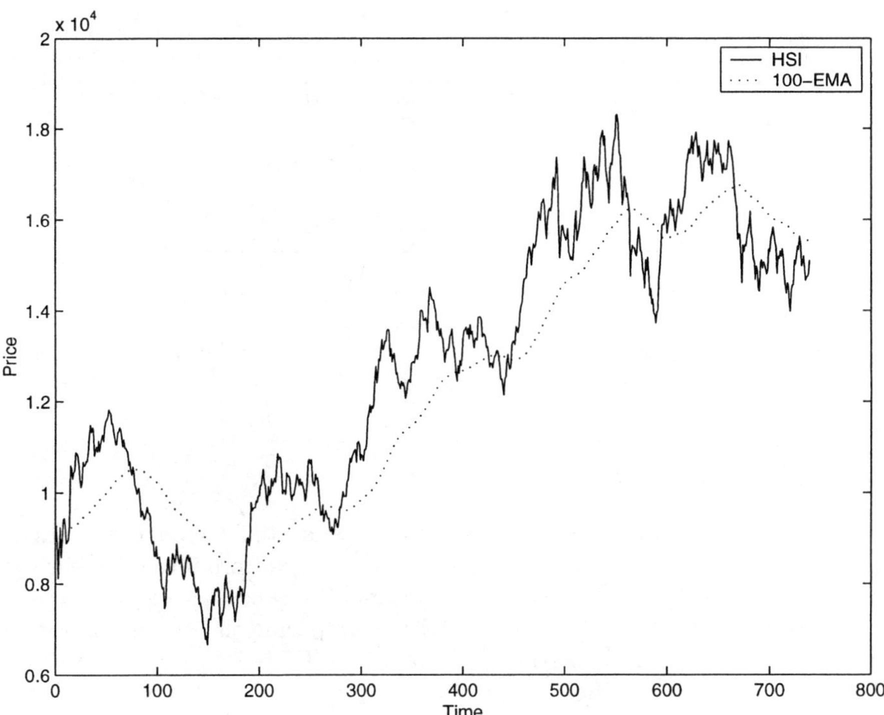

Fig. 3. HSI and 100 days' EMA

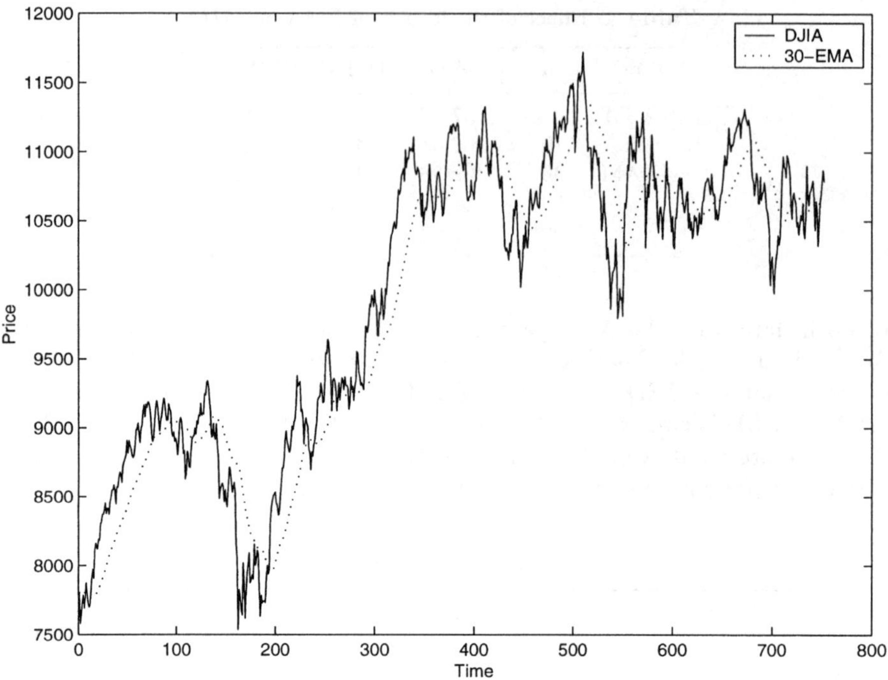

Fig. 4. DJIA and 30 days' EMA

Table 5. ASD and AAM

data set	ASD	AAM		ratio
		n	Δ	
HSI	182.28	100	20.80	0.114
DJIA	79.95	30	15.64	0.196

4.2.1.2 FASM and FAAM

For the fixed margin setting, we set the margin width to 0.03, i.e. $u(\mathbf{x})+d(\mathbf{x}) = 0.03$, for both data sets. The up margin, $u(\mathbf{x})$, ranges from 0 to 0.03, each increments is 0.0075. For these setting, we obtain the results in Table 6 for data set HSI and in Table 7 for data set DJIA. Comparing the corresponding results of non-fixed margin settings (Table 3 and Table 4) with the results of fixed margin settings (Table 6 and Table 7), we observe that the predictive performance of non-fixed margin settings is better than that of the fixed margin cases generally. From Table 6 and Table 7, we can see that the MAE is in a wide range. This means that using a fixed margin setting may have bad predictive result which gives more risk.

Table 6. Fixed margin results on HSI

$u(\mathbf{x})$	$d(\mathbf{x})$	MAE	UMAE	DMAE
0	0.03	259.32	43.37	215.95
0.0075	0.0225	233.28	66.21	167.07
0.0150	0.0150	220.50	94.07	126.43
0.0225	0.0075	216.87	126.96	89.91
0.03	0	227.17	167.34	59.83

Table 7. Fixed margin results on DJIA

$u(\mathbf{x})$	$d(\mathbf{x})$	MAE	UMAE	DMAE
0	0.03	99.97	17.00	82.97
0.0075	0.0225	90.42	25.24	65.18
0.0150	0.0150	86.70	35.46	51.24
0.0225	0.0075	87.61	48.47	39.14
0.03	0	93.24	64.30	29.94

4.2.2 AR Model

Here, we use the AR model with order four to predict the prices of HSI and DJIA; hence, we can compare the AR model with NASM, NAAM in SVR with the same order. The results are listed in the Table 8. We can see that NASM and NAAM are superior to AR model with same order.

Table 8. Results on AR(4)

data set	MAE	UMAE	DMAE
HSI	217.75	105.96	111.79
DJIA	88.74	46.36	42.38

4.2.3 RBF network

The RBF network we used is implemented in NETLAB [18]. We perform the one-step ahead prediction to predict the prices of HSI and DJIA. More specifically, we set the effect of hidden units to 3, 5, 7, 9 and set other parameters as default. The corresponding results are listed in Table 9 for HSI, in Table 10 for DJIA respectively. Comparing these two tables with Table 3 and Table 4, we can see that NASM and NAAM are also better than the RBF network.

Table 9. Effect of number of hidden units on HSI

# hidden	MAE	UMAE	DMAE
3	386.65	165.08	221.57
5	277.83	128.92	148.91
7	219.32	104.15	115.17
9	221.81	109.46	112.35

Table 10. Effect of number of hidden units on DJIA

# hidden	MAE	UMAE	DMAE
3	88.31	44.60	43.71
5	98.44	48.46	49.98
7	90.53	46.22	44.31
9	87.23	44.09	43.14

5 Discussion and Conclusion

In this work, we propose to use non-fixed and asymmetrical margin (NAAM) setting in the prediction of HSI and DJIA. From the experiments, we make the following observations:

1. Comparing NAAM with the case of NASM which just uses the standard deviation, we find that by adding the momentum to set the margin we can reduce the predictive downside risk. We may also improve the accuracy of our prediction by selecting a suitable length of EMA.
2. The selection of the length of EMA may depend on the volatility of the financial data. A long term EMA may be suitable for a higher volatility financial time series. A short term EMA may be suitable for the opposite case.
3. In SVR, non-fixed margin settings (NAAM and NASM) are better than the fixed margin settings (FAAM and FASM). Using a fixed margin setting may have more risk, which results in poor performance.
4. The SVR algorithm with NASM and NAAM outperforms the AR model with the same order.
5. The SVR algorithm with NASM and NAAM is also better than the RBF network.

In our work, how to find more suitable parameters easily, i.e., C and β, for a specific data set is still a problem. In addition, we just consider the momentum term to trace the changing trend of the stock market here. Other more general or robust methods are still needed to be explored and to be applied in the margin settings in the order of capturing the valuable information of stock market promptly.

Acknowledgement

The work described in this work was partially supported by a grant from the Research Grants Council of the Hong Kong Special Administration Region, China.

References

1. D. E. Baestaens. *Neural Network Solutions for Trading in Financial Markets.* London: Financial Times: Pitman Pub., 1994.
2. K. Bennett and E. Bredensteiner. Duality and Geometry in SVM Classifiers. In P. Langley, editor, *Proc. of Seventeenth Intl. Conf. on Machine Learning*, pages 57–64, San Francisco, 2000. Morgan Kaufmann.
3. B. E. Boser, I. Guyon, and V. N. Vapnik. A Training Algorithm for Optimal Margin Classifiers. In *Computational Learing Theory*, pages 144–152, 1992.
4. G. E. P. Box and G. M. Jenkins. *Time-Series Analysis, Forecasting and Control.* San Francisco: Holden-Day, third edition, 1994.
5. C. Burges and D. Crisp. Uniqueness of the SVM Solution. In S. A. Solla, T. K. Leen, and K. R. Müller, editors, *Advances in Neural Information Processing Systems*, volume 12, pages 223–229, Cambridge, MA, 2000. MIT Press.
6. Chih-Chung Chang and Chih-Jen Lin. LIBSVM: a Library for Support Vector Machines (version 2.31), 2001.
7. C. Chatfield. *The Analysis of Time Series: An Introduction.* Chapman and Hall, fifth edition, 1996.
8. C. Chatfield. *Time-Series Forecasting.* Chapman and Hall/CRC, 2001.
9. B. Cheng and D. M. Titterington. Neural Networks: A Review from a Statistical Perspective. *Statistical Science*, 9:2–54, 1994.
10. R. O. Duda and P. E. Hart. *Pattern Classification and Scene Analysis.* New York: Wiley, London; New York, 1973.
11. C. W. J. Granger and A. P. Andersen. *Introduction to Bilinear Time Series.* Göttingen: Vandenhoeck and Ruprecht, 1978.
12. C. W. J. Granger and R. Joyeux. An Introduction to Long-Memory Time Series Models and Fractional Differencing. *Journal of Time Series Analysis*, 1, 1980.
13. S. Haykin. *Neural Networks : A Comprehensive Foundation.* Upper Saddle River, N. J.: Prentice Hall, 2nd edition, 1999.
14. A. K. Jain, M. N. Murty, and P. J. Flynn. Data Clustering: A Review. *ACM Computing Surveys*, 31(3):264–323, 1999.
15. S. Mukherjee, E. Osuna, and F. Girosi. Nonlinear Prediction of Chaotic Time Series Using Support Vector Machines. In J. Principe, L. Giles, N. Morgan, and E. Wilson, editors, *IEEE Workshop on Neural Networks for Signal Processing VII*, pages 511–519. IEEE Press, 1997.
16. K. R. Müller, S. Mika, G. Rätsch, K. Tsuda, and B. Schölkopf. An introduction to Kernel-Based Learning Algorithms. *IEEE Transactions on Neural Networks*, 12:181–201, 2001.
17. K. R. Müller, A. Smola, G. Rätsch, B. Schölkopf, J. Kohlmorgen, and V. Vapnik. Predicting Time Series with Support Vector Machines. In W. Gerstner, A. Germond, M. Hasler, and J. D. Nicoud, editors, *ICANN*, pages 999–1004. Springer, 1997.

18. Ian T. Nabney. *Netlab: Algorithms for Pattern Recognition*. Springer, London; New York, 2002.

19. D. F. Nicholls and A. Pagan. Varying Coefficient Regression. In E.J. Hannan, P.R. Krishnaiah, , and M.M. Rao, editors, *Handbook of Statistics*, volume 5, pages 413–449, North Holland, Amsterdam, 1985.

20. B. Schölkopf Pai-Hsuen Chen, Chih-Jen Lin. A Tutorial on ν-Support Vector Machines. Technical report, National Taiwan University, 2003.

21. M. B. Priestley. *Spectral Analysis and Time Series*. New York: Academic Press, London, 1981.

22. J. R. Quinlan. Induction of Decision Trees. *Machine Learning*, 1:81–106, 1986.

23. Baldev Raj and Aman Ullah. *Econometrics: A Varying Coefficients Approach*. New York: St. Martin's Press, 2nd edition, 1981.

24. B. Schölkopf, A. Smola, R. Williamson, and P. Bartlett. New Support Vector Algorithms. Technical Report NC2-TR-1998-031, GMD and Australian National University, 1998.

25. A. Smola, N. Murata, B. Schölkopf, and K.-R. Müller. Asymptotically Optimal Choice of ε-Loss for Support Vector Machines. In *Proc. of Seventeenth Intl. Conf. on Artificial Neural Networks*, 1998.

26. A. Smola and B. Schölkopf. A Tutorial on Support Vector Regression. Technical Report NC2-TR-1998-030, NeuroCOLT2, 1998.

27. E. H. Tay and L. J. Cao. Application of Support Vector Machines to Financial Time Series Forecasting. *Omega*, 29:309–317, 2001.

28. H. Tong. *Non-Linear Time Series*. Clarendon Press, Oxford, 1990.

29. T. B. Trafalis and H. Ince. Support Vector Machine for Regression and Applications to Financial Forecasting. In *Proceedings of the IEEE-INNS-ENNS International Joint Conference on Neural Networks (IJCNN2000)*, volume 6, pages 348–353. IEEE, 2000.

30. V. N. Vapnik. *The Nature of Statistical Learning Theory*. Springer, New York, 1995.

31. V. N. Vapnik, S. Golowich, and A. Smola. Support Vector Method for Function Approximation, Regression Estimation and Signal Processing. In M. Mozer, M. Jordan, and T. Petshe, editors, *Advances in Neural Information Processing Systems*, volume 9, pages 281–287, Cambridge, MA, 1997. MIT Press.

32. C. Watkins. *Learning from Delayed Rewards*. PhD thesis, King's College, Cambridge, England, 1989.

33. Haiqin Yang. Margin Variations in Support Vector Regression for the Stock Market Prediction. Master's thesis, Chinese University of Hong Kong, 2003.

34. Haiqin Yang, Laiwan Chan, and Irwin King. Support Vector Machine Regression for Volatile Stock Market Prediction. In Hujun Yin, Nigel Allinson, Richard Freeman, John Keane, and Simon Hubbard, editors, *Intelligent Data Engineering and Automated Learning — IDEAL 2002*, volume 2412 of *LNCS*, pages 391–396. Springer, 2002.

35. Haiqin Yang, Laiwan Chan, and Irwin King. Margin Settings in Support Vector Regression for the Stock Market Prediction, 2003. To be submitted.

36. Haiqin Yang, Irwin King, and Laiwan Chan. Non-fixed and Asymmetrical Margin Approach to Stock Market Prediction Using Support Vector Regression. In *International Conference on Neural Information Processing — ICONIP 2002*, volume 3, pages 1398–1402, 2002.

A Method for Applying Neural Networks to Control of Nonlinear Systems

Jinglu HU and Kotaro HIRASAWA

Graduate School of Information, Production and Systems
Waseda University
Hibikino 2-7, Wakamatsu, Kitakyushu 808-0135, Japan
{jinglu, hirasawa}@waseda.jp

Summary. This chapter discusses a new method for applying neural networks to control of nonlinear systems. Contrast to a conventional method, the new method does not use neural network directly as a nonlinear controller or nonlinear prediction model, but use it indirectly via an ARX-like macro-model, in which neural network is embedded. The ARX-like model incorporating neural network is constructed in such a way that it has similar linear properties to a linear ARX model. The nonlinear controller is then designed in a similar way as designing a controller based on a linear ARX model. Numerical examples are used to illustrate the usefulness of the new method.

Key words: Nonlinear system, nonlinear control, neural networks, linear control theory, ARX model

1 Introduction

Neural networks have recently attracted much interest in system control community because they can learn any nonlinear mapping [5, 16, 4, 20]. Many approaches have been proposed to apply neural networks to control of general nonlinear systems [14, 20, 15, 21]. Although neural networks are universal approximators [7], there are two major criticisms on using neural network models; one is that they do not have useful interpretations in their parameters, especially for multilayer perceptron (MLP) networks [1]; the other is that they do not have structures favorable to applications such as controller design and system analysis [15, 3]. Because of the nonlinearity, many neural network based control approaches have to use two neural networks: one for representing the system and the other used as a controller. A typical example is the direct inverse control with specialized training principle [16, 18]. A question arises here of whether "we can develop a designing scheme that the two neural networks: the one used for model and the one used for controller

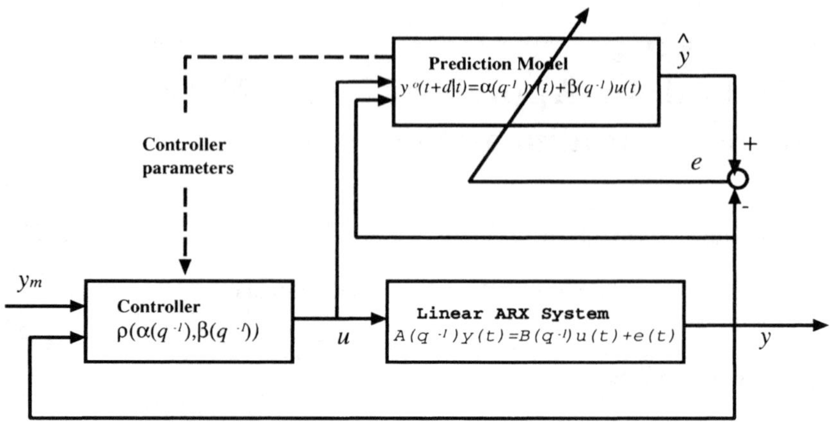

Fig. 1. A typical linear ARX model based control system where prediction model and controller sharing their parameters

share the same parameters?". If so, one only needs to train the one used for modeling which is usually easier to realize, especially in the sense of off-line training.

On the other hand, in linear cases it seems that there is no problem for sharing parameters between models and controllers. Figure 1 shows a well-known control system based on linear ARX model [6]. In the control system, the ARX prediction model and the controller share the same parameters. We know that the linearity of ARX model for input variables make sharing parameters possible. This motivates us to develop an ARX like neural network based model that not only has flexible representation ability like a conventional neural network, but also has the similar linearities like a linear ARX model. In such method, neural network plays only supportive role to synthesize and tune feedback controller autonomously instead of replacing conventional controller. By combining the existing results in the literature of control theory and neural networks, the new method is expected to be more acceptable by engineers and practitioners [21].

It has been shown that a general nonlinear system can be represented as an ARX like regression form by using mathematical transformations such as Taylor expansion [10]. Such an ARX macro-model has "state dependent coefficients". The "state dependent coefficients" are then parameterized by using a multi-input and multi-output (MIMO) neural network. The model obtained in this way is called quasi-ARX model. The quasi-ARX model is then further transformed into one linear in the input variables by introducing an extra input variable.

The chapter is organized as follows: Section 2 describes the problem. Section 3 introduces a quasi-ARX prediction model. Section 4 develops a dual loop learning algorithm for parameter estimation. Section 5 describes the con-

352

trol system based on the quasi-ARX prediction model. Section 6 gives some numerical examples to illustrate the usefulness of the proposed method. Finally, Section 7 presents discussions and conclusions.

2 Problem Description

2.1 Systems

Let us consider a single input single output nonlinear time-invariant system whose input-output relation described by

$$y(t) = g(\varphi(t)) + e(t) \tag{1}$$

$$\varphi(t) = [y(t-1) \dots y(t-n_y)\, u(t-1) \dots u(t-n_u-d+1)]^T \tag{2}$$

where $y(t)$ is the output at time t $(t = 1, 2, \dots)$, $u(t)$ the input, $\varphi(t)$ the regression vector with known order (n_y, n_u), d the known integer time delay, $e(t)$ the disturbance, and $g(\cdot)$ the unknown nonlinear function. It is further assumed that

1) $g(\cdot)$ is a continuous function, and at a small region around $\varphi(t) = 0$, it is C^∞ continuous;
2) the system (1) is stable. That is, for bounded input $u(t)$, the output $y(t)$ is bounded.

2.2 Problems

We consider a minimum variance control with a criterion function defined by

$$J(t+d) = \frac{1}{2}\left[(y(t+d) - y^*(t+d))^2 + \lambda u(t)^2\right] \tag{3}$$

where $y^*(t)$ is reference signal, and λ is weighting factor for the control input. The problem is to design a controller

$$u(t) = \rho(\Omega, \xi(t)) \tag{4}$$

where Ω and $\xi(t)$ are the parameter and regression vectors that will be defined later, by minimizing the criterion function (3) with

$$\frac{\partial J(t+d)}{\partial u(t)} = 0. \tag{5}$$

It is well known from linear control theory, see e.g., G.C. Goodwin and K.S. Sin (1984) [6], that if a prediction model for the system is linear in the input variable $u(t)$, a controller can be easily obtained by solving (5). Unfortunately, a conventional neural network model is nonlinear in the variable $u(t)$. To solve this problem, instead of using neural network directly as a model, we embed neural network in an ARX like macro-model and construct a nonlinear prediction model that is linear in the variable $u(t)$.

3 Quasi-ARX Prediction Model

3.1 Regression Form Representation

It has been shown that a general nonlinear system described by (1) can be represented in a regression form [9, 10].

Performing Taylor expansion to the unknown nonlinear function $g(\varphi(t))$ in (1) on a small region around $\varphi(t) = 0$

$$y(t) = g(0) + g'(0)\varphi(t) + \frac{1}{2}\varphi^T(t)g''(0)\varphi(t) + \cdots + e(t) \tag{6}$$

where the prime denotes differentiation with respect to $\varphi(t)$, and introducing the notation

$$y_0 = g(0)$$

$$\theta(\varphi(t)) = \left(g'(0) + \frac{1}{2}\varphi^T(t)g''(0) + \ldots\right)^T$$

$$= [a_{1,t} \cdots a_{n_y,t} \ b_{0,t} \cdots b_{n_u-1,t}]^T,$$

we have a regression form of the system described by (1)

$$y(t) = y_0 + \varphi^T(t)\theta(\varphi(t)) + e(t) \tag{7}$$

where the coefficients $a_{i,t} = a_i(\varphi(t))$, $b_{i,t} = b_i(\varphi(t))$ are nonlinear function of $\varphi(t)$.

On the other hand, one needs to predict $y(t)$ by using the input-output data available up to time $t - d$ in a prediction model. For this consideration, we hope that the coefficients $a_{i,t}$ and $b_{i,t}$ are calculable using the input-output data up to time $t-d$. To do so, let us replace iteratively $y(t-i)$, $i = 1, ..., d-1$ in the expressions of $a_{i,t}$ and $b_{i,t}$ with their predictions

$$y(t - i) \Rightarrow \hat{g}(\hat{\varphi}(t - i)), \ i = 1, ..., d - 1 \tag{8}$$

where $\hat{g}(\cdot)$ is a predictor, $\hat{\varphi}(t-i)$ whose elements $y(t-k)$, $i+1 < k \leq d-1$ are replaced by their predictions, and define the new expressions of the coefficient by

$$a_{i,t} = \tilde{a}_{i,t} = \tilde{a}(\phi(t - d)), \quad b_{i,t} = \tilde{b}_{i,t} = \tilde{b}(\phi(t - d))$$

where $\phi(t - d) = q^{-d}\phi(t)$ and $\phi(t)$ is a vector define by

$$\phi(t) = [y(t) \ldots y(t - n_y + 1) \, u(t) \ldots u(t - n_u - d + 2)]^T. \tag{9}$$

And q^{-1} is a backward shift operator, e.g. $q^{-1}u(t) = u(t - 1)$.

3.2 ARX Like Macro-Model

Let us introduce two polynomials $A(q^{-1}, \phi(t))$ and $B(q^{-1}, \phi(t))$ based on the coefficients $a_{i,t}$ and $b_{i,t}$, defined by

$$A(q^{-1}, \phi(t)) = 1 - a_{1,t}q^{-1} - \dots - a_{n_y,t}q^{-n_y} \tag{10}$$

$$B(q^{-1}, \phi(t)) = b_{0,t} + b_{1,t}q^{-1} + \dots + b_{n_u-1,t}q^{-n_u+1}. \tag{11}$$

We can then express the system (1) by an ARX macro-model

$$A(q^{-1}, \phi(t-d))y(t) = y_0 + B(q^{-1}, \phi(t-d))q^{-d}u(t) + e(t). \tag{12}$$

Furthermore, for a system described by (12), we have the following theorem for the d-step prediction, which is similar in form to that in linear case.

Theorem 1. *For a system described by (12), the d-step-ahead prediction, $y^o(t+d|t, \phi(t))$, of $y(t)$ satisfies*

$$y^o(t+d|t, \phi(t)) = y_\phi + \alpha(q^{-1}, \phi(t))y(t) + \beta(q^{-1}, \phi(t))u(t) \tag{13}$$

where

$$y^o(t+d|t, \phi(t)) = y(t+d) - F(q^{-1}, \phi(t))e(t+d),$$
$$y_\phi = F(q^{-1}, \phi(t))y_0,$$
$$\alpha(q^{-1}, \phi(t)) = G(q^{-1}, \phi(t))),$$
$$= \alpha_{0,t} + \alpha_{1,t}q^{-1} + \dots + \alpha_{n_y-1,t}q^{-(n_y-1)},$$
$$\beta(q^{-1}, \phi(t)) = F(q^{-1}, \phi(t))B(q^{-1}, \phi(t)),$$
$$= \beta_{0,t} + \beta_{1,t}q^{-1} + \dots + \beta_{n_u+d-2,t}q^{-n_u-d+2},$$

and $G(q^{-1}, \phi(t))$, $F(q^{-1}, \phi(t))$ are unique polynomials satisfying

$$F(q^{-1}, \phi(t))A(q^{-1}, \phi(t)) = 1 - G(q^{-1}, \phi(t))q^{-d} \tag{14}$$

Proof: Let us denote the polynomials $F(q^{-1}, \phi(t))$ and $G(q^{-1}, \phi(t))$ by

$$F(q^{-1}, \phi(t)) = 1 + f_{1,t}q^{-1} + \dots + f_{d-1,t}q^{-d+1}$$
$$G(q^{-1}, \phi(t)) = g_{0,t} + g_{1,t}q^{-1} + \dots + g_{n_y-1,t}q^{-n_y+1}$$

For a given small interval time Δt, the coefficients $a_{i,t+d}$, $b_{i,t+d}$, $f_{i,t}$ and $g_{i,t}$ can be seen as constants. Therefore, from the division algorithm of algebra [2], there exist unique polynomials such that (14) is satisfied for any given time t. Replacing t with $t+d$ in (12), we have

$$A(q^{-1}, \phi(t))y(t+d) = y_0 + B(q^{-1}, \phi(t))u(t) + e(t+d). \tag{15}$$

Then multiplying (15) on the left by $F(q^{-1}, \phi(t))$, and using (14), we have

$$y(t+d) - F(q^{-1}, \phi(t))e(t+d)$$
$$= F(q^{-1}, \phi(t))y_0 + G(q^{-1}, \phi(t))q^{-d}y(t+d)$$
$$+ F(q^{-1}, \phi(t))B(q^{-1}, \phi(t))u(t). \tag{16}$$

Finally, using the definition of $y^o(t+d|t, \phi(t))$ and y_ϕ, we then get

$$y^o(t+d|t, \phi(t)) =$$
$$y_\phi + G(q^{-1}, \phi(t))y(t) + F(q^{-1}, \phi(t))B(q^{-1}, \phi(t))u(t). \tag{17}$$

This proves the theorem.

3.3 Linearity for the Input Variable $u(t)$

The prediction model (13) is a general one that is nonlinear in the variable $u(t)$, because the coefficients y_ϕ, $\alpha_{i,t}$ and $\beta_{i,t}$ are functions of $\phi(t)$ whose elements contain $u(t)$. To solve this problem, we introduce an *extra variable* $x(t)$ and replace the variable $u(t)$ in $\phi(t)$ with an unknown nonlinear function $\tilde{\rho}(\xi(t))$, where

$$\xi(t) = [y(t) \ ... \ y(t - n_1 + 1) \ x(t+d) \ ...$$
$$x(t - n_3 + d + 1) \ u(t-1) \ ... \ u(t - n_2)]^T$$

including the extra variable $x(t+d)$ as an element. Then we have

$$y^o(t+d|t, \xi(t)) = y_\xi + \alpha(q^{-1}, \xi(t))y(t) + \beta(q^{-1}, \xi(t))u(t) \tag{18}$$

where y_ξ is y_ϕ whose variable $u(t)$ is replaced by $\tilde{\rho}(\cdot)$. Moreover, we typically choose $n_1 = n_y$, $n_2 = n_u + d - 2$, $n_3 = 1$, which gives

$$\xi(t) = [y(t) \ ... \ y(t - n_y + 1) \ x(t+d) \ u(t-1) \ ... \ u(t - d + 2)]^T.$$

Obviously, in a control system, the extra variable $x(t+d)$ can be replaced with the reference signal $y^*(t+d)$. Introducing

$$\Phi(t) = [1 \ y(t)...y(t - n_y + 1) \ u(t)...u(t - n_u - d + 2)]^T,$$
$$\Theta_\xi = [y_\xi \ \alpha_{0,t} \ ... \ \alpha_{n_y-1,t} \ \beta_{0,t} \ ... \ \beta_{n_u+d-2}]^T,$$

we finally express the ARX-like macro-model (18) by

$$y^o(t+d|t, \xi(t)) = \Phi^T(t)\Theta_\xi. \tag{19}$$

3.4 Incorporation of Neural Network

The macro-model (19) is not feasible at this stage because the elements of Θ_ξ are unknown nonlinear function of $\xi(t)$, which must be parameterized. In contrast with our previous work [10], in the new method we will parameterize

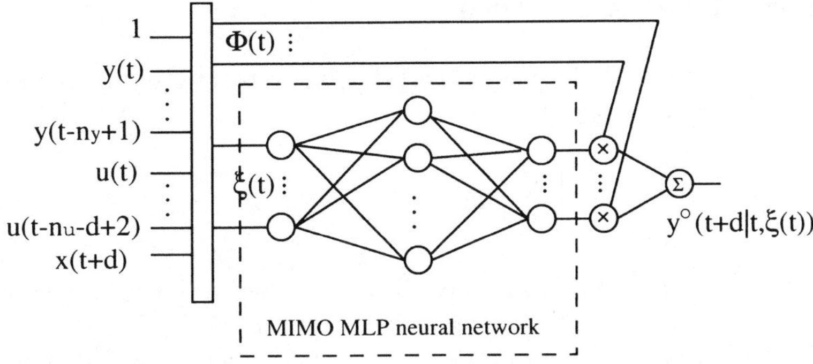

Fig. 2. The quasi-ARX prediction model incorporating neural network.

the elements of Θ_ξ using neural network. A significant advantage of using neural network is that it can be used to deal with higher dimensional problems.

Parameterizing Θ_ξ with a MIMO neural network, the quasi-ARX prediction model is expressed by

$$\mathcal{M} : y^o(t + d|t, \xi(t)) = \Phi^T(t)\mathcal{N}(\xi(t), \Omega) \tag{20}$$

where $\mathcal{N}(\cdot, \cdot)$ is a 3-layer neural network with n input nodes, M sigmoid hidden nodes and $n + 1$ *linear* output nodes[1]. Figure 2 shows the quasi-ARX prediction model incorporating neural network.

Let us express the 3-layer neural network by

$$\mathcal{N}(\xi(t), \Omega) = W_2\Gamma(W_1\xi(t) + B) + \theta \tag{21}$$

where $\Omega = \{W_1, W_2, B, \theta\}$, $W_1 \in \mathcal{R}^{M \times n}$, $W_2 \in \mathcal{R}^{(n+1) \times M}$ are the weight matrices of the first and second layers, $B \in \mathcal{R}^{M \times 1}$ is the bias vector of hidden nodes, $\theta \in \mathcal{R}^{(n+1) \times 1}$ is the bias vector of output nodes, and Γ is the diagonal nonlinear operator with identical sigmoid elements σ (i.e., $\sigma(x) = \frac{1-e^{-x}}{1+e^{-x}}$). Then the quasi-ARX prediction model (19) is expressed in a form of

$$\mathcal{M} : \quad y^o(t + d|t, \xi(t)) = \Phi^T(t) \cdot W_2\Gamma(W_1\xi(t) + B) + \Phi^T(t)\theta. \tag{22}$$

The quasi-ARX prediction model consists of two parts: the second term of the right side of (22) is a linear ARX prediction model part, while the first term is a nonlinear part. Therefore, in the quasi-ARX prediction model the bias of output nodes θ describes a linear approximation of the object system. This makes θ distinctive to other parameters. This feature allows us to use a dual loop leaning algorithm for the estimation.

---

[1] The number of input node is $n = \dim(\xi(t)) = n_y + n_u + d - 1$, the number of output node is equal to $\dim(\Phi(t)) = n + 1$.

4 Hierarchical Algorithm

Identification of the prediction model (22) can be performed by minimizing a criterion function defined by

$$E_N = \frac{1}{N} \sum_{t=1}^{N} [y(t+d) - y^o(t+d|t, \xi(t))]^2 \tag{23}$$

where $y(t)$ is the system output, and N the number of data. There are many estimation algorithms available for the identification [19, 11, 12]. However, it has been found that these methods are not so efficient for the quasi-ARX prediction model as for the conventional neural networks.

On the other hand, from (22) we can see that the quasi-ARX prediction model consists of two parts: linear part and nonlinear part, especially the parameter vector θ describes a linear approximation. Therefore, it is natural to introduce a dual-loop scheme where θ is estimated in a different way to other parameters.

1) Loop I: Estimation of θ

In Loop I, the parameter vectors W_1, W_2, and B are fixed and treated as constant vectors. The parameter vector θ can therefore be estimated as a parameter vector of linear model defined by

$$\mathcal{SM}1: \quad z_L(t) = \Phi^T(t)\theta \tag{24}$$

where $z_L(t)$ is the output of linear ARX submodel, and its real values are calculated by

$$z_L(t) = y(t+d) - \Phi^T(t)W_2\Gamma(W_1\xi(t) + B). \tag{25}$$

Since (24) is a linear model, the well-known recursive least squares (RLS) algorithm [13] can be applied for θ estimation.

2) Loop II: Estimation of W_1, W_2 and B

In Loop II, the parameter vector θ is fixed and treated as a constant vector. Then W_1, W_2 and B are estimated as parameter vectors of a nonlinear submodel, defined by

$$\mathcal{SM}2: \quad z_N(t) = \Phi^T(t)W_2\Gamma(W_1\xi(t) + B) \tag{26}$$

where $z_N(t)$ is the output of nonlinear submodel and its real values are calculated by

$$z_N(t) = y(t+d) - \Phi^T(t)\theta. \tag{27}$$

A backpropagation (BP) algorithm widely used in neural network community is applied to this estimation.

It follows from the above considerations that a hierarchical training algorithm consisting of Loop I and Loop II, and executing the two loops alternatively, can be described by the following four steps:

Step 1, set $\theta = 0$, and assign *small* initial values to W_1, W_2, and B;

Step 2, calculate $z_L(t)$ by (25), then estimate θ for the linear submodel $\mathcal{SM}1$ by using a RLS algorithm;

Step 3, calculate $z_N(t)$ by (27), then estimate W_1, W_2 and B for the nonlinear submodel $\mathcal{SM}2$. This is realized by using the well-known BP algorithm, *but the BP is only performed for a few epochs L in order to avoid overfitting*;

Step 4, stop if pre-specified conditions are met, otherwise go to *Step 2* and repeat the estimation of θ, and W_1, W_2, B.

In the above dual-loop estimation algorithm, *Step 2* is a RLS estimation where there is no problems such as local minimum or overfitting. And the change of θ on this step may be considered as noise to the followed estimation on *Step 3*. The noise may help BP algorithm to get out of local minima [17]. Furthermore, reducing the BP estimation step L in each iteration, the role of BP estimation is reduced. This is found to have a role to solve the overfitting problem to some extent. For small noisy data sets, the value of L should be small, see [8] for more details.

5 Controller Design

For a given system to be controlled, the controller design includes two steps: the step for identifying quasi-ARX prediction model, and the step for deriving and implementing control law.

5.1 Identifying Quasi-ARX Prediction Model

It is obviously that identifying the quasi-ARX prediction model for a given unknown nonlinear system can be implemented on-line similar to the case of linear adaptive control. However, if input-output data from the system are available, the identification can be implemented off-line. This is preferable from a viewpoint of convergence and stability.

Besides input variable $u(t)$, there is an extra input variable $x(t)$ in the quasi-ARX prediction model. If control system is designed for a specific known reference signal $y^*(t)$, then the reference signal $y^*(t)$ can be used as the extra input variable $x(t)$ in the identification, otherwise, a random signal may be used. When random signal is used, the identified quasi-ARX prediction can theoretically deal with any reference signal with reasonable control accuracy.

The identified quasi-ARX prediction model, expressed by

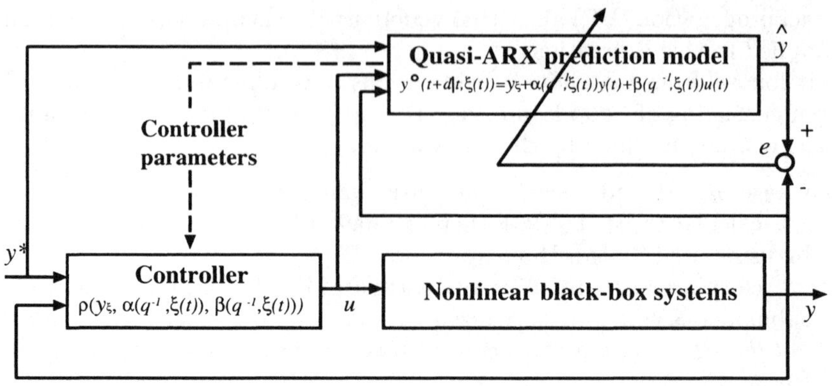

Fig. 3. Quasi-ARX model based control system for unknown nonlinear systems.

$$y^o(t+1|t, \xi(t)) = \Phi^T(t)\hat{\Theta}_\xi \tag{28}$$

where $\hat{\Theta}_\xi = [\hat{y}_\xi\, \hat{\alpha}_{0,t} \ldots \hat{\alpha}_{n_y-1,t}\, \hat{\beta}_{0,t} \ldots \hat{\beta}_{n_u+d-2,t}]^T$, will then be used for controller design.

5.2 Deriving Control Law

In the case where a conventional neural network is used as prediction model, a controller can not be derived directly from an identified model because of the nonlinearities. However, the quasi-ARX model is linear in the input variable $u(t)$. Taking this advantage, we may derive directly a controller by solving (5)

$$u(t) = \frac{\hat{\beta}_{0,t}}{\hat{\beta}_{0,t}^2 + \lambda} \left\{ \left[(\hat{\beta}_{0,t} - \hat{\beta}(q^{-1}, \xi(t)))q\right] u(t-1) \right.$$
$$\left. + y^*(t+1) - \hat{\alpha}(q^{-1}, \xi(t))y(t) - \hat{y}_\xi \right\}. \tag{29}$$

In the control law (29), as an element of $\xi(t)$, the extra input variable of the prediction model $x(t)$ is replaced by reference signal $y^*(t)$. Figure 3 shows the quasi-ARX model based control system for unknown nonlinear systems. A feature of the control system is that the prediction model and the controller share their parameters.

6 Numerical Simulations

6.1 Identification Examples

We first carry out a simulation to illustrate the effectiveness of the hierarchical estimation algorithm by applying the algorithm to identify a simulated nonlinear system described by

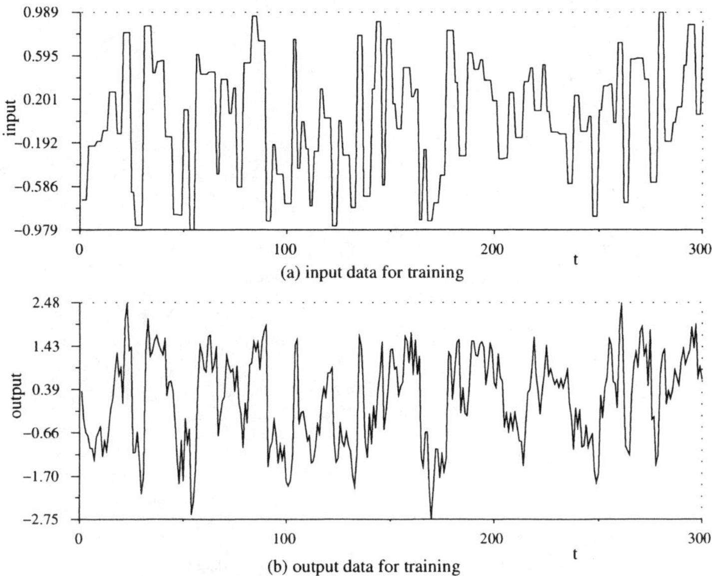

(a) input data for training

(b) output data for training

Fig. 4. Input-output data set for training.

$$y(t) = \frac{\exp(-y^2(t-2) * y(t-1))}{1 + u^2(t-3) + y^2(t-2)}$$

$$+ \frac{\exp(0.5 * (u^2(t-2) + y^2(t-3))) * y(t-2)}{1 + u^2(t-2) + y^2(t-1)}$$

$$+ \frac{\sin(u(t-1) * y(t-3)) * y(t-3)}{1 + u^2(t-1) + y^2(t-3)}$$

$$+ \frac{\sin(u(t-1) * y(t-2)) * y(t-4)}{1 + u^2(t-2) + y^2(t-2)} + u(t-1) + e(t) \tag{30}$$

where $e(t) \in (0, 0.1)$ is white Gaussian noise. We record 600 input-output data sets by exciting the system with a random input sequence. The first 300 data set is used as training data, while the second 300 data set is using as testing data. The training data set is depicted in Fig.4.

To identify the system, we use the following quasi-ARX model

$$y(t + d) = \Phi^T(t)W_2\Gamma(W_1\phi(t) + B) + \Phi^T(t)\theta \tag{31}$$

with $n_y = 4$, $n_u = 3$, $M = 8$, $d = 1$. The model is slight different to the one described by (22). It is nonlinear for the input variable $u(t)$ and do not contain the extra input $x(t)$. The model contains 127 parameters to be adjusted. First, we trained the model by using a fast BP algorithm modified from "trainbpx.m" for MATLAB Neural Network Toolbox. We found that a BP algorithm has much higher probability to be stuck at a local minimum than the case used for

361

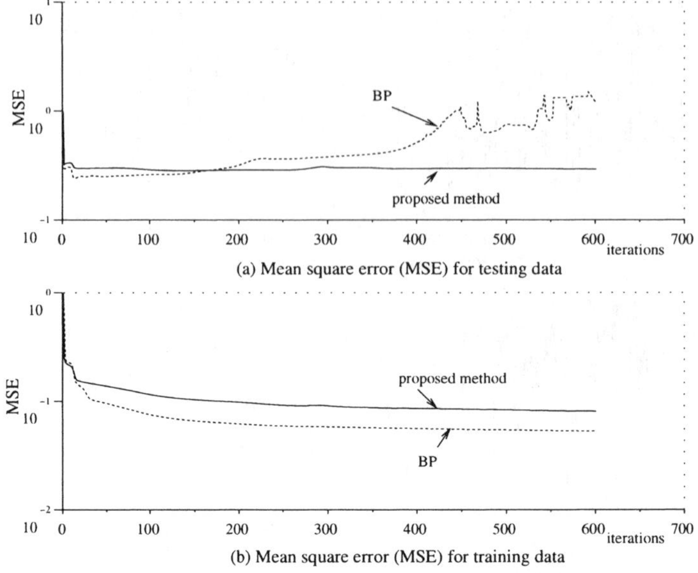

Fig. 5. Mean squares error. (a) for testing data; (b) for training data. The dashed lines show the results of BP algorithm, and solid line the results of the proposed hierarchical algorithm.

a conventional neural network; a mechanism for randomly resetting learning rate is introduced to overcome the difficulty. On the other hand, when using the proposed hierarchical algorithm, no local minimum problem appeared in our simulations. This means that the proposed hierarchical training algorithm can more and less solve local minimum problem because of the dual-loop structure.

Figure 5 shows the mean square errors of the BP (dashed lines), and the proposed algorithm (solid lines) for (a) testing data, and (b) training data. We can see that using the BP has overfitting problem although it has smaller training error, while this overfitting problem is solved by using the proposed hierarchical algorithm, where comparative small L ($L = 40$) was used because the data sets are rather short and noisy.

To show how well the system has been identified, we excite the system and the identified model using an input sequence described by

$$u(t) = \begin{cases} \sin(\frac{2\pi t}{250}) & t \leq 500 \\ 0.8\sin(\frac{2\pi t}{250}) + 0.2\sin(\frac{2\pi t}{25}) & t > 500 \end{cases}$$

in noise free case. Figure 6 shows the simulation of the model trained using the proposed hierarchical algorithm (dashed line). The solid line shows the

362

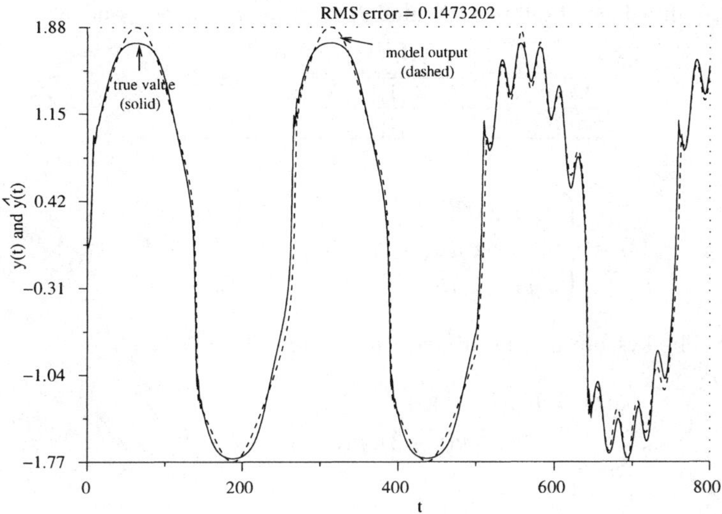

Fig. 6. Simulation of the model on testing data in the case where the model is training using the proposed algorithm.

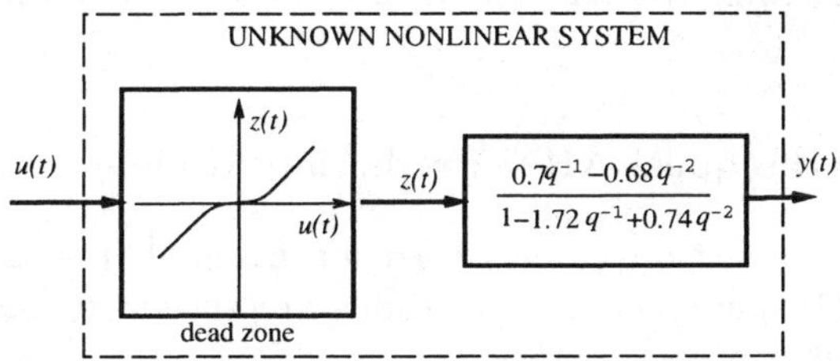

Fig. 7. A unknown nonlinear system with dead zone in the input.

system output. We can see that the model represents the system quite well, where RMS (root mean square) error is 0.1473202.

6.2 Control Examples

Example 1: The system to be controlled is assumed to have a dead zone, shown in Fig.7. The linear part of the system is described by

$$G(q^{-1}) = \frac{0.7q^{-1} - 0.68q^{-2}}{1 - 1.72q^{-1} + 0.74q^{-2}}, \tag{32}$$

while the nonlinear element is a dead zone described by

363

Table 1. Specifications of quasi-ARX models used for *Example 1* and *Example 2*

	n_y	n_u	d	M	MLP	NoP
Example 1	2	2	1	15	$N_{4\times15\times5}$	155
Example 2	3	2	1	20	$N_{5\times20\times6}$	246

$$
z(t) = \begin{cases}
u(t) - 0.75 & \text{if } u(t) > 1; \\
0.25\,\mathrm{sign}(u(t)) \times u^2(t) & \text{if } |u(t)| \le 1; \\
u(t) + 0.75 & \text{if } u(t) < -1
\end{cases}
$$

Example 2: The system is a nonlinear one governed by

$$
y(t) = f[y(t-1), y(t-2),
$$
$$
y(t-3), u(t-1), u(t-2)] \tag{33}
$$

where

$$
f[x_1, x_2, x_3, x_4, x_5] = \frac{x_1 x_2 x_3 x_5 (x_3 - 1) + x_4}{1 + x_2^2 + x_3^2}.
$$

The above two systems are treated as unknown nonlinear ones. No prior information concerning system nonlinearity is used in the following simulations.

1) Identifying the quasi-ARX Prediction Models

The quasi-ARX prediction models used for *Example 1* and *Example 2* are described by (22), and their specifications are shown in Tab.1, where NoP denotes the number of parameters.

To identify the prediction models, from each system we record 1000 input-output data sets by exciting the system with a kind of random input sequence. Figure 8 shows the first 300 data sets taken from the *Example 1*. Note that there is an extra input variable in the models, which is corresponding to the reference signal in a control system. A kind of random sequence shown in Fig.8(c) is used for this input variable in the simulations.

The hierarchical algorithm described in Section 4 is used to estimate model parameters. For each example, 200 iterations are carried out, where BP estimation step L is 500. When a noisy input-output data set is used for estimation, one should decrease the value of L and increase the total iterations. Figure 9 shows the RMS error for the estimations; solid line is the result of *Example 1*, and dashed line the result of *Example 2*.

2) Controlling the Systems

The identified quasi-ARX prediction models are then applied to control the systems via the control law described by (29), where λ is chosen to be 0.001.

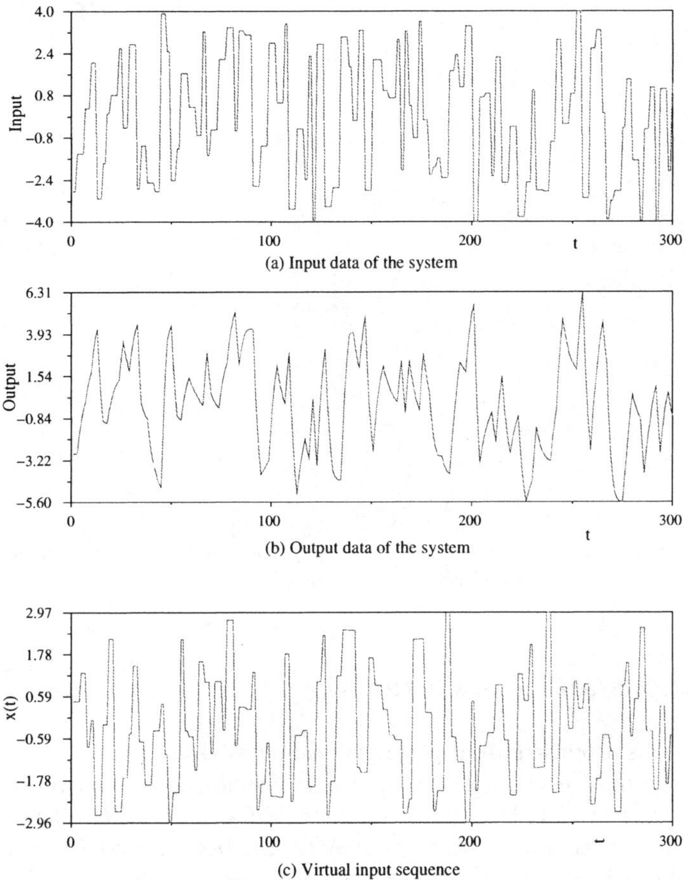

Fig. 8. Data for identification, (a) and (b) Input-output data taken from the system; (c) A kind of random input sequence used for the extra input variable

The two systems in *Example 1* and *Example 2* are rather nonlinear. Linear ARX prediction model has been found to be not able to control the systems well. Because of page limitation, we only show the results using the quasi-ARX prediction model. Figure 10 and 11 show the control results for *Example 1* and *Example 2* by using the proposed quasi-ARX prediction model, respectively. Although the reference signals have suddenly changes at $t = 100$ and $t = 150$, the controlled systems can track the reference signal quite well; the RMS control errors are 0.158 and 0.050, respectively.

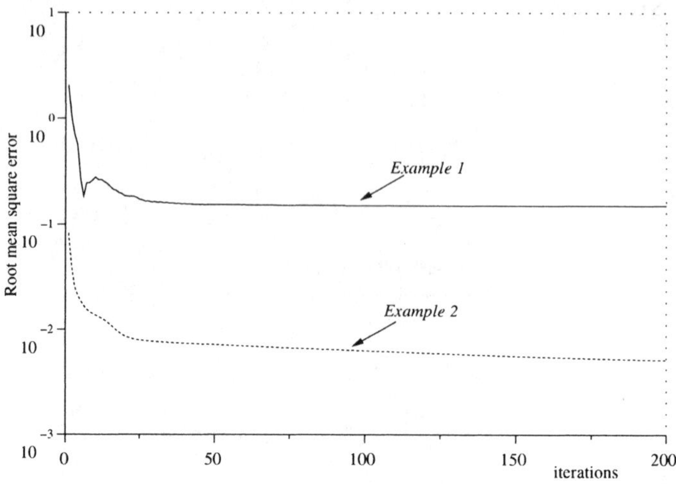

Fig. 9. Root mean square errors for estimation of the quasi-ARX prediction models for *Example1* (solid line) and *Example 2* (dashed line).

7 Discussions and Conclusions

A new method is proposed for applying neural networks to control of nonlinear systems. In the new method, neural networks are not used directly as prediction model or controller, but indirectly used via an ARX-like macro-model. This is realized by constructing a quasi-ARX prediction model linear in the input variables, where neural network is incorporated. A distinctive feature of the new method is that the neural network used in prediction model and the one used in controller share the same parameters. This allows one only needs to identify the prediction model to design a control system. The identification can be done off-line, which is usually easier from a convergence and stability point of view. A dual loop algorithm has been introduced for this off-line identification.

Obviously, after implementing the control system based on the identified quasi-ARX prediction model, an adaptive control algorithm may be applied to adjust the parameters of prediction model in order to increase control accuracy or to deal with a time-variant system. However, a robust adaptive algorithm is crucial for such applications. Further studies are needed.

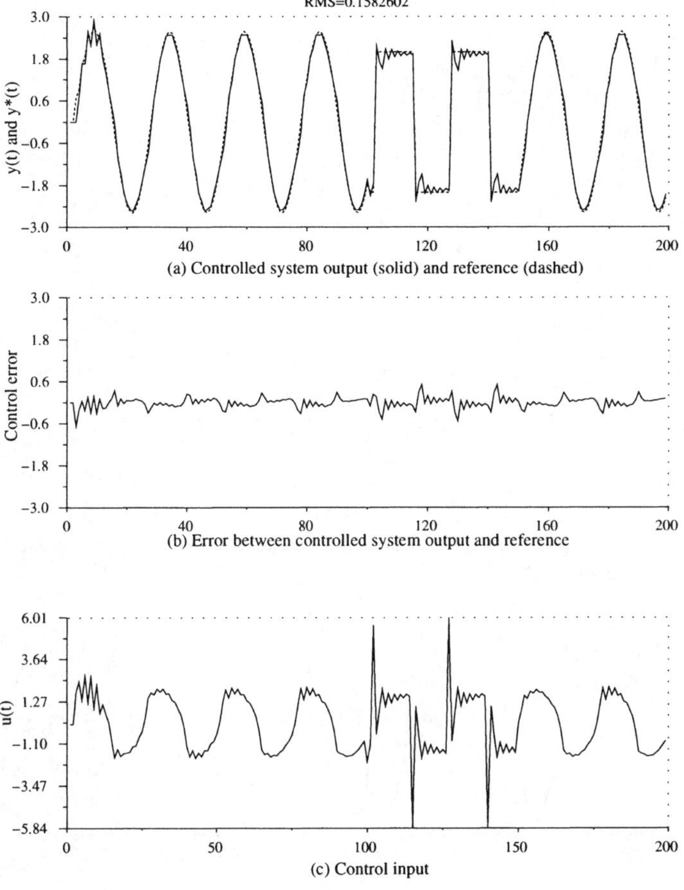

Fig. 10. Control results for *Example 1*.

References

1. J.M. Benitez, J.L. Castro, and I. Requena. Are artificial neural networks black boxes? *IEEE Trans on Neural Networks*, 8(5):1156–1164, 1997.
2. D.M. Burton. *Introduction to Modern Abstract Algebra*. Addison-Wesley, Reading, Mass, 1967.
3. C.L. Chen and W.C. Chen. Self-organizing neural control system design for dynamic processes. *INT. J. SYSTEMS SCI.*, 24(8):1487–1507, 1993.
4. F.C. Chen and H.K. Khalil. Adaptive control of a class of nonlinear discrete time systems using neural networks. *IEEE Trans. on Automatic Control*, 40(5):791–801, 1995.
5. S. Chen and S.A. Billings. Neural networks for nonlinear dynamic system modeling and identification. *INT. J. Control*, 56(2):319–346, 1992.
6. G.C. Goodwin and K.S. Sin. *Adaptive Filtering Prediction and Control*. Prentice-Hall, Inc., 1984.

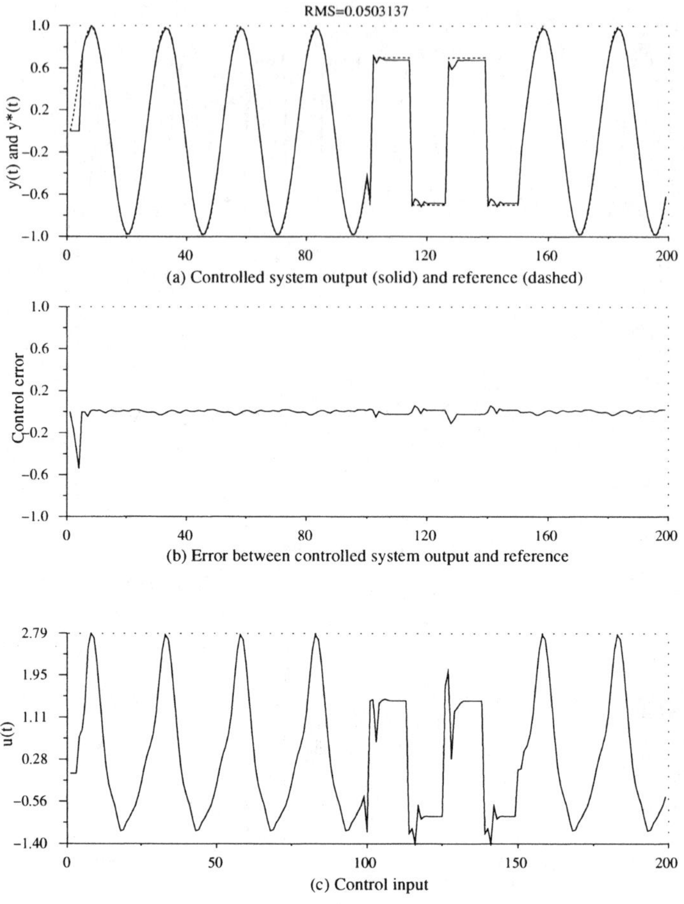

Fig. 11. Control results for *Example 2*.

7. K. Hornik. Multilayer feedforward networks are universal approximators. *Neural Networks*, 2:359–366, 1989.

8. J. Hu and K. Hirasawa. A hierarchical method for training embedded sigmoidal neural networks. In *Lecture Notes in Computer Science 2130 (Proc. of ICANN'01)*, pages 937–942, 8 2001.

9. J. Hu, K. Kumamaru, and K. Hirasawa. Quasi-ARMAX modeling approaches to identification and prediction of nonlinear systems. In *Proc. of the 12th IFAC Symp. on Identification (Santa Barbara)*, 6 2000. (CDROM: FrMD3-2).

10. J. Hu, K. Kumamaru, and K. Hirasawa. A quasi-ARMAX approach to modeling of nonlinear systems. *International Journal of Control*, 74(18):1754–1766, 12 2001.

11. R.A. Johansen. On Tikhonov regularization, bias and variance in nonlinear system identification. *Automatica*, 33:441–446, 1997.

12. L. Ljung. *System Identification: Theory for the User, Second Edition.* Prentice-Hall PTR, Upper Saddle River, N.J., 1999.

13. L. Ljung and T. Söderström. *Theory and Practice of Recursive Identification.* MIT press, Cambridge, Mass, 1983.

14. W.T. MillerIII, R.S. Sutton, and P.J. Werbos. *Neural Networks for Control.* The MIT Press, Massachusetts, 1990.

15. K.S. Narendra and S. Mukhopadhyay. Adaptive control using neural networks and approximate models. *IEEE Trans. on Neural Networks*, 8(3):475–485, 1997.

16. K.S. Narendra and K. Parthasarathy. Identification and control of dynamical systems using neural networks. *IEEE Trans. on Neural Networks*, 1(1):4–27, 1990.

17. Oliver Nelles. *Nonlinear System Identification.* Springer, Berlin, 2001.

18. M. Norgaard, O. Ravn, N.K. Poulsen, and L.K. Hansen. *Neural Networks for Modeling and Control of Dynamic Systems.* Springer, London, 2000.

19. J. Sjöberg and L. Ljung. Overtraining, regularization and searching for minimum with application to neural networks. *INT. J. Control*, 62(6):1391–1407, 1995.

20. J. Sjöberg, Q. Zhang, L. Ljung, A. Benveniste, B. Deglon, P.Y. Glorennec, H. Hjalmarsson, and A. Juditsky. Nonlinear black-box modeling in system identification: a unified overview. *Automatica*, 31(12):1691–1724, 1995.

21. Y. Zhang and J. Wang. Recurrent neural networks for nonlinear output regulation. *Automatica*, 37(8):1161–1173, 2001.

Robot Manipulator Control via Recurrent Neural Networks

Luis J. Ricalde[1], Edgar N. Sanchez[1], and Jose P. Perez[2]

[1] CINVESTAV, Unidad Guadalajara,
   Apartado Postal 31-438, Plaza La Luna,
   Guadalajara, Jalisco C. P. 45091, Mexico,
   e-mail: lricalde@gdl.cinvestav.mx, sanchez@gdl.cinvestav.mx
[2] School of Physics and Mathematics,
   Universidad Autonoma de Nuevo Leon (UANL),
   Monterrey, Mexico

Abstract. This chapter presents the application of neural networks to robot manipulator control. The main methodologies, on which the approach is based, are recurrent neural networks and the recent introduced technique of inverse optimal control for nonlinear systems. The proposed controller structure is composed of a neural identifier and a control law defined by using the inverse optimal control approach. The proposed new control scheme is applied via simulations to control a robot manipulator model where friction terms are included.

Keywords: Neural networks, Trajectory tracking, Adaptive control, Lyapunov function, Stability analysis.

1.1   Introduction

Artificial neural networks, computational models of the brain, are widely used on engineering applications due to their ability to estimate the relation between inputs and outputs from a learning process. Motivated by the seminal paper [12], there exists a continuously increasing interest in applying neural networks to identification and control of nonlinear systems. Most of these applications use feedforward structures [3], [6]. Recently, recurrent neural networks are being developed; as an extension of the static neural networks capability to aproximate nonlinear functions, recurrent neural networks can aproximate nonlinear systems. They allow more efficient modeling of the underlying dynamical systems [15]. Three representative books [23], [16] and [14] have reviewed the application of recurrent neural networks for nonlinear system identification and control. In particular, [23] uses off-line learning, while [16] analyzes adaptive identification and control by means of on-line learning, where stability of the closed-loop system is established based on the Lyapunov function method. In [16], the trajectory tracking problem is reduced to a linear model following problem, with application to DC electric motors. In [14], analisys of Recurrent Neural Networks for identification, estimation and control are developed, with applications on chaos control, robotics and chemical processes.

Control methods that are applicable to general nonlinear systems have been intensely developed since the early 1980's. Main approaches include, for example, the use of differential geometry theory [8]. Recently, the passivity approach has generated increasing interest for synthesizing control laws [4]. An important problem for these approaches is how to achieve robust nonlinear control in the presence of unmodelled dynamics and external disturbances. In this direction, there exists the so-called H_∞ nonlinear control approach [1]. One major difficulty with this approach, alongside its possible system structural instability, seems to be the requirement of solving some resulting partial differential equations. In order to alleviate this computational problem, the so-called inverse optimal control technique was recently developed, based on the input-to-state stability concept [10].

On the basis of the inverse optimal control approach, a control law for generating chaos in a recurrent neural network was designed in [17]. In [18] and [19], this methodology was modified for stabilization and trajectory tracking of an unknown chaotic dynamical system. The proposed adaptive control scheme is composed of a recurrent neural identifier and a controller, where the former is used to build an on-line model for the unknown plant and the latter, to ensure that the unknown plant tracks the reference trajectory. In this chapter, we further improve the design by adequating it to systems with less inputs than states. The approach is based on the methodology developed in [18] and [19], in which the control law is optimal with respect to a well-defined cost functional.

Robot manipulators present a practical challenge for control purposes due to the nonlinear and multivariable nature of their dynamical behaviour. Motion control in joint space is the most fundamental task in robot control; it has motivated extensive research work in synthetizing different control methods such as fuzzy computed torque control [11], PI+PD fuzzy control [21] and static neural network control [2]. An important problem for developing control algorithms is that most robot models neglect practical aspects such as actuator dynamics, sensor noise, and friction. which if are not considered in the design may cause performance deterioration. In this chapter, we apply the approach developed in [20] for robot manipulator control including friction terms.

1.2　Recurrent High Order Neural Networks

Artificial neural networks have become an useful tool for control engineering thanks to their applicability on modelling, state estimation and control of complex dynamic systems. Using neural networks, control algorithms can be developed to be robust to noise, uncertainties and modelling errors.

Neural Network consists of a number of interconnected processing elements or neurons. The way in wich the neurons are interconnected determines its structure. For identification and control, the most used structures are:

Feedforward networks. In feedforward networks, the neurons are grouped into layers. Signals flow from the input to the output via unidirectional connections. The network exhibits high degree of conectivity, contains one or more hidden layers of neurons and the activation function of each network is smooth, generally a sigmoidal function.

Recurrent networks. In a recurrent neural network, the outputs of the neuron are fedback to the same neuron or neurons in the preceding layers. Signals flow in forward and backward directions.

Artificial Recurrent Neural Networks are mostly based on the Hopfield model [5]. These networks are considered as good candidates for nonlinear systems applications which deal with uncertainties and are attractive due to their easy implementation, relatively simple structure, robustness and the capacity to adjust their parameters on line.

In [9], high-order recurrent neural networks (RHONN) are defined as

$$\dot{x}_i = -a_i x_i + \sum_{k=1}^{L} w_{ik} \prod_{j \in I_k} y_j^{d_j(k)}, i = 1, ..., n \tag{1.1}$$

where x_i is the ith neuron state, L is the number of high-order connections, $\{I_1, I_2, ..., I_L\}$ is a collection of non-ordered subsets of $\{1, 2, ..., m+n\}$, $a_i > 0$, w_{ik} are the adjustable weights of the neural network, $d_j(k)$ are nonnegative integers, and y is a vector

$$\begin{aligned} y &= \left[y_1, .., y_n, y_{n+1}, ..., y_{n+m} \right]^\mathsf{T} \\ &= [S(x_1), ..., S(x_n), S(u_1), ..., S(u_m)]^\mathsf{T} \end{aligned} \tag{1.2}$$

with $u = [u_1, u_2, ..., u_m]^\mathsf{T}$ being the input to the neural network, and $S(\cdot)$ is a smooth sigmoid function formulated by $S(x) = \frac{1}{1+\exp(-\beta x)} + \varepsilon$, where β is a positive constant and ε is a small positive real number.

Clearly, system (1.1) allows the inclusion of high-order terms. By defining a new vector,

$$z(x, u) = [z_1(x, u), ..., z_L(x, u)]^\mathsf{T} = \left[\prod_{j \in I_1} y_j^{d_j(1)}, ..., \prod_{j \in I_L} y_j^{d_j(L)} \right]^\mathsf{T} \tag{1.3}$$

system (1.1) can be rewritten as

$$\dot{x}_i = -a_i x_i + \sum_{k=1}^{L} w_{ik} z_k(x, u) \tag{1.4}$$

$$\dot{x}_i = -a_i x_i + w_i^\mathsf{T} z_i(x, u) \qquad i = 1, ..., n$$

where $w_i = [w_{i,1}...w_{i,L}]^\mathsf{T}$.

In this work, we assume that the input enters the neural network directly. We are particularly interested in cases where the inputs are less than the states.

For this purpose, we further rewrite system (1.1) as

$$\dot{x}_i = -a_i x_i + w_i^\top z_i(x) + \Omega u_i, \ i = 1, ..., n \quad \Omega \in R^{nxm} \tag{1.5}$$

Then, we reformulate (1.4) in a matrix form and obtain

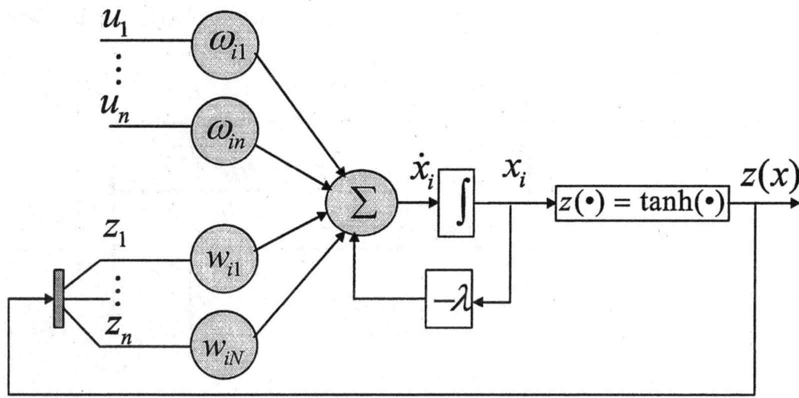

Fig. 1.1. Recurrent neural network scheme

$$\dot{x} = Ax + Wz(x) + \Omega u \tag{1.6}$$

where $x \in \Re^n$, $W^* \in \Re^{n \times L}$, $z(x) \in \Re^L$, $u \in \Re^n$, and $A = -\lambda I$, with $\lambda > 0$.
A descriptive diagram of a recurrent neural network is seen in Fig. 1.1 .

In the following, we use a slight modification for the RHONN given as

$$\dot{x} = Ax + W\Gamma z(x) + \Omega u \tag{1.7}$$

$$\Gamma = \begin{pmatrix} \gamma_1 & 0 & \cdot & \cdot & & 0 \\ 0 & \gamma_2 & \cdot & \cdot & \cdot & 0 \\ \cdot & \cdot & \cdot & \cdot & \cdot & \cdot \\ \cdot & \cdot & \cdot & \cdot & \cdot & \cdot \\ \cdot & \cdot & \cdot & \cdot & \gamma_{L-1} & 0 \\ 0 & 0 & \cdot & \cdot & 0 & \gamma_L \end{pmatrix}$$

$$\gamma_i > 0, \quad i = 1....L$$

Note that these γ parameters play the role of the step for the gradient-descent
algorithm defined below.

1.3 Inverse Optimal Control

Many control applications have to deal with nonlinear processes in presence
of uncertainties and disturbances. All these phenomena must be considered
for controller design in order to obtain the desired closed loop performance.

Under the assumption that all the states are available for measurement, we develop an adaptive neural control for uncertain disturbed nonlinear systems in order to track a trajectory. A recurrent neural network is used in order to modelize the unknown plant. The weight adaptation law and the robust feedback controller are obtained using Control Lyapunov Functions [9]. The controller is shown to be optimal with respect to a meaningful cost functional via the inverse optimal control approach[10]. The control scheme is displayed in Fig. 1.2.

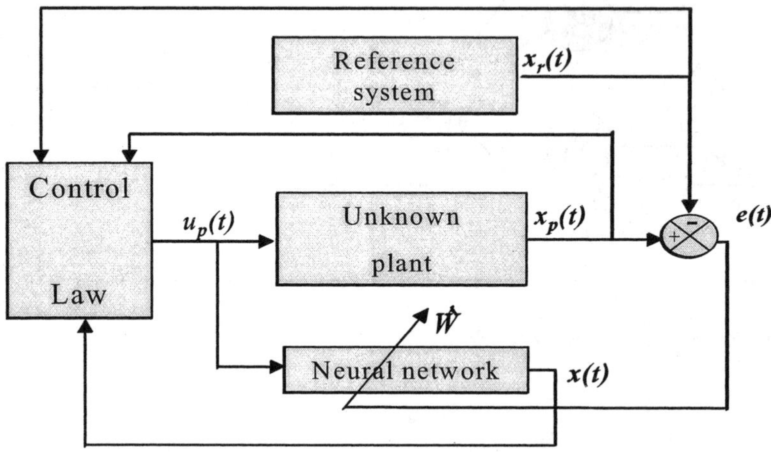

Fig. 1.2. Recurrent neural control scheme

This section closely follows [10], and [22]. As stated in [22], optimal stabilization guarantees several desirable properties for the closed loop system, including stability margins. In a direct approach, we would have to solve the Hamilton-Jacobi-Bellman (HJB) equation which is not an easy task. Besides, the robustness achieved is largely independent of the particular choice of functions $l(x) > 0$ and $R(x) > 0$. This motivated to pursue the development of design methods which solve the inverse problem of optimal stabilization. In the inverse approach, a stabilizing feedback is designed first and then shown to optimize a cost functional of the form

$$J = \int_0^\infty \left(l(x) + u^T R(x)u \right) dt \tag{1.8}$$

The problem is inverse because the functions $l(x)$ and $R(x)$ are a posteriori determined by the stabilizing feedback, rather than a priori chosen by the design.

374

A stabilizing control law $u(x)$ solves an inverse optimal problem for the system

$$\dot{x} = f(x) + g(x)u \tag{1.9}$$

if it can be expressed as

$$u = -k(x) = -\frac{1}{2}R^{-1}(x)\left(L_g V(x)\right)^\top, \qquad R(x) = R^\top(x) > 0 \tag{1.10}$$

where $V(x)$ is a positive semidefinite function, such that the negative semi-definiteness of $\dot{V}$ is achieved with the control $u = -\frac{1}{2}k(x)$. That is

$$\dot{V} = L_f V(x) - \frac{1}{2}L_g V(x)k(x) \leq 0 \tag{1.11}$$

When the function $-l(x)$ is set to be the right hand side of

$$l(x) = -L_f V(x) + \frac{1}{2}L_g V(x)k(x) \geq 0 \tag{1.12}$$

then $V(x)$ is a solution of the HJB equation

$$l(x) + L_f V(x) - \frac{1}{4}\left(L_g V(x)\right)R^{-1}(x)\left(L_g V(x)\right)^T = 0 \tag{1.13}$$

The inverse optimal control approach is based on the concept of equivalence between the input to state stability (ISS) and the solution of the H_∞ non-linear control problem [10]. Using Lyapunov control functions (CLF's) a stabilizing feedback is designed first and then shown to be optimal with respect to a cost functional that imposes penalties on the error and the control input. This approach provides robust controllers where robustness is obtained as a result of the control law optimality, which is independent of the cost functional [22].

1.4   Trajectory Tracking Analysis

The unknown nonlinear plant is defined as

$$x_p = f_p(x_p) + g_p(x_p)u \tag{1.14}$$

We propose to modelize this unknown nonlinear plant by the recurrent neural network,

$$\dot{x}_p = \dot{x} + w_{per} \tag{1.15}$$
$$= Ax + W^*\Gamma z(x) + (x - x_p) + \Omega u$$

where $x_p, x, z(x) \in \Re^n$, $W^*, \Gamma \in \Re^{n \times n}$, $u \in \Re^{mx1}$, $\Omega \in R^{nxm}$, and $w_{per} = x - x_p$ represents the modelling error, with W^* being the unknown values of the neural network which minimize the modelling error.

The reference trajectory is given by

$$\dot{x}_r = f_r(x_r, u_r), \quad x_r \in \Re^n \tag{1.16}$$

and we define the tracking error as

$$e := x_p - x_r \tag{1.17}$$

Its time derivative is

$$\dot{e} = Ax + W^* \Gamma z(x) + w_{per} + \Omega u - f_r(x_r, u_r) \tag{1.18}$$

Now, we proceed to add and subtract the terms $\hat{W}\Gamma z(x_r)$, Ae, and $\Omega\alpha_r(t, \hat{W})$, so that

$$\dot{e} = Ae + W^* \Gamma z(x) + \Omega u +$$
$$\left(-f_r(x_r, u_r) + Ax_r + \hat{W}\Gamma z(x_r) + x_r - x_p + \Omega\alpha_r(t, \hat{W})\right) \tag{1.19}$$
$$-Ae - \hat{W}\Gamma z(x_r) - \Omega\alpha_r(t, \hat{W}) - Ax_r - x_r + x + Ax$$

Note that the plant will track the reference signal, even in the presence of uncertainties, if there exists a function $\alpha_r(t, \hat{W})$ such that

$$\alpha_r(t, \hat{W}) = \left(\Omega^T \Omega\right)^{-1} \Omega^T (f_r(x_r, u_r) - Ax_r - \hat{W}\Gamma z(x_r) - (x_r - x_p)) \tag{1.20}$$

Next, assume that (1.20) holds, and let define

$$\tilde{W} = W^* - \hat{W} \tag{1.21}$$
$$\tilde{u} = u - \alpha_r(t, \hat{W}), \tag{1.22}$$

so that (1.19) is reduced to

$$\dot{e} = Ae + \tilde{W}\Gamma z(x) + \hat{W}\Gamma \left(z(x) - z(x_r)\right) + (A + I)(x - x_r) - Ae + \Omega\tilde{u}$$
$$= Ae + \tilde{W}\Gamma z(x) + \hat{W}\Gamma(z(x) - z(x_p) + z(x_p) - z(x_r)) \tag{1.23}$$
$$+ (A + I)(x - x_p + x_p - x_r) - Ae + \Omega\tilde{u}$$

Then, by defining

$$\tilde{u} = u_1 + u_2 \tag{1.24}$$

with

$$u_1 = \left(\Omega^T \Omega\right)^{-1} \Omega^T (-\hat{W}\Gamma(z(x) - z(x_p)) - (A + I)(x - x_p)) \tag{1.25}$$

equation (1.23) reduces to

$$\dot{e} = Ae + \tilde{W}\Gamma z(x) + \hat{W}\Gamma z(x_p) - z(x_r) + (A + I)(x_p - x_r) - Ae + \Omega u_2 \tag{1.26}$$

Moreover, by taking into account that $e = x_p - x_r$, equation (1.26) can be rewritten as the following equation:

$$\dot{e} = (A + I)e + \tilde{W}\Gamma z(x) + \hat{W}\Gamma(z(x_p) - z(x_r)) + \Omega u_2 \tag{1.27}$$

Therefore, the tracking problem reduces to a stabilization problem for the error dynamics (1.27). To solve this problem, we next apply the inverse optimal control approach.

1.4.1 Tracking Error Stabilization

Once (1.27) is obtained, we consider its stabilization. Note that $(e, \tilde{W}) = (0,0)$ is an equilibrium point of the undisturbed autonomous system. For stability analysis, we define the candidate Lyapunov function as

$$V = \frac{1}{2}\|e\|^2 + \frac{1}{2}tr\left\{\tilde{W}^\top \tilde{W}\right\} \tag{1.28}$$

Its time derivative, along the trayectories of (1.27), is

$$\dot{V} = e^\top(A+I)e + e^T\tilde{W}\Gamma z(x) + e^\top \Omega u_2$$
$$+ e^\top \hat{W}\Gamma\left(z(x_p) - z(x_r)\right) + tr\left\{\dot{\tilde{W}}^\top \tilde{W}\right\} \tag{1.29}$$

As in [16], we propose the following learning law:

$$tr\left\{\dot{\tilde{W}}^\top \tilde{W}\right\} = -e^\top \Gamma \tilde{W} z(x) \tag{1.30}$$

Then, we substitute (1.30) into (1.29), to obtain

$$\dot{V} = -(\lambda - 1)e^\top e + e^\top \Gamma \hat{W}\phi_z(e, x_r) + e^\top \Omega u_2 \tag{1.31}$$

where

$$\phi_z(e, x_r) = z(x_p) - z(x_r) = z(e + x_r) - z(x_r)$$

Next, we consider the following inequality

$$X^\top Y + Y^\top X \leq X^\top \Lambda X + Y^\top \Lambda^{-1} Y \tag{1.32}$$

which holds for all matrices $X, Y \in \Re^{n x k}$ and $\Lambda \in \Re^{n x n}$ with $\Lambda = \Lambda^\top > 0$ [15].

Applying (1.32) to $e^T \Gamma W\phi(e, x_r)$ with $\Lambda = I$, we obtain

$$\dot{V} \leq -(\lambda - 1)e^\top e + \frac{1}{2}e^\top e + \frac{1}{2}\left(\phi_z(e, x_r)\right)^\top \Gamma\left(\hat{W}^\top \hat{W}\right)\Gamma\phi_z(e, x_r) + e^\top u_2$$

$$\dot{V} \leq -(\lambda - 1)e^\top e + \frac{1}{2}e^\top e + \frac{1}{2}\|\hat{W}\|^2 \|\Gamma\|^2 \|\phi_z(e, x_r)\|^2 + e^\top u_2 \tag{1.33}$$

where $\|\hat{W}\|$, $\|\Gamma\|$ are any matrix norm for $\hat{W}$ and Γ.

Since $\phi_z(e, x_r)$ is Lipschitz with respect to e, then, there exists a positive constant L_ϕ such that

$$\phi_z(e, x_r)^\top \leq L_\phi \|e\| \tag{1.34}$$

Hence (1.33) can be rewritten as

$$\dot{V} \leq -(\lambda - 1)e^T e + e^T(\frac{1}{2} + \frac{1}{2}L_{\phi_z}^2 \|\hat{W}\|^2 \|\Gamma\|^2)e + e^\top \Omega u_2 \tag{1.35}$$

To this end, we define the following control law:

$$u_2 = -\mu\ \Omega_s(1 + L_{\phi_z}^2 \|\hat{W}\|^2 \|\Gamma\|^2)e \tag{1.36}$$

$$\triangleq -\beta\left(R(e,\hat{W})\right)^{-1}(L_g V)^{\top}\ ,\quad \mu > \frac{1}{2}$$

with scalars $R(e,\hat{W}) > 0, \beta > 0$.

Substituting (1.36) in (1.35) yields

$$\dot{V} = -((\lambda - 1) - (\mu - \frac{1}{2})(1 + L_{\phi_z}^2 \|\hat{W}\|^2 \|\Gamma\|^2)) \|e\|^2 \tag{1.37}$$

$$< 0 \qquad\qquad \forall e, \hat{W} \neq 0$$

Now, let us consider the following lemma

Lemma 1. *[7] For scalar valued functions,*

i) A function f(t) which is bounded by below and not increasing has a limit when $t \to \infty$.

ii) Consider the non negative scalar functions f(t), g(t) defined for all $t \geq 0$. If $f(t) \leq g(t)$, $\forall t \geq 0$ y $g(t) \in L_p$ then $f(t) \in L_p$ for all $p \in [1, \infty]$.

If $e = 0$, and $\hat{W} \neq 0$, our tracking goal is achieved. Then, we proceed to prove the boundedness of the on-line weights.

Since $\dot{V}$ is a negative semidefined function, not increasing and bounded by below, from Lemma 1 we have

$$\lim_{t \to \infty} V \to V_\infty \tag{1.38}$$

Hence, V_∞ exists and it is bounded, then we have

$$\lim_{t \to \infty} \hat{W} \to \hat{W}_\infty, \tag{1.39}$$

Hence $\hat{W}_\infty$ exists and it is bounded

From (1.37), we conclude that for all $e, \hat{W} \neq 0$,

$$\lim_{t \to \infty} e(t) = 0. \tag{1.40}$$

Finally, the control law, which affects the plant and the neural netwok, is given by

$$u = \tilde{u} + \alpha_r(t, \hat{W}) \tag{1.41}$$

$$= \Omega_s(-\hat{W}\Gamma(z(x) - z(x_p)) - (A + I)(x - x_p)$$

$$-\mu\left(1 + L_{\phi_z}^2 \|\hat{W}\|^2 \|\Gamma\|^2\right)e + f_r(x_r, u_r) - Ax_r - \hat{W}\Gamma z(x_r) - x_r + x_p)$$

This control law guarantees asymptotic stability of the error dynamics and therefore ensures tracking to the reference signal.

1.4.2 Optimization with Respect to a Cost Functional

Once the problem of finding the control control law which stabilizes (1.27) is solved, we proceed to formulate a cost functional defined by

$$J(\tilde{u}) = \lim_{t \to \infty} \left\{ 2\beta V + \int_0^t \left(l(e, \hat{W}) + u_2^\top R(e, \hat{W}) u_2 \right) d\tau \right\} \tag{1.42}$$

where the Lyapunov function solves the Hamilton-Jacobi-Bellman family of partial derivative equations parametrized with $\beta > 0$ as

$$l(e, \hat{W}) + 2\beta L_f V - \beta^2 L_g V R(e, \hat{W})^{-1} L_g V^\top = 0 \tag{1.43}$$

Note that $2\beta V$ in (1.42) is bounded when $t \to \infty$, since by (1.37) V is decreasing and bounded from below by $V(0)$. Therefore, $\lim\limits_{t \to \infty} V(t)$ exists and is finite.

Recall that in [10], we need $l(e, \hat{W})$ to be positive define and radially unbounded with respect to e. Here, from (1.43) we have

$$l(e, \hat{W}) = -2\beta L_f V + \beta^2 L_g V R(e, \hat{W})^{-1} L_g V^\top \tag{1.44}$$

Substituting (1.36) into (1.44) and then applying (1.32) to the second term on the right side of $L_f V$, we have

$$l(e, \hat{W}) \geq (\lambda - 1)\|e\|^2 + (\mu - 1)\left(1 + L_\phi^2 \|\Gamma\|^2 \|W\|^2\right) \|e\|^2 \tag{1.45}$$

Since we select $\lambda > 1$ and $\mu > 1$, we ensure that $l(e, \hat{W})$ satisfies the condition of being positive definite and radially unbounded. Hence, (1.42) is a cost functional.

The integral term in (1.42) can be written as

$$l(e, \hat{W}) + u_2^\top R(e, \hat{W}) u_2 = -2\beta \left(L_f V\right) + 2\beta^2 \left(L_g V\right) \left[R\left(e, \hat{W}\right)\right]^{-1} (L_g V)^T \tag{1.46}$$

The Lyapunov time derivative is defined as

$$\dot{V} = L_f V + L_g V u \tag{1.47}$$

and substituting (1.36) in (1.47), we obtain

$$\dot{V} = L_f V + L_g V \left[-\beta \left(R(e, \hat{W})\right)^{-1}\right] (L_g V)^\top$$

Then, multiplying $\dot{V}$ by -2β we have

$$-2\beta \dot{V} = -2\beta \left(L_f V\right) + 2\beta^2 \left(L_g V\right) \left[R\left(e, \hat{W}\right)\right]^{-1} (L_g V)^T$$

Hence

$$l(e, \hat{W}) + u_2^\top R(e, \hat{W}) u_2 = -2\beta \dot{V} \tag{1.48}$$

Replacing (1.48) in the cost functional (1.42), we obtain

$$\begin{aligned}
J(u_2) &= \lim_{t \to \infty} 2\beta V - 2\beta \int_0^t \dot{V} d\tau \\
&= \lim_{t \to \infty} \{2\beta V(t) - 2\beta V(t) + 2\beta V(0)\} \\
&= 2\beta V(0)
\end{aligned} \tag{1.49}$$

The cost functional optimal value is given by $J^* = 2\beta V(0)$. This is achieved by the control law (1.36).

The obtained results can be summarized in the following theorem.

Theorem 1. *For the unknown nonlinear system (1.14), modelized by the recurrent high order neural network (1.15), whose weights are on-line adapted by (1.30), the control law (1.41)*
i) guarantees asymptotic stability of the tracking error.
ii) minimizes the cost functional defined by (1.42).

1.5   Simulation Results for Robot Trajectory Tracking

In this application, the robust adaptive control law is sinthetized in order to overcome the effects of disturbances and uncertain variables such as friction forces.

Friction in robotic arms is an important caracteristic. It is affected by many factors such as lubrication, velocity, and applied forces. Friction terms are classified as static friction, dry friction, viscous friction, exponential friction and drag friction [2]. In general, friction models are considered as a combination of the Coulomb and viscous friction.

The Coulomb friction can be defined as

$$f(\dot{q}) = f_c sgn(\dot{q}) \tag{1.50}$$

where $\dot{q}$ is the angular velocity of the link and f_c is a constant parameter .

The viscous friction is considered proportional to the velocity as follows

$$f(\dot{q}) = f_v sgn(\dot{q}) \tag{1.51}$$

where f_v is a constant parameter.

Then, we can parametrize the friction term as

$$f(\dot{q}) = f_v \dot{q} + f_c sgn(\dot{q}) \tag{1.52}$$

These aproximations are very useful for practical cases, but there may exists some applications which require a more complex combination of the friction parameters.

In order to test the applicability of the proposed control scheme, we consider the trajectory tracking problem for a robot manipulator model. The dynamics of a 2-link rigid robot arm, with a friction term and with torque control input can be written as

$$D\left(q\right)\ddot{q} + C\left(q,\dot{q}\right)\dot{q} + G\left(q\right) + f\left(\dot{q}\right) = \tau \tag{1.53}$$

where $D\left(q\right)$ is a positive definite and symmetric inertia matrix, $C\left(q,\dot{q}\right)$ is the matrix containing the effect of centripedal and Coriolis forces, and $G\left(q\right)$ contains the gravitational torques. These elements are defined as follows,

$$q = \begin{bmatrix} q_1 \\ q_2 \end{bmatrix}, \quad \dot{q} = \begin{bmatrix} \dot{q}_1 \\ \dot{q}_2 \end{bmatrix}, \quad \ddot{q} = \begin{bmatrix} \ddot{q}_1 \\ \ddot{q}_2 \end{bmatrix}$$

$$D\left(q\right) = \begin{bmatrix} D_{11} & D_{12} \\ D_{21} & D_{22} \end{bmatrix}, \tag{1.54}$$

$$D_{11} = m_1 l_{c1}^2 + m_2 \left(l_1^2 + l_{c2}^2 + 2l_1 l_{c2} \cos\left(q_2\right)\right) + I_{zz1} + I_{zz2}$$

$$D_{12} = m_2 \left(l_{c2}^2 + l_1 l_{c2} \cos\left(q_2\right)\right) + I_{zz2}$$

$$D_{21} = D_{12}$$

$$D_{22} = m_2 l_{c2}^2 + I_{zz2}$$

$$C\left(q,\dot{q}\right) = \begin{bmatrix} -2m_2 l_1 l_{c2} \dot{q}_2 sen\left(q_2\right) & -m_2 l_1 l_{c2} \dot{q}_2 sen\left(q_2\right) \\ m_2 l_1 l_{c2} \dot{q}_1 sen\left(q_2\right) & 0 \end{bmatrix}$$

$$G\left(q\right) = \begin{bmatrix} m_1 g l_{c1} \cos(q_1) + m_2 g \left(l_1 \cos(q_1) + l_{c2} \cos(q_1 + q_2)\right) \\ m_2 g l_{c2} \cos(q_1 + q_2) \end{bmatrix}$$

$$\tau = \begin{bmatrix} \tau_1 \\ \tau_2 \end{bmatrix}$$

Coulomb and viscous friction have been considered to model the friction torque $f\left(\dot{q}\right)$,

$$f\left(\dot{q}\right) = \begin{bmatrix} f_{v1} \dot{q}_1 + f_{c_1} sgn\left(\dot{q}_1\right) \\ f_{v2} \dot{q}_2 + f_{c_2} sgn\left(\dot{q}_2\right) \end{bmatrix} \tag{1.55}$$

where

$$f_{v1} = 2.288 \ \text{N s}$$
$$f_{v2} = 0.175 \ \text{N s}$$
$$f_{c1} = -8.049 \ \text{N m}$$
$$f_{c2} = 1.734 \ \text{N m}$$

We consider the example studied in [11], as the nonlinear unknown plant. The parameters for this robot model, are the following:

$$
\begin{array}{llll}
m_1 = 23.902 & \text{kg} & m_2 = 1.285 & \text{kg} \\
l_1 = 0.45 & \text{m} & l_2 = 0.45 & \text{m} \\
l_{c1} = 0.091 & \text{m} & l_{c2} = 0.048 & \text{m} \\
I_1 = 1.226 & \text{kg m}^2 & I_2 = 0.093 & \text{kg m}^2
\end{array}
$$

Fig. 1.3 shows a picture of this robot (This picture is included with permission of Prof. Victor Santibañez, Instituto Tecnologico de La Laguna, Torreon, Mexico). We try to force this manipulator to track a reference signal given

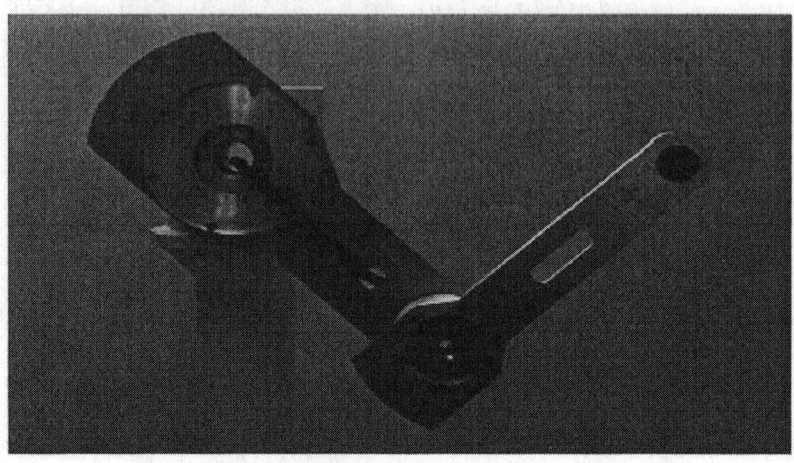

Fig. 1.3. Robotic manipulator

by

$$q_{r1} = 1.57 + 0.78(1 - e^{-2t^3}) + 0.17(1 - e^{-2t^3})\sin{(w_1 t)} \qquad (1.56)$$
$$q_{r2} = 1.57 + 1.04(1 - e^{-1.8t^3}) + 2.18(1 - e^{-1.8t^3})\sin{(w_2 t)}$$

where w_1 and w_2 are the frequencies of the desired trajectories for link 1 and link 2, respectively. This trajectory imposes large velocity and acceleration values to the manipulator. For the simulation, $w_1 = 15$ rad/s and $w_2 = 3.5$ rad/s are used. Furthermore, we consider an external torque disturbance modelized as a pulse train shown in Fig.1.4. We select the initial position at $q_1 = -90°$ and $q_2 = 0°$, which corresponds to an equilibrium point.

We use the dynamical neural network (1.15) to modelize the manipulator

$$\text{with} \quad A = -100I \, , I \in R^{4x4},$$
$$k = 0.35, \ \Gamma = 0.5I,$$
$$z(x_i) = (\tanh(kx_i))$$
$$\Omega = \begin{pmatrix} 0\ 0\ 1\ 0 \\ 0\ 0\ 0\ 1 \end{pmatrix}^T$$

For the control law, we select $\mu = 85$. The time evolution for the angles and applied torques to the links are shown in Figs. 1.5–1.8. As can be seen, trajectory tracking is successfully obtained.

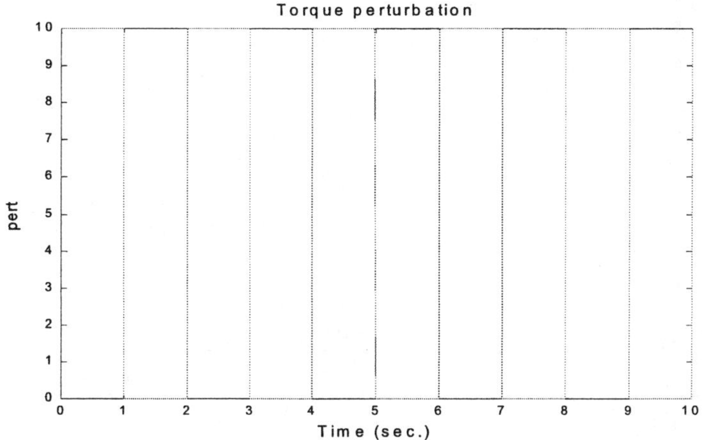

Fig. 1.4. Torque perturbation applied to link 1

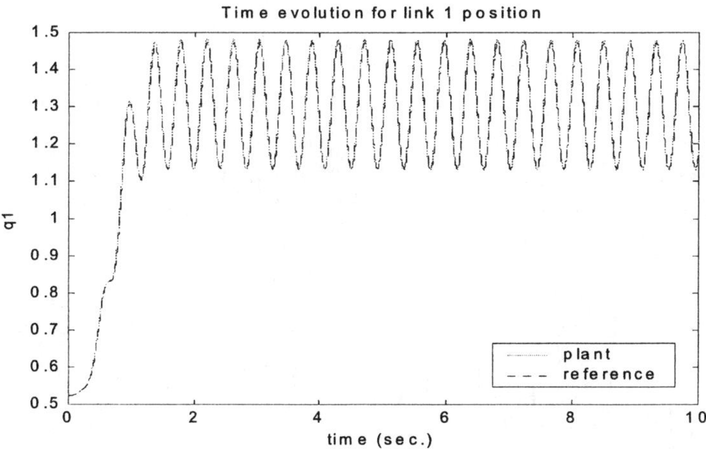

Fig. 1.5. Time evolution for the position of link 1

We can see that the Recurrent Neural Controller ensures rapid convergence of the system outputs to the reference trajectory. The controller is robust in presence of disturbances applied to the system. Another important issue of this approach related to other neural controllers, is that most neural controllers are based on indirect control, first the neural network identifies the unkown system and when the identification error is small enough, the control is applied. In our approach, direct control is considered, the learning laws for the neural networks depend explicitly of the tracking error instead

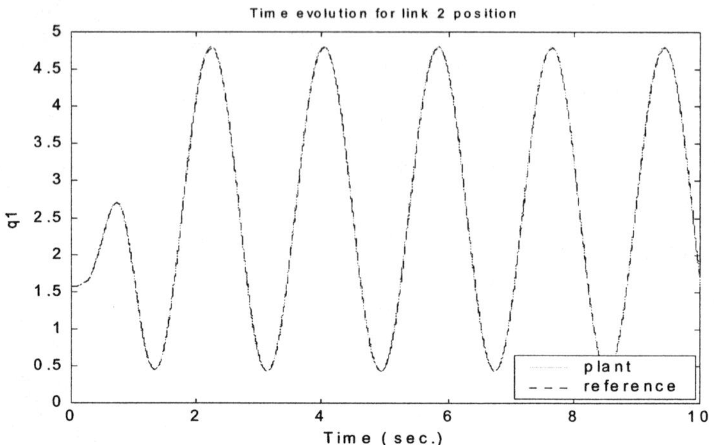

Fig. 1.6. Time evolution for the position of link 2

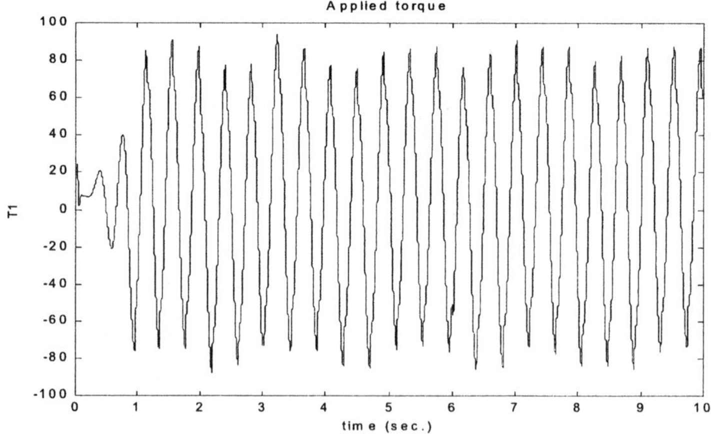

Fig. 1.7. Torque applied to link 1

of the identification error. This approach results in faster response of the system.

Acknowledgements.- We thank the support of CONACYT, Mexico, Project 32059A. The first author thanks the support of Centro de Ensenanza Tecnica Industrial (CETI Guadalajara).

384

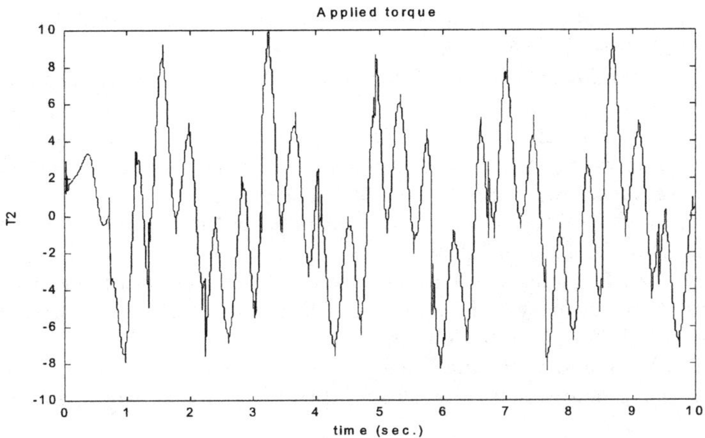

Fig. 1.8. Torque applied to link 2

1.6 Conclusions

We have extended the adaptive recurrent neural control previously developed in [18], [19] and [20] for trajectory tracking control problem in order to consider less inputs than states. Stability of the tracking error is analized via Lyapunov control functions and the control law is obtained based on the inverse optimal control approach. A robot model with friction terms and unknown external disturbances is used to verify the design for trajectory tracking, with satisfactory performance. Research along this line will continue to implement the control algorithm in real time and to further test it in a laboratory enviroment.

References

1. Basar T.and P. Bernhard, *H - Infinity Optimal Control and Related Minimax Design Problems*, Birkhauser, Boston, USA, 1995.
2. Harris C. J., T. H. Lee and S. S. Ge, *Adaptive Neural Network Control of Robotic Manipulators* ,World Scientific Pub., 1999.
3. M. M. Gupta and D. H. Rao (Eds.), *Neuro-Control Systems, Theory and Applications*, IEEE Press, Piscataway, N.J., USA, 1994.
4. Hill D. J. and P.Moylan, "The Stability of nonlinear dissipative systems", *IEEE Trans. on Auto. Contr.*, vol. 21, 708-711, 1996.
5. Hopfield J., "Neurons with graded responses have collective computational properties like those of two state neurons" *Proc. Nat. Acad. Sci.*, USA, 1984, 81, pp. 3088-3092.
6. K. Hunt, G. Irwin and K. Warwick (Eds.), *Neural Networks Engineering in Dynamic Control Systems*, Springer Verlag, New York, USA, 1995.

7. Ioannou P. A. and J. Sun, *Robust Adaptive Control,* Prentice Hall, Upper Saddle River, New Jersey, USA, 1996.

8. Isidori A., *Nonlinear Control Systems*, 3rd Ed., Springer Verlag, New York, USA, 1995.

9. Kosmatopoulos E. B., M. A. Christodoulou and P. A. Ioannou , "Dynamical neural networks that ensure exponential identification error convergence", *Neural Networks*, vol. 1, no. 2, pp 299-314, 1997.

10. Krstic M. and H. Deng, *Stablization of Nonlinear Uncertain Systems*, Springer Verlang, New York, USA, 1998.

11. Llama M. A., R. Kelly and V. Santibañez, "Stable Computed torque of Robot manipulators via fuzzy self-tuning", *IEEE Trans. on Systems, Man. and Cybernetics*, Vol. 30, 143-150, February, 2000.

12. Narendra K. S. and K Parthasarathy, "Identification and control of dynamical systems using neural networks", *IEEE Trans. on Neural Networks*, vol. 1, no. 1, pp 4-27, 1990.

13. Pham D. T. and X. Liu, *Neural Networks for Identification, Prediction and Control*, Springer Verlag, London, 1995.

14. Poznyak A. S., E. N. Sanchez and W. Yu, *Differential Neural Networks for Robust Nonlinear Control*, Worl Scientific, USA, 2000.

15. Poznyak A. S., W. Yu, E. N. Sanchez and J. P. Perez, "Nonlinear adaptive trajectory tracking using dynamic neural networks", *IEEE Trans. on Neural Networks*, vol. 10, no 6, pp 1402-1411, Nov. 1999.

16. Rovitahkis G. A. and M. A. Christodoulou, *Adaptive Control with Recurrent High-Order Neural Networks*, Springer Verlag, New York, USA, 2000.

17. Sanchez E. N., J. P. Perez and G. Chen, "Using dynamic neural control to generate chaos: An inverse optimal control approach", *Int. J. of Bifurcation and Chaos*, 2001.

18. Sanchez E. N., J. P. Perez, L.Ricalde and G. Chen, "Trajectory tracking via adaptive neural control", *Proceedings of IEEE Int. Symposium on Intelligent Control*, Mexico City, pp. 286-289, September 2001.

19. Sanchez E. N., J. P. Perez, L. Ricalde and G. Chen, "Chaos production and synchronization via adaptive neural control", *Proceedings of IEEE Conference on Decision and Control*, Orlando, Fl, USA, December 4-7, 2001.

20. Sanchez E. N., J. P. Perez and L. Ricalde, "Recurrent neural control for robot trajectory tracking", *Proceedings of International Federation of Automatic Control*, Barcelona Spain, July, 2002.

21. Sanchez E. N. and V. Flores, "Real-Time fuzzy PI+PD control for an under-actuated robot", *Proc. of 2002 IEEE International Workshop on Intelligent Control*, Vancouver, B.C., Canada, October 2002.

22. Sepulchre R. , M. Jankovic and P. V. Kokotovic, "*Constructive nonlinear control*", Springer, New york, USA, 1997.

23. Suykens K. , L. Vandewalle and R. De Moor, *Artificial Neural Networks for Modelling and Control of Nonlinear Systems*, Kluwer Academic Publishers, Boston, USA,1996.

Gesture Recognition Based on SOM Using Multiple Sensors

Masumi Ishikawa

Department of Brain Science and Engineering, Graduate School of Life Science
and Systems Engineering, Kyushu Institute of Technology
2-4 Hibikino, Kitakyushu 808-0196, Japan
ishikawa@brain.kyutech.ac.jp

Abstract. Gesture recognition is important, because it is a useful communication
medium between humans and computers. In this paper we use multiple sensors,
i.e., PSD cameras for detecting LEDs attached on a body and DataGloves for both
hands. One of the major difficulties in gesture recognition is temporal segmentation
from continuous motion. We use training samples which are manually segmented
and labeled as prior knowledge. A self-organizing map(SOM) is constructed based
on training samples. Test gestural data are segmented by systematic search to obtain
the best match with reference vectors on a competitive layer. A comparative study is
done between the use of a single SOM and 3 SOMs for representing spatio-temporal
information obtained from PSD cameras and DataGloves.

Keywords: gesture, recognition, self-organizing map, sensor, segmentation,
spatio-temporal information

1 Introduction

Gesture recognition is needless to say important, because it is a useful communication medium between humans and computers[2][3][4][5][10][11]. It is
widely known that efficient processing of spatio-temporal information such
as recognition of gestures is one of the central issues in current information
processing technology.

Methods for the recognition of gestures or sign languages are roughly
classified into three categories. In the first category a mouse or a stylus is
used[11][20]. They are practical media for human computer interaction, but,
with the exception of a 3-D mouse, they are too restrictive, because gestures
in this category are limited to be 2-D.

In the second category users wear sensors such as DataGloves[10], Data-
Suit and LEDs. They have difficulties in wearing them in practical situation,

because they tend to prevent flexible motion and are too expensive for daily use.

In the third category an image processing technique reconstructs 3-D shapes of arms and hands from 2-D image[12]. It has such difficulties as feature extraction, occlusion and ill-posedness. Required facilities are inexpensive, but computational cost for the recognition tend to be large. To overcome these difficulties, the use of background knowledge such as a model of arms and hands is effective[14][19].

There are four major difficulties in gesture recognition. First difficulty is that a *sequence* of shapes of arms and hands rather than a *snapshot* of arms and hands has to be treated for gesture recognition. Second difficulty is temporal segmentation of a sequence of shapes of arms and hands. i.e., determination of the start and the end of each gesture. This difficulty is mainly due to temporally continuous nature of gestures. Third difficulty is spatio-temporal variation of gestures; even a gesture of a particular word of one person is subject to infinite variation both spatially and temporally. This variation is prevalent in pattern recognition, but in many cases pattern recognition has dealt with only spatial variation such as character recognition or temporal variation such as speech recognition. Fourth difficulty is that the amount of resulting spatio-temporal gestural data is enormous. This necessitates feature extraction from dynamic scene images for efficient information processing.

The present paper focuses on the first three difficulties, while disregarding the last one. To solve the last difficulty we use multiple sensors as an effective shortcut to feature extraction[7].

There have been a few existing methods for the recognition of gestures. The first one is a hidden Markov model(HMM), which has been extensively used for speech recognition[16][17]. Its advantages are that variation of temporal length of gestures can be handled naturally by transition probability matrices. On the other hand, HMM needs a lot of segmented and labeled samples for estimating transition probability matrices and other model parameters. Furthermore, multidimensional gestural data must be discretized in advance for use in HMMs.

The second method is the use of continuous dynamic programming(CDP) for matching a given sequence of gestural data with a set of standard gestural data sequence[21]. This matching method, called the spotting recognition in speech recognition, makes "frame-wise" recognition possible.

The third method is a self-organizing map(SOM), which has the following advantages. Firstly, clustering can be done without supervised signals, i.e., class labels. Although class labels for training samples are needed, they are not used in constructing a SOM, but used only for determining the class boundaries on SOM after training. Secondly, it is applicable to data with arbitrary number of classes. Thirdly, visualization of resulting self-organizing maps is quite effective in understanding gestural data, which is hardly available in other methods. Fourthly, discretization of data is not necessary in contrast to HMMs. On the other hand, data for SOM should have a constant dimensionality, which is generally not satisfied in temporal data.

The present paper adopts SOMs for gesture recognition due to above advantages. SOM is also expected to ameliorate the third difficulty of inherent spatio-temporal variation of gestures. We must also devise a method for solving the difficulty of the constant dimensionality requirement for inputs to SOMs.

Section 2 presents the basics of SOM. Section 3 describes measurement of gestural data and training samples. Section 4 provides a solution to handling temporal data by SOM, which corresponds to the first difficulty in gesture recognition. Section 5 explains occlusions and temporal adjustments in gesture recognition. Section 6 presents a method for segmentation and recognition, which provides a solution to the second difficulty in gesture recognition. A comparison is carried out between the use of one self-organizing map and 3 self-organizing maps for representing spatio-temporal information obtained from PSD cameras and DataGloves. Sections 7 and 8 explain experimental results. This is followed by conclusions and discussions.

2 Basics of SOM

Kohonen proposed a method for creating self-organizing maps of given samples in an unsupervisory way[9]. Let an input vector be $\mathbf{x} \in R^n$ and the reference vector i on a competitive layer be $\mathbf{w}_i \in R^n$.

For a given input vector, the distance between the input vector, $\mathbf{x}$, and each reference vector, $\mathbf{w}_i$, is calculated. The winner neuron, $\mathbf{w}_c$, is determined by,

$$\|\mathbf{x} - \mathbf{w}_c\| = \min_i \{\|\mathbf{x} - \mathbf{w}_i\|\} \tag{1}$$

A Euclidian distance measure is adopted in Eq.(1).

The next step is the modification of neighboring reference vectors as follows.

$$\mathbf{w}_i(t+1) = \mathbf{w}_i(t) + h_{ci}(t)[\mathbf{x}(t) - \mathbf{w}_i(t)] \tag{2}$$

where t is a time step and $h_{ci}(t)$ is a neighborhood function. The step function in Eq.(3) or the Gaussian function in Eq.(4) is used as a neighborhood function.

$$h_{ci}(t) = \begin{cases} \alpha(t) : i \in N_c \\ 0 \quad\; : i \notin N_c \end{cases} \tag{3}$$

$$h_{ci}(t) = \alpha(t) \cdot exp\left(-\frac{\|\mathbf{r}_c - \mathbf{r}_i\|^2}{2\sigma^2(t)}\right) \tag{4}$$

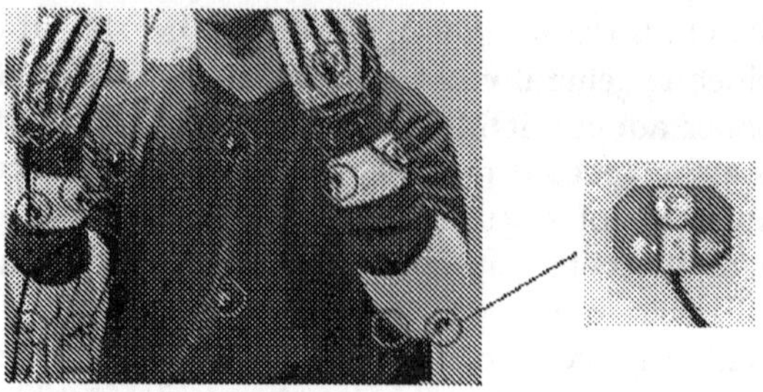

Fig. 1. A subject wearing 16 LEDs and DataGloves

where N_c is a set of suffixes in a neighborhood region, r_c is the location of the winner neuron on a competitive layer, r_i is the location of the competitive neuron, w_i, on a competitive layer, $\alpha(t)$ is a learning rate at time t, and $\sigma(t)$ is the standard deviation of the Gaussian function at time t.

After learning reference vectors become ordered in the sense that similar reference vectors are located nearby on a competitive layer. The resulting self-organizing map provides class boundaries determined by the distance between adjacent competitive neurons. These boundaries classify not only input vectors for training but also novel input vectors.

3 Gesture measurement and training samples

16 LEDs are attached on a body to measure its shape as in Fig. 1; one LED is at the back of the hand, 3 LEDs at the wrist, 2 LEDs at the elbow, 1 LED at the shoulder, 1 LED at the breast, and 1 LED at the belly. Each LED emits infrared for 100 microseconds. Two cameras equipped with PSD(Position Sensitive Device) are used to measure the locations of LEDs. Measurement of emitted infrared by two PSD cameras makes calculation of the 3-D position of each LED possible, provided appropriate calibration is done in advance. The fact that each LED emits infrared at different timing makes this calculation straightforward.

Two DataGloves are used to measure the shapes of both hands. Each DataGlove has 10 sensors to measure the angles of the first and the second joints of 5 fingers.

Shapes of arms and hands are measured every 33 msec. In case of occlusion, PSD cameras cannot measure infrared from occluded LEDs, hence interpolation is necessary to estimate their locations. Spline interpolation and Newton's interpolation are carried out, the details of which will be presented in section 5.

For the recognition of temporary *continuous* gestures, prior knowledge is indispensable, because in continuous gestures there is no clue to decide the start and the end of each gesture. Each gesture is a sequence of data on the shapes of arms and hands. Segmentation and labeling of training data are carried out manually by an experimenter with the help of video recording and are used as prior knowledge. They are used for training of a self-organizing map.

4 Temporal data in SOM

There have been two approaches in self-organization for temporal or sequence data. One is to regard data at each time as an input to SOM. The other is to regard temporal or sequence data as an input to SOM.

In the former approach, the winner neuron is determined at each time for the corresponding input data, hence temporal data are represented by a trajectory on a competitive layer. This is easy to implement, but it only transforms a trajectory in a high dimensional input space into that on a two dimensional competitive layer. An additional procedure is required accordingly to extract feature from a trajectory on a competitive layer.

Chappell et al. proposed a slight modification of the former approach, i.e., use of a leaky integrator to calculate the distance between an input vector and a reference vector[1]. In this way past inputs are taken into account in calculating the distance.

Salmela et al. proposed to use a two-dimensional binary map composed of winner neurons for recognizing isolated spoken numbers independent of speakers[15]. A resulting two-dimensional binary map is given to a multilayer Perceptron as an input. This method provides a feature vector with constant length based on spoken numbers of different duration. However, it cannot discriminate two words with the same set of phonemes but with different order, because a *sequence* of phonemes are not taken into account.

The latter approach generates a feature vector of fixed size from temporal data of varying length. There are two extremes in this approach. One extreme is to compress temporal data by calculating weighted averages. Kangas proposed to obtain weighted average of temporal data by using an exponentially decreasing function[8]. Mozer also proposed to obtain weighted averages of temporal data by using the gamma function or the Gaussian function[13]. One advantage is that the effect of noise is decreased by calculating weighted averages. However, since it compresses information too much, even necessary information tends to be lost in this process.

The other extreme in the latter approach is to form an input vector by simply concatenating temporal data as an input to SOM[8]. Since the dimension of a resulting input vector inevitably varies, it cannot be used as it is. Furthermore, computational cost tend to be huge due to large dimensionality of resulting input vectors.

In this paper we propose to adopt midway between the two extremes by compressing an input vector to a fixed length by interpolation. For this

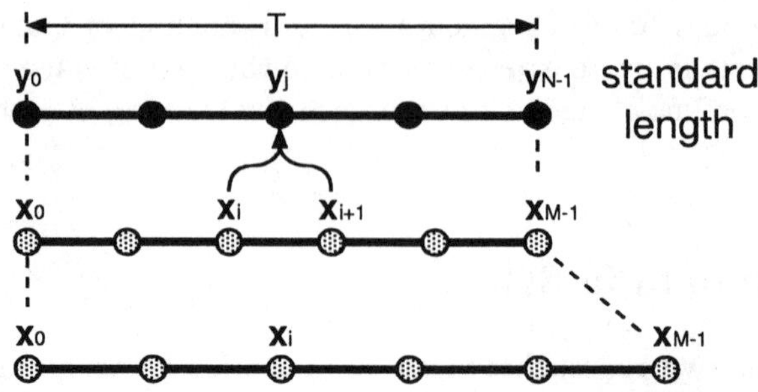

Fig. 2. Temporal normalization by linear interpolation

purpose linear interpolation and Gaussian interpolation are compared in the following subsections.

4.1 Linear interpolation

Fig. 2 illustrates how temporal data are transformed into those with the standard length. A value at a new sampling point is estimated by linear interpolation as in Eq.(5).

$$
\mathbf{y}_j = \begin{cases} \mathbf{x}_0, \; j=0 \\[2ex] \dfrac{((i+1)A - jB)\mathbf{x}_i + (jB - iA)\mathbf{x}_{i+1}}{A} \\[1ex] \qquad\qquad , \; 0 < j \leq N-1 \end{cases} \tag{5}
$$

where

$$
A = \frac{T}{M-1} \tag{6}
$$

$$
B = \frac{T}{N-1} \tag{7}
$$

$$
i = \max_k\{kA < jB\}, \; k = 0, 1, ..., M-1 \tag{8}
$$

where T is the standard length of a sequence.

Ishikawa et al. used this method for the recognition of gestures based on data obtained from a DataGlove[5]. It is straightforward to see that linear interpolation is easily affected by noise and temporal variation.

4.2 Gaussian interpolation

Eq.(9) gives Gaussian temporal adjustment and Fig. 3 illustrates its basic principle[6].

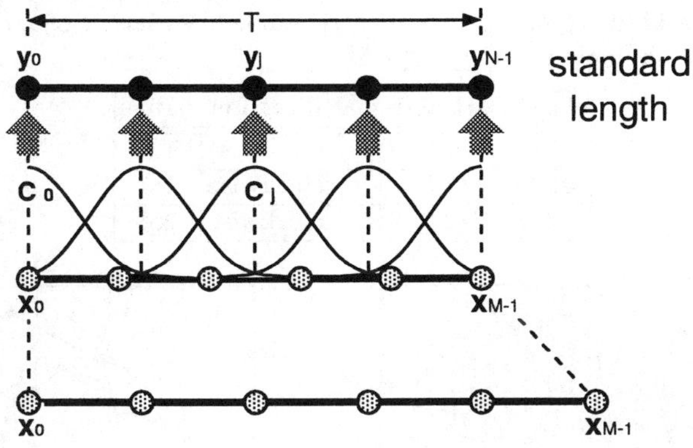

Fig. 3. Temporal normalization by Gaussian functions

$$\mathbf{y}_j = \frac{\displaystyle\sum_{i=0}^{N-1} c_j(i)\mathbf{x}(i)}{\displaystyle\sum_{i=1}^{N-1} c_j(i)} \tag{9}$$

$$c_j(i) = \exp\{-\frac{\{(i-j)A\}^2}{2\sigma^2}\} \tag{10}$$

$$A = \frac{T}{N-1} \tag{11}$$

where T is the standard length of a sequence.

Eq.(9) calculates Gaussian weighted averages of temporal data at various points in time by using multiple Gaussian functions to compress information of temporal data.

4.3 Experimental results

We consider here 3 concentric circular paths: class r1 is a path with radius 1.0, class r2 is a path with radius 1.5, and class r3 is a path with radius 2.0[6]. Table 1 and Fig. 4(a) describe 3 classes. Fig. 4(b) illustrates examples of paths for 3 classes.

Either linear interpolation or Gaussian interpolation is applied, and resulting sequences are used as inputs to SOM with parameters in Table 2. Tables 3 and 4 indicate that a sequence of length 2 is sufficient in case of Gaussian interpolation, but even a sequence of length 5 is not sufficient in case of linear interpolation to classify 3 concentric circular paths. This result clearly indicates that Gaussian interpolation retains features of the original sequence

Table 1. Three kinds of paths of concentric circles. Variation of a radius is ± 0.5, and variation of an angle is $10° \le \theta \le 45°$.

class	start and goal	direction	radius
r1	(1.0, 0)	clockwise	1.0
r2	(1.5, 0)	clockwise	1.5
r3	(2.0, 0)	clockwise	2.0

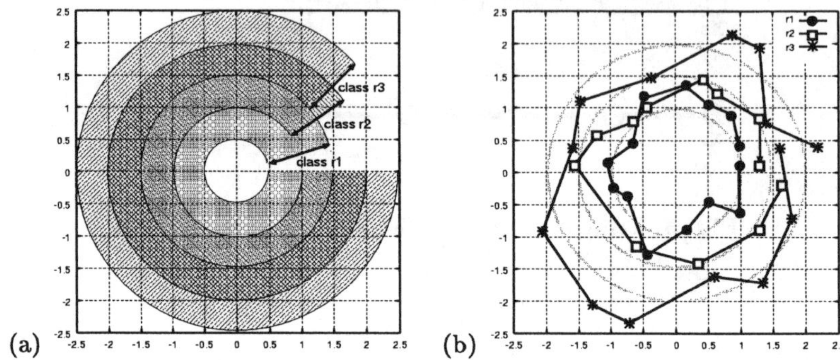

Fig. 4. Concentric circular paths. (a) Feasible regions of 3 concentric circles. (b) Examples of paths for 3 classes.

more efficiently than linear interpolation. This is attributed to the fact that Gaussian interpolation is more robust to noise and spatio-temporal variation than linear interpolation.

Three other examples, i.e., circular paths, handwriting identification and recognition of finger sign language, also show that Gaussian interpolation is superior[6]. These are omitted here due to space limitation. These examples well demonstrate the superiority of Gaussian interpolation over linear interpolation.

Table 2. Parameters for self-organization for concentric circles. "$n \to m$" indicates that the radius of neighborhood linearly decreases from n to m during 1st(or 2nd) stage of learning.

	1st stage	2nd stage
learning rate	0.4	0.02
no. of iterations	1000	5000
radius of neighborhood	$4 \to 3$	$3 \to 1$
map size	10×10	

Table 3. Performance of self-organization for concentric circles by Gaussian interpolation. RR stands for recognition rate and MQE stands for the mean quantization error.

length	training		test	
	RR(%)	MQE	RR(%)	MQE
1	94.17	0.042	81.83	0.056
2	96.67	0.157	92.17	0.201
3	98.50	0.216	93.33	0.271
4	97.50	0.242	96.00	0.300
5	97.67	0.270	94.17	0.331

Table 4. Performance of self-organization for concentric circles by linear interpolation. RR stands for recognition rate and MQE stands for the mean quantization error.

length	training		test	
	RR(%)	MQE	RR(%)	MQE
1	85.67	0.054	70.83	0.068
2	88.67	0.214	82.33	0.274
3	92.33	0.384	83.33	0.475
4	92.67	0.504	85.50	0.621
5	92.00	0.604	85.67	0.733

5 Occlusions and temporal adjustments

When either or both of the PSD cameras have insufficient amount of infrared from an LED due to occlusion, its 3-D position cannot directly be calculated. To make occlusion less frequent, we use multiple LEDs around the wrist and the elbow. Nevertheless, occlusions do occur and we have to use interpolation to overcome this difficulty.

Table 5 indicates that the average error is smaller when either of the cameras have normal inputs than when both cameras have insufficient inputs. It also shows that the longer the interval of occluded data is, the larger the average error of location estimation is. Table 6 compares the Newton's interpolation and the spline interpolation. It clearly indicates that the spline interpolation is superior to the Newton's interpolation in terms of the average error. Based on these results, we decide to adopt the spline interpolation in this paper.

Even after adjustment, there is a slight difference of time intervals of data acquisition from PSD cameras and DataGloves. Time intervals of the former and the latter are 34.38msec and 34.43msec, respectively. We apply linear interpolation in Fig. 2 for this adjustment. Concatenation of gestural data from PSD cameras and DataGloves of both hands, and temporal normalization above provide a fixed length input to SOM. Training of a self-organizing

Table 5. Comparison of the average error and the standard deviation of error for the Newton's interpolation(6th order). *"Both"* stands for the case when an LED is occluded from both cameras. *"A"* stands for the case when an LED is occluded only from camera A. *"B"* stands for the case when an LED is occluded only from camera B. Values in parentheses are the standard deviation of error.

no. of occluded data	*Both*	*A*	*B*
3	1.344(1.046)	0.589(0.579)	0.694(0.731)
7	2.708(2.082)	1.194(1.263)	1.356(1.392)
10	4.578(3.711)	1.946(2.424)	2.538(2.413)

Table 6. The average error and the standard deviation of error for the Newton's interpolation(6th order) and the spline interpolation(3rd order) when an LED is occluded from both cameras. Values in parentheses are the standard deviation of error.

no. of occluded data	Newton	spline
3	1.344(1.046)	1.209(0.930)
7	2.708(2.082)	1.819(1.417)
10	4.578(3.711)	2.289(1.844)

map gives clustering of gestural samples. Each neuron on the resulting SOM is attached a label of its closest training sample[18].

6 Segmentation and recognition

Candidates for the starting and the end of each gesture are systematically generated, and Euclidean distances between the candidates and reference vectors are calculated. The candidate with the minimum Euclidean distance is considered to provide the correct segmentation. At the same time the corresponding reference vector provides a label for the candidate. It is to be noted that segmentation and recognition are determined simultaneously. Modification based on prior information on the length of each gesture is also introduced. A procedure for the segmentation and the recognition is:

Step 1: Calculate frequency distribution of the length of each gesture based on training samples. The resulting frequency distribution, $P(m)$, is regarded as prior probability distribution, where m is a gestural label. The frequency distribution is approximated by the Gaussian function.

$$P(m) = \frac{1}{\sqrt{2\pi}\sigma_m} \exp(-\frac{(x - \mu_m)^2}{2\sigma_m^2}) \tag{12}$$

where x is the number of time steps, μ_m is the average length of gesture m, and σ_m is the standard deviation of gesture m.

Step 2: Calculate the minimum Euclidean distance between a given candidate and all the reference vectors. For each candidate we determine the winner neuron. Among winner neurons with label m, let the minimum distance be $r_{min}^{(m)}$ and the maximum distance be $r_{max}^{(m)}$.

Step 3: Estimate the conditional probability for each gesture.

$$P(r|m) = \exp(-k \frac{r - r_{min}^{(m)}}{r_{max}^{(m)} - r_{min}^{(m)}}) \tag{13}$$

where r is a Euclidean distance and k is a parameter converting normalized Euclidean distance into probability.

Step 4: Segmentation and recognition are simultaneously done by maximizing the following joint probability.

$$P(m, r) = P(m)P(r|m) \tag{14}$$

7 Experimental results(single SOM)

The following 10 sentences are examples of a sign language here.

(1) What is your research theme?
(2) My research theme is recognition of a sign language.
(3) Do you understand a sign language?
(4) I like baseball.
(5) I don't like baseball.
(6) I like soccer.
(7) I don't like soccer.
(8) I want to talk with you.
(9) I want to walk.
(10) I want to talk with you while walking.

Fifteen words are included in the above sentences. We decompose *talk* into two primitive gestures. We also add *no action*(both hands being placed on the knee). Accordingly we use 17 primitive gestures in total.

Two subjects give gestures of the above 10 sentences with hands and arms for obtaining training and test data. The numbers of training samples for subject 1 and 2 are 488 and 138, respectively. The number of time steps is normalized to 5. PSD data are composed of 16 LEDs and each LED has 3 dimensions. Therefore, the number of dimensions for one primitive gesture is 340(=16x3x5+10x2x5). Data are normalized between 0 and 1. Table 7 shows parameters for training a SOM. The percentage of the correct recognition for training data is 91.86%, and the mean quantization error(MQE) is 0.8310. Fig. 8 illustrates the resulting SOM.

The numbers of test samples for subject 1 and subject 2 are 30 and 20, respectively. Table 8 indicates the recognition performance. In the experiment

Table 7. Parameters for training a single SOM.

	1st stage	2nd stage
learning rate	0.4	0.02
no. of iterations	10000	100000
radius of neighborhood	15 → 5	5 → 1
map size	30 × 30	

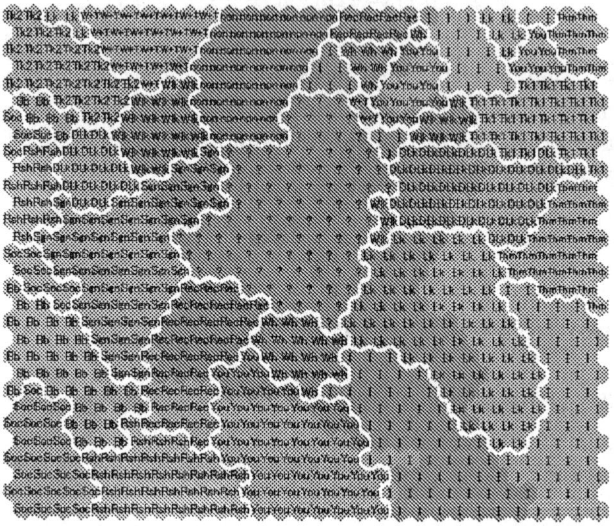

Fig. 5. Resulting SOM for 2 subjects. Labels on the map are: you(You), I(I), research(Rsh), theme(Thm), recognition(Rec), sign language(Sgn), question mark(?), like(Lk), dislike(DLk), talk(Tk1, Tk2), baseball(Bb), soccer(Soc), walk(Wlk), what(Wh), talk while walking(W+T) and no action(non).

Table 8. Recognition performance of test data using a single SOM. NoC and NoW stand for the number of correct recognition and the number of wrong recognition, respectively.

NoC	NoW	correct ratio(%)	MQE
136	39	79.43	0.9733

$k = 10$ is used, since it gives the best empirical result. MQE=0.9733 corresponds to error of 0.258° for a DataGlove and error of 11.72 for each PSD (-2047~2047), which are both fairly small. Closer investigation reveals that segmentation is sometimes done where no actual gesture exists. This might be due to integrating data from 3 sensors, which tend to make salient features from one sensor obscure.

Table 9. Parameters for training 3 SOMs.

		PSD		right DataGlove	
learning stage	1st st.	2nd st.		1st st.	2nd st.
learning rate	0.4	0.02		0.4	0.02
no. of iterations	10000	50000		5000	20000
	left DataGlove				
learning stage	1st st.	2nd st.			
learning rate	0.4	0.02			
no. of iterations	2000	10000			

Table 10. Recognition performance for training data using 3 SOMs

	right-hand	left-hand	PSD
recognition rate(%)	87.56	99.50	90.10
MQE	290.41	153.26	1645.55

8 Experimental results(3 SOMs)

The reason for using 3 SOMs is that gestures such as *I* and *you* use only the right hand, hence data of the left hand might disturb the recognition. Another reason is that computation time can be reduced by using 3 SOMs, because SOM for the left hand could be trained on a smaller competitive layer due to smaller variation of the motion of the left hand. Table 9 shows parameters for 3 SOMs. Figures 6, 7 and 8 illustrate the resulting SOMs. Table 10 shows the recognition performance. It is to be noted that data are not normalized in contrast to the previous section. PSD data are -2047~2047 and DataGlove data are 0~900.

Test is done either by multiplying the joint probabilities in Eq.(14) or by considering only Euclidean distances.

8.1 Multiplication of joint probabilities

Probability is calculated by multiplying three joint probabilities obtained from PSD, right-hand and left-hand.

$$P_{B \cdot R \cdot L}(m_{B \cdot R \cdot L}, r_{B \cdot R \cdot L}) = \tag{15}$$
$$P_B(m_B, r_B) * P_R(m_R, r_R) * P_L(m_L, r_L)$$

Among candidates of gestures the one with the maximum probability is the estimated one. Since value of k influences the performance of recognition, recognition performance for various k's are evaluated as shown in Table 11. Table 11 indicates that $k_B > k_R > k_L$ shows the best performance. In case of $k_R > k_B > k_L$ the sequence of the same right-hand gestures such as *I* and *baseball* seems to deteriorate the recognition performance. This suggests better

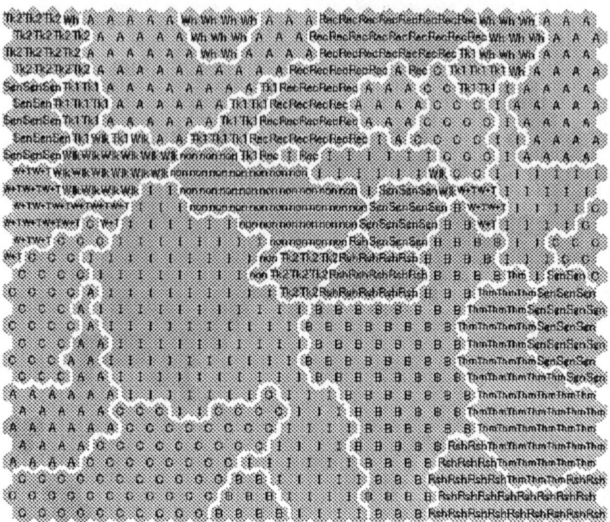

Fig. 6. Resulting SOM for PSD. Labels on the map are: you(A), question mark(A), baseball(B), soccer(B), like(C), dislike(C), I(I), recognition(Rec), sign language(Sgn), talk(Tk1, Tk2), research(Rsh), theme(Thm), walk(Wlk), what(Wh), talk while walking(W+T) and no action(non).

Table 11. Recognition performance for various values of k_R, k_L, and k_B. NoC and NoW stand for the number of correct recognition and the number of wrong recognition, respectively.

k_R	k_L	k_B	NoC	NoW	recognition rate(%)
10	10	10	117	53	68.82
10	1	5	106	64	62.35
5	1	10	122	42	73.49

performance by increasing k_B. Compared with the results by a single SOM, the number of obtained segmentation is decreased, suggesting that unnecessary segmentation can be avoided by using multiple SOMs.

8.2 Euclidean distance

Previous experiments have considered not only Euclidean distance but also prior information on the length of gestures. In cases where the variance of the length of gestures is large, recognition rate tend to deteriorate. This suggests a method using only Euclidean distance.

For this purpose we calculate the following normalized distance, $\hat{r}_i$, corresponding to each sensor.

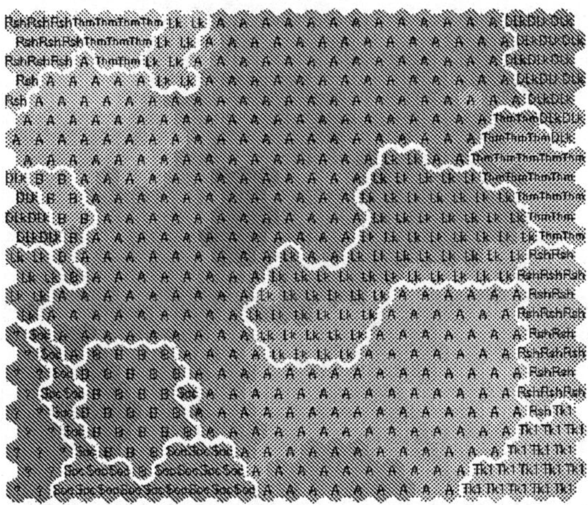

Fig. 7. Resulting SOM for the right DataGlove. Labels on the map are: you(A), I(A), sign language(A), baseball(A), recognition(A), what(A), walk(B), talk while walking(B), soccer(Soc), like(Lk), dislike(DLk), question mark(?), research(Rsh), theme(Thm), and talk1(Tk1). "talk1" stands for the first half of "talk." "talk2"(the latter half of "talk") and "no action" do not appear on the map.

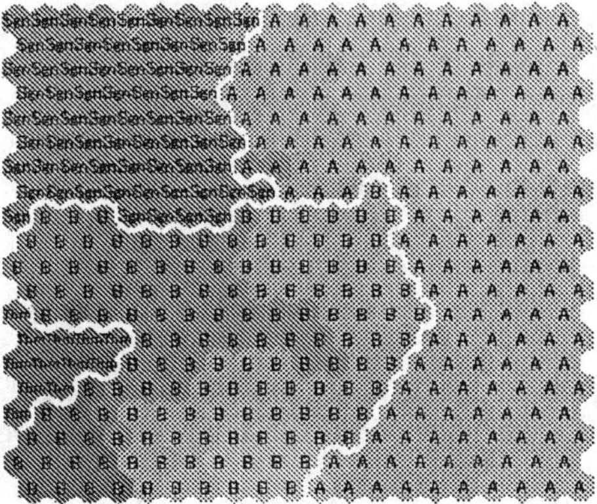

Fig. 8. Resulting SOM for the left DataGlove. Labels on the map are: baseball(A), research(A), soccer(A), talk while walking(B), talk2(B), theme(Thm) and sign language(Sgn). The rest of the words do not appear on the map.

$$\hat{r}_i = \frac{r_i - r_{min}^{(m)}}{r_{max}^{(m)} - r_{min}^{(m)}} \qquad i = B, R, L \qquad (16)$$

For each sensor we generate 3 best candidates. If some candidate is supported by all the sensors, i.e., small $\hat{r}_i$, this is considered to be a good estimate.

Table 12. Recognition performance for test data using only Euclidean distances. NoC and NoW stand for the number of correct recognition and the number of wrong recognition, respectively.

NoC	NoW	recognition rate(%)
119	53	69.20

Table 12 shows the performance of recognition. The recognition rate of 69.20% is rather low. This is attributed to the failure to recognize gestures, notwithstanding the existence of gestures.

9 Conclusions and discussions

Generally speaking there are two major components in gesture recognition from spatio-temporal data. One is feature extraction, and the other is segmentation and recognition. This present paper has focused on the latter.

We have proposed to recognize temporary continuous gestures using PSD cameras and DataGloves. Various difficulties such as temporal segmentation and temporal adjustment are solved. Self-organizing maps are used as a basic clustering method.

Although we have conjectured that multiple SOMs would perform better than a single SOM, experimental results so far indicate the superiority of the latter. To improve the performance of the former, various improvements need to be done. In case of multiplication of joint probabilities, more detailed search of the value of k, which converts Euclidean distance into probability, would be necessary. In case of Euclidean distance, more candidates should be reserved for recognition. These are left for immediate future studies.

In the present study of using a single SOM the dimension of an input vector is 340, which is obviously very large. We think that angle information instead of 3-D position should be used; angle information contribute to the decrease of computational cost, and also is advantageous for not being sensitive to the size of a subject.

Segmentation in the present study requires large computation time due to exhaustive search. We have also done research to accelerate this search; total computation time could be reduced down to about 5% of exhaustive search by efficient pruning of search tree. This result will be presented in the near future.

We have skipped the task of feature extraction by introducing sensors. We have currently been doing a research on estimating the direction of gaze from dynamic scene images. This will be combined with dynamic image processing techniques in the future.

The recognition rate obtained so far is far from satisfactory. It is also expected that as the number of gestures increases, the recognition rate deteriorates. This suggests inherent difficulty of gesture recognition. Although various improvements based on the above would be possible, these alone would not realize very high recognition rate. Top down information such as syntax and semantics information would help further improve the recognition rate. Of course, this is not an easy task as can be seen from speech recognition.

This study was supported by Special Coordination Funds for Promoting Science and Technology from the Ministry of Education, Culture, Sports, Science and Technology of Japan. I would also like to express my appreciation to former students at my Lab., Mr. Naohiro Sasaki, Mr. Hiroshi Suenaga and Satoru Tanaka, for their contributions to this study.

References

1. Chappell, G. J. and Taylor, J. G. (1993). The temporal Kohonen map, Neural Networks, Vol.6, pp.441-445.
2. Harling, P.A. et al. Eds., (1996). Progress in Gestural Interaction: Proceedings of Gesture Workshop'96, Springer.
3. Wachsmuth, I, Froelich, M. Eds., (1997). Gesture and Sign Language in Human-Computer Interaction, International Gesture Workshop, Bielefeld, Germany, Springer.
4. Braffort, A. et al. Eds., (1999). Gesture-Based Communication in Human Computer Interaction: International Gesture Workshop, GW'99, Lecture Notes in Computer Science, VOL.1739, Springer-Verlag.
5. Ishikawa, M. (2000). Recognition of hand-gestures based on self-organization using a DataGlove, Australian Journal of Intelligent Information Processing Systems, Vol.6, No.2, pp.65-71.
6. Ishikawa, M. and Suenaga, H. (2001). Self-organization for temporal data of varying length, ICONIP2001, Shanghai, China, pp.247-252.
7. Ishikawa, M. and Sasaki, N. (2002). Gesture recognition based on SOM using multiple sensors, 9th International Conference on Neural Information Processing(ICONIP2002), Singapore, pp.1300-1304.
8. Kangas, J. (1990). Time-delayed self-organizing maps, IJCNN-90, Vol.2, pp.331-336, San Diego, CA.
9. Kohonen, T., (2001). Self-organizing Maps, 3rd Ed., Springer.
10. Kurokawa, T, Morichi, T. and Watanabe, S. (1993). Bidirectional translation between sign language and Japanese for communication with deaf-mute people, Advances in Human Factors/Ergonomics, 19B, pp.1109-1114.
11. Kurtenbach, G. and Hulteen, E.A. (1990). Gestures in human-computer communication, The Art of Human-Computer Interface Design, Addison-Wesley, pp.309-317.
12. Lee, H-J. and Chen, Z., (1985). Determination of 3D human body postures from a single view, Computer Vision, Graphics, and Image Processing, Vol.30, pp.148-168.
13. Mozer, M. C. (1994). Neural net architectures for temporal sequence processing, in A. Weigend and N, Gershenfeld (Eds.), Time Series Prediction : Forecasting the Future and Understanding the Past, pp.243-264, Addison Wesley.
14. O'Rourke, J. and Badler, N.I. (1980), Model-Based Image Analysis of Human Motion Using Constraint Propagation, IEEE Transactions on Pattern Analysis and Machine Intelligence, Vol. PAMI-2, No.6, November, pp.522-536.

15. Salmela, P., Kuusisto, S., Saarinen, J., Laurila, K. and Haavisto, P. (1996). Isolated spoken number recognition with hybrid of self-organizing map and multilayer Perceptron, Proceedings of the International Conference on Neural Networks, (ICNN'96), Vol.4, pp.1912-1917.

16. Rabiner, L.R. and Juang, B.H. (1986). An introduction to hidden Markov models, IEEE ASSP Magazine, January 1986, pp.4-16.

17. Rabiner, L.R. (1989). A tutorial on hidden markov models and selected applications in speech recognition, Proceedings of the IEEE, vol.77, No.2, pp.257-285.

18. Ritter, H. and Kohonen, T. (1989). Self-organizing semantic maps, Biological Cybernetics, vol.61, pp.241-254.

19. Rohr, K. (1994), Towards Model-Based Recognition of Human Movements in Image Sequences, CVGIP: Image Understanding, Vol.59, No.1, pp.94-115.

20. Rubine, D. (1991). Specifying gestures by example, Computer Graphics, Vol.25, No.4, pp.329-337.

21. Takahashi, K., Seki, S., and Oka, R. (1994), Spotting recognition of human gestures from motion images, Time-Varying Image Processing and Moving Object Recognition, 3, in Cappellini, V. (Ed.), Proceedings of the 4th International Workshop, pp.65-72, Elsevier.

Enhanced phrase-based document clustering using Self-Organizing Map (SOM) architectures

M. Hussin [a], J. Bakus [b], and M. Kamel[b]

[a] Department of Computer Science and Automatic Control, University of Alexandria, Alexandria, Egypt
[b] PAMI Lab, University of Waterloo, Canada
{mfarouk,jbakus,mkamel}@uwaterloo.ca

Summary. Availability of large full-text document collections in electronic form has created a need for tools and techniques that assist users in organizing these collections. Document clustering is one of the popular methods used for this purpose. The Self-organizing map (SOM), an unsupervised algorithm for clustering and topographic mapping, has shown promising results in this task. Most of the existing SOM techniques rely on a "bag of words" document representation. Each word in the document is considered as a separate feature, ignoring the word order. In this chapter we investigate the use of phrases rather than words as document features for document clustering. We present a phrase grammar extraction technique, and use the extracted phrases as features in two different document clustering algorithms, self-organizing map (SOM) and hierarchical self-organizing map (HSOM). We present results of clustering documents from the REUTERS corpus and show an improvement in the clustering performance evaluated using the entropy and F-measure.

Key words: Text mining, Document clustering, Self-organizing map, Phrase extraction

1 Introduction

The growing number of text information available on the internet and other sources is raising interest in automatic organization of this data. A number of ma-chine learning techniques have been proposed to deal with this issue, and one such technique is document clustering. The objective of document clustering is to identify inherent groupings of text documents, such that given a set of documents it is able to separate them into a number of clusters, where each cluster contains documents about similar topics. Document clustering has been used in a number of applications. In information retrieval systems, it has been used to improve the precision and recall performance [1], and as an efficient way to find similar documents [2]. It has also been proposed in browsing document collections [3], organizing the results of a search engine query [4], and automatically generating hierarchical grouping of documents [5]. One of the successful approaches to document clustering is to use self-organizing map (SOM) neural networks. The SOM network is a special type of neural network that can learn patterns from complex, multi-dimensional data and transform them into visually organized clusters. The theory of the SOM network is motivated by the observation of the operation of the brain. Various human sensory impressions are neurologically mapped into the brain such that spatial or other relations among stimuli correspond to spatial relations among the neurons organized into a two-dimensional map [6, 7, 8]. The main function of SOM networks is to map the input data from an high dimensional space to a lower (usually one or two) dimensional plot, while maintaining the original topological relations. The physical location of points on the map shows the relative similarity between the points in the original space. SOM is well suited to text input, because it works very well on noisy highly dimensional data, which is the characteristic of text data. A number of SOM networks have been successfully applied to the document clustering problem. Some examples include the ET-Map [9], SOMLib [10] using hierarchical maps, WEBSOM [11, 12] focused on very large corpora, and hierarchical SOMART [13] which uses a combination of SOM and adaptive resonance theory (ART) to generate the clusters. Representation model is very important in the text clustering. While clustering methodology has evolved to allow computation of large high dimensional data sets by dimensionality reduction [14] and optimization of the clustering process [12], the representation model based on the "bag of words" paradigm essentially remained the same [15]. In this paradigm, the entire document is represented as a list of all (or a subset of) the words found in the document, ignoring the order or context of the words. In order to improve the quality of document clustering algorithm, this chapter presents a document clustering approach that captures some of the context of the words by using phrases rather than individual words as the features. We used the hierarchical phrase grammar [16] to extract frequently occurring phrases, and use this as feature in two different SOM based document clustering architectures. One of the architectures is the original SOM map, and other is a hierarchical version of the SOM. The phrases are extracted utilizing a grammar technique using statistical collocations. The grammar is constructed from a training corpus by iteratively merging a pair of adjacent words tokens to form a rule. Each rule can be further merged together in subsequent iterations to form longer phrases. The merging is performed according to mutual information association measure. To extract the phrases, the grammar rules are applied in the order of the association weights to a new document. The reminder of this chapter is organized as follows. Section 2 describes related work. Section 3 presents the document clustering architecture. Section 4 proposes the vector space model, and section 5 describes phrase extraction from text using hierarchical grammar. Section 6 describes document clustering using the two different architectures of SOM. Quality measures of document clustering are presented in section 7. In Section 8 we report the experimental results and their evaluation. The conclusion is given in section 9.

2 Related Work

The representation of documents is an important issue in the machine learning tasks, such as document clustering or document classification. The representation should be as compact as possible in order to allow efficient processing of large document collections, yet it should contain all of the relevant information. SOM based document clustering methods usually use the vector space model [17] to represent document objects, where each document is represented as a vector of features. Each feature in this vector is assigned a weight in several ways: binary, term frequency, or term frequency/inverse document frequency (tf/idf) [18] depending on the application. In most cases, the "bag of words" feature representation is used, where each feature corresponds to one word in the document, however a number of attempts have been made to extend the representation to include phrases rather than words. Zamir and Etzioni [19] proposed a new phrase model called suffix tree for a document clustering. Using this model, the clustering algorithm finds common phrases between two documents by building a tree of phrases and then compares the branches. This tree representation was extended by Hammouda in [20, 21] to a graph representation. The phrase based representation has also been successfully used in document classification. Mladenic and Gorbelnik [22] enumerate all possible phrases up to a length of five words and use feature extraction to prune out the irrelevant phrases. The remaining phrases are used in a document classification task using a Naive Bayes classifier. Furnkranz et al. [23] use the Ripper [24] algorithm to extract phrases and use these in a document classifier. In the above work, phrase based text classification is based on a particular classifier. Caropresso et at. [25] investigate phrase based text classification independent of the classifier used.

3 Document Clustering Architecture

In this section, an overview of the document clustering architecture will be presented. Document clustering can be viewed as a process of grouping the set of documents into a set of clusters, as shown in Figure 1

3.1 Feature extraction

It is the process of transforming the documents collection into a suitable representation to be enable a further analysis by the clustering algorithm. In the preprocessing step, the document set is first cleaned by removing stop-words and then applying a stemming algorithm that converts different word forms into a similar canonical form. The stop-words are frequent words that carry no information (i.e. pronouns, prepositions, conjunctions etc.). By word stemming we mean the process of suffix removal to generate word stems. This is done to group words that have the same conceptual meaning, such as walk, walker, walked, and walking. In the indexing step, a number of different strategies have been suggested over the years of information retrieval research. The most commonly used document representation is the so called vector

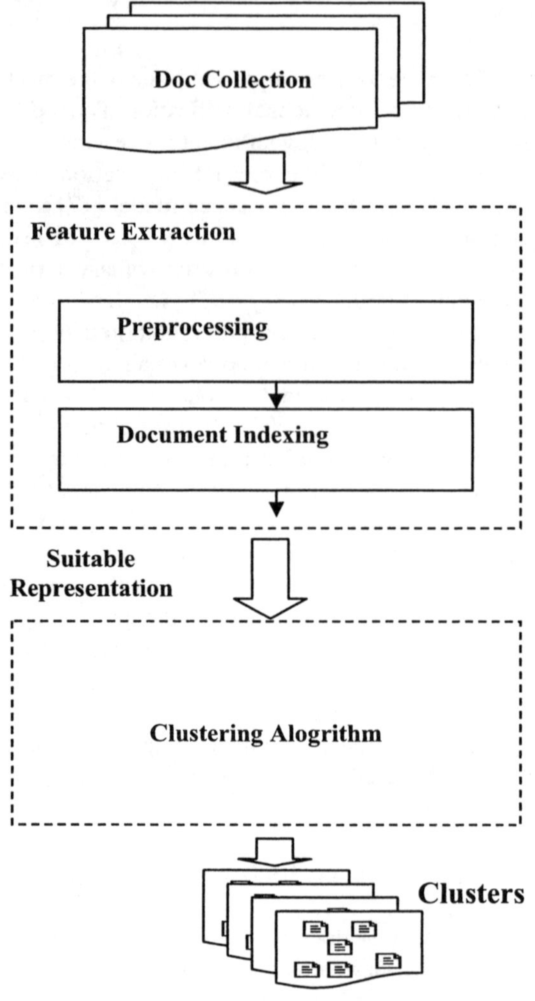

Fig. 1. Cluster Architecture

space model, where each document is represented as a vector of features, the detailed description of the vector space model will be given in section 4.

3.2 Clustering Algorithm

It takes the suitable representation of documents collection as input and output the clustered data. There are three main categories of clustering : Hierarchical clustering, partitional clustering, and unsupervised neural networks.

Hierarchical clustering

A hierarchical clustering algorithm creates an hierarchical decomposition of the given set of data objects. Depending on the direction of building the hierarchy we can identify two methods of hierarchical clustering: Agglomerative and Divisive. The agglomerative approach is the most commonly used in hierarchical clustering.

- **Agglomerative Hierarchical Clustering (bottom-up):** This method starts with the set of objects as individual clusters, then, at each step merges the most two similar clusters. This process is repeated until a minimal number of clusters have been reached, or, if a complete hierarchy is required then the process continues until only one cluster is left. The method is very simple but needs to specify how to compute the distance between two clusters. Three commonly used methods for computing this distance are listed in table 1,

Method	Distance Function
Single Linkage	$\| T - S \| = \min_{x \in T; y \in S} \| x - y \|$
Complete Linkage	$\| T - S \| = \max_{x \in T; y \in S} \| x - y \|$
Average Linkage	$\| T - S \| = \dfrac{\sum_{x \in T; y \in S} \| x - y \|}{|S|.|T|}$

Table 1. Three different distance function in agglomerative hierarchical clustering.

where T, and S are two clusters.

- **Divisive Hierarchical Clustering (top-down):** This method recursively partitioning the whole data objects as one cluster until singleton sets are reached. In this case, we need to decide at each step, which cluster to split and how to perform the split. A simpler way would be to choose the largest cluster to split, splitting a cluster requires the decision of which objects go to which sub-cluster. One method is to find the two sub-clusters using K-means, resulting in a hybrid technique called bisecting K-means [2].

Partitional clustering

Partitional clustering algorithms create a one level un-nested partitioning of the data sets. If K is the desired number of clusters, then partitional approaches typically find all K clusters at once. The most known class of partitional clustering algorithms are the k-means algorithm and its variants. K-means starts by randomly selecting k seed cluster means, then assigns each object to its nearest cluster mean. The algorithm then iteratively recalculates the cluster means and new object memberships. The process continues up to a certain number of iterations, or when no changes are detected in the clusters means.

Unsupervised neural networks

A most successful approach to document clustering is to use unsupervised artificial neural networks. Interestingly, neural networks are highly suited to textual in-put, being capable of identifying structure of high dimensions within a body of natural language text. Neural networks work best with data that contains noise, has a poorly understood structure and changing characteristics. The self-organizing map (SOM) is one of the most versatile unsupervised neural network architectures. It is capable of ordering high-dimensional statistical data in such a way that similar input items are grouped spatially close to one another. In this chapter we used the SOM as our clustering algorithm, more details will be given in section 6. The Adaptive Reasonance Theory (ART) is another unsupervised neural network architectures used in clustering. It appears to be particularly suitable because of their well-defined interfaces as well as features that most other networks lack. In specific, ART networks have the ability to create new output nodes (i.e. category) dynamically, and do not suffer from the problem of forgetting previously learned categories if the environment changes. They too, however, can only develop input categories at a given level of specificity, which depends on a global parameter called vigilance.

4 Vector Space Model for Document Representation

Most document clustering methods use the vector space model to represent document objects, where each document is represented by a word vector. Typically, the input collection of documents is represented by a word-by-document matrix A

$$A = (a_{ik}),$$

where a_{ik} is the weight of word i in document k. There are three different main weighting schemes used for the weight a_{ik}. Let f_{ik} be the frequency of word i in document k, N the number of documents in the collection, and n_i the total number of times word i occurs in the whole collection. Table 2 summarizes the three weighting schemes that will be used in our experiments.

5 Hierarchical Phrase Grammar

The objective of the phrase extraction procedure is to find a pair of adjacent words (bigram) that tend to occur frequently together. Given a bigram $< w_i; w_j >$ where the word w_i is followed by the word w_j that tend to occur frequently together, the two words are merged and replaced by a new word w_k. The merge of the bigram is determined by an association measure of the bigram determined from a training corpus. We use a mutual information association measure [26]. The first step is to create a frequency table of all the words and bigrams present in a training corpus. The frequency count of a word w_i and a bigram $< w_i; w_j >$ is given as $N(w_i)$ and $N(w_i, w_j)$ respectively. The total number of words and bigrams in the corpus

Method	Weighting Scheme	Comments
Binary	$a_{ik} = \begin{cases} 1 & \text{if} f_{ik} > 0 \\ 0 & \text{otherwise} \end{cases}$	The simplest approach is to let the weight be 1 if the word occurs in the document and 0 otherwise.
Term Frequency	$a_{ik} = f_{ik}$	Another simple approach is to use the frequency of the word in the document.
Term Frequency Inverse Document Frequency	$a_{ik} = f_{ik} * \log\left(\frac{N}{n_i}\right)$	A well-known approach for computing word weights, which assigns the weight to word i in document k in proportion to the number of occurrences of the word in the document, and in inverse proportion to the number of documents in the collection for which the word occurs at least once.

Table 2. Three different weighting schemes used in vector space model.

is given as N_w and N_b respectively. The mutual information association measure is defined as the measure of the amount of information that one event contains about another event. Given two events i and j, then the mutual information between them is given as:

$$I(i,j) = log_2 \frac{p(i,j)}{p(i)p(j)} \tag{1}$$

where $P(i)$ and $P(j)$ are the event probabilities and $P(i,j)$ is the joint probability. Using the training corpus frequency counts, the mutual information association measure between words w_i and w_j is given as:

$$A_{MI}(w_i, w_j) = log_2 \frac{N(w_i, w_j)/N_b}{N(w_i)N(w_j)/N_w^2} \tag{2}$$

Using the mutual information, association weight is calculated for each of the bigrams in the training corpus. The bigram $< w_i; w_j >$ with the highest positive weight is replaced in the entire corpus with a new non-terminal symbol w_k, such that this new symbol may form bigrams in subsequent operations. The bigram $< w_i; w_j >$, new symbol w_k , and association weight $A(w_i, w_j)$ are stored in the grammar table. Table 2 shows an example grammar for the phrase "new york stock exchange". The frequency counts and association weights for all the bigrams are recalculated to reflect the merge operation and the procedure is repeated until there are no more bigrams with a positive weight. The grammar generation algorithm is as follows:

1. Make a frequency table of all the words and bigrams in the training corpus.
2. From the frequency table calculate the association weight for each bigram.
3. Find a bigram $< w_i; w_j >$ with the maximum positive association weight $A(w_i, w_j)$. Quit the algorithm if none found.

4. In the training corpus, replace all the instances of the bigram $< w_i; w_j >$ with a new symbol w_k. This is now the new corpus.
5. Add $< w_i; w_j >$, w_k, and $A(w_i, w_j)$ to the grammar.
6. Update the frequency table of all the words and bigrams.
7. Go to step 2.

The merge operation creates a new symbol w_k that can be used in subsequent merges, and therefore arbitrary long phrases can be extracted. Similarly, each of the merged bigrams expands to a text phrase of two or more words. Table 3 shows a list of the rules that make the phrase "new york stock exchange". To extract the phrases, a similar iterative algorithm is used. First, a list of bigrams from the input word sequence is collected and followed by similar iterative bigram merging operation. Rather than calculating the association weights from the occurrence counts, the weights extracted from the training corpus are used. As a result, only bigrams that have been merged in the training can be merged during the phrase extraction. The phrase extraction is given as follows:

1. Make a list of all the bigrams in the input sequence.
2. Find a bigram $< w_i; w_j >$ with the maximum association weight $A(w_i, w_j)$ in the list of rules.
3. In the sequence, replace all the instances of the bigram $< w_i; w_j >$ with a new symbol w_k (taken from the rule). This is now the new sequence.
4. For each replaced instance output the expanded phrase.
5. Update the list of all the bigrams in the sequence.

Associated Weight	Bigram	New Symbol	Expanded Phrase
7.6	(new,york)	w_1	"new york"
5.2	(stock,exchange)	w_2	"stock exchange"
8.5	(w_1, w_2)	w_3	"new york stock exchange"

Table 3. Resulting grammar from the phrase "new york stock exchange" using the mutual information association measure.

6 Self-Organizing Map (SOM) and Hierarchical SOM

The self-organizing map consists of a layer of input units each of which is connected to a grid of output units. These output units are arranged in some application dependent topological order. The notion of the input layer is used more or less for historical reasons to have some compatibility to other architectures of artificial neural networks. Input units take the input in terms of a feature vector, and propagate it to the output units. Each of the output units in turn is assigned a weight vector of the same dimensionality as the input data.

The learning algorithm of self-organizing maps can be seen as an extension to the competitive learning paradigm. Pragmatically speaking, during each of the learning steps the unit with the highest activation, i.e. the best matching unit, with respect to a randomly selected input vector is adapted in a way that it will exhibit even higher activation with respect to this input in future. Additionally, the units in the neighborhood of the best matching unit are also adapted to exhibit higher activation with respect to the given input. The learning algorithm of self-organizing maps can be described as follows:

1. Random selection of one input-vector x.
2. Determination of the best matching unit i by using the Euclidean distance measure.

$$i :\| w_i - x \| \leq \| w_j - x \| \; \forall j \in O$$

   In this formula w_i and w_j denote the weight vectors assigned to unit i j in output space O respectively.
3. Adaptation of the weight vectors w_j in the neighborhood of the best matching unit i.

$$\Delta w_j = \varepsilon * \delta(i,j) * (x - w_j) \forall j \in O,$$

   where ε is learning-rate, and δ is the function to guarantee that the larger the distance between units i and j, i.e. $\| i - j \|$ in the output space O, the smaller the adaptation of the weight vector w_j.
4. Repeat steps (1) through (3) until no more changes to the weight vectors are observed.

The strength of the adaptation is determined with respect to parameter ε, the neighborhood relation between the best matching unit i and the output unit j which is currently under consideration, and the distance between input vector x and weight vector w_j assigned to unit j.

To guarantee convergence of the map the learning-rate as well as the adapted neighborhood have to shrink in the course of time. In other words, the adaptation of the weights as well as the set of units that are subject to adaptation decrease gradually over the training. For more detailed description of the learning algorithm consult [7].

The key idea of the hierarchical SOM is to build a hierarchically organized SOM neural network with layered architecture where each layer consists of a number of independent SOMs as shown in Figure 2. More precisely, for each map unit in one layer, a SOM is added to the next layer, based on the size of unit "number of documents in the unit".

This network is trained sequentially from the first layer downwards along the hierarchy until the SOM networks of the last layer are reached. Hence, as soon as the first layer SOM has reached its stable state, training continues with the SOMs in the second layer based on the size of each unit in the first SOM. In this layer, however, each map is trained only by using the patterns mapped onto the corresponding unit of the layer above. Moreover, on the transition from one layer to the next, the input

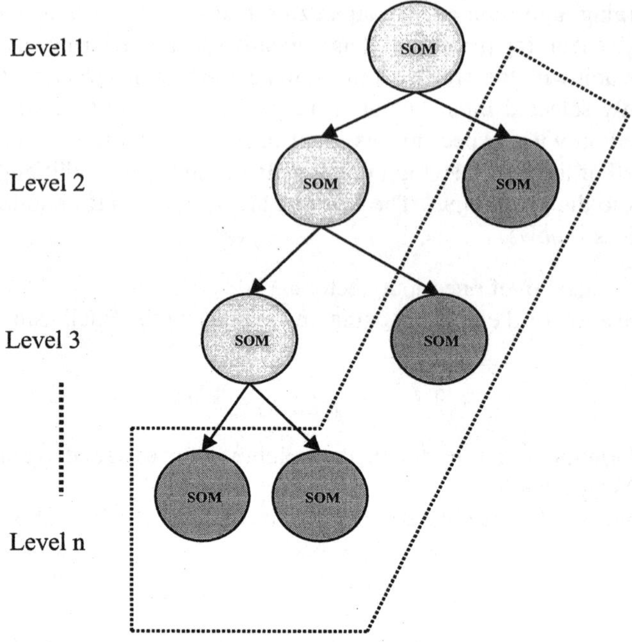

Fig. 2. Architecture of a hierarchical SOM.

patterns may be shortened by omitting those components that are equal in the patterns mapped onto the same unit. This feature is of special importance when dealing with text as the underlying input data because first text documents are represented in a very high dimensional feature space and second, some of these features are common to each text belonging to a particular topic, i.e. cluster. Thus, the pattern reduction inherent to hierarchical SOM with high quality cluster output is expected to play an important role during the learning process. The recursive algorithm for document clustering using hierarchical SOM is summarized in Table 4. A highly valuable property of hierarchical SOM is the remarkable speed-up in the training process as compared to SOM.

7 Quality Measures

An important issue of document clustering is how to measure the performance of the clustering algorithm. Many measures have been used, each of which has been designed to evaluate some aspect of the categorization performance of a system. Two measures are widely used in text mining literature to evaluate the quality of the clustering algorithms: cluster entropy and F-measure [2]. Both of these techniques rely on labelled test corpus, where each document is assigned to a class label. The mea-

Given a set of documents S,
1. Prepare the vector space representation of set S.
2. Apply SOM technique to partition S into m units, $(s_1, s_2, \cdots, s_m)$.
3. Loop
    For each unit of s_i
      If size of unit s_i in suitable dimension Then
        output the cluster.
      Else
        Repeat from 1 to 3 on s_i
      Endif
    Until s_i is finished.

Table 4. Algorithm for hierarchical SOM document clustering.

sures compare the resulting clusters to the labelled classes, and measure the degree to which the documents from same classes are assigned to the same clusters.

The cluster entropy uses the entropy concept from information theory and measures the "homogeneity" of the clusters. Lower entropy indicates more homogeneous cluster and vice versa. Consider the results of a clustering experiment, and let $P(i, j)$ be the probability that a document has a class label i and is assigned to cluster j. Then the entropy E_j for a cluster j is given as:

$$E_j = -\sum_i P(i, j) log_2 P(i, j) \tag{3}$$

The total entropy for a set of clusters is calculated as the sum of entropies for each cluster weighted by the cluster size:

$$E = \sum_j \frac{n_j}{n} E_j \tag{4}$$

where n_j is the number of documents in cluster j and n is the total number of documents.

The F-measure combines the precision and recall concepts from information retrieval, where each class is treated as the desired results for the query, and each cluster is treated as the actual results for the query. To evaluate the queries, precision and recall for class i and cluster j is given as:

$$\begin{aligned} Recall(i, j) &= \frac{n_{ij}}{n_j} \\ Precision(i, j) &= \frac{n_{ij}}{n_i} \end{aligned} \tag{5}$$

where n_{ij} is the number of documents with class label i in cluster j, n_i is the number of documents with class label i, and n_j is the number of documents in cluster j, and n is the total number of documents. The F-measure for class i and cluster j is given as:

$$F(i, j) = 2 \left[\frac{Recall(i, j) Precision(i, j)}{Recall(i, j) + Precision(i, j)} \right] \tag{6}$$

The F-measure for all the clusters is given as a weighed sum of the maximum F-measures for each class as:

$$F = \sum_i \frac{n_i}{n} \max_j F(i, j) \tag{7}$$

In the test corpus, some documents have multiple classes assigned to them. This does not affect the F-measure, however the cluster entropy can no longer be calculated. Instead of cluster entropy, we define the class entropy [27], which measures the homogeneity of a class rather than the homogeneity of the clusters. For every cluster j and class i, we compute $P(j \mid i)$, the probability that a document is assigned to cluster j given that it belongs to class i. Using these probabilities, the entropy for a class i is given as:

$$E_i^* = -\sum_j P(j \mid i) \log_2 P(j \mid i) \tag{8}$$

We use the conditional probabilities, rather than joint probabilities (as in Eq. 3), because the probabilities are normalized over the summation, i.e. $\sum_j P(j \mid i) = 1$. Therefore, Eq. 8 is a true entropy expression, while Eq. 3 is not a true entropy expression, because $\sum_i P(i, j) \neq 1$. The overall class-entropy E^* is calculated as a sum of the entropies E_i^*, weighted by the class probabilities:

$$E^* = \sum_i \frac{n_i}{n} E_i^* \tag{9}$$

where n_i is the number of document in class i, and n is the total number of documents.

8 Experimental Results

In order to test the effectiveness of the phrase features compared to the word features, we used the REUTERS [28] test corpus. This is a standard text clustering corpus, composed of 21578 news articles. The documents in the Reuters collection were collected from Reuters newswire in 1987. From this corpus, 1000 documents were selected and used as the test set to be clustered. From the remainder, 10000 documents were selected as the training corpus for the phrase extraction grammar. Each document is processed by removing a set of common words using a "stopword" list, and the suffixes are removed using a Porter stemmer. The summary of data sets used in this chapter is shown in Table 5.

Two different cases were investigated: individual words as features, and extracted phrases as features. For each case, three different representations were used: bi-nary, term frequency, and tf/idf. The documents were clustered using the SOM, and HSOM techniques implemented using the SOM-PAK package developed by Kohonen et al. [29], and a square map was used with different sizes ranging from 4 units (2×2) to 100 units (10×10).

The configurations of these document clustering techniques was as follows:

Data Set	Source	Number of documents	Features dimension
Word	Reuters-21578	1000	7293
Phrase	Reuters-21578	1000	16462

Table 5. Summary description on data sets.

- The HSOM consists of two layers the first layer is SOM with dimension 2×2 using 0.02 learning rate, and the second layer was organized from giving each resulting unit in the first SOM layer another SOM using 0.02 learning rate, with different dimensions 2, 3×3, 3×4, 4×4, 4×5, 5×5, 5×6, and 6×6. The resulted number of clusters is 18, 27, 36, 48, 60, 75, 90, 108 respectively.
- The SOM used with dimensions 2×2, 3×3, 4×4, 5×5, 6×6, 7×7, 8×8, 9×9, and 10×10, and 0.02 learning rate, which results in the number of clusters as 4, 9, 16, 25, 36, 49, 64, 81, and 100 respectively.

Number of clusters	Data Set				Improvement	
	Word		Phrase		Class entropy	F-measure
	Class entropy	F-measure	Class entropy	F-measure		
4	1.19	0.38	1.24	0.44	4.20%	16.19%
9	2.09	0.43	1.95	0.41	-6.69%	-5.72%
16	2.49	0.37	2.37	0.42	-4.81%	12.08%
25	2.86	0.36	2.86	0.43	0.00%	19.68%
36	3.25	0.31	3.16	0.39	-2.76%	26.59%
49	3.46	0.27	3.45	0.39	-0.28%	42.83%
64	3.64	0.27	3.64	0.38	0.00%	38.68%
81	3.86	0.28	3.77	0.29	-2.33%	1.67%
100	3.97	0.31	3.82	0.30	-3.77%	-2.06%

Table 6. Comparison of the Class entropy and F-measure using binary representation and clustered by SOM technique.

The class entropy and F-measure results are shown in Tables 6-11 and Figures 3-8 for the three different document representations applied to SOM and HSOM techniques. In most cases with SOM and HSOM the phrase representation performed better in the clustering task than the word representation with different number of output clusters. The average reduction of the class entropy using the SOM technique and the use of phrases rather than words is 1.8%, 2.5%, and 4.2% for binary, term frequency, and tf/idf respectively. The corresponding improvement in the F-measure is 16.7%, 10.6%, and 4.1% for binary, term frequency, and tf/idf respectively. Similarly, the average reduction of the class entropy using HSOM and phrases instead of words is 40.1%, 59.4%, and 65.1% for binary, term frequency, and tf/idf respectively, while the corresponding improvement in F-measure is 10.3%, 5.2%, and 12.4%.

Number of clusters	Data Set				Improvement	
	Word		Phrase		Class entropy	F-measure
	Class entropy	F-measure	Class entropy	F-measure		
4	1.16	0.38	1.17	0.44	0.86%	14.65%
9	2.20	0.34	2.12	0.44	-3.63%	29.06%
16	2.70	0.32	2.63	0.40	-2.59%	24.63%
25	3.07	0.31	3.01	0.36	-1.95%	16.49%
36	3.45	0.29	3.34	0.31	-3.18%	6.66%
49	3.65	0.29	3.62	0.31	-0.82%	7.88%
64	3.93	0.28	3.71	0.27	-5.59%	-2.27%
81	4.09	0.25	3.99	0.26	-2.44%	3.72%
100	4.23	0.27	4.11	0.26	-2.83%	-5.20%

Table 7. Comparison of the Class entropy and F-measure using term frequency representation and clustered by SOM technique.

Number of clusters	Data Set				Improvement	
	Word		Phrase		Class entropy	F-measure
	Class entropy	F-measure	Class entropy	F-measure		
4	1.07	0.46	1.00	0.46	-6.54%	-0.88%
9	2.03	0.40	1.93	0.44	-4.92%	11.02%
16	2.53	0.37	2.40	0.45	-5.13%	20.34%
25	2.89	0.37	2.75	0.40	-4.84%	7.80%
36	3.25	0.37	3.02	0.34	-7.07%	-8.99%
49	3.34	0.38	3.25	0.34	-2.69%	-10.35%
64	3.51	0.33	3.53	0.36	0.56%	8.64%
81	3.77	0.32	3.54	0.34	-6.10%	7.36%
100	3.83	0.31	3.81	0.31	-0.52%	1.82%

Table 8. Comparison of the Class entropy and F-measure using tf/idf representation and clustered by SOM technique.

Number of clusters	Data Set				Improvement	
	Word		Phrase		Class entropy	F-measure
	Class entropy	F-measure	Class entropy	F-measure		
16	2.07	0.38	1.27	0.42	-38.64%	8.22%
25	2.11	0.38	1.27	0.42	-39.81%	9.48%
36	2.13	0.39	1.28	0.42	-39.90%	9.53%
49	2.15	0.39	1.28	0.43	-40.46%	10.68%
64	2.15	0.39	1.28	0.43	-40.46%	10.18%
81	2.15	0.39	1.28	0.43	-40.46%	11.35%
100	2.16	0.39	1.28	0.43	-40.74%	11.86%
108	2.15	0.39	1.28	0.43	-40.46%	11.09%

Table 9. Comparison of the Class entropy and F-measure using binary representation and clustered by HSOM technique.

Number of clusters	Data Set				Improvement	
	Word		Phrase		Class entropy	F-measure
	Class entropy	F-measure	Class entropy	F-measure		
16	2.58	0.40	1.28	0.41	-50.38%	2.93%
25	3.00	0.40	1.28	0.41	-57.33%	1.78%
36	3.19	0.40	1.28	0.41	-59.87%	3.73%
49	3.24	0.40	1.29	0.42	-60.18%	4.69%
64	3.30	0.40	1.29	0.43	-60.90%	5.75%
81	3.40	0.41	1.29	0.43	-62.05%	5.97%
100	3.41	0.40	1.29	0.44	-62.17%	8.38%
108	3.43	0.41	1.29	0.44	-62.39%	8.50%

Table 10. Comparison of the Class entropy and F-measure using term frequency representation and clustered by HSOM technique.

Number of clusters	Data Set				Improvement	
	Word		Phrase		Class entropy	F-measure
	Class entropy	F-measure	Class entropy	F-measure		
16	2.67	0.41	1.1	0.47	-58.80%	14.41%
25	2.85	0.42	1.1	0.48	-61.40%	13.04%
36	3.08	0.42	1.1	0.48	-64.28%	14.11%
49	3.22	0.42	1.1	0.48	-65.83%	12.31%
64	3.33	0.42	1.1	0.48	-66.96%	13.66%
81	3.39	0.43	1.1	0.48	-67.55%	12.50%
100	3.41	0.44	1.1	0.48	-67.74%	9.470%
108	3.47	0.44	1.1	0.48	-68.29%	9.78%

Table 11. Comparison of the Class entropy and F-measure using tf/idf representation and clustered by HSOM technique.

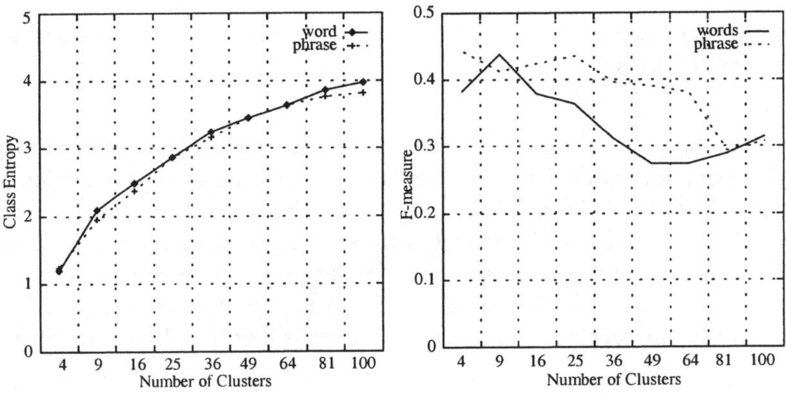

Fig. 3. Class entropy and F-measure for 1000 document set, using binary representation and clustered by SOM technique.

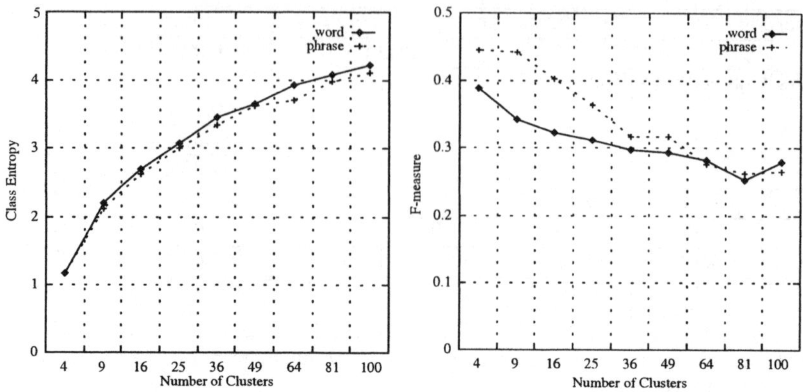

Fig. 4. Class entropy and F-measure for 1000 document set, using term frequency representation and clustered by SOM technique.

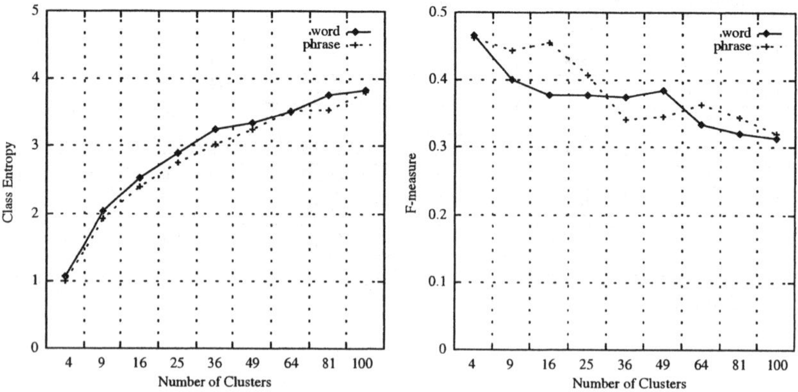

Fig. 5. Class entropy and F-measure for 1000 document set, using tf/idf representation and clustered by SOM technique.

As shown in Figure 9, the HSOM clustering technique also performed better than the SOM clustering technique. Using the words as features, the average reduction of the class entropy of HSOM over SOM is 45.6%, 19.4%, and 11% for binary, term frequency, and tf/idf respectively, and the corresponding increase in the F-measure is 43%, 63%, and 40%. Similarly, using phrases as features, the average reduction of the class entropy is 66.5%, 68.6%, and 71.1% for binary, term frequency, and tf/idf respectively, while the improvement in the F-measure is 48.6%, 67.4%, and 51.5% respectively.

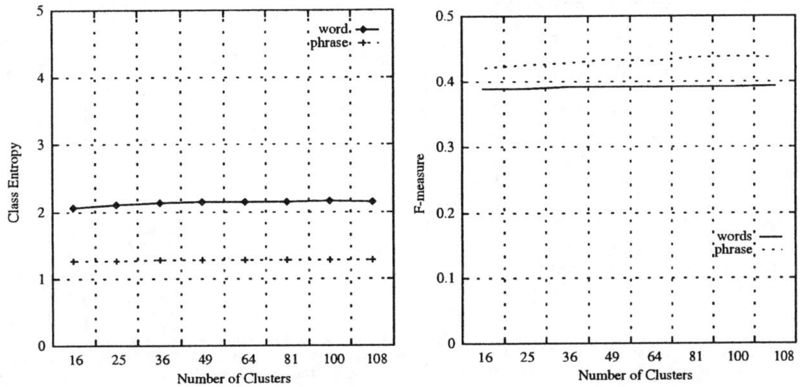

Fig. 6. Class entropy and F-measure for 1000 document set, using binary representation and clustered by HSOM technique.

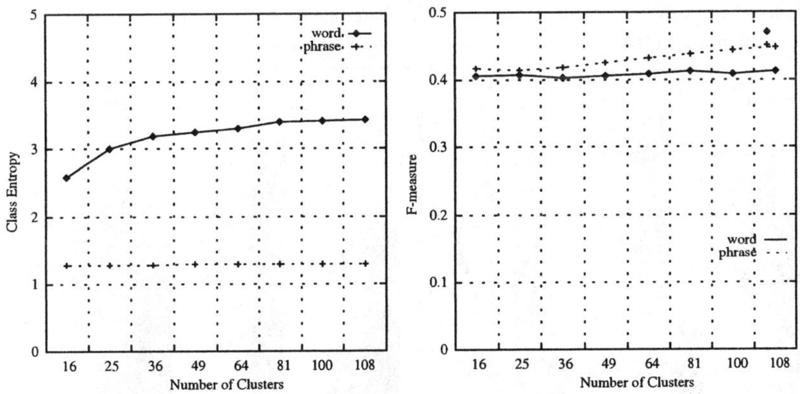

Fig. 7. Class entropy and F-measure for 1000 document set, using term frequency representation and clustered by HSOM technique.

9 Conclusion

In this chapter, we presented two different types of self-organizing map architectures SOM and HSOM and tested them using both word features and phrase features. The clustering was performed using two stages, where in the first stage phrases are extracted using grammar based on mutual information and used to generate a vector space for document representation, at the second stage vector space model applied to the SOM and HSOM document clustering techniques. The performance was evaluated by testing the phrase based document clustering on the REUTERS test corpus, and comparing it to word based document clustering using both class entropy and F-measure. The experimental results show that the phrase based features achieved

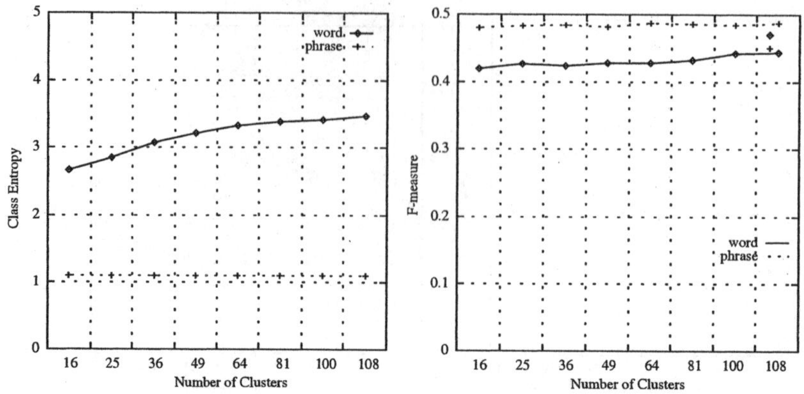

Fig. 8. Class entropy and F-measure for 1000 document set, using tf/idf representation and clustered by HSOM technique.

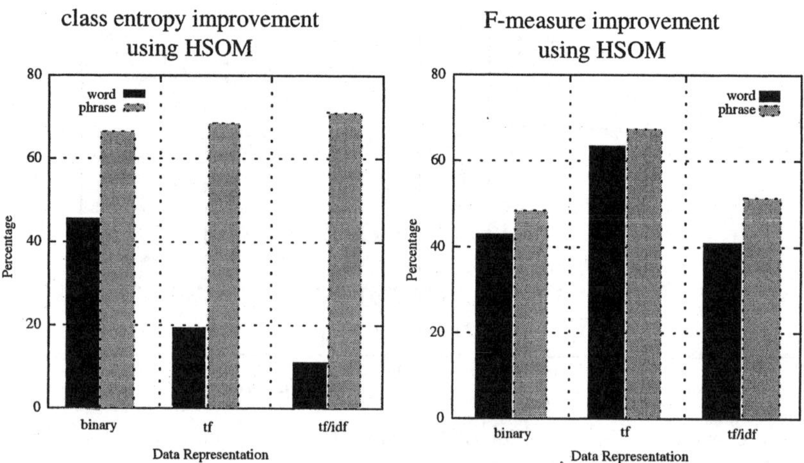

Fig. 9. Class entropy and F-measure improvement using HSOM compared with SOM in both phrase and word data representation.

a better quality clustering than word in both SOM and HSOM techniques. It also demonstrate an improvement in the clustering performance of HSOM over that of SOM.

References

1. C. Van Rijsbergen, "Information Retrieval", Butterworth, London, UK, 1979.

2. M. Steinbach, G. Karypis, and V. Kumar, "A comparison of document clustering techniques", KDD'2000, Workshop on Text Mining, 2000.

3. D. Cutting, D. Karger, J. Pedersen, and J. Tukey, "Scatter/gather: A cluster-based approach to browsing large document collections", In SIGIR'92, pp. 318-329, 1992.

4. O. Zamir, O. Etzioni, O. Madani, and R. Karp, "Fast and intuitive clustering of web documents", In KDD'97, pp. 287-290, 1997.

5. D. Koller and M. Sahami, "Hierarchically classifying documents using very few words", In Proceedings of the 14th International Conference on Machine Learning (ICML), pp. 170-178, 1997.

6. T. Kohonen, "Cybernetic Systems: Recognition, Learning, Self-Organization", In: E. Caianiello, and G. Musso (Eds.), Research Studies Press Ltd., Letchworth, Herfordshire, UK, pp. 3, 1984.

7. T. Kohonen, "Self-Organization and Associative Memory", Springer-Verlag, Berlin, 1989.

8. T. Kohonen, "Self-organizing maps", Springer-Verlag, Berlin, 1995.

9. H. Chen, C. Schuffels, and R. Orwing, "Internet categorization and search: A machine learning approach", Journal of Visual Communication and Image Representation, 7, pp. 88-102, 1996.

10. A. Rauber, and D. Merkl, "The SOMLib digital library system", Proceedings of the Third European Conference on Research and Advanced Technology for Digital Libraries (ECDL'99), Paris, France, September 22-24. Springer-Verlag, 1999.

11. T. Honkela, "Self-organizing maps in natural language processing", Doctoral Dissertation, Helsinki University of Technology, Espoo, Finland, 1997.

12. T. Kohonen, S. Kaski, K. Lagus, J. Salojärvi, V. Pattero, and A. Saarela, "Organization of a massive document collection", IEEE Transactions on Neural Networks, Special Issue on Neural Networks for Data Mining and Knowledge Discovery, 11(3), pp. 574-585, 2000.

13. M. Hussin, and M. Kamel, "Document clustering using hierarchical SOMART neural network", In Proceedings International Joint Conference on Neural Network (IJCNN'03), Portland, Oregon, USA, July, 2003.

14. S. Kaski, "Dimensionality reduction by random mapping: Fast similarity computation for clustering", Proceedings of International Joint Conference on Neural Network (IJCNN'98), Vol. 1, pp. 413-418, 1998.

15. D. Pullwitt, "Integrating contextual information to enhance SOM-based text document clustering", Journal of Neural Networks, 15, pp. 1099-1106, 2002.

16. J. Bakus, M. Kamel, and T. Carey, "Extraction of text phrases using hierarchical grammar". In Proceedings of the 15th conference of the Canadian Society for Computational Studies of Intelligence, pp. 319-324, 2002.

17. G. Salton, C. Yang, and A. Wong, "A vector-space model for automatic indexing", Communications of the ACM, 18(11), pp. 613-620, 1975.

18. G. Salton, and C. Buckley, "Term weighting approaches in automatic text retrieval", Technical Report 87-88, Cornell University, Department of Computer Science, Ithaca, New York, USA, 1987.

19. O. Zamir, and O. Etzioni, "Web document clustering: A feasibility demonstration". In SIGIR'98, pp. 46-54, 1998.

20. K. Hammouda, and M. Kamel, "Phrase-based Document Similarity based on an Index Graph Model", In Proceedings of 2002 IEEE International Conference on Data Mining (ICDM02), Maebashi, Japan, pp. 203-210, 2002.

21. K. Hammouda, and M. Kamel, "Document Similarity Using a Phrase Indexing Graph Model", Knowledge and Information Systems, In Press, 2003.

22. D. Mladenić, and M. Grobelnik, "Word sequences as features in text-learning", In Proceedings of the 17th Electrotechnical and Computer Sciences Conference (ERK-98), Ljubljana, Slovenia, pp. 145-148, 1998.

23. J. Furnkranz, T. Mitchell, and E. Riloff, "A case study in using linguistic phrases for text categorization on the WWW". In Proceedings of the 1st AAAI Workshop on Learning for Text Categorization, pp. 5-12, 1998.

24. W. Cohen, and Y. Singer, "Context-sensitive learning methods for text categorization", ACM Transactions on Information Systems, 17(2), pp. 141-173, 1999.

25. M. Caropreso, S. Matwin, and F. Sebastiani, "A learner-independent evaluation of the usefulness of statistical phrases for automated text categorization". In: A. Chin (Ed.), "Text Databases and Document Management: Theory and Practice", Hershey, USA, Idea Group Publishing, pp. 78-102, 2001.

26. C. Manning, and H. Schutze, "Foundations of Statistical Natural Language Processing", The MIT Press, Cambridge, Massa-chusetts, USA, 1999.

27. J. Bakus, M. Hussin, and M. Kamel, "A SOM-Based Document Clustering using Phrases", In Proceedings of the 9th International Conference on Neural Information Processing (ICONIP'02), Novemeber 18-22, Singapore, pp. 2212-2216, 2002.

28. D. Lewis, and M. Ringuette, "Comparison of two learning algorithms for text categorization". In Proceedings of the 3rd Annual Symposium on Document Analysis and Retrieval (SDAIR'94), pp. 81-93, 1994.

29. T. Kohonen, J. Kangas, and J. Laaksonen, "SOM-PAK: the self-organizing map program package ver.3.1", SOM programming team of Helsinki University of Technology, Apr. 1995.

Discovering gene regulatory networks from gene expression data with the use of evolving connectionist systems

Nikola K. Kasabov[1] and Dimiter S. Dimitrov[2]

1: Knowledge Engineering and Discovery Research Institute, Auckland University of Technology, Private Bag 92006, Auckland 1020, New Zealand; nkasabov@aut.ac.nz
2: National Cancer Institute NCI – Frederick, NIH, Washington DC, dimitrov@helix.nih.gov

Abstract. The paper describes the task of model creation of gene networks from time course RNA gene expression data (a reverse engineering problem). It introduces a novel method for this purpose based on evolving connectionist systems (ECOS). A case study is used to illustrate the approach. Gene regulatory networks (GRN), once constructed, can be potentially used to model the behaviour of a cell or an organism from their initial conditions. The problem of modelling GRN of brain cells (neurons) in their relation to the functioning of the brain and the creation of neuro-genetic systems is discussed and potential solutions are outlined.

Keywords: evolving connectionist systems; gene regulatory networks; neuro-genetic systems; Leukemia cell line; rule extraction.

1. Evolving processes in molecular biology and the problem of gene network discovery

In a single cell, the DNA, the RNA and the protein molecules evolve and interact in a continuous way. At the cell level evolving are all the metabolic processes, the cell growing, the cell division, etc. [1,2,3,17]. This interaction can be represented as a complex genetic regulatory network (GRN) of genes connected to each other so that the connections represent this interaction [4]. Genes can trigger other genes to over-express, or to become under-expressed, or may not have a direct relation at all.

The following issues are related to the problem:

• It is assumed that a GRN describes the regulatory interaction between genes;

• It is assumed that reverse engineering – from gene expression data to GRN, is appropriate to apply;

• It is assumed that gene expression data reflect the underlying GRN;

• If there are co-expressed genes over time – either one regulates the other, or both are regulated by same other genes;

• The time unit of interaction needs to be defined;

• Appropriate data need to be obtained;

• A validation procedure needs to be used;

• A correct interpretation of the models may generate new biological knowledge.

The problem of predicting the behavior of cells is much more complex than taking a snapshot of expression values of the genes from the RNA as it has been used in many studies [4]. Genes may very in their expression level over time manifesting complex dynamics of the cell processes. By measuring the expression of all genes over time, we can make a step towards finding some relationships between the genes and inferring Gene Regulatory Networks (GRN) that govern the underlying interaction between the genes [1].

In a single cell, the DNA, the RNA and the protein molecules interact in a continuous way during the process of the RNA transcription from DNA (genotype), and the subsequent RNA to protein (phenotype) translation. A single gene interacts with many other genes in this process, inhibiting, directly or indirectly, the expression of some of them, and promoting others at the same time. This interaction can be represented as a GRN. A simple example of such network is given in fig.1.

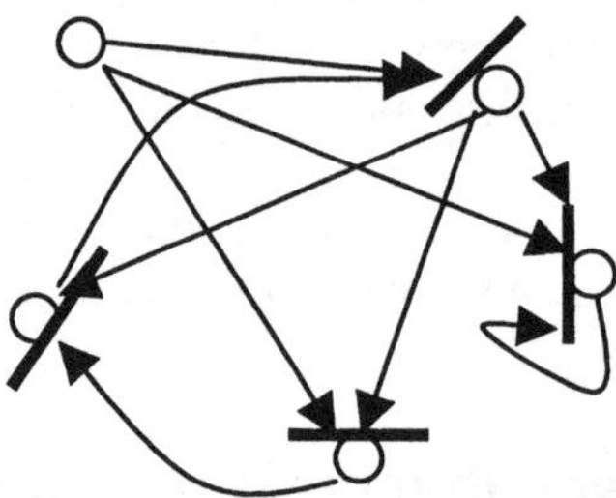

Figure 1. A simplified gene regulatory network GRN, where the nodes represent genes, gene clusters, or proteins. The links represent the relationship between the clusters in consecutive time moments.

A significant challenge for information scientists and biologists is to create computational models of GRN from both dynamic data (e.g. gene expression data of thousands of genes over time, and also from protein data) and from static data (e.g. DNA), under different external inputs (diseases, drugs, etc.). A large amount of both static and dynamic gene expression data is available from public domain databases (http://www.ncbi.nlm.nih.gov/, http://www.genome.ad.jp/kegg/ (www.ebi.ac.uk/microarray). Collecting both static and time course gene expression data from up to 30,000 genes is now a common practice in many biological, medical and pharmaceutical laboratories.

Several approaches have been introduced so far for the problem of genetic network discovery and modeling as presented briefly in the next section.

2. GRN models – a brief review

An extended review of the literature on the existing models for modeling GN is presented in [4].

There are several types of GN representation, some of them listed below:

• Boolean GRN (using Kauffman boolean networks), where boolean vectors represent the state of the genes at every time point, i.e. values of 1 or 0; this representation is too simplistic and is imprecise [5];

• Bayesian and regression networks – posterior probabilities of state transition are represented in the model [13, 14];

• Connectionist networks (genes are represented as neurons and the interaction between them – as weighted connections [20,21,24,26];

• Fuzzy connectionist networks - fuzzy representation is used to represent the transition in a connectionist GRN network [24];

Several methods have been introduced for reverse engineering in order to derive a GRN of the above representation from data:

• Deriving gene relations from MEDLINE abstracts [19];

• Analytical modeling – formulas are derived from gene data [10,15];

• Correlation analysis of gene data to find correlations between gene expression over time [12].

• Cluster analysis – genes are clustered based on their expression [7,8,9] and then linked based on functional similarity;

• Evolutionary computation – GRN are evolved from gene data based on a fitness function [8, 11,16];

• Connectionist techniques (neural networks) are used to learn a GRN from data [20,21,26].

Despite of the existence of these methods, the problem of the genetic network discovery has not been solved so far. One of the reasons is that the processes are too complex for the existing computational models. Generally speaking, modeling genetic networks requires that the models evolve both its structure and functionality in time. A potential approach to apply to this task is the *evolving connectionist systems (ECOS)* approach as presented and applied in this paper.

3. Evolving connectionist systems

Evolving connectionist systems are multi-modular, connectionist architectures that facilitate modelling of evolving processes and knowledge discovery [22-25]. An evolving connectionist system may consist of many evolving connectionist modules.

An evolving connectionist system is a neural network that operates continuously in time and adapts its structure and functionality through a continuous interaction with the environment and with other systems according to: (i) a set of parameters P that are subject to change during the system operation; (ii) an incoming continuous flow of information with unknown distribution; (iii) a goal (rationale) criteria (also subject to modification) that is applied to optimise the performance of the system over time.

The set of parameters P of an ECOS can be regarded as a chromosome of "genes" of the evolving system and evolutionary computation can be applied for their optimisation.

The evolving connectionist systems presented in [22-25] have the following specific characteristics: (1) they evolve in an open space, not necessarily of fixed dimensions; (2) they learn in on-line, pattern mode, incremental learning, fast learning - possibly by one pass of data propagation; (3) they learn in a life-long learning mode; (4) they learn as both individual systems, and evolutionary population systems; (5) they have evolving structures and use constructive learning; (6) they learn locally and locally partition the problem space, thus allowing for a fast adaptation and tracing the evolving processes over time; (7) they facilitate different kinds of knowledge extraction, mostly combined memory based, statistical and symbolic rule knowledge.

The evolving connectionist models presented in [22-25] are knowledge-based models, facilitating Zadeh-Mamdani fuzzy rules (EFuNN, HyFIS), Takagi-Sugeno fuzzy rules (DENFIS), on-line cluster formation (for example − the evolving clustering method ECM).

Fig.2 shows a simplified version of an evolving fuzzy neural network (EFuNN) [2] that facilitates the extraction of rules of the type of Zadeh-Mamdani; an example is given below:

IF x1 is High (0.7) and x2 is Low (0.8) THEN y is Medium (0.9), number of examples accommodated in the rule is 45; radius of the cluster covered by the rule is 0.5.

Each rule node captures one fuzzy rule that can be extracted at any time of the operation of the system. A rule links a cluster of data from the input space to a cluster of data from the output space and can be further interpreted as a piece of local problem space knowledge.

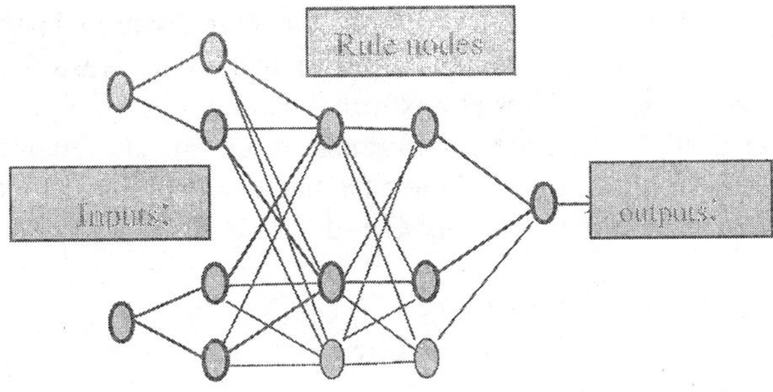

Fig. 2. A simplified version of EFuNN (from [24])

Another type of ECOS – DENFIS [25] learn Takagi-Sugeno fuzzy rules of the form of:

IF x1 is High (0.7) and x2 is Low (0.8) THEN y=0.5 +3.7x1 + 4.5x2, number of examples accommodated in the rule is 15; the area of the cluster covered by the rule is [0.5, 0.7]. The rules can contain non-linear function instead of a linear one [25].

Each evolving connectionist system consists of three main parts:
(1)   Pre-processing and feature evaluation part
(2)   Connectionist modelling part
(3)   Knowledge acquisition part

4. Evolving connectionist systems for GRN modeling and discovery

Genes are complex structures and they cause dynamic transformation of one substance into another during the life of an individual, as well as of the human population over many generations. When genes are "in action", the dynamics of the processes in which a single gene is involved are complex, as this gene interacts with many other genes, proteins, and is influenced by many environmental and developmental factors.

Modelling these interactions, learning about them and extracting knowledge, is a major goal for the scientific area of computational molecular biology and bioinformatics. The whole process of the expression of genes and the production of proteins, and back to the genes, is an evolving process.

Microarray data can be used to evolve an ECOS with inputs being the expression level of a certain number of selected genes (e.g.100) and the outputs being the expression level of the same genes at the next time moment as recorded in the data [26]. After an ECOS is trained on time course gene expression data, rules are extracted and linked between each other in terms of time of their creation in the model, thus representing the GRN. The rule nodes in an EfuNN capture clusters of input genes that are related to the output genes at next time moment.

The extracted rules in an EFuNN structure for example, represent the relationship between the gene expression of a group of genes G(t) at a time moment t and the expression of the genes at the next time moment G(t+dt),   e.g.:

IF g13(t) is High (0.87) and g23(t) is Low (0.9)
THEN g87 (t+dt) is High (0.6) and g103(t+dt) is Low

Through modifying a threshold for rule extraction (see [24]) one can extract stronger or weaker patterns of relationship [26].

Using EfuNNs allows for learning dynamic GRN, so that on-line, incremental learning of a GRN is possible as well as adding new inputs/outputs (new genes) to the GRN.

Another ECOS that can be applied to GRN modeling is DENFIS [24,26]. A set of DENFIS models can be trained, one for each gene g_i so that input vectors are the expression   vectors G(t) of the selected genes at the time moment t and the output is the expression of a single gene at the next time moment t+1: g_i(t+dt). DENFIS allows for a dynamic partitioning of the input space.

Takagi-Sugeno fuzzy rules, that represent the relationship between each gene gi and the rest of the genes, are extracted from each DENFISi model, e.g.:

If   g1 (t) is (0.63   0.70   0.76) and
        g2 (t) is (0.71   0.77   0.84) and
        g3 (t) is (0.12   0.2   0.34) and
        g4 (t) is (0.59   0.66   0.72)
then g4 (t+1) =   1.84 - 1.26 g1 - 1.22g2 + 0.58g3 - 0.03 g4,

In the above representation triangular fuzzy membership functions are used to represent the expression of of the genes in the input problem space (with the left side, the center and the right side of the triangle shown in the rules).

If Gaussian membership functions are used, the type of the rules extracted will be of the form of:

If   g1 (t) is (0.70   0.23) and
        g2 (t) is (0.77   0.13) and
        g3 (t) is (0.2   0.01) and
        g4 (t) is (0.66   0.15)
then g4(t+1) =   1.84 - 1.26 g1 - 1.22g2 + 0.58g3 - 0.03 g4,

where the membership functions defined by their Gaussian shape, center and standard deviation.

5. A case study of GRN modeling with the use of ECOS

Retinoic acid and other reagents can induce differentiation of cancer cells leading to gradual loss of proliferation activity and in many cases death by apoptosis. Elucidation of the mechanisms of these processes may have important implications not only for our understanding of the fundamental mechanisms of cell differentiation but also for treatment of cancer. We studied differentiation of two sub-clones of the leukemic cell line U937 induced by retinoic acid [27]. These sub-clones exhibited highly differential expression of a number of genes including c-Myc, Id1 and Id2 that were correlated with their telomerase activity – the PLUS clones had about 100-fold higher telomerase activity than the MINUS clones [27]. It appears that the MINUS clones are in a more "differentiated" state. The two sub-clones were treated with retinoic acid and samples were taken before treatment (time 0) and then at 6 h, 1, 2, 4, 7 and 9 days for the plus clones and until day 2 for the minus clones because of their apoptotic death. The gene expression in these samples was measured by Affymetrix gene chips that contain probes for 12,600 genes. To specifically address the question of telomerase regulation we selected a subset of those genes that were implicated in the telomerase regulation and used ECOS for their analysis.

The task is to find the gene regulatory network $G=\{g1,g2,g3,g_{rest-},g_{rest+}\}$ of three genes $g1$=c-Myc, $g2$=Id1, $g3$=Id2 while taking into account the integrated influence of the rest of the changing genes over time denoted as g_{rest-} and g_{rest+} representing respectively the integrated group of genes which expression level decreases over time (negative correlation with time), and the group of genes which expression increases over time (positive correlation with time).

Groups of genes g_{rest-}, g_{rest+} were formed for each experiment of PLUS and MINUS cell line, forming all together four group of genes. For each group of genes, the average gene expression level of all genes at each time moment was calculated to form a single aggregated variable g_{rest}.

Two EfuNN models, one for the PLUS cell, and one – for the MINUS cell, were trained on 5-gene input vector data, the expression level of the genes $G(t)$ at time moment t, and 5 gene output vectors – the expression level $G(t+1)$ of the same genes recorded at the next time moment. Rules were extracted from the trained structure that describe the transition between the gene states in the problem space. The rules are given in appendix and their transition in time is represented as graphs on fig. 3a,b.

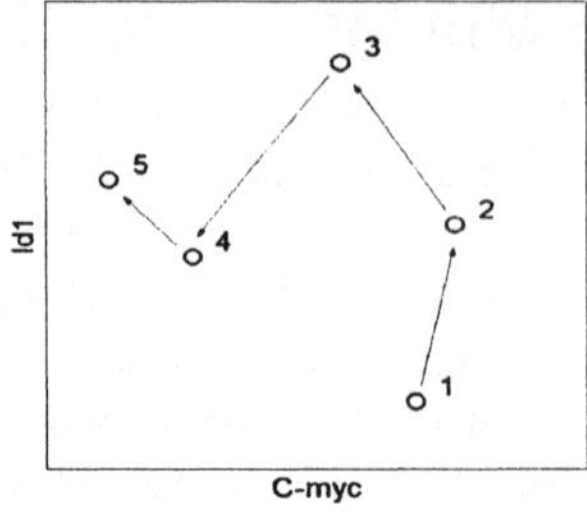

Fig. 3a. The genetic regulatory network extracted from a trained EfuNN on time course gene expression data of genes related to telomerase of the PLUS leukemic cell line U937. Each point represents a state of the 5 genes used in the model, the arrows representing (rules) transitions of the states.

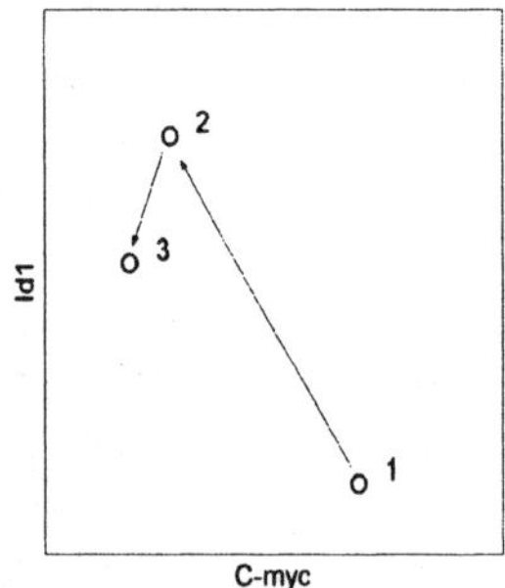

Fig. 3b. The regulatory network of three time steps for the MINUS cell line represented in the 2D space of the expression level of the first two genes – c-Myc and Id1

The shown transitional diagrams can be potentially used to predict the state of the selected genes in time from an initial state of the cell (the initial expression of the selected genes in a cell).

6. Conclusions and future directions

Modeling gene regulatory networks is a challenging task due to the complexity of the biological processes in the cells. It becomes even more complicated when the GRN is related to the functioning of a neuronal cell. And this seems to be the only way to truly model neurons and neural networks of the brain and discover some patterns that can explain fatal diseases.

432

ECOS are suitable techniques for modeling GRN in an adaptive, evolving way, with more data added when such becomes available. Using the extracted rules form an ECOS GRN model of a cell, one can attempt to simulate the development of the cell from initial state G(t=0), through time moments in the future, thus predicting a final state of the cell.

Future directions include a more rigorous analysis of the theoretical limits of ECOS, building multi-modular systems of multiple sources of information, building large ECOS to model complex gene/protein complexes, building large scale adaptive decision support systems that consists of hundreds and thousands of adaptive modules.

Modeling of GRN of cancerous cells, brain cells (neurons) and of stem cells, for the prediction of their functioning in future time moments, have a tremendous potential for medical applications and for knowledge discovery.

7. References

[1] Baldi, Bioinformarics – A Machine Learning Approach, 2001.

[2] L. Hunter, Artificial intelligence and molecular biology. Canadian Artificial Intelligence, No 35, Autumn 1994.

[3] B. Sobral, Bioinformatics and the future role of computing in biology. In: From Jay Lush to Genomics: Visions for animal breeding and genetics, 1999

[4] H. De Jong, Modeling and simulation of genetic regulatory systems: a literature review, Journal of Computational Biology, vol.9, No.1, 67-102, 2002

[5] T. Akutsu, S. Miyano, and S. Kuhara, "Identification of genetic networks from a small number of gene expression patterns under the boolean network model," Pacific Symposium on Biocomputing, vol. 4, pp.17-28, 1999

[6] S. Ando, and E. Sakamoto, and H. Iba, "Evolutionary Modelling and Inference of Genetic Network," Proceedings of the 6th Joint Conference on Information Sciences, March 8-12, pp.1249-1256, 2002.

[7] P. D'Haeseleer, S. Liang, and R. Somogyi, "Genetic network inference: from coexpression clustering to reverse engineering,", Bioinformatics, vol. 16, no. 8, pp.707-726, 2000.

[8] G. Fogel and D. Corne (eds), *Evolutionary Computation for Bioinformatics*: Morgan Kaufmann Publ., 2003.

[9] S. Kauffman, "The large scale structure and dynamics of gene control circuits: an ensemble approach," Journal of Theoretical Biology, vol. 44, pp.167-190, 1974.

[10] K. W. Kohn, and D. S. Dimitrov, "Mathematical Models of Cell Cycles," Computer Modeling and Simulation of Complex Biological Systems}, 1999

[11] J. R. Koza, W. Mydlowec, G. Lanza, J. Yu, M. A. Keane, "Reverse Engineering of Metabolic Pathways from Observed Data using Genetic Programming," Pacific Symposium on Biocomputing, vol. 6, pp.434-445, 2001

[12] A. Lindlof, and B. Olsson, "Could Correlation-based Methods be used to Derive Genetic Association Networks?," Proceedings of the 6thJoint Conference on Information Sciences, March 8-12, pp.1237-1242, 2002.

[13] M. Kato, T.Tsunoda, T.Takagi, Inferring genetic networks from DNA microarray data by multiple regression analysis, Genome Informatics, 11, 118-128, 2000

[14] S.Gomez, S.Lo, A.Rzhetsky, Probabilistic prediction of unknown metabolic and signal-transduction networks, Genetics 159, 1291-1298, November 2001

[15] Liang, S. Fuhrman, and R. Somogyi, REVEAL: A general reverse engineering algorithm for inference of genetic network architectures," Pacific Symposium on Biocomputing, vol. 3, pp.18-29, 1998.

[16] Mimura, and H. Iba, "Inference of a Gene Regulatory Network by Means of Interactive Evolutionary Computing," Proceedings of the 6th Joint Conference on Information Sciences, March 8-12, pp.1243-1248,2002.

[17] P. A. Pevzner, Computational Molecular Biology: An Algorithmic Approach, MIT Press, 2000.

[18] R. Somogyi, S. Fuhrman, and X. Wen, "Genetic network inference in computational models and applications to large-scale gene expression data," Computational Modeling of Genetic and Biochemical Networks, in: J. Bower and H. Bolouri (eds.), {MIT} Press, pp.119-157, 1999.

[19] Z. Szallasi, "Genetic Network Analysis in Light of Massively Parallel Biological Data Acquisition," Pacific Symposium on Biocomputing, vol. 4, pp.5-16, 1999

[20] J. Vohradsky, "Neural network model of gene expression," The FASEB Journal, vol. 15, March, pp.846-854, 2001.

[21] J. Vohradsky, "Neural model of gene network," Journal of Biological Chemistry, vol. 276, pp.36168-36173, 2001[22] N. Kasabov, "ECOS: A framework for evolving connectionist systems and the ECO learning paradigm", Proc. of ICONIP'98, Kitakyushu, Japan, Oct. 1998, IOS Press, 1222-1235.

[23] N. Kasabov, Evolving fuzzy neural networks for on-line supervised/unsupervised, knowledge–based learning, IEEE Trans. SMC – part B, Cybernetics, vol.31, No.6, 902-918, December 2001.

[24] N. Kasabov, Evolving connectionist systems: Methods and Applications in Bioinformatics, Brain study and intelligent machines, Springer, London, New York, Heidelberg, 2002.

[25] N. Kasabov and Q. Song, DENFIS: Dynamic, evolving neural-fuzzy inference systems and its application for time-series prediction, IEEE Trans. On Fuzzy Systems, vol.10, No.2, 144-154, April 2002.

[2] N. Kasabov and D. Dimitrov, "A method for gene regulatory network modelling with the use of evolving connectionist systems," Proc. ICONIP'2002 - International Conference on Neuro-Information Processing, Singapore, 2002, IEEE Press

[27] X.Xiao, Phogat, S., Sidorov, I.A., Yang, J., Horikawa, I., Prieto, D., Adelesberger, J., Lempicki, R., Barrett, J.C., and Dimitrov, D.S. Identification and characterization of rapidly dividing U937 clones with differential telomerase activity and gene expression profiles: Role of c-Myc/Mad1 and Id/Ets proteins. Leukemia, 2002, 16:1877-1880

Acknowledgement

We would like to acknowledge Dr X.Xiao, Dr I.Sidorov, and Dr R.Lempicki for preparing the data at NCI at Frederick and Dr Q.Song for assisting with the experiments on the prepared data.

Appendix A. Some of the gene regulatory rules extracted from the PLUS cell line ECOS model.

Denotation: the type of the rules is: IF G(t) THEN G(T+1); [1],[2],[3],[4],[5] denote the 5 genes used in the model; 1,2 and 3 denote Small, Medium and High expression level as a fuzzy membership function; the number attached to it is the membership degree, for example [1] (2 0.299)(3 0.701) means that gene 1 is expressed at a medium level with a membership degree of 0.299 and at a High level with a degree of 0.701.

Rule 1: if   [1] (2 0.299) (3 0.701) and
              [2] (1 0.909) (2 0.091) and
              [3] (1 0.070) (2 0.930) and
              [4] (2 0.683) (3 0.317) and
              [5] (1 0.731) (2 0.269)
    then   [1] (2 0.091) (3 0.909) and
              [2] (1 0.798) (2 0.202) and
              [3] (1 0.048) (2 0.952) and
              [4] (2 0.439) (3 0.561) and
              [5] (1 0.838) (2 0.162)

Rule 2: if   [1] (2 0.091) (3 0.909) and
              [2] (2 0.961) (3 0.039) and
              [3] (2 0.955) (3 0.045) and
              [4] (2 0.559) (3 0.441) and
              [5] (1 0.836) (2 0.164)
    then   [1] (2 0.622) (3 0.378) and
              [2] (1 0.231) (2 0.769) and
              [3] (1 0.909) (2 0.091) and
              [4] (2 0.896) (3 0.104) and
              [5] (1 0.355) (2 0.645)
Rule 3:
     if   [1] (2 0.691) (3 0.309) and
              [2] (2 0.091) (3 0.909) and
              [3] (1 0.909) (2 0.091) and
              [4] (1 0.174) (2 0.826) and
              [5] (1 0.341) (2 0.659)
    then   [1] (1 0.311) (2 0.689) and
              [2] (1 0.909) (2 0.091) and
              [3] (1 0.244) (2 0.756) and
              [4] (2 0.091) (3 0.909) and
              [5] (1 0.909) (2 0.091)

Appendix B. Gene regulatory rules extracted from the MINUS cell ECOS model (same denotation as in Appendix A above is used):

Rule 1:
if   [1] (2 0.091) (3 0.909) and
        [2] (1 0.909) (2 0.091) and
        [3] (2 0.091) (3 0.909) and
        [4] (2 0.604) (3 0.396) and
        [5] (2 0.983) (3 0.017)

then [1] (2 0.091) (3 0.909) and
        [2] (2 0.091) (3 0.909) and
        [3] (2 0.996)   and
        [4] (2 0.091) (3 0.909) and
        [5] (1 0.909) (2 0.091)

Rule 2:
if   [1] (1 0.583) (2 0.417) and
        [2] (2 0.091) (3 0.909) and
        [3] (1 0.909) (2 0.091) and
        [4] (2 0.091) (3 0.909) and
        [5] (1 0.909) (2 0.091)

then [1] (1 0.840) (2 0.160) and
        [2] (1 0.909) (2 0.091) and
        [3] (2 0.091) (3 0.909) and
        [4] (1 0.641) (2 0.359) and
        [5] (2 0.810) (3 0.190)

Rule 3:
if   [1] (1 0.909) (2 0.091) and
        [2] (2 0.757) (3 0.243) and
        [3] (1 0.114) (2 0.886) and
        [4] (1 0.909) (2 0.091) and
        [5] (2 0.091) (3 0.909)

then [1] (1 0.909) (2 0.091) and
        [2] (1 0.508) (2 0.492) and
        [3] (1 0.909) (2 0.091) and
        [4] (1 0.909) (2 0.091) and
        [5] (2 0.091) (3 0.909)

Experimental Analysis of Knowledge Based Multiagent Credit Assignment

Ahad Harati,   Majid Nili Ahmadabadi

Control and Intelligent Processing Center of Excellence and AI & Robotics Lab., Dep. of Electrical and Computer Engineering, University of Tehran, Tehran, Iran

School of Intelligent Systems, Institute for Studies on Theoretical Physics and Mathematics, Tehran, Iran

harati@ece.ut.ac.ir,   mnili@ut.ac.ir

Abstract Multiagent Credit Assignment is one the major problems in realization of multiagent reinforcement learning. Since the environment usually is not intelligent enough to qualify individual agents in a cooperative team, it is very important to develop some methods for assigning individual agent credits when just single group reinforcement is available. Depending on the type of cooperation, role of the agents can be complementary or redundant. We consider this as the group's task type and call the former case AND-type and the latter OR-type task.

In this research an approach based on agents' learning history and knowledge is proposed to solve Multiagent Credit Assignment problem. The main idea behind this approach is that, the more knowledgeable agents make fewer mistakes. Therefore, it is more probable that behavior of less knowledgeable agents be the reason of team failure.

Normal Expertness, Relative Normal Expertness and Certainty are introduced as measures of agents' knowledge. Implementing one of these measures, Knowledge Evaluation Based Credit Assignment (KEBCA) is used to judge agents' actions and assign them proper credits.

In this paper, we study the effects of task type (AND-type and OR-type tasks) in solving the Multiagent Credit Assignment problem and, using a simulated environment, discuss the result of applying KEBCA in each case. In addition, performance of the system is examined in presence of some uncertainties in the environment, functioning as noise in agents' actions.

Keywords Multiagent Credit Assignment, Cooperative Parallel Task, Feedback distribution, Reinforcement Sharing.

1 Introduction

Reinforcement Learning (*RL*) [1][2][3] is widely used in multiagent systems to realize cooperation [4]. Each *RL* agent should be provided with a proper reinforcement as a scalar feedback of its performance. Since cooperative multiagent systems usually receive a team reinforcement signal as a measure of total system performance, this major problem arises: Having a global team credit, how to assign suitable credit to each agent in the team. This is called *Inter-agent Credit As-*

signment [5] or *Multiagent Credit Assignment (MCA)* [6][7] in general and *Reward Sharing* [4][8] in some special cases where there is no punishment.

As mentioned in [4], solving this problem properly enhances speed and quality of learning. However, in systems with independent learning modules, the role of *MCA* is even more serious and important. Having independent learning modules is usual in many practical applications where the total state-space of problem is huge. In such situations, the *MCA* is not just an option to reach faster and more qualitative learning, but it is a vital problem for learning in multiagent systems.

In most of existing researches, some ad-hoc strategies are used to solve the *MCA* problem. Mostly, it is assumed that local reinforcements can be obtained for each agent. Another common assumption is considering rich state information for the distributed modules.
These approaches both fail when face the practical tasks and conditions. In many cases, the environment critic is not intelligent enough to asses the role of agents separately in a proper way. In addition the task structure does not provide a local judgment criterion. So, when the extension of local state information is not desirable or possible, a different approach must be taken to solve the *MCA* problem.

In this research, a parallel team configuration and a one-step deterministic task for Q-learning [9] agents are considered that let us study the effects of different *MCA* methods more easily. In fact, considering a one-step task eliminates the effects of temporal credit assignment for each agent. In addition, each agent observes just the portion of environment it needs to perform its own task (i.e. the task is completely parallel) and selects an individual action accordingly. Agents' actions are combined to form the group action. This group action is then evaluated by the environment and, as a result, it returns a team reinforcement signal. It is assumed that a critic agent has the duty of distributing the team reinforcement among the Q-learning team members, see Figure 1.

In our approach, the critic uses some measures of the agents' histories that denote their amount of knowledge. Then, it tries to guess the role of each agent in the group outcome. This is done by judging each agent separately. It is assumed that, the critic can access internal information of the learning agents, like their Q-cell values. The critic doesn't know the task partially or totally and cannot learn it also. Moreover, it is taken that the critic has no control on the environment and

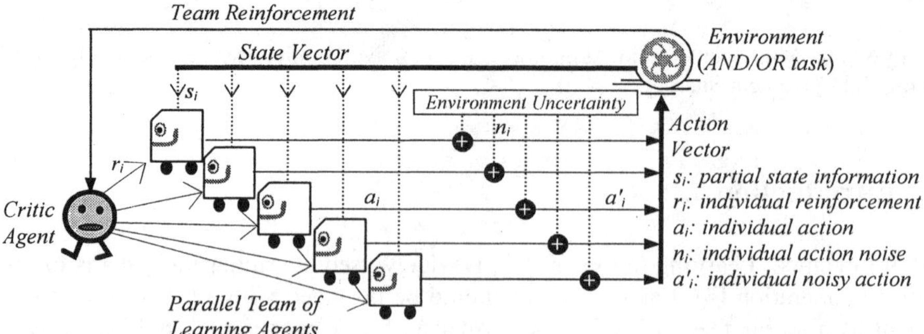

Figure 1 An overview of the critic agent's task

can neither plan the training scenarios for the agents nor decide about the respective state transitions.

In [6][10][11] we introduced some knowledge evaluation measures and the basic idea of *Knowledge Evaluation Based Credit Assignment*. Based on this idea, a credit assignment method for a team of agents with a parallel and deterministic AND-type task is developed. In this paper we extend our *MCA* methods for both AND-type and OR-type tasks. In addition the effects of environment noise on the quality of our credit assignment methods are discussed.

In the following section Q-learning is briefly introduced to clarify the agents' learning model and algorithm. Then the *MCA* problem is stated more precisely. Some related researches are reviewed in section 4. In the fifth section our approach is presented. Then the implemented environment uncertainty is explained. Test bed is explained in the seventh section. After that, the simulation results are reported and discussed. Future researches and a conclusion of this paper are given in the last section.

2 Q-Learning

Reinforcement learning is originally obtained as a model of animals' learning method. In this method, the learner perceives something about the *state* of its environment and, based on a predefined criterion, chooses an *action*. The action changes the world's state and the agent receives a scalar *reward* or *reinforcement*, indicating the goodness of its new state. After receiving the reward or the punishment, it updates the learnt strategy based on a learning rate and some other parameters.

In one-step Q-learning algorithm [23] the external world is modeled as a Markov decision process with discrete time and finite states. Next to each action, the agent receives a scalar reward or reinforcement.

The state-action value table, the Q-table, which estimates the long-term discounted reward for each state-action pair, determines the learned policy of the agent. Given the current state s_t and the available actions set (*Actions*), a Q-learning agent selects action a_i with the probability $p(a_i \mid s_t)$ given by the Boltzmann distribution:

$$p(a_i \mid s_t) = \frac{e^{Q(s_t,a_i)/\tau}}{\sum_{k \in Actions} e^{Q(s_t,a_k)/\tau}} \qquad (1)$$

where τ is the temperature parameter and adjusts randomness of the decision. The agent executes the action, receives an immediate reward r, observes the next state s_{t+1} and updates $Q(s_t,a_i)$ as:

$$Q(s_t,a_i) \leftarrow (1-\alpha)Q(s_t,a_i) + \alpha(r + \gamma V(s_{t+1})) \qquad (2)$$

where $0 \le \alpha \le 1$ is the learning rate, $0 \le \gamma \le 1$ is a discount parameter and $V(s)$ is given by:

$$V(s) = \max_{b \in Actions} Q(s, b) \qquad (3)$$

Q-table is improved gradually and the agent learns when it searches the state space.

3 Problem Statement

Credit Assignment (*CA*) is one of the most important problems in *RL*. It can be classified into three categorizes: *Structural* [1][3], *Temporal* [12], and *Multiagent Credit Assignment* (*MCA*). The first two categories are somehow well studied in the literature while *MCA* is quite a new problem in both machine learning and multiagent domains.

The main problem of *MCA* is how to assign a suitable credit to each agent in a team of cooperative agents. This problem arises when multi learning agents are used to perform a cooperative task and there is no well defined goal for each agent. In such situations, just a single team reinforcement signal is provided by the task nature (environment critic). Therefore, individual agents' role in the team outcome is to be determined. Doing this, it is possible to qualify each agent with a proper individual reinforcement. To study *MCA* in more detail, different group configurations and task types are considered.

A group of agents can cooperate to perform a serial or parallel task. The task is serial if the agents are serially selected for trying to solve the problem and the environment state is changed by the previous ones. In a parallel task, the environment changes caused by the other agents do not affect the state of acting agent. In such cases, each agent faces just its portion of task.

MCA in a serial task is studied in [4][8] and [13]. In serial tasks, agents' actions come after each other, so the problem is slightly similar to temporal credit assignment [12]. In this paper, a parallel task is implemented in order to separate the *MCA* problem from other involved subjects and have a chance to study the effects of introduced methods more closely.

The agents' task type also highly affects the *MCA* methods. In fact, the team credit is determined according to a combination of partial solutions proposed by the agents (individual agents' actions). In practical cases, this combination can take many complicated forms. At one limit, it is needed that all agents perform their job successfully to obtain a desirable group outcome. In other words, agents' roles are complementary and no one can compensate others' fault. Such tasks are called AND-type tasks here. At the other limit, one can consider a group of agents with redundant roles. In such configuration, a single success in the team is enough to obtain a desired outcome at the group level. Such tasks are called OR-type tasks here. In this type of tasks, each agent can completely amend faults of its teammates.

Practical situations are usually a combination of AND-type and OR-type tasks. The *MCA* problem for such task types can be solved using the basic solutions to be discussed for AND-type and OR-type tasks provided that, the task structure is known.

4 Related Works

As discussed in [7], till now sufficient attention is not paid to *MCA*. Therefore, some simple techniques have been implemented to overcome it, or strictly speaking, to ignore it.

If the agents can sense each other, a uniform credit distribution leads to acceptable results. The reason is that, each agent can learn others' roles, situations or behaviors. Many researches assume this condition, e.g. [14][15]. Also if it is possible to compute local reinforcement for each agent, as in [8], then inter agent *CA* is not needed at least for local outcomes. In [8] it is shown that, using just global rewards slows down the learning progress.

In [16], a Distributed RL algorithm is proposed for applications like traffic flow, based on distributing representation of the value function across nodes. In this research it is assumed that, local rewards can be computed for each node directly based on its sensory information and actions. In order to reach higher levels of cooperation, each node considers the value function of its two neighbors. This work differs from ours, as it tries to optimize total team performance (or global reward function) by using local reinforcements. We are trying to compute individual reinforcements when just a global feedback at the team level is available.

In some cases, the team totally forms a single *RL* unit. Market Based *RL* Systems [17][18][19] are some examples of such systems. In Market Based *RL*, the credit is distributed among different agents as money or wealth. In such systems, each agent does not have a separate *RL* for itself, since each agent has just one action. Hence, all agents together form a single *RL* system. So, this wealth distribution is considered as structural *CA* and no inter agent *CA* is performed.

Tan used uniform distribution of credit between two predators cooperating to catch a prey [15]. He showed that, when the predators ignore each other, the agents couldn't learn the task properly.

Solving the inter agent *CA* problem (even partially) results enhancements in quality and speed of learning in most of the cases. As it is shown in [20], vicarious rewards improve average team performance. Tile world [21] is used in this research to experimentally show the effects of proposed approach. The main idea is assigning extra reward to the agents that are near the agent who finds food.

This observation is studied from cooperation point of view. In fact, the researchers intended to propose a method for creating cooperative behavior in multiagent learning systems. Here, we are interested to study their work from *CA* point of view. Assigning vicarious rewards is actually judging agents' actions. In other words, with this approach we accept that when one of the agents succeeds (a piece of food is eaten) more than just the last agent must be rewarded. Distance to the food is the selected criterion for judging the importance of agents' roles. However, such criterion is not generally available and this seems more like an ad-hoc approach from the *CA* point of view.

The same concept is used in [22]. This research assumes that positive reinforcement is consistent across the agents. This assumption holds in the studied multi robot platform, because of the spatial and temporal locality of the shared information. In other words, when two robots are at nearby the same place at the same time, they are likely to receive same reinforcement for the same actions [22].

In [4][8] it is shown that, rewarding all effective agents, not only the last one, can speed up learning as the agents can have more chance to learn the task. This kind of reward sharing among the agents can only have meaning in serial tasks with no punishments and results improvements in group learning in Profit Sharing [19][23]. In this research no judgment is done for evaluating agents' actions.

As shown in [13], better results are obtained if a criterion more suitable than temporal difference is used for distributing the team reinforcement signals among the learning agents. In this research, different types of expertness are used in order to distribute the team reinforcement among serially activated agents. A deterministic task is considered and two methods are proposed for distributing the reinforcement signals. First method, called *Direct*, is useful when there is a kind of hierarchy among the agents' roles. In this method, more expert agents receive more rewards in the case of success and more punishments when the team fails. In contrast, the second method, called *Inverse,* is applicable to teams of homogenous agents with similar and balanced roles. This method assigns larger rewards and smaller punishments to the expert agents as it is less probable that they make mistakes. These ideas are implemented via weighting agents' roles according to their expertness. Hence, this work is also a kind of reinforcement sharing as the signs of feedback for all agents are the same.

MCA is studied in [24] for coordinated group tasks at design time. This paper discusses the way of determining agents' reward functions such that local greedy policies lead to desirable group outcome. We are interested in solving *MCA* for independent agent at run-time.

In [6][10][11] *MCA* in deterministic AND-type tasks is discussed. It is shown that, knowledge evaluation is a suitable approach to such problem when there is some reliable history of the agents' learning. For realization of this idea, different measures of agents' knowledge are proposed and their functionalities were evaluated using some introduced indexes. In addition, it is experimentally shown that the proposed methods perform much better than discussed intuitive strategies for the *MCA*. In this paper, we extend our approach to both AND-type and OR-type tasks and study the role of task type on *MCA*. In addition, performance of the system is examined in presence of some uncertainties in the environment, functioning as noise in agents' actions.

5 Our Approach to Multiagent Credit Assignment

5.1 Measures

To assign individual agents' credits, the critic agent needs some criteria to judge each agent, because the information content of the team reinforcement signal received from the environment is not sufficient [7]. Measures of the agents' knowledge are suitable for such purpose, since it's needed to approximate the probability of correctness of agents' decisions. If this approximation is used to criticize some agents in the team properly, each individual agent can learn its task. In this paper, three such measures, called *Normal Expertness*, *Relative Normal Expertness* and *Certainty,* are discussed in the next parts.

5.1.1 Non-Relative Criteria

Using history of received reinforcements, some measures of agents' *expertness* are calculated [25]. Here, with the same idea, *Normal Expertness* is used to show haw much an action is probable to be the best action or correct one in a state. Therefore in this paper, *Normal Expertness* for every Q-cell is used as credit assignment criterion[1] . In fact to act independent of reward and punishment signal magnitude, a slightly modified version of expertness is implemented. Normal Expertness for a state-action is defined as the number of rewards minus number of punishments that has been assigned to it since the start of learning. For state s and action a of ith agent it is calculated as:

$$E_i^{Nrm}(s,a) = N_r(s,a) - N_p(s,a) \qquad (4)$$

where $N_r(s,a)$ and $N_p(s,a)$ are its number of rewards and punishments respectively, received from start of learning till now. Since this measure is computed for each Q-cell, it denotes how much this state-action is experienced positively. So it will be a good estimate for future outcomes of that action. It is called non-*relative* criterion, since it just depends on single state-action pair.

5.1.2 Relative Criteria

Considering history of other possible actions in a state leads to more informative measures. Two of such measures are introduced here. They are called relative criteria since they depend on more than one Q-cell.

The first one, called *Relative Normal Expertness*, is simply achieved by considering relative value of *Normal Expertness* against the other possible actions in the same state. In other words, for state s and action a, of ith agent it is calculated based on $E_i^{Nrm}(s,a)$ as:

---

[1] In [25] *Expertness* is calculated for the Q-table.

$$RE_i^{Nrm}(s,a) = \frac{e^{E_i^{Nrm}(s,a)}}{\sum_{a' \in Actions_i'} e^{E_i^{Nrm}(s,a')}} \qquad (5)$$

The second relative criterion is called *Certainty*. It is defined as the action selection probability used by the agent to choose its current action. Since our Q-learning agents use Boltzmann exploration method for their action selection, *Certainty* for i^{th} agent is computed as:

$$C_i(s,a) = \frac{e^{Q_i(s,a)/T_i}}{\sum_{a' \in Actions_i'} e^{Q_i(s,a')/T_i}} \qquad (6)$$

where s and a are its state and selected action at current trial respectively and T is a small enough temperature that must be determined by the amount of uncertainty in the environment and other parameters of learning. It is worth mentioning that this measure can express the history of agent's learning regardless of its action selection mechanism. It considers the current temperature of each agent (T_i), and also it is affected by all of learning parameters because of the usage of Q-cells.

Normally, Q-learning agents explore the environment and exploit their knowledge with different probabilities to find better possible solutions and to gain more reward respectively. If the agent is exploring, it is not certain enough about the outcome of its action and it is more probable to select wrong actions. On the other hand, when the agent is exploiting its knowledge, it is more likely to select proper actions provided that its knowledge is correct. So, *Certainty* can be used as a credit distribution criterion to judge correctness of agents' actions.

5.2 Credit Assignment Methods

5.2.1 Optimal Credit Assignment

In AND-type tasks team rewards and in OR-type tasks team punishments can safely be assigned to the all group members. Hence if the critic agent knows about the task type of agents, the *MCA* problem is reduced to *Punishment Distribution* [7] and *Reward Distribution* in AND-type and OR-type tasks respectively. Therefore as the best theoretically possible solution to *MCA* problem, *Optimal Punishment Distribution (OPD)* and *Optimal Reward Distribution (ORD)* are defined for AND-type and OR-type tasks respectively in noise free environments. *OPD (ORD)* is method of *MCA* in AND-type (OR-type) tasks used by optimal critic. It is assumed that optimal critic can evaluate actions of each agent individually regardless of other group members' actions. Although this is not possible in practice, we use it as a best case to evaluate our methods.

5.2.2 Intuitive Strategies

No Punishment (*NP*) [7] is one of the intuitive strategies for *MCA* in AND-type tasks that ignores punishment distribution and uses just team rewards for training the agents. Hence, all trials with the team punishments are ignored and no reinforcement is assigned to the agents. *No Reward* (*NR*) is a similar approach in OR-type tasks. In this method team punishments are assigned to all individual agents and trials with team rewards are ignored.

5.2.3 Knowledge Evaluation Based Credit Assignment

If a reliable learning history of the agents is available, the critic agent can use this information to decide about agents' actions and assign them the proper reinforcement. We call this approach *Knowledge Evaluation Based Credit Assignment* (*KEBCA*). Some result of *KEBCA* for AND-type tasks is reported in [6][10][11]. Using suitable measures of agents' knowledge, such as *Normal Expertness*, *Relative Normal Expertness* and *Certainty*, the critic agent acts according to one the following algorithms based on what information it may have.

In *Case Study I*, it is assumed that the critic agent receives both the environment reinforcement signal and the number of agents with wrong actions (w). All agents are rewarded (punished) if the team receives a reward (punishment) signal when the task is AND-type (OR-type). Otherwise, the critic agent sorts the agents according to a measure of their knowledge in a decreasing manner. Then, it punishes w agents from end of the list and rewards the remaining ones. Clearly, this method is a bit risky; especially in the initial stages of learning when the number of Q-cells visited for the first time is relatively large. But using the number of wrong actions, *MCA* is robust enough to handle some rare mistakes.

In *Case Study II*, number of agents with wrong actions is not known. The critic agent rewards (punishes) all agents if the team is rewarded (punished) in AND-type (OR-type) tasks as in the previous case. However, when the team gets a punishment (reward), it doesn't know how many agents must be rewarded and how many of them must be punished. In fact, it is more likely that the group with lower knowledge selects wrong actions. Therefore, if the critic partitions the set of learning agents into two reasonable sections based on a measure of agents' knowledge, then the problem can be somehow solved.

Considering lack of critic information in some cases, a more effective method for *MCA* is obtained if the critic does not try to guess outcome of agents' actions and act just considering the previous events confidently. Therefore, to lower the risk of wrong reinforcement assignments, the critic must partition the set of agents into three sections. The middle part is the area where the selected measure cannot judge the correctness of agents' actions firmly enough. So the related agents ignore that trial. To do this, two thresholds are computed for each knowledge measure. Here threshold values are selected such that critic does not risk and acts fully rational according to the previous trials. For further discussions see [7].

However, the final algorithm is highly affected with task conditions. Using thresholds is suitable when task is deterministic and there is exactly one correct action in each state while in more realistic tasks, it may be better to use a cluster-ing method for classifying agents into reward deserved and punishment deserved groups. Such idea is discussed in [6].

5.3 Evaluation Indexes

For studying the effects of *MCA* methods, some evaluation indexes are defined. In this paper five such indexes are used: *Correctness, Performance, Efficiency, Learning ratio* and *Group Success Ratio* (*GSR*).

Correctness is defined as the ratio of correct assignments to the number of agents. It measures how the critic can guess suitable credits for all agents in each trial. An assignment is correct if the assigned credit has not the reverse sign of agents' real credit. In other words *Correctness* at trial t (C_t) is defined as follows.

$$C_t = \frac{|\{i\,|\,1 \le i \le N, Cr_i = RealCr_i \text{ or } Cr_i = Undefined\}|}{N} \tag{7}$$

N is the number of learning agents, Cr_i is the credit assigned to *agent_i*, $RealCr_i$ is its real credit and $Cr_i = Undefined$ denotes the cases when no credit is assigned[2]. An *MCA* method (the critic) is considered rational if its *Correctness* is always one.

Efficiency denotes how much the critic agent uses the learning trails. In other words, this index shows the ratio of number of assigned credits (correct or incor-rect) to the number of assignments made by the optimal critic. It is calculated at trail t as:

$$E_t = \frac{|\{agent\,|\,its\ assigned\ credit\ is\ not\ Undefined\}|}{N} \tag{8}$$

Performance shows the effect of used *MCA* method on the group success. In other words, it is a scaled group performance index that denotes the ratio of num-ber of correct individual actions to the number of all agents' actions. This index is a function of the critic performance, the learning method, and its parameters. It qualifies the group in terms of its members and regardless of its task type. *Per-formance* is calculated as:

$$P_t = \frac{|\{agent\,|\,its\ action\ is\ correct\}|}{N} \tag{9}$$

Learning ratio shows how much the team has learned its task. For each agent, it is the ratio of learnt states to the total number of them. In this research, the task is deterministic and designed such that only one action is the correct action in each state. Therefore, this index is calculated by considering greedy action selection policy. In other words, a state is considered as a learnt state if the correct action has the single maximum Q-value among possible actions. *Learning ratio* of the

---

[2] The agent ignores its learning trial and doesn't update its Q-table when its individual re-inforcement signal is *undefined*.

team is the average of learning ratio of all learning agents. This is a good index to judge how well the critic acts, as it is less sensitive to the agent action selection method when compared with *Performance*. Formally, it is given by:

$$Lr_i = \frac{\left|\{s \mid \arg\max_a \{Q(s,a)\} \text{ is unique correct action}\}\right|}{n_s}$$

$$Lr_t = \sum_{i=1}^{N} Lr_i \Big/ N \tag{10}$$

where Lr_i is the *Learning ratio* of ith agent and Lr_t is the group *Learning ratio* at time t.

Finally *Group Success Ratio* (*GSR*) is the ratio of team rewards (successes) in past 100 trials. Against *Performance*, this index qualifies the group as a whole. When R_t is the team reinforcement at time t, it is calculated as:

$$GSR_t = \frac{\left|\{t' \mid t - 100 < t' \leq t; R_{t'} = Reward\}\right|}{100} \tag{11}$$

6 Uncertain Environments and Noise Effects

In many applications, there is uncertainty in successfully performing the decided commands. For example, a mobile robot decides to go straight forward but wheel slippages deviates it from the desired direction. If the robot cannot understand this deviation, it evaluates its decision wrongly. Such environment uncertainties are modeled here as noise in the environment.

6.1 Noise

MCA and the proposed methods are sensitive to the environment reinforcement signals as this signal is the only feedback the system receives. Therefore, a simulated noise is added to the reinforcement signal in order to create the most critical conditions and evaluate the system behavior in such situations. More precisely, the added noise randomly changes the agents' selected actions. This in turn will lead to an unreliable reinforcement signal as the agents do not know about such digression.

For example when there is 10% noise, the agent's action is changed to the predecessor action with the probability of 5% and to the successor one in a circular order with the same probability. After applying such noises to the agents' actions, the environment criticizes the group action which is constructed by noisy individual actions and assigns the team a reinforcement signal accordingly. In this paper, results of *MCA* for two different amount of noise (10% and 30%) are reported and discussed.

Since noise is not filtered by the critic agent in any case, it is the duty of agents to reduce or eliminate its effects. Therefore, in these experiments, the role of learning rate is crucial and it must be kept small enough for learning convergence. In fact, the optimal value for this parameter depends on the value of uncertainty in the environment (noise). Here, to be able to compare the result of different *MCA* methods, a unique learning rate is used for all reported experiments. In addition, in order to be able to compare learning scenarios with different lengths (number of trials), learning rate is kept constant during the learning period.

6.2 Credit Assignment

Although *ORD* and *OPD* were initially defined in noise free environments, they can be used in presence of noise too, with some modification in their definition[3]. In such conditions, *ORD* and *OPD* are optimal just from credit assignment point of view not from noise effect compensation. So these methods do not assign real credits to the agents in the presence of noise any more and it is probable that a correct action be punished or vice versa.

NR and *NP* (the most intuitive and simplest methods) and the proposed approach (*KEBCA)* are also employed in noisy environments in their original form. It will be shown that, *KEBCA* along with a proper knowledge measure (like *Certainty*) can handle bounded amount of noise using a suitable learning rate. Sensitivity of *NR, NP* and *KEBCA* methods to the learning rate depends on the agents' initial knowledge. This will be discussed in more detail latter.

7 Test Bed

For testing the explained ideas in practice, a group of five homogeneous agents with a parallel configuration and a one-step task is considered. It is assumed that the environment qualifies the team performance as a whole and the critic agent assigns credits due to one of the explained measures.

Two 5-digit set of numbers are given to the team to calculate their sum. Each digit which is less than five is given to one of the agents as its state (5x5 states). The agents don't know the summation and should learn it. Therefore, nine actions are defined for each agent. Each action for each state results in one digit from zero to eight.

When simulating an AND-type task, the environment rewards the group if all of the agents perform their summations correctly, otherwise it punishes the team. In the case of OR-type tasks, the team is rewarded when there is at least one successful agent. The team is punished if all of the agents do their summation wrong.

Since this task is one-step, the discount factor of Q-learning algorithm (γ) [9] is zero. The reward and punishment are 10.0 and -10.0 respectively. Q-cells are uni-

---

[3] In a noisy system, the optimal critic observes the noisy effect of each agent's action on the environment.

448

formly initialized with the value of 0.1. Learning rate (α) and temperature (T) are kept constant at 0.7 and 5.0 respectively during the learning period. Constant temperature bounds *Performance* of the individual agents at 87% when optimal Q-cell values are achieved.

Before the start of group learning in simulation of AND-type tasks, each agent is given some individual learning opportunities[4]. Having different number of individual learning trials (20, 30, 100, 100 and 300 trials), the group is learnt 37-38% of its task in average. The group learning is started without any initial knowledge in the other cases.

8 Simulation Results

In this section, the results of using introduced *KEBCA* method and knowledge evaluation measures for the AND-type task are discussed first. Then, the best fitted knowledge evaluation measure is selected and used for credit assignment in the OR-type task. The effects of task type on *MCA* are discussed in this subsection as well. At the end, *Performance* of implementing *KEBCA* in the noisy environment is evaluated.

The reported results are obtained by averaging the records of at least thirty separate simulations. In order to filter high frequency fluctuations in the results caused by the agents' exploration, a moving average with factor of 100 is calculated for each graph except for *Learning ratio* and *GSR* that are smooth enough.

8.1 *KEBCA* and Knowledge Evaluation Measures (AND-Type Task)

In Figure 2, the *Correctness* is shown for all criteria. In this experiment, real credits are given to the agents to make an ideal case for comparison. It can be seen that for case study I, *Correctness* of *Certainty* and *Relative Normal Expertness* are similar and as relative measures, better than *Normal Expertness* that is non-relative. This is an expected result since they both consider all possible actions in a state that obviously leads to a more informative criterion for predicting outcome of selected action.

An interesting fact in this figure is that, *Correctness* of *Certainty* and *Normal Expertness* is exactly 100% in case study II for the entire learning period. This shows that they act fully rational as conservative threshold values are selected. But, this selection leads to a lower *Efficiency*.

---

[4] In AND-type tasks in *Case Study II* team punishments are not used for training inexperienced states. Hence without any initial knowledge, the scarcity of success (team reward) in initial stages makes the learning period very long. Since in *Case Study I* and OR-type tasks the team could learn from failures (team punishments), the role of initial knowledge is much less significant.

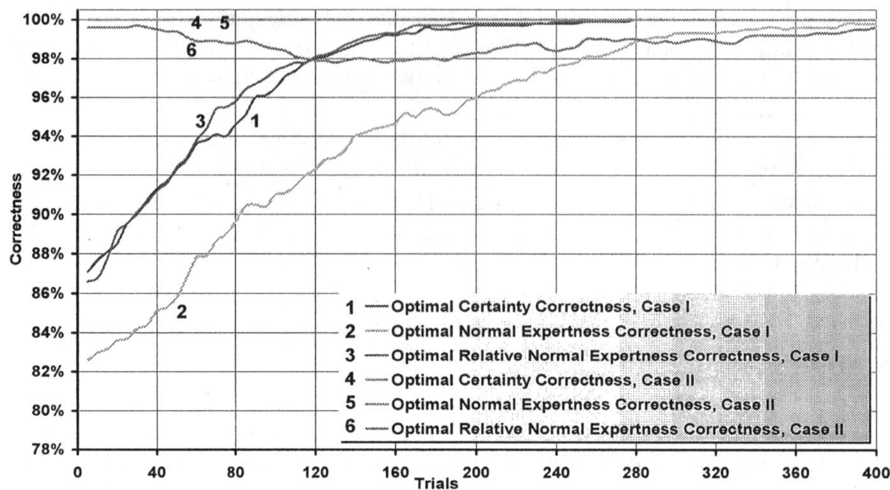

Figure 2 *Correctness* of criteria in ideal conditions, AND-type task

In Figure 3 *Performance* of applying different methods are compared. Results of giving agents their real credits are also shown for comparison. Some noticeable observations in this figure are as follows:

First, all of methods have a positive slope in their *Performance* curve. So they all can lead to reasonable fast group learning. On the other hand, the results of case study I, is near optimal case. This shows that, knowing the number of wrong decisions in the group is very helpful. Finally, *Certainty* acts best among the others. The reason is that with the current parameters, it can fit to the thresholds better. In other words, it can express the agent's history of experiences better than the others.

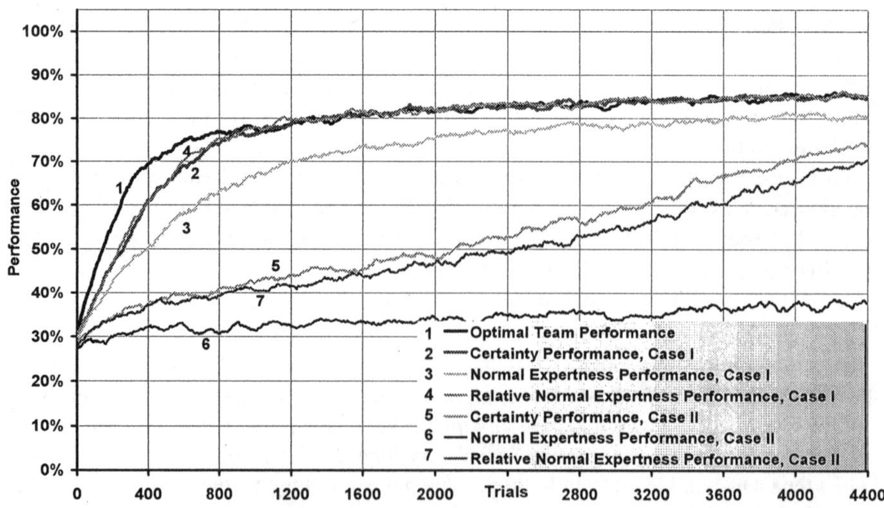

Figure 3 Group *Performance* for different criteria, AND-type task

450

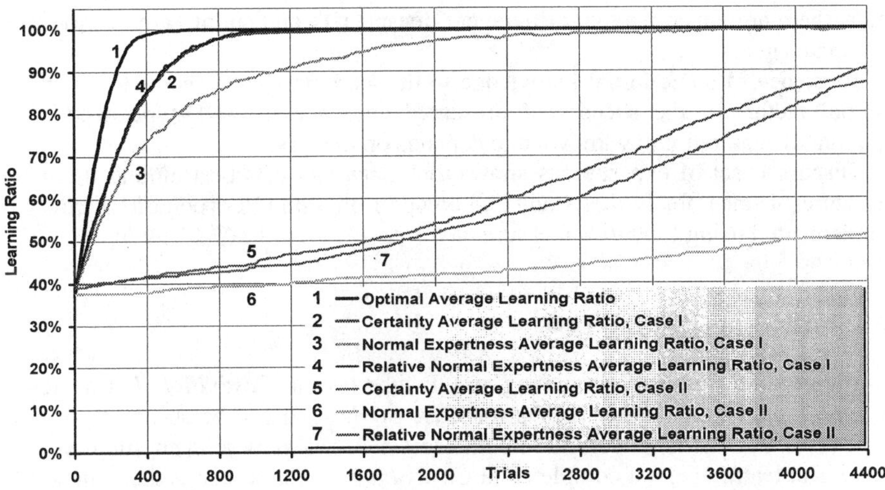

Figure 4 Learning ratio for credit assignment criteria, AND-type task

A problem with *Normal Expertness* that makes it act worse than the other criteria is that, the mentioned procedure of setting thresholds can't be done for this measure completely; as it simply ignores alternatives of the selected action. Therefore all conditions must rely on the history of action itself.

In Figure 4, *Learning ratio* is shown for different credit assignment methods and for the optimal critic case. This ratio is the average among all group agents. Due to the initial agents' knowledge obtained via individual learning, this ratio starts from 38-39% and grows during the learning period. These results completely match with the group performances in Figure 3.

Finally Figure 5 shows the *Efficiency* of criteria in case study II. As mentioned, this parameter is 100% for all criteria in case study I, as all agents get positive or negative credits and no trial is ignored. But in case study II, depending on thresh-

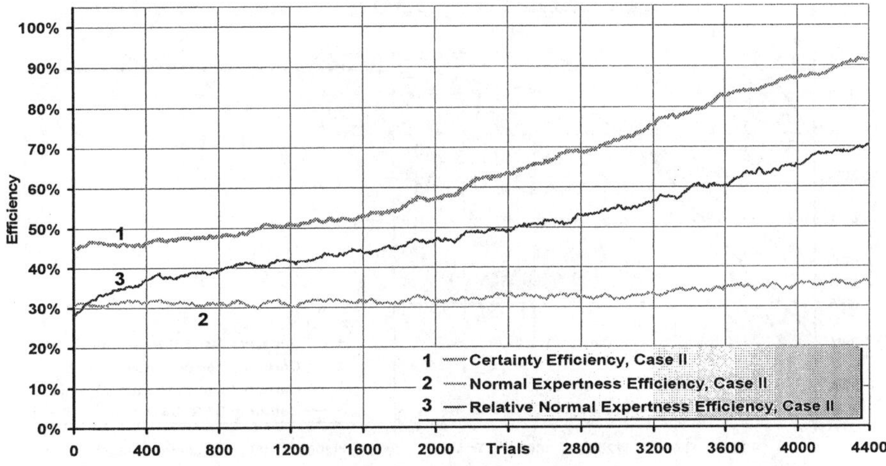

Figure 5 Efficiency of different credit assignment criteria, AND-type task

451

olds, there are situations that critic agent doesn't risk and simply ignores the trial for some agents.

Existence of some initial knowledge in the learning agents seems to be an essential factor for the success of presented approach in AND-type tasks. The amount of required initial knowledge depends on the task.

Discussed set of experiments shows that *Certainty* acts best among the other measures. Hence, for further study and comparison with OR-type tasks *KEBCA* is implemented using *Certainty* measure.

8.2 *KEBCA* in OR-Type Tasks

In Figure 6 the result of simulating an OR-type task in *Case Study I* is depicted. Figure 7 shows the similar results for *Case Study II*. It can be seen that, in OR-type tasks *KEBCA* is successful in both cases even when there is no initial knowledge. But learning is not completed in *Case Study II*. The reason is that when *GSR* reaches one, there is no team punishment anymore. In fact, receiving team rewards, the critic agent cannot judge inexperienced state-actions. Consequently, no new Q-cell is learnt and just already learnt Q-cells are reinforced. In such conditions, further learning of the agents has no cost from the group outcome point of view as the team performs its job correctly.

Correctness is not shown in Figure 7 as it is one in the entire learning period. The reason is that in deterministic environments *KEBCA* acts fully rational in *Case Study II* [7]. In Figure 6 for *Case Study I*, it can be seen that although *Correctness* is not one and *MCA* is a bit risky, but providing the critic agent with the number of wrong actions in the team, the agents can learn their job efficiently.

In Figure 8 result of using an optimal critic is shown. In optimal method, the presence of other agents has no effect on each individual agent learning. Therefore, this figure is applicable to both AND-type and OR-type tasks. *Efficiency* and *Correctness* are always 100% for an optimal critic. Comparing Figure 8 with

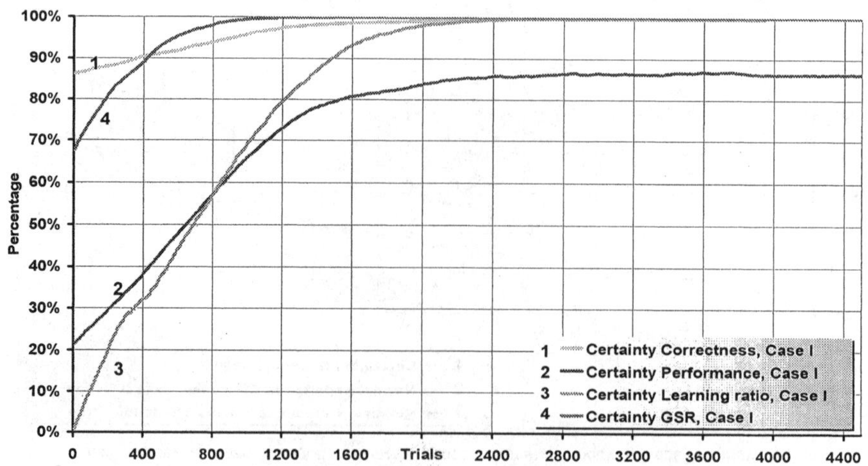

Figure 6 Results of *KEBCA* in OR-type task, Case Study I

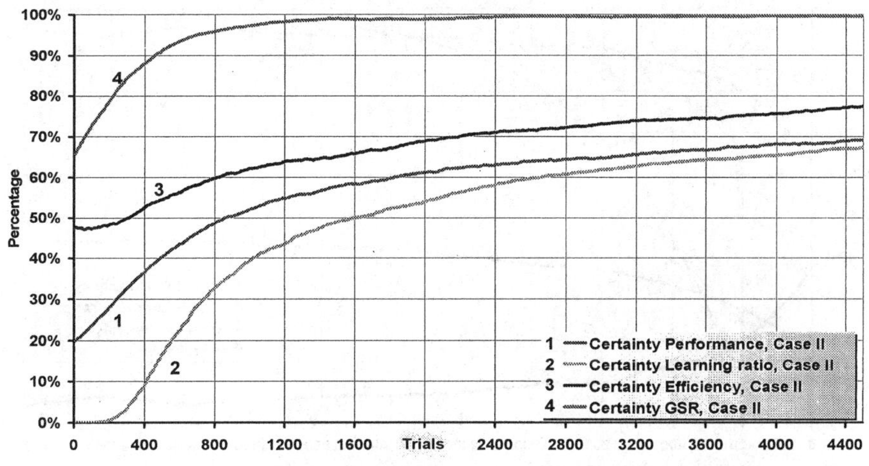

Figure 7 Results of KEBCA in OR-type task, Case Study II

Figure 6, Figure 7, and Figure 9 (*NR* case), it is observed that *MCA* in OR-type tasks is more efficient in the start of learning.

8.3 *KEBCA* in a Noisy Environment

Figure 10 shows *Performance* of *KEBCA* and *NP* strategy in AND-type tasks. In these experiments, the agents have some initial knowledge. As mentioned, implemented learning parameters bounds *Performance* to 87%. In Figure 10, it can be seen that *KEBCA* reaches this limit in a noise free environment. Even with 10% noise, *Performance* is acceptable after 4000 trials, although with a slower growth. In this figure, three carves of *NP* are completely below the *KEBCA* ones. This shows the effectiveness of our approach even when there are relatively large amounts of noise in the environment.

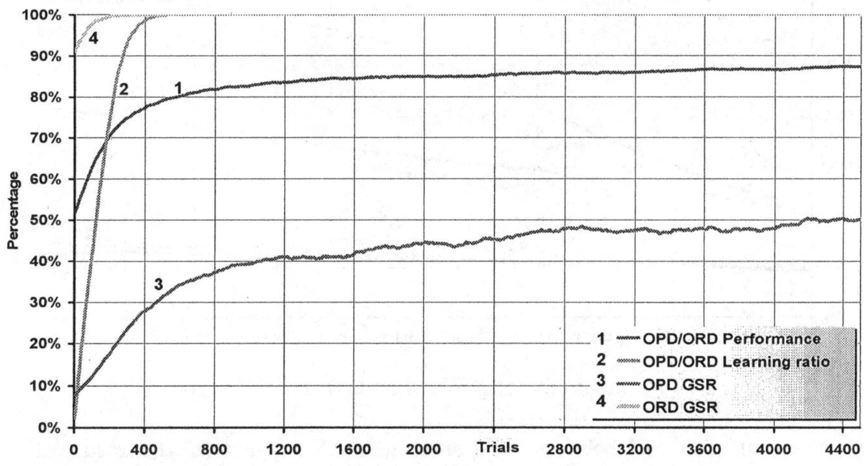

Figure 8 Outcome of MCA using OPD/ORD

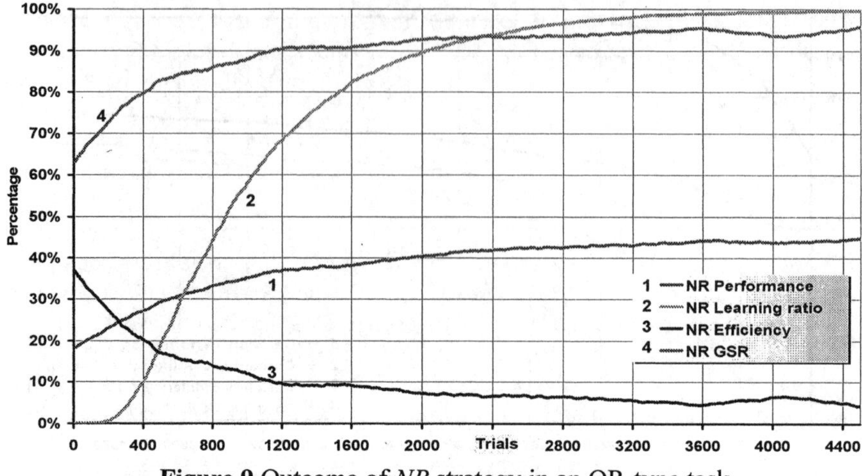

Figure 9 Outcome of *NR* strategy in an OR-type task

In Figure 11, an OR-type task is simulated with agents with no initial knowledge. It can be seen that, in all conditions *KEBCA* acts better than *NR*. In the presence of noise, *Performance* drops after a temporary growth at the start of group learning. This is due to the fact that, the agents loose their gained knowledge after some trials. The reason for this observation is that, the learning rate is not small enough[5]. In fact, the agents must use a smaller learning rate to handle uncertainty in the team reinforcements by averaging their knowledge over more experiences.

Performance curves of *OPD* and *ORD* are depicted in Figure 12. A comparison of this figure and Figure 10 shows that, *KEBCA* for agents with initial knowledge performs better than the optimal critic in AND-type tasks in the noisy envi-

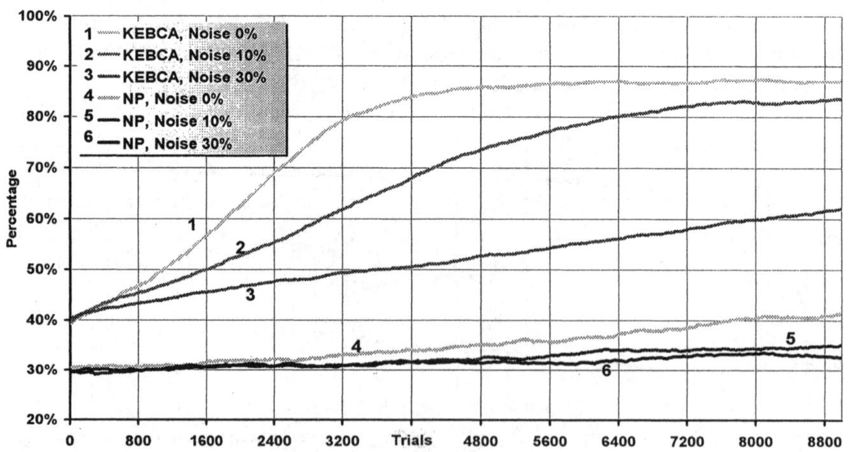

Figure 10 *Performance* of *KEBCA* and *NP* in AND-type task

---

[5] It is noteworthy that, such behavior is not observed in AND-type tasks. The reason will be discussed in the next section.

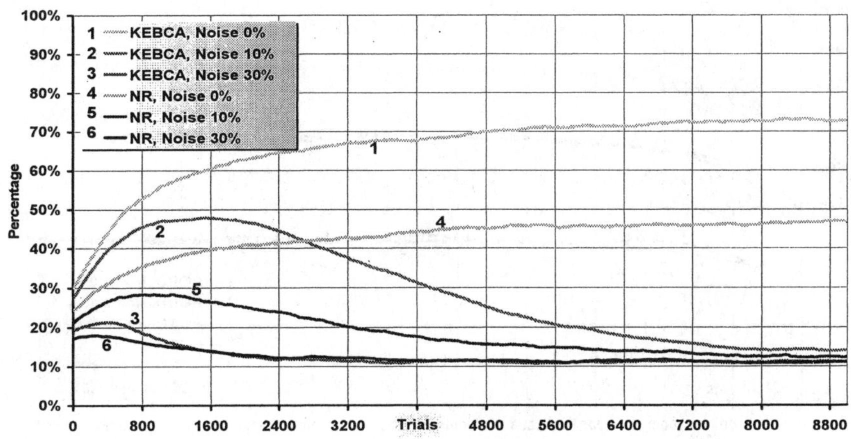

Figure 11 *Performance* of *KEBCA* in OR-type task

ronments. But, optimal critic (*ORD*) performs better than *KEBCA* in OR-type tasks (compare Figure 11 and Figure 12).

Figure 12 shows very little difference between *ORD* and *OPD* as the agents' initial knowledge in *OPD* is obtained by the agents in *ORD* after 110 trails[6].

9 Discussion

9.1 Noise Free Environments

The experiments show that, *MCA* is highly affected by the task type. Each of the two mentioned task types has its own features and difficulties for the critic agents.

At the start of learning in OR-type tasks, the critic agent can use the team punishments to lead the group before *GSR* reaches one. If there is any necessity that all agents reach to a reasonable level of individual performance, the critic agent can control the agents' temperature. At the start of learning, the agents must use a large enough temperature to provide the critic agent with the informative environment reinforcements (team punishments). Using the measures of agents' knowledge, the critic agent can determine the time that the desired learned knowledge is obtained and then the agents must decrease their temperature to exploit their knowledge. In fact, with controlling the exploration versus exploitation in this way, the critic agent can train the agents completely. The cost of this strategy is low *GSR* during the learning. Hence in OR-type tasks, the role of initial knowledge is not very important. In addition, for the same reason, different *MCA* methods act similar to each other.

---

[6] It worth reminding that, *ORD* and *OPD* are the same methods and the observed differences are just due to the difference in the agents' initial knowledge.

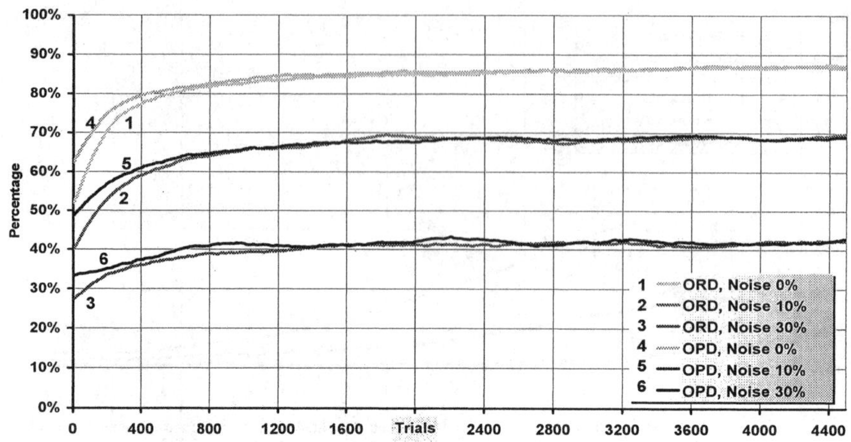

Figure 12 *Performance* of *OPD* and *ORD*

In contrast, in AND-type tasks the initial group learning phase is a big challenge to the critic agent when the agents have no initial knowledge. This difficulty is due to the fact that, the agents are similar in their amount of knowledge and judgment is hard. Also, informative feedbacks (team rewards) are rare. Another problem with this type of task is that, *GSR* is much less than *Performance*. In fact, the group always acts weaker than a single agent and this is more evident in more populated groups.

9.2 Noisy Environments

Presence of noise generally affects the team learning negatively, but the amount of this effect depends on the *MCA* method and the task type.

If there is some reliable (noise free) history of learning (initial knowledge), then *KEBCA* is preferable as it intelligently processes the new received reinforcements. In fact, *KEBCA* acts like a noise filter in such conditions. This is observed considering *Correctness* (not reported in this paper) of different *MCA* methods.
On the other hand, with the increase of uncertainty, the learning rate must be readjusted (reduced). The sensitivity to unsuitable or unadjusted learning rate in *KEBCA* is less than in the optimal case as a result of mentioned noise filtering behavior when confident initial knowledge is available.

When there isn't such initial knowledge, things get changed. The reliability of rewards in AND-type tasks and punishments in OR-type tasks is less than the individual reinforcement signals in a similar single agent scenario. In addition, the noise filtering characteristic isn't present any more as the gathered knowledge is affected by noise and is not completely reliable. Missing this advantage and having fewer individual agent reinforcements in *KEBCA* –which mean weaker feedback– results in a higher sensitivity to the learning rate.
This condition is observable in Figure 11. A notable fact in this figure is that, *Performance* drops after a temporal growth in the first learning trials. This drawback

456

and non-monotony in the learning curves imply that, the learning rate is high. This means the learning rate of 0.7, which works well with ORD and OPD, is too high in the case of *KEBCA* to reach a reasonable stable condition. Therefore, the learning system needs a smaller learning rate to obtain proper rules from its less informative experiences. Results of simulations with the learning rate of 0.5 (not reported in this paper) confirm this conclusion. In fact, with this more suitable learning rate, better results – even comparable with results in a noise free environment – are obtained.

10 Conclusion and Future Works

It is discussed that, Multiagent Credit Assignment (*MCA*) is a very basic and important problem in groups of cooperative independent reinforcement learners. In such groups, a common reinforcement signal must be distributed among the team members, when the role of each agent is not known clearly. For doing this, a suitable criterion must be provided to estimate the role of each agent in the team performance and to judge if an agent has done a wrong action, especially when there is no local judgment criterion. In this paper, three such criteria, named *Certainty*, *Normal Expertness* and *Relative Normal Expertness*, were introduced. The introduced criteria are based on the idea of using agents' learning history or knowledge to judge their actions. Hence, they are used as measures of agents' knowledge. It was shown that, this approach enhances speed and quality of learning and *Certainty* performed best among the others. It was also observed that, having extra information, like the number of wrong actions in the team, can be used to produce much faster and better learning.

It is discussed that, task type completely affects the scenario of learning. Generally, *MCA* is harder when facing parallel AND-type tasks and the presence of some initial knowledge seems necessary for a reasonable learning speed. The main problem with OR-type tasks is that, the amount of attainable knowledge by the agents is bounded. However, this problem can be eliminated using the agents' exploration strategy properly.

It is observed that, *KEBCA* has a noise filtering capability if the agents have some reliable initial knowledge. Therefore, relative to individual learning, the presented method results in a faster and more qualitative learning.

In this paper, the proposed methods are tested in deterministic one-step tasks as the first step in solving the very complicated *MCA* problem. In fact, with the explained scenario, we somehow tried to approach the *MCA* problem in a new way. It seems that extending *KEBCA* to multi step tasks is straight forward, but the extension to non-deterministic environments needs more revisions and efforts. However, in all conditions any available local information can be used beside the proposed measures to judge agents' actions.

We are now studying the more general cases of *MCA* in non-deterministic environments as these environments add new challenges to this problem. In addition, mathematical analysis of our approach and testing it in more practical problems are among our current research.

References

[1]   R. S. Sutton, and A. G. Barto, *Reinforcement Learning: An Introduction*, MIT Press, Cambridge, MA, 1998.

[2]   R. S. Sutton (editor), "Machine Learning: Special Issue on Reinforcement Learning", *Machine Learning,* Vol. 8, May 1992.

[3]   L. P. Kaelbling, M. L. Littman, and A. W. Moore, "Reinforcement Learning: A Survey", *Journal of Artificial Intelligence Research*, pp. 237-285, May 1996.

[4]   K. Miyazaki, and S. Kobayashi, "Rationality of Reward Sharing in Multi-agent Reinforcement Learning", *Second Pacific Rim Int. Workshop on Multi-Agents*, 1999, pp. 111-125.

[5]   S. Sen, and G. Weiss, "Learning in Multiagent Systems", In G. Weiss (Editor), *Multiagent Systems*, MIT Press, Cambridge, MA, Second Edition, 2000, pp. 259-298.

[6]   A. Harati, and M. Nili Ahmadabadi, "A New Approach to Credit Assignment in a Team of Cooperative Q-Learning Agents", *Proc. IEEE Conf. Systems, Man & Cybernetics (SMC'2002),* Hammamet, Tunisia, Oct. 2002

[7]   A. Harati, "*A Study on Credit Assignment among Cooperative and Independent Reinforcement Learning Agents*", Master Thesis in Persion, University of Tehran, Tehran, Iran, Jul. 2003.

[8]   S. Arai, K. Miyazaki, and S. Kobayashi, "Multi-agent Reinforcement Learning for Crane Control Problem: Designing Rewards for Conflict Resolution", *in Proc. of the Fourth Int. Symposium on Autonomous Decentralized Systems*, 1999, pp. 310-317.

[9]   C. Watkins, J. Christopher, and P. Dayan, "Q-Learning", *Technical note in* [2], pp. 55-68, May 1992.

[10] A. Harati, and M. Nili Ahmadabadi, "Multiagent Credit Assignment in a Team of Cooperative Q-Learning Agents with a Parallel Task", *Proc. First Eurasian Workshop on Agents for Information Management,* Shiraz, Iran, Oct. 2002, pp. 301-305.

[11] A. Harati, and M. Nili Ahmadabadi, "Certainty and Expertness-Based Credit Assignment for Cooperative Q-Learning Agents with an AND-Type Task", *Proc. 9th Int. Conf. Neural Information Processing (ICONIP'2002),* Nov. 2002, pp. 306-310.

[12] R. S. Sutton, *Temporal Credit Assignment in Reinforcement Learning*, PhD Dissertation, University of Massachusetts, Amherst, MA, USA, Feb. 1984.

[13] M. A. Abbasi, M. Nili Ahmadabadi, and M. Asadpour, "An Analysis on the Effects of Reinforcement Distribution in a Distributed Multi-Agent Team", *in Proc. 4th Iranian Conf. Computer Engineering (CSICC),* Tehran, Feb 2002, pp. 236-243.

[14] S. Arai, K. Sycara, and T. R. Payne, "Multi-agent Reinforcement Learning for Scheduling Multiple-Goals", *Proc. 4th Int. Conf. Multi-Agent Systems (ICMAS' 2000),* 2000, pp. 359-360.

[15] M. Tan, "Multi Agent Reinforcement Learning Independent vs. Cooperative Agents", *Proc. 10th Int. Conf. Machine Learning,* Amherst, MA, USA, Jun. 1993, pp. 330-337.

[16] J. Schneider, W. K. Wong, A. Moore, and M. Riedmiller, "Distributed Value Functions", *Proc. 16th Int. Conf. Machine Learning (ICML'99),* Bled, Slovenia, July 1999.

[17] J. H. Holland, "Properties of the bucket brigade", *Proc. Int. Conf. Genetic Algorithms*, Hillsdale, NJ, 1985.

[18] J. Schmidhuber, *"Evolutionary Principles in Self-Referential Learning, or on Learning How to Learn: the Meta-Meta... Hook"*, Diploma thesis, Institut für Informatik, Technische Universität München, 1987.

[19] J. J. Grefenstette, "Credit Assignment in Role Discovery Systems Based on Genetic Algorithms", *Machine Learning*, Vol. 3, pp. 225-245, 1988.

[20] K. Irwig, and W. wobcke, "Multi-Agent Reinforcement Learning with Vicarious Rewards", *Linköping Electronic Articles in Computer and Information Science*, Vol. 4 (1999), No. 34, http://www.ep.liu.se/ea/cis/1999/034, Dec. 30, 1999.

[21] D. N. Kinny, and M. P. Georgeff, "Commitment and Effectiveness of Situated Agents", *Proc. 12th Int. Joint Conf. Artificial Intelligence*, 1991, pp. 82-88.

[22] M. J. Matarić, "Using Communication to Reduce Locality in Multi-Robot Learning", *Proc. 14th National Conf. Artificial Intelligence*, 1997, pp. 643-648.

[23] S. Arai, K. Sycara, and T. R. Payne, "Experience Based Reinforcement Learning to Acquire Effective Behavior in a Multi Agent Domain", *Proc. of 6th Pacific Rim Int. Conf. Artificial Intelligence*, Springer-Verlag, 2000, pp. 125-135.

[24] D. H. Wolpert, K. Tumer, "An Introduction to Collective Intelligence", *NASA Technical Report:NASA-ARC-IC-99-63, to appear in J. M. Bradshaw, ed, handbook on agent technology*.

[25] M. Nili Ahmadabadi, and M. Asadpour, "Expertness Based Cooperative Q-Learning", *in IEEE Trans. On SMC*, Part B, Vol. 32, No. 1, Feb. 2002, pp. 66-76.

Implementation of Visual Tracking System using Artificial Retina Chip and Shape Memory Alloy Actuator

W.C. Kim[1], M. Lee[1], J.K. Shin[1], H.S.Yang[2]

[1]School of Electrical Engineering & Computer Science, Kyungpook National University, 1370 Sankyuk-Dong, Puk-Gu, Taegu 702-701, Korea
[2]Department of Computer Science, Korea Advanced Institute of Science and Technology, 373-1 Gusong-Dong, Yusong-Gu, Taejon 305-701, Korea

Abstract—We implemented a visual tracking system using an artificial retina chip and the shape memory alloy actuator. A foveated complementary metal oxide silicon (COMS) retina chip for edge detection was designed and fabricated for an image sensor of the developed system, and the shape memory alloy actuator was used for mimicking the roles of the ocular muscles to track a moving object. Also, we proposed a new computational model that mimics the functional roles of our brain organs for generating the smooth pursuit eye movement. In our model, a neuromorphic model for the medial temporal (MT) cell generates motion energy, and the medial superior temporal (MST) cell is considered to generate an actuating signal so that the developed active vision system smoothly pursues the moving object with similar dynamics to the motion of our eyeball during the smooth pursuit. Experimental results show that the developed system successfully operates to follow the edge information of a moving object.

Keywords—Visual tracking, artificial retina chip, smooth pursuit eye movement, shape memory alloy actuator

1 Introduction

The cerebral control system in the brain that directs the eye towards an object for viewing is as significant as the system that interprets the visual signals received from the eye [1]. In the same way as the cerebral control system, the most important role of the active vision system is to direct an artificial visual receptor such as a charge-coupled device (CCD) camera toward an interesting object in the visual field as a human vision system. The ability of the present active vision system for achieving this function is less powerful and inefficient than that of the human vision system, which has motivated us to mimic mechanism of human eye movements.

In this paper, we focus on modeling and developing an active vision system that mimics the smooth pursuit eye movement of a human being, which is one of the crucial features of the human visual system. Ringach proposed a tachometer feed-

460

back model for smooth pursuit eye movements [2]. His model imitates the dynamics during smooth pursuit eye movement, but he did not consider a motion energy calculation of the moving object. Dickie and his colleagues explained the role of cortical area medial superior temporal (MST) and also proposed a simple smooth eye-head pursuit model [3]. Recently, Grossberg presented a mathematical model for smooth pursuit eye movement including the roles of MST and MT cells [4]. We propose a new computational model that mimics the functional roles of our brain organs for generating the smooth pursuit eye motion. In our model, a neuromorphic model for the MT cell generates motion energy, and the MST cell is considered to generate an actuating signal so that the developed active vision system moves to the target with similar dynamics to the motion of our eyeball during the smooth pursuit. In a conventional image processing system using a CCD type camera, the computational cost is quite high, and the processing speed is still not fast for real-time application [5]-[7]. The retina is superior to the digital image processing systems because human retina uses a parallel processing with space variant structure [8]-[12]. Using a space variant structure, the advantage is the selective reduction of image data. This strategy gives good resolution in the center, while keeping a wide view field [13]-[15]. In our study, a foveated (log-polar) CMOS retina chip for edge detection has been designed and fabricated. Pixel distribution on the surface of retina chip follows the log-polar transformation having more resolution in the center than in the periphery. The log-polar transformation has been widely used for efficient image construction and analysis in the retina topic structure [13]. For the tracking work, we use a shape memory actuator (SMA) for mimicking the roles of ocular muscles to track a desired object. In robotics, the shape memory alloy actuator has been frequently adopted in substitution for artificial muscle.

A wide variety of topics for retinal chips has been researched by many people during last decades. Some outcomes of studies are general purpose image sensor with primitive signal processing such as the edge detection chip [8], the retinal implant [16] and the motion detection chips [17]-[19]. Most of these studies are mainly related with the design of new type of an image sensor without considering of a system level design such as an active vision system. In our study, the retina chip with edge detection function is applied to design of active vision system for visual tracking.

Section 2 describes the developed retina chip with foveated structure and edge detection function, and experimental results for the retina chip will be also included in Section 2. In Section 3, the modeling of MT and MST cells is explained, and the shape memory alloy actuator is explained in Section 4. The hardware implementation and experimental results will be followed in Section 5. Conclusions and further works will be discussed in Section 6.

2 Foveated Retina Chip for Edge Detection

2.1 Architecture and Operational Principle of the Edge Detection Chip

The functions of photoreceptor, horizontal cell and bipolar cell in the human retina should be transformed to an equivalent electrical circuit [8]. In biological process, the photoreceptors in the retina receive a visual stimulus and transform an optical signal into an electrical signal. The horizontal cell spatially smoothes the transformed optical signal, while the bipolar cell yields the differential signal, which is the difference between the optical signal and the smoothed signal. By the output signal of the bipolar cell, an abrupt difference from negative to positive or positive to negative at the edge can be obtained [11]. These are implemented in CMOS circuits.

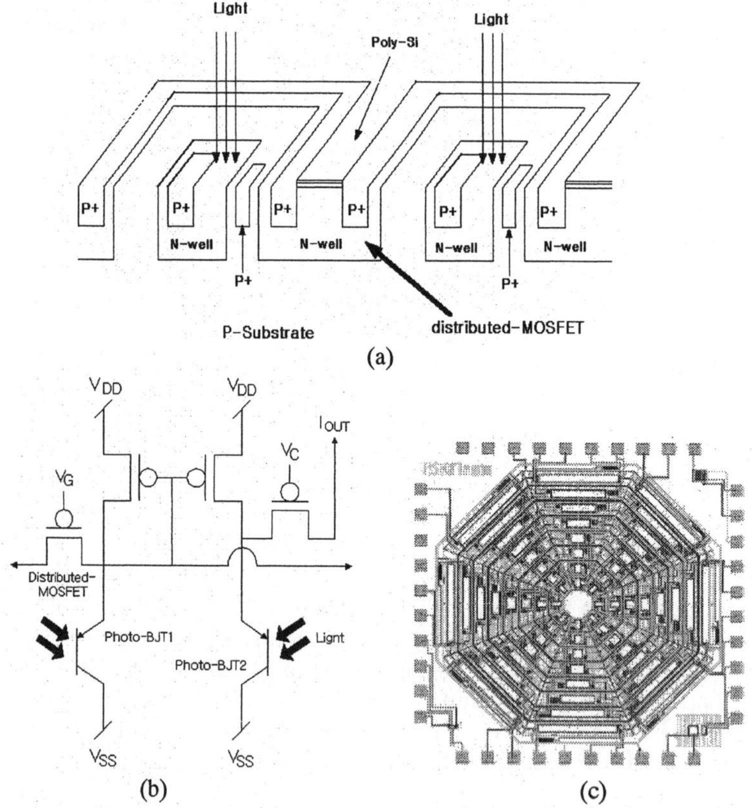

Fig. 1. Unit pixel and layout: A pixel plays three main roles: the sensing, the differentiation and the smoothing. (a) Cross-sectional view, (b) Equivalent circuit, (c) Layout.

According to the edge detection mechanism, we constructed the pixel for edge detection as shown in Fig. 1 with two photo-BJTs (photoreceptor), one current mirror circuit (bipolar cell), distributed-MOSFET (horizontal cell) and one MOS transistor [20]. The foveated retina chip has eight circular arrays whose unit circular array consists of eight pixels and radius grows linearly from the center. This linear growth of the radius means a linear growth also for a unit pixel area. The growth factor depends on the shape of the cell, etc. The resolution of the foveated retina chip is 8×8 (8 circumferences with eight pixels each), which is a total of 64 photo cells. A minimum pixel size of $100\mu m \times 100\mu m$ has been achieved and the chip occupies an area of 3mm×3mm. We aimed to confirm the capability of the log-polar typed edge detection chip, so we used a small number of pixels for the chip. The log-polar type must increase the pixel size according to radius, so the scaling is needed for each circumference [14]. If the size of photo-BJT in each ring was increased according to a fixed scaling rate and the peripheral components such as mirror circuit were not increased and maintained the same size, the output for the same input light was not constant. The result of SPICE simulation is shown in Fig. 2(a). The reason is that the larger photo-BJT makes the larger photocurrent for the same light intensity; therefore the modification of a peripheral component is needed as in a larger aspect ratio. To avoid this problem, we used the same scaling ratio of the photo-BJT scaling factor for the MOS aspect ratio in a mirror circuit [15]. The channel length of MOS was fixed for minimum value and only channel width was increased. Simulation results for channel width scaling are shown in Fig. 2(b). The output is constant for the same incident light regardless of pixel size. Therefore, output current responds to only the real edge of input image.

2.2 Experiments and Results

The electrical and optical characteristics of the fabricated chip have been measured. Fig. 3 shows the variation of measured current distribution of I_{out} in the fabricated chips when the object was moved from the left side to the right. The dark cells in Figs. 3 (b), (d), (f), (h) and (j) represent the highest current level ($3\mu A$), and the white cells are the case of middle current level ($-0\mu A$). The gray cells are the case of lowest current level ($-2\mu A$). If the object covers a specific cell area of the chip, the covered cells inside an object and the cells outside an object have the white cells. The boundary cells inside an object are dark cells and the boundary cells outside an object are gray cells. The edge information is extracted from the boundaries that maximize the current difference between exposed pixels and covered pixels against an object. The edge is represented by binary information. Thus, we can detect edge information by a threshold that is implemented by a simple comparator using the current levels. Fig. 3 (k) shows one of examples for edge detection of the foveated retina chip, when the object covers the retina chip as shown in Fig. 3 (e). As shown in Fig. 3 (k), we can detect edge information at both boundaries such as between the second cell and the first cell and between the fifth cell and the sixth cell. As shown in Fig. 3, the fabricated retina chip successfully detects the edge information. Since the foveated retina chip has a small resolution such as 8 x 8, it is very difficult to control the light intensity and resultantly the current output contains high interference as shown in Fig. 3.

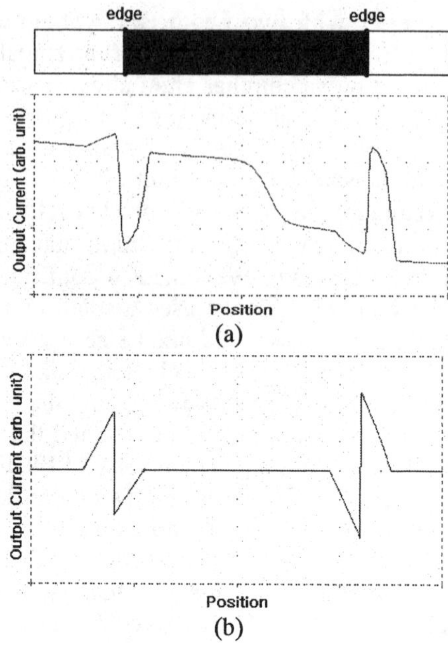

Fig. 2. SPICE simulation for channel width scaling: (a) Output current without scaling, (b) Output current with scaling.

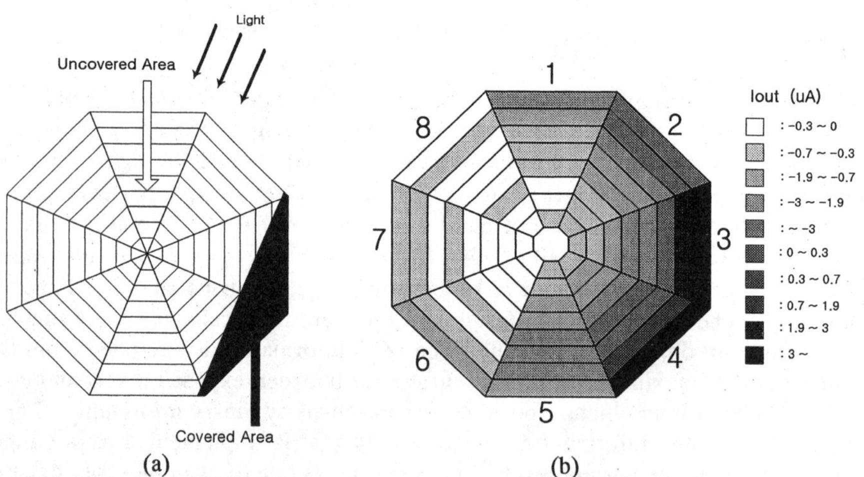

Fig. 3. Measured distributions of output currents I_{out}: when the light was incident upon the left half of the chip, the output current flows. (a), (c), (e), (g) and (i) show several object patterns on the foveated retina chip. (b), (d), (f), (h) and (j) show the output currents of foveated retina chip for each object pattern. For the edge confirmation, (k) shows that the edge information is detected at the boundary, which is both the interval of cell 1 and cell 2 and the interval of cell 5 and cell 6.

464

Light
Uncovered Area
Covered Area
(c)
1
2
3
4
5
6
7
8
Iout (uA)
: -0.3 ~ 0
: -0.7 ~ -0.3
: -1.9 ~ -0.7
: -3 ~ -1.9
: ~ -3
: 0 ~ 0.3
: 0.3 ~ 0.7
: 0.7 ~ 1.9
: 1.9 ~ 3
: 3 ~
(d)
Light
Uncovered Area
Covered Area
(e)
1
2
3
4
5
6
7
8
Iout (uA)
: -0.3 ~ 0
: -0.7 ~ -0.3
: -1.9 ~ -0.7
: -3 ~ -1.9
: ~ -3
: 0 ~ 0.3
: 0.3 ~ 0.7
: 0.7 ~ 1.9
: 1.9 ~ 3
: 3 ~
(f)
Light
Uncovered Area
Covered Area
(g)
1
2
3
4
5
6
7
8
Iout (uA)
: -0.3 ~ 0
: -0.7 ~ -0.3
: -1.9 ~ -0.7
: -3 ~ -1.9
: ~ -3
: 0 ~ 0.3
: 0.3 ~ 0.7
: 0.7 ~ 1.9
: 1.9 ~ 3
: 3 ~
(h)

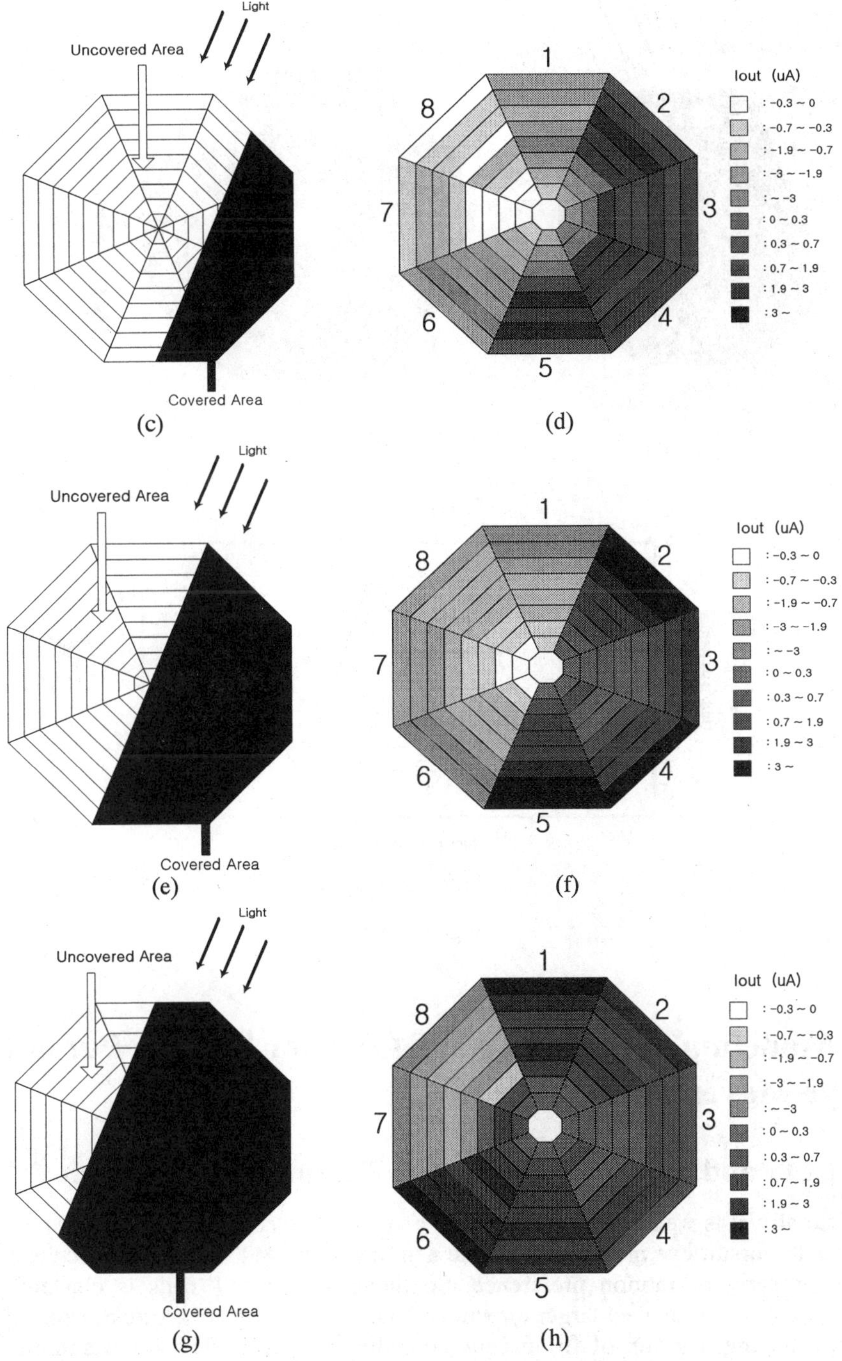

465

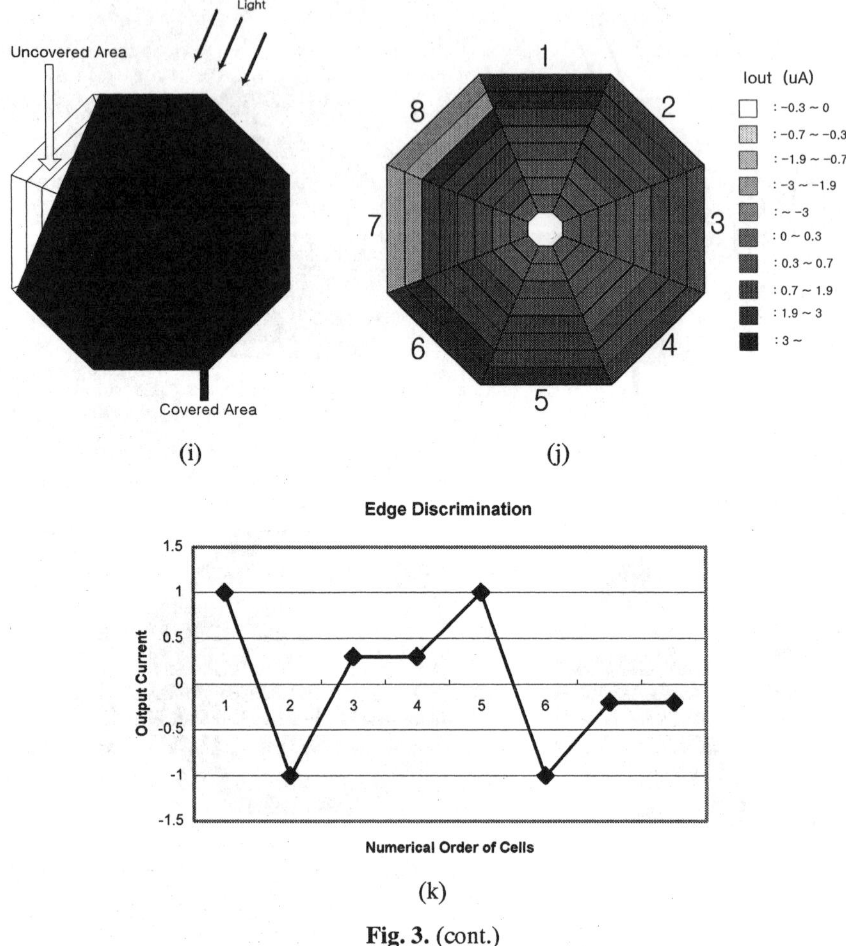

Fig. 3. (cont.)

3 Modeling of MT and MST Cells for Smooth Pursuit Eye Movement

3.1 MT and MST Cells for Smooth Pursuit Eye Movement

Generally, it is well known that the MT and the MST cells are essential organs for smooth pursuit eye movements. Some neurons in the MT and the MST cells have the property of motion preference. So the group of MT cells is clustered or weighted to a preferred target direction. Thus, we can use an artificial neural network having a group of firing cells according to a specific motion stimulus to model the MT cells. Moreover, in the MT area, the motion signals are segregated into a target and a background [21]-[22]. If there exists a moving object, edge information of the moving object can be detected by the retina chip. Then, the center of gravity of the edge cells on the retina chip is regarded as the target, and the others are considered as background. The MT area is connected with the MST area, which receives the motion energy of moving object. The MST cells provide a neural signal to move the muscle of the eyeball using the efference copy signal and the output signal of MT cells. In the modeling of an MST cell, we used the tachometer feedback model, which is a linear model of smooth pursuit eye movement based on an efference copy signal of eye acceleration [2]. Based on biological facts, when the velocity of a moving object is below 3°/sec, the human eye is in fixation, and if the velocity exceeds the fixation boundary, the eye moves to follow a moving object smoothly. In our simulation, the distance between current fixation point and the center value of a moving target is obtained by Euclidean distance.

3.2 Modeling of MT and MST cells

The model of smooth pursuit eye movement follows to the principal feature of the MT cells. At first, group activities of MT cells are modeled on the self-organizing feature map (SOFM) [23]-[24]. The outputs of self-organizing feature map emulate the neural outputs of MT cells. The SOFM generates an output signal weighted to preferred target direction. We consider the two-dimensional feature map with two inputs, in which inputs represent the centroid of an edge for a moving object. The trained SOFM represents a different cell activation of MT area according to the motion. The centroid is obtained by the center of gravity of the cells that is related with the boundary between the covered cells and the uncovered cells. It is used as a target for smooth pursuit of a moving object. The output emulating the activity of MT cells is organized into 64×64 sized maps. In the training of SOFM, we consider 120 cases θ_k for each 3 degrees of angle on polar coordinates for direction, and eight cases R_i, that is a point on the radial axis for distance between successive stimuli on retina space. We assumed that the retina space consists of 64 pixels for horizontal and vertical resolution, respectively. As a result, the magnitude of R_i should be

$$i \times (radius\,of\,retina\,space)/8, \quad where\ i = 1,2,3\cdots 8\ and\ radius\,of\,retina\,space\ is\ 32 \tag{1}$$

The origin of coordinates is center of fovea. Training sets are 960 points, which are combinations of 120 angle cases every eight radial distance cases on polar retina space. A training set is the centroid of the edges of a moving object that may come out on the retina space. Outputs of trained SOFM are different to the direction and velocity of the moving stimulus. The human eye moves according to on/off set mechanism. When the velocity of a moving object reaches the criterion $3°/sec$ of on/off set, the human eye initiates the smooth pursuit eye movement [3]. In order to reflect these mechanisms, we adopt the winner-takes-all model to model the onset and offset functions. The winner-takes-all model consists of two dimensions; one is for direction selectivity with 120 cases for each 3 degrees of angle and the other is for different velocities with eight cases for each difference of centroid between successive frames.

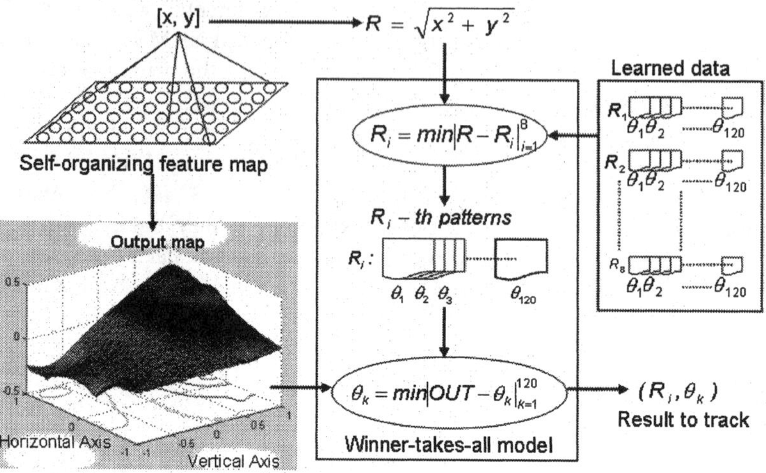

Fig. 4. Modeling of MT cell with SOFM and winner take all

Fig. 4 shows the proposed model for emulating the roles of MT cells. After the training of SOFM, the outputs of self-organizing feature map are used for constructing the weight values for the winner-takes-all model. As shown in Fig. 4, when a test input with (x, y) coordinate is given, the velocity is estimated by the comparison of Euclidean distance for centroid difference of successive frames and eight reference levels (R_i) of velocity. A reference level with the minimum distance represented by $min(|R - R_i|),\ i = 1,2,3,\dots\dots 8$ is regarded as an estimated velocity. Then, the direction is estimated by computation of matching degree between weight values for each node of the winner-takes-all model and the present outputs of the feature map. The matching degree is obtained by minimizing $|OUT - \theta_k|,\ k = 1,2,3,\dots\dots 120$, where the OUT is an output of SOFM to current input and θ_k is one of 120 directional maps for the estimated distance R_i. The winner node with the largest matching score is selected for a directional estimation. We can select a winner node among 960 nodes, and the winner node interprets the motion energy. Finally, the next pursuit point is $[R_i, \theta_k]$ on the polar coordinates of the retina.

In the MST modeling, we used the tachometer feedback model to maintain some properties of the MST cell as shown in Fig. 5. The tachometer feedback model emphasizes the maintenance of pursuit dynamics relatively to an efference copy in smooth pursuit eye movement [2]. In this model, the input is the target velocity $\overline{T}(t)$ in the visual field, and the output is the eye velocity $\overline{E}(t)$. In the experiments on primates, it is known that the delay of efference copy affects the pursuit dynamics. The tachometer feedback model successfully shows some features that

468

the stability and the oscillation depend on feedback delay in tracking. Total delay of the tachometer feedback model is proportional to the eye oscillation and the damping ratio in the pursuit movement [2].

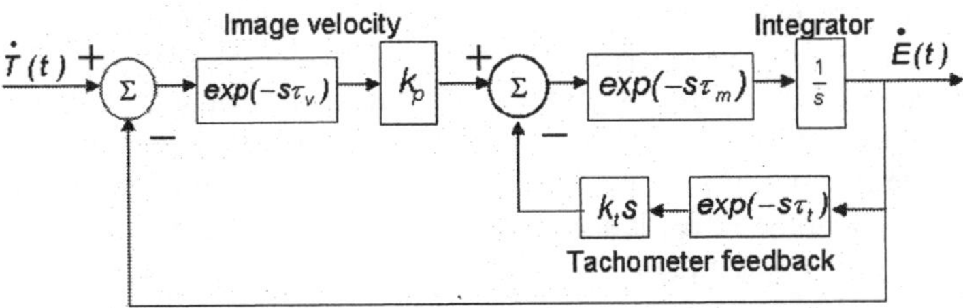

Fig. 5. Modeling of MST cell with tachometer feedback model: The tachometer feedback model shows some features, which are the stability and the oscillation depending on feedback delay in smooth pursuit eye movement. We assumed all parameters, where k_p, k_t, τ_v, τ_m and τ_t are 10, 0.7, 0.07, 0.03 and 0.03, respectively.

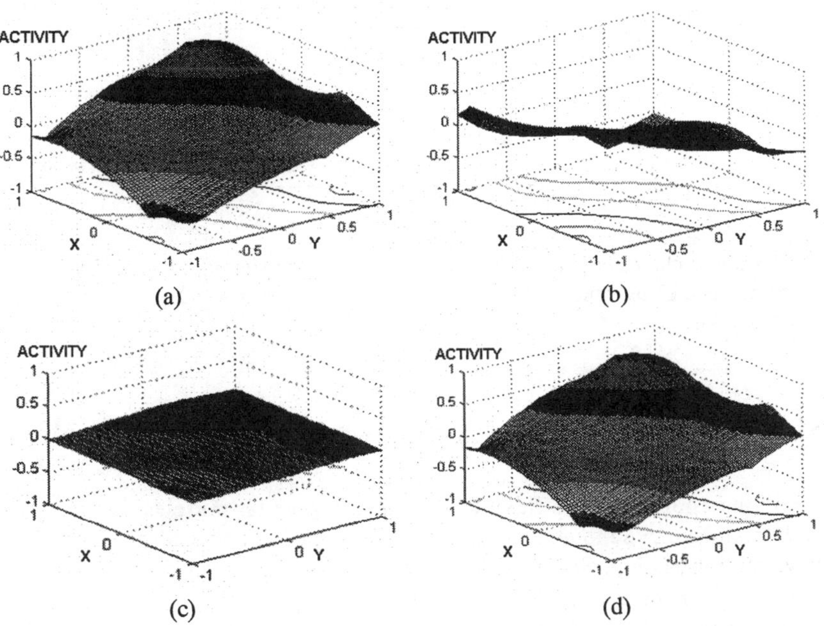

Fig. 6. Trained results for self-organizing feature map: The trained feature maps are organized differently according to object velocity and moving direction. (a) Output at speed=64 and angle=$\pi/2$, (b) Output at speed=64 and angle=$3\pi/2$, (c) Output at speed=8 and angle=$\pi/2$, (d) Output at speed=64 and angle=$\pi/2$

3.3 Computer Simulation Results

In the training of the MT model, we used 960 points that consist of 120 directional cases and eight cases for velocity. Through the training, we can get a feature map imitating the activities of MT cells. Fig. 6 (a) to (d) show several examples of outputs of the SOFM according to the various directions and velocities of the moving stimulus. In the Fig. 6, the 'X' and 'Y' indicate the axes on each point of retina coordinates. The '*Activity*' on the graph symbolizes a response of MT cells. Consequently, we can have 960 independent feature maps, which are a standard pattern to be compared with an on-line output of SOFM for deciding a tracking location. The Fig. 7 shows the characteristics of response of the MST model coupled with the MT model. In the Matlab simulation, dynamic response of the proposed system using the tachometer feedback model is as follows; the percentage overshoot is 15.3411[%], the rising time is 0.0825[sec] and the settling time is 0.4494[sec].

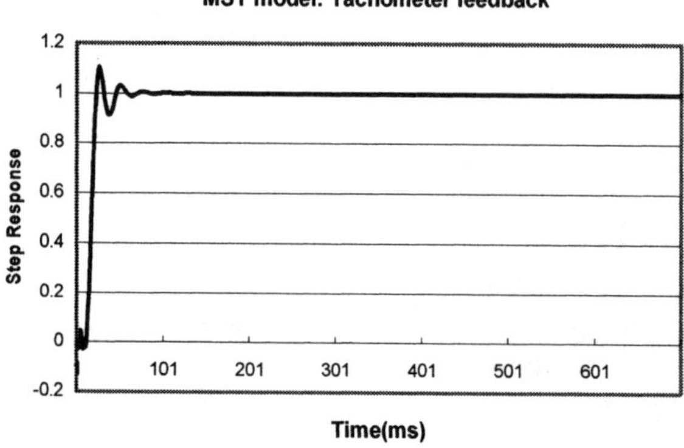

Fig. 7. Dynamic characteristics of the MST model: In the simulation of tachometer feedback, the percentage overshot is 15.3411[%], the rising time is 0.0825[sec] and the settling time is 0.4494[sec].

4 Shape Memory Alloy Actuator for Ocular Muscles

Eye motion is guided by the flexible movement of ocular muscles. In order to imitate a human-like eye movement, we adopt the shape memory alloy to the actuator to drive the retina chip. Since the SMA actuator among various muscle-like actuators is easy to handle, we can manipulate the shape memory alloy using simple electrical circuit. It means that the shape memory alloy actuator saves labor in mimicking the movement feature of the ocular muscles.

The shape memory alloys are applied to the group of metallic materials that demonstrate the ability to return to some previously defined shape or size when subjected to the appropriate thermal procedure. Generally, these materials can be plastically deformed at some relatively low temperature, and upon exposure to some higher temperature will return to their shape prior to the deformation. Therefore, we can control the length of the shape memory alloy actuator using a pulse width modulation (PWM) signal that modulates the current flowing into the actuator for heating. In this research, the used shape memory alloy is the NiTi wire, which is the compound of nickel-titanium. The NiTi alloys have greater shape memory strain (up to 8% versus 4 to 5% for the copper-base alloys) with the maximum deformation to 90°/C. Table 1 shows the specification of NiTi in detail.

Table 1. Specification of shape memory alloy

D [μm]	R [Ω/m]	RC [mA]	TCS [sec]	TRS [sec]	TCR [cycle/min]	AST [°C]	AFT [°C]	RST [°C]	RFT [°C]
100	150	180	0.5	0.7	50	88	98	72	62

D Wire diameter, *R* Resistance, *RC* Recommended current, *TCS* Typical contraction speed, *TRS* Typical relaxation speed, *TCR* Typical cycle rate, *AST* Activation start temp, *AFT* Activation finish temp, *RST* Relaxation start temp, *RFT* Relaxation finish temp.

In order to control the motion of SMA actuator, we use a simple proportional and derivative (PD) controller and a linearized model of a shape memory alloy but NiTi has a hysteresis effect [25]. The precision control of hysteresis is an open issue, but the PD controller gives the tolerable control performance. Moreover, it is hard to find an engineering model of shape memory alloy due to nonlinear attribute. The linearized model of a shape memory alloy simplifies the complicated characteristic. Fig. 8 shows the feedback block diagram including the linearized model of shape memory alloy actuator and the PD controller.

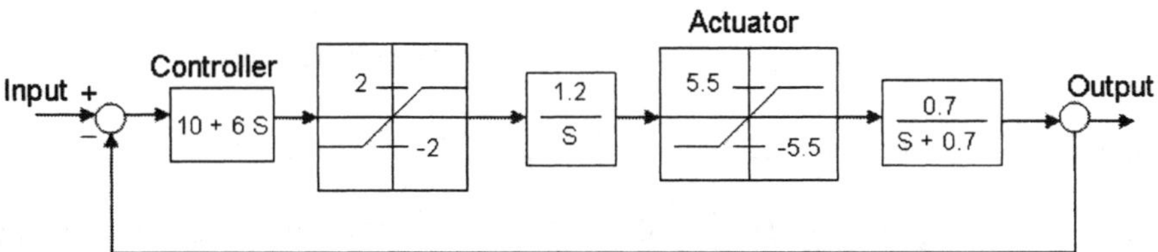

Fig. 8. Block diagrams of shape memory alloy actuator and PD controller

5 Hardware Implementation and Experimental Results

We developed a simple active vision system based on the retina chip, the smooth pursuit eye movement model and the shape memory alloy actuator. Fig. 9 shows the overall block diagram. The edge information of a target transmits from the foveated artificial retina chip to the MT model. The output of MT model transfers to the MST block that is the tachometer feedback model. The motor signal calculated in the MST model is given to the PD controller and finally to shape memory actuators. The retina chip can focus on an object in visual field, and also successfully follows an moving object by developed system.

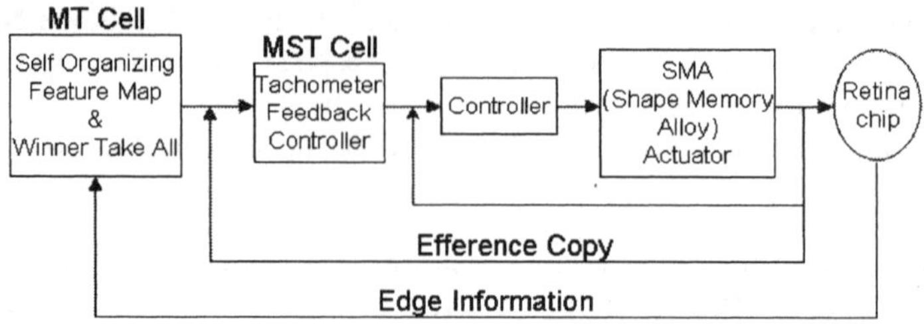

Fig. 9. Block diagram of the overall system using artificial retina chip, shape memory alloy actuator and smooth pursuit eye movement model

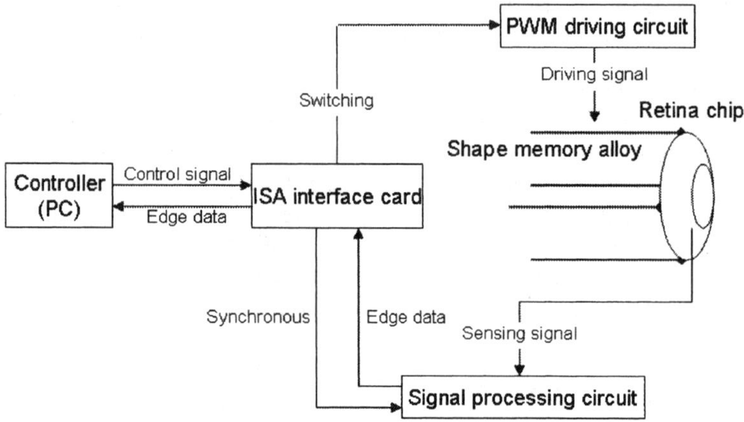

Fig. 10. System configuration of the developed active vision system

As shown in Fig. 10, the developed system has four SMA wires to show the four directional activities: the left side, the right side, an upper direction and a downward direction. The main control unit of the developed system is a personal computer (IBM PC). The other devices consist of the PWM drive circuit, the industry standard architecture (ISA) interface [26] and the signal processing circuits. The PWM circuit for driving shape memory alloy actuator supplies the regulated current, which is the source of a heating energy to deform the length of the shape memory alloy actuator. The ISA interface card is designed to interface with the personal computer and the periphery devices. This ISA interface aids both the PWM circuit in driving the retina chip and the signal processing circuit in extracting the edge information. The signal processing circuits consist of some comparators for detecting edge in retina chips. Fig. 11 shows the real picture for the developed system. The maximum field of view is 20° for both panning and tilting, and the moving speed is $30^{\circ}/\sec$. We could get the eight bit resolution of movement, such as upward, downward, left, right and slanting directions. According to light stimulus detected in the retina chip, our system can follow the moving edge, which is caused by the change in light stimulus. The distance between the light source and artificial retina chip is fixed as 0.3[m].

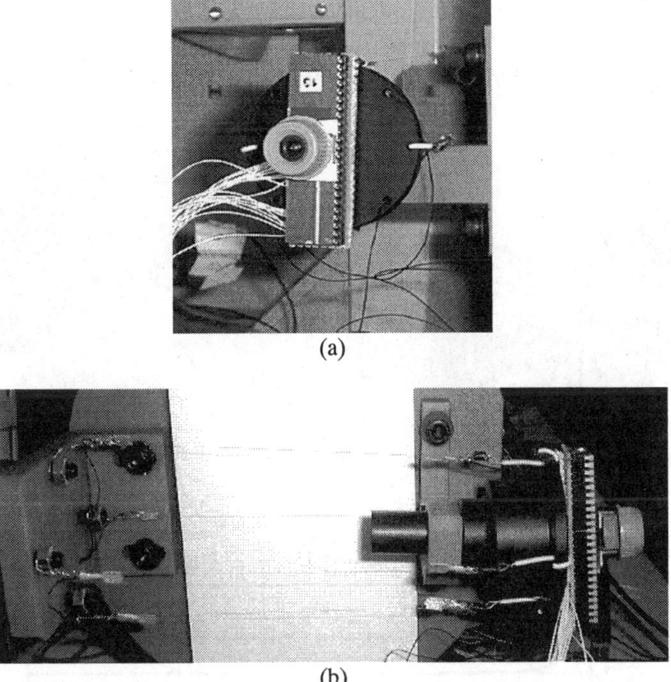

(a)

(b)

Fig. 11. Real configuration of the developed system: (a) Front view, (b) Shape memory alloy actuator with the artificial retina chip.

As shown in Fig. 12, the light source is shown to the left or right, the developed active vision system follows the light source successfully. The experimental results show that the proposed system operates to follow the moving edge of a slowly moving object. On the quantitative evaluation, the actual measurement was completed, and the quantitative data was compared with that of computer simulation. The data acquisition was done by a linear potentiometer and an eight bit analog-to-digital converter with 1[ms] sampling time. Computer simulation and experimental results are shown in Fig. 12 (c) and (d), respectively. The response of the system shows a slower phase than the theoretical one of computer simulation. In the simulation, the tachometer feedback model shows the dynamic response that the percentage overshot is 15.3411[%], the rising time is 0.0825[sec] and the settling time is 0.4494[sec]. For the SMA actuator based on the PD controller, the percentage overshoot is 10.1815[%], the rising time is 0.0987[sec] and the settling time is 0.6377[sec]. The implemented system shows a little slow response that the percentage overshoot is 1.0256[%], the rising time is 0.1630[sec] and the settling time is 0.1960[sec]. It is due to the linear control for non-linear SMA actuator. The square error between simulation and experimental results are shown in Fig. 12 (e).

The large error in transient dynamic occurred mostly by inaccurate modeling result of the SMA actuator. However, the developed smooth pursuit system follows the edge movement according to a moving object. Fig. 12 (f) shows the target cell location according to the specific visual stimulus as shown in Fig. 12 (b). Target location is the cell with 4th ring from the center and $\pi/4$ angle for direction in retina space. Fig. 12 (g) shows the firing SOFM map among the 960 trained feature maps according to the visual stimulus shown in Fig. 12 (b).

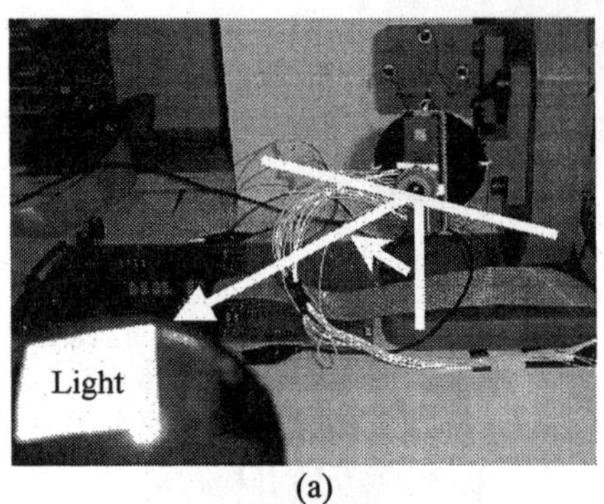

(a)

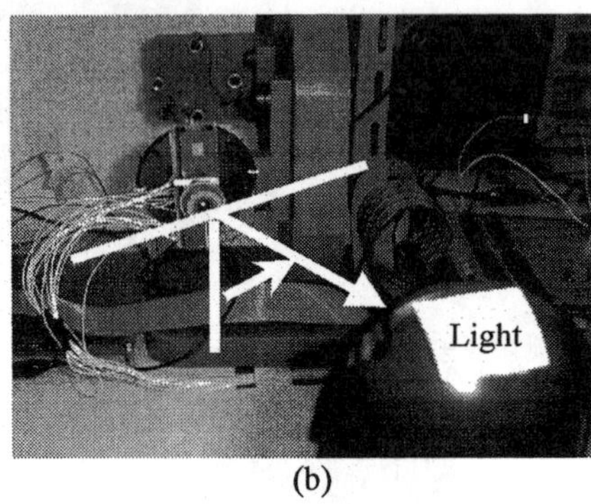

(b)

Computer Simulation: PD contorller & SMA model

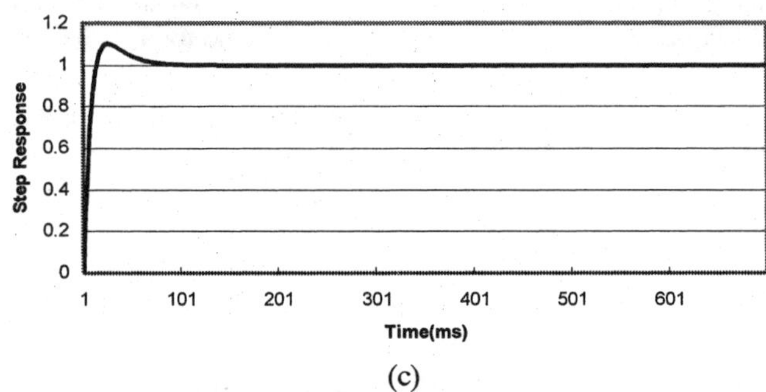

(c)

Fig. 12. Experimental results of the developed system on smooth pursuit movement: (a) Retina chip's movement on the light-source to the right, (b) Movement of the light-source situated at the left side in visual field, (c) Computer simulation on the SMA actuator's response based upon PD controller. In the time-domain response, the percentage overshoot is 10.1815[%], the rising time is 0.0987[sec] and the settling time is 0.6377[sec]. (d) Measurement on the changes of actuator in shape deformation under movement. For the measured data, the percentage overshoot is 1.0256[%], the rising time is 0.1630[sec] and the settling time is 0.1960[sec]. (e) Verification of developed system under the square-error measure. (f) Target location according to the visual stimulus shown in (b). (g) The activated feature map according to the visual stimulus shown in (b).

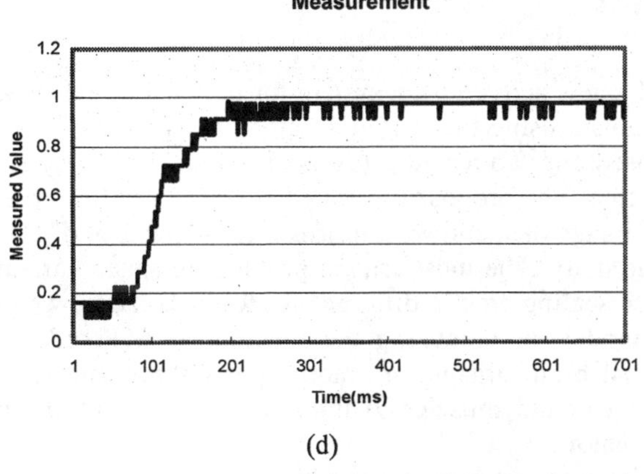

(d)

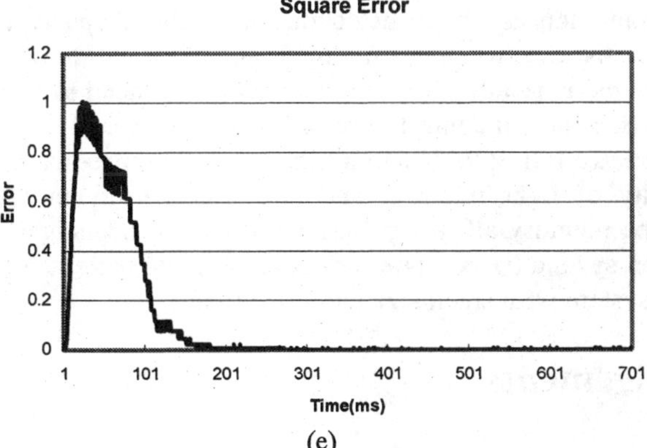

(e)

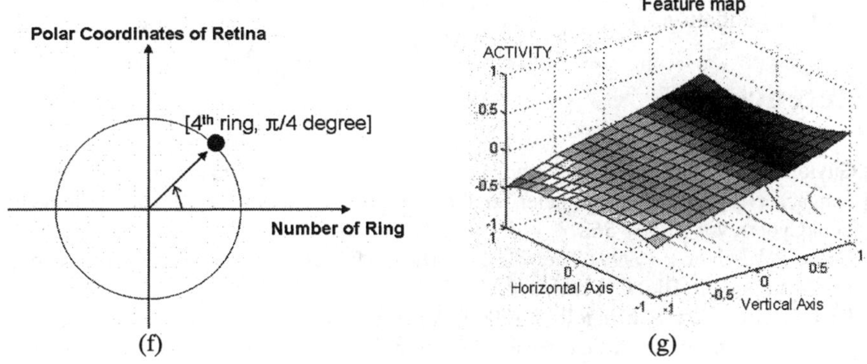

(f)   (g)

Fig. 12. (cont.)

6 Conclusions

We implemented a new active vision system for mimicking the smooth pursuit eye movement mechanism using the artificial retina chip and shape memory alloy actuator. We proposed and fabricated a foveated retina chip for edge detection with increasing pixel area. The proposed foveated structure has the advantage of selective reduction of image data and we confirmed the capability of foveated structure retina chip in our study. The most critical problem in space variant retina chip is the pixel response scaling among different sized pixels. Also, we proposed a new smooth pursuit model for emulating the function of MT cells and MST cells, which are essential brain organs for smooth pursuit eye movement. In order to mimic the roles of ocular muscles to track a desired target, we used the shape memory alloy actuator.

Due to lack of image resolution in the fabricated retina chip, it is difficult to use in real application such as object detection. Also, the shape memory alloy has nonlinear characteristic between heating input and deformation output. In order to control the shape memory alloy actuator accurately, we need to consider a nonlinear controller such as neural controller or fuzzy controller.

Accordingly, we are trying to design a higher resolution retina sensor with local adaptation ability of light intensity and also considering an effective control method for shape memory alloy actuator. Also, we are investigating the application of the vision system based upon the retina chip for pattern recognition or intelligent vision system with small size and low price.

Acknowledgements

This research was funded by the Brain Science & Engineering Research Program of the Ministry of Korea Science and Technology and the grant No. R05-2003-000-11399-0(2003) from the Basic Research Program of the Korea Science & Engineering Foundation.

References

[1] Guyton AC (1991) Textbook of medical physiology. WB Saunders Company, USA
[2] Ringach DL (1995) A tachometer feedback model of smooth pursuit eye movements. Biol Cybern vol 73:561-568
[3] Dicke PW, Thier P (1999) The role of cortical MST in a model of combined smooth eye-head pursuit. Biol Cybern vol 80:71-84
[4] Pack C, Grossberg S, Mingolla E (2000) A neural model of smooth pursuit control and motion perception by cortical area MST. Journal of Cognitive Neuroscience, Technical Report CAS/CNS-TR-99-023
[5] Gruss A, Carley LR, and Kanade T (1991) Integrated sensor and range finding analog signal process. IEEE J Solid-State Circuits vol 26:184-191
[6] Mead CA (1989) Analog VLSI and Neural Systems. Addison-Wesley

[7] Maruyama M, Nakahira H, Araki T, Sakiyama S, Kitao Y, Aono K, Yamada H (1990) An image signal multiprocessor on a single chip. IEEE J Solid-State Circuits vol 25:1476-1483

[8] Ikeda H, Tsuji K, Asai T, Yonezu H, Shin JK (1998) A novel retina chip with simple wiring for edge extraction. IEEE Photonics Technology Letters vol 10:261-263

[9] Kobayashi H, White JL, Abidi AA (1991) An active resistor network for Gaussian filtering of images. IEEE J Solid-State Circuits vol 26:738-748

[10] Mead CA, Mahowald MA (1988) A silicon model of early visual processing. Neural Networks vol 1:91-97

[11] Wu CY, Chiu CF (1995) A new structure of the 2-D silicon retina. IEEE J Solid-State Circuits vol 30:890-897

[12] Yu PC, Decker SJ, Lee HS, Sodini CG, Wyatt JL (1992) CMOS resistive fuses for image smoothing and segmentation. IEEE J Solid-State Circuits vol 27:545-553

[13] Boldue M, Levine MD (1998) A review of biologically motivated space-variant data reduction models for robotic vision. Computer Vision and Image Understanding vol 69:170-184

[14] Pardo F, Boluda JA, P´erez JJ, Felici S, Dierickx B, Scheffer D (1996) Response properties of a foveated space-variant CMOS image sensor. Proceeding ISCAS-96 vol 1, pp 373–376

[15] Pardo F, Dierickx B, Scheffer D (1997) CMOS foveated image sensor: signal scaling and small geometry effects. IEEE Transactions on Electron Devices vol 44 no 10:1731-1737

[16] Li L and Yagi T (2001) For the development of a retinal implant. Proceeding ICONIP, vol 3, pp1518-1523

[17] Andreou AG, Strohbehn K, and Jenkins RE (1991) Silicon retina for motion computation. Proceeding IEEE International Symposium on Circuits and Systems

[18] Delbruck T (1993) Silicon retina with correlation-based, velocity-tuned pixels. IEEE Transactions on Neural Networks vol 4 no 3:529-541

[19] Torralba AB and Herault J (1999) An efficient neuromorphic analog network for motion estimation. IEEE Transaction on Circuits and Systems-I: special issue on bio-inspired processors and CNNs for vision vol 46(2)

[20] Wu CY, Jiang HC (1999) An improved BJT-based silicon retina with tunable image smoothing capability. IEEE Transactions on Very Large Scale Integration (VLSI) Systems vol 7 no 2:241-248

[21] Born RT, Groh JM, Zhao R, Lukasewycz SJ (2000) Segregation of object and background motion in visual MT: Effects of microstimulation on eye movements. Neuron vol 26:725-734

[22] Krauzlis RJ, Zivotosky AZ, Miles FA (1999) Target selection for pursuit and saccadic eye movements in human. Journal of Cognitive Neuroscience vol 11:641-649

[23] Kohonen T (1990) The self-organizing map. Proceeding IEEE vol 78 no 9, pp1464-1480

[24] Haykin S (1999) Neural Networks. Prentice Hall, pp 443-483

[25] Choi BJ, Lee YJ (1998) Motion control of a manipulator with SMA actuators. Proceeding KACC, pp 220-223

[26] Tompkins WJ, Webster JG (1988) Interfacing sensors to the IBM PC, Prentice Hall

Printing: Strauss GmbH, Mörlenbach
Binding: Schäffer, Grünstadt